Plants of the Western Boreal Forest & Aspen Parkland

Written by

Derek Johnson
Linda Kershaw
Andy MacKinnon
Jim Pojar

with contributions from

Trevor Goward & Dale Vitt

First printed in 1995 5 4 3 2 1
Printed in Canada

The Publisher
Lone Pine Publishing
206, 10426-81 Ave.
Edmonton, Alberta
Canada T6E 1X5

202A-1110 Seymour Street
Vancouver, British Columbia
Canada V6B 3N3

16149 Redmond Way, #180
Redmond, Washington
USA 98052

Canadian Cataloguing in Publication Data
Main entry under title:
Plants of the western boreal forest and aspen parkland

Includes bibliographical references and index.
ISBN 1-55105-058-7

1. Botany—Canada, Western—Handbooks, manuals, etc. 2. Plants—Identification—Handbooks, manuals, etc. I. Johnson, J. D.
QK203.W4P53 1995 581.9712 C95-910653-7

Editor-in-Chief: Glenn Rollans
Botanical Editor: Linda Kershaw
Design: Bruce Timothy Keith
Editors: Roland Lines, Jennifer Keane
Layout, Maps & Production: Greg Brown, Bruce Timothy Keith
Separations: Elite Lithographers Co. Ltd., Edmonton, Alberta, Canada
Printing: Quebecor Jasper Printing Inc., Edmonton, Alberta, Canada
Cover Design: Jang & Willson
Front Cover Photo: Daryl Benson
Back Cover Background: Dean Dragich
Back Cover Inset Photos: Derek Johnson, David Mussell, Jim Pojar

Funding for this publication was provided in part by the Canadian Forest Service under the Canada-Alberta Partnership Agreement in Forestry.

Natural Resources Canada | Ressources naturelles Canada
Canadian Forest Service | Service canadien des forêts

The publisher gratefully acknowledges the assistance of Alberta Community Development and the Department of Canadian Heritage, and the financial support provided by the Alberta Foundation for the Arts.

Table of Contents

List of Keys and Illustrations

Keys

Illustrations

ACKNOWLEDGEMENTS

We'd like to thank some of the people who helped to produce this book. Special thanks to Trevor Goward for his detailed review of the lichens section. Bernard Goffinet also provided comments on the lichens section and Dale Vitt reviewed the bryophytes. In Alaska, Barbara Murray reviewed the lichens and bryophytes, and Alan Batten provided comments on the vascular plants sections. The vascular plants text was also reviewed by Vern Harms (in Saskatchewan) and Norm Kenkel (in Manitoba). Also in Manitoba, Robin Marles provided information on the ethnobotany of the boreal forest and reviewed the introduction. Thank you all for your comments and corrections.

We would also like to acknowledge the contribution of the editors and contributors of two previous publications, which provide continuity to this book: *Plants of Northern British Columbia* (including Ray Coupé, George Argus, Frank Boas, Craig DeLong, George Douglas, Trevor Goward, Andy MacKinnon, Jim Pojar, Rosamond Pojar and Anna Roberts; 1992) and *Plants of Coastal British Columbia* (including Paul Alaback, Joe Antos, Trevor Goward, Ken Lertzman, Andy MacKinnon, Jim Pojar, Rosamund Pojar, Andrew Reed, Nancy Turner and Dale Vitt; 1994).

Many individuals took the photographs used in this guide: Blaine Andrusek, Gerry Allen, Joe Antos, Frank Boas, Robin Bovey, Adolf Ceska, Blake Dickens, Robert Frisch, Ian Corns, Trevor Goward, Julie Hrapko, Derek Johnson, Linda Kershaw, Ron Long, Bill Merilees, Del Meidinger, M. Naga, Robert Norton, George Otto, Jim Pojar, Anna Roberts, Rob Scagel, Don Thomas, Terry Thormin, Nancy and Robert Turner, Trygve Steen, Dale Vitt, Cliff Wallis, Cleve Wershler, Kathleen Wilkinson, Dave Williams and J.M. Woollett. Photo credits are listed on p. 381.

Line drawings have been used from Britton and Brown (1913), and with kind permission from the following: British Columbia Ministry of Forests (including original line drawings by Shirley Salkeld); Columbia University Press (Crum and Anderson, 1981; Schuster 1969; Thomson 1984); Trevor Goward; Lone Pine Publishing (MacKinnon et al. 1992; Pojar and MacKinnon, 1994); John Maywood; Royal B.C. Museum (Brayshaw, 1976, 1985; Douglas, 1982, 1995; Schofield, 1992; Szczawinski, 1962; Taylor, 1963, 1973, 1974a, 1974b, 1983); Stanford University Press (Hulten, 1968); University of Michigan Press (Crum 1973, 1988); University of Washington Press (Frye and Clark, 1937; Hitchcock et al. 1955–69); Wm. C. Brown Co. Publishers (Hale 1979); National Museum of Canada (Porsild and Cody, 1980). Line drawing credits are listed on pp. 381–82.

Many thanks to staff at Lone Pine Publishing for their assistance in putting this guide together: Bruce Keith and Greg Brown for their hard work in design and layout; Roland Lines for his help with editing; and Glenn Rollans and Shane Kennedy for managing the project. Thanks are also due to staff at the Northern Forestry Centre of the Canadian Forest Service for their assistance, particularly Brenda Laishley (editing) and Martin Siltanen (preparation of map and general sounding board).

Financial assistance in production of this guide has been provided by the Canadian Forest Service under the Canada-Alberta Partnership Agreement in Forestry.

Derek Johnson, Linda Kershaw, Andy MacKinnon and Jim Pojar

ABOUT THIS GUIDE

The boreal forest is a vast patchwork of interesting and distinctive ecosystems — from aspen forests, jack pine/lichen forests, and old growth spruce/fir forests, to waterlogged peatlands, dry rock outcrops and grasslands. The plants of these ecosystems are the subject of this book.

This guide is designed for anyone interested in the plants of the western boreal forest and aspen parkland. The guide presents a simple yet thorough account of all major land-plant groups, and some aquatics, occurring in this diverse but relatively little-known region. While most other guides describe plants occurring within particular states or provinces, we have tried to present information on the entire western boreal bioregion, following ecological and climatic, rather than political, boundaries.

We wrote this guide with several types of users in mind: residents who want to know more about their natural surroundings; students, scientists and resource specialists who require an up-to-date reference on the plants of this region; and travellers who need a relatively simple and easy-to-use guide. When technical terms are unavoidable, they are defined in the glossary or explained in the introductions to plant groups.

The boreal forest region is large and diverse and so is its flora. This guide includes descriptions of the most common vascular plants, mosses, liverworts and lichens native to or naturalized in the forested boreal region west of roughly 92° W. Many of the common plants of the boreal forest are found from Alaska to Newfoundland and south to the aspen parkland region. Our primary focus is the closed boreal forest stretching across Manitoba, Saskatchewan, Alberta, northeastern British Columbia, and the southern District of Mackenzie, plus the northwestern boreal forest, in the Yukon, Northwest Territories and Alaska. However, it also includes many eastern boreal species.

Most primary entries include at least one colour photograph accompanied by line drawings to illustrate habit (that is, what the whole plant looks like) or details, where required. Keys are included to assist in separating some of the larger or more difficult groups.

Although this guide is fairly comprehensive, definitive identification of some groups of species will require consulting more technical references. For example, more than 50 species of sedge occur commonly in this region, and sometimes a dissecting microscope is required to distinguish their key features. The information provided in this guide is insufficient to clearly identify all of the sedges that can be found in the region, but the references listed at the end of this book can guide interested readers to more comprehensive technical manuals on sedges, grasses, willows and other difficult groups.

No other guide specifically covers the plants of the western boreal region, so working with these plants has required consulting several technical manuals for areas within this region. For vascular plants these included Hultén (1968) and Welsh (1974) for Alaska; Porsild and Cody (1980) for the Northwest Territories; Moss (1983) for Alberta; Looman and Best (1979) for the prairie provinces; Scoggan (1957) for Manitoba; Gleason and Cronquist (1963) and Britton and Brown (1913) for eastern areas; and Scoggan (1978–79) for Canada. Primary sources of information for non-vascular plants included the following: Vitt et al. (1988) for all groups; Conard and Redfearn (1979), Frye and Clark (1937) and Schuster (1969) for the liverworts; Conard and Redfearn (1979), Crum (1976), Crum and Anderson (1981), Ireland (1982), Lawton (1971), Schofield (1992) for the mosses; and Brodo and Hawksworth (1977), Goffinet and Hastings (1994), Hale (1979) and Thomson (1967, 1979, 1984) for the lichens. Some non-technical guides are available for various parts of the region (e.g. MacKinnon et al. 1992), but none has accurate habitat and range information for the entire area.

The usual habitat of a species in the boreal forest/aspen parkland is described under the heading 'Where Found,' along with its distribution in this region. Many boreal plants are found all around the world. These species are noted as 'circumpolar,' at the end of the distribution notes.

The 'Notes' section describes similar and/or closely related species that are also found in this region. It also provides information about aboriginal uses of the plant, historical uses in other places, natural history, plantlore, and/or etymology (name derivation).

HOW TO USE THIS GUIDE

This guide is organized so that species most likely to be confused with one another appear in the same section. Trees, shrubs, wildflowers, insectivorous, saprophytic and parasitic plants, aquatics, grasses and grass-like plants, ferns and relatives, bryophytes and lichens comprise the major sections in the book. Within these broad sections, we have organized plants by large families and by groups of smaller families.

For most readers, the quickest way to identify an unfamiliar plant using this guide is to start by browsing through the illustrations to find those species or groups of species the plant most closely resembles. The colour key to the wildflowers (pp. 77–80) may help to guide you to different groups of non-woody plants. Once you think you have found your plants, read the written description carefully. Sometimes, it may not be clear which characteristics provide the most consistent distinctions between or among closely related species. To complicate matters further, the appearance of any individual plant varies with its age, the time of year, the weather, soil types, disturbances, or other factors affecting growing conditions. The existence of local variants can also add to the difficulty of identifying a species.

For the most perplexing groups, especially those with closely related species, a technical key is necessary for positive identification. We have provided keys, with specialized terms kept to a minimum, for most major plant groups in this guide. These are simple, two-branched keys that rely on vegetative and floral characteristics. When flowers are necessary to distinguish species or genera, and all that remains on the specimen is the fruit or flowers of the previous year, a bit of imagination and reconstruction may be necessary to determine the path to follow in the key. When confusion arises at a branch in the key, examine both options, then determine which looks more plausible for your specimen.

The most common and widespread species are described in the greatest detail. Less common or localized species are discussed in the 'Notes' section of more common, closely related species. Rare species, or those whose range is limited in the region, are not generally included. For example, we have excluded several species that occur only near the eastern or northern limit of the western boreal forest, but are common in other regions.

Plant Names

Plants in this guide are listed with both common and scientific names. Scientific names are generally more widely accepted than common names. However, species concepts and names are not always stable or universal. The bog orchids, for example, are placed in the genus *Habenaria* in this book, while other works may refer to this genus as *Platanthera*.

We used the following sources for names of plants: common names largely follow Ealey (1992) for vascular plants; Glime (1989; 1991; 1992, 1993 a,b; 1994 a,b,c) for mosses; MacKinnon et al. (1992) and Pojar and MacKinnon (1994) for liverworts; and Goward et al. (1994) for lichens. Scientific names largely follow Scoggan (1978–1979) for vascular plants, Ireland et al. (1987) for mosses, Stotler and Crandall-Stotler (1977) for liverworts, and Egan (1987) for lichens. All names are listed alphabetically in the index at the end of this guide. Alternative scientific or common names may be used in other books about the plants of this region, so we have included some of these in the text and index.

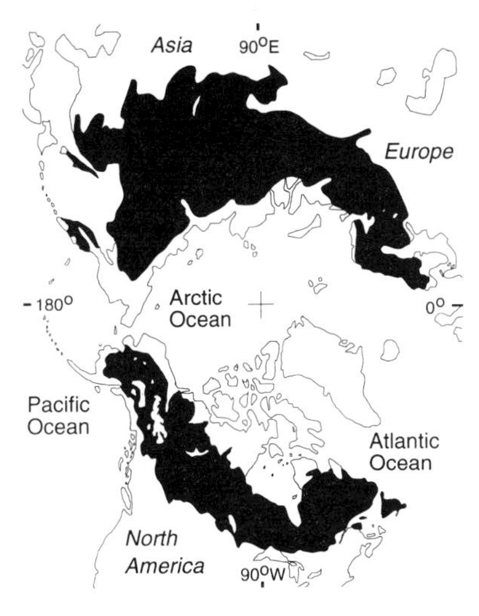

The Circumpolar Boreal Forest

THE REGION

The boreal zone is a broad northern circumpolar belt that, in places in North America, spans 10° of latitude. Boreal forest and taiga are terms often used interchangeably. Taiga is a Russian word originally applied to the broad ecotone between subarctic forest and tundra in Eurasia. Its use has been expanded to include subarctic forests in both Eurasia and North America.

The boreal forest of North America stretches from Alaska and the Rocky Mountains eastward to the Atlantic Ocean. To the north, it is bounded by the treeline. In the south, it is bounded by the aspen parkland, a transition zone to the prairie grasslands, and farther west by the great arc of the North American Cordillera. The boreal forest occupies more than 60% of the total area of the forests of Canada and Alaska. It is generally dominated by coniferous trees, mainly pine, spruce, fir and tamarack, the dominant species varying regionally and with local habitat. Broad-leaved trees, such as aspen and poplar, occur either in pure stands or mixed with conifers. These species become increasingly dominant towards the southern edge of the region, in the aspen parkland, as the northern forest grades to prairie.

This guide covers species of forested and non-forested habitats in the western boreal and aspen parkland regions.

Regional Variation

Boreal forest or taiga extends as a continuous band across North America. It covers about 28% of North America north of Mexico, making it the most extensive vegetation formation or biome on the continent. Although it presents a relatively uniform aspect over a vast area, and has comparatively low plant species diversity, the boreal forest is not the same everywhere. Several ecological classification schemes have been applied to the taiga, in an attempt to sort out the latitudinal and regional variation.

Rowe (1972) delimits 3 boreal subregions: forest, forest and barren (to the north), and forest and grass or aspen parkland (to the south). Similarly, Elliot-Fisk (1988) divides the boreal forest into 3 structural units or formations: closed forest, lichen woodland, and forest-tundra ecotone. The northern treeline separates the lichen woodland from the forest-tundra ecotone (or sometimes the tundra proper).

Other classifications combine physiography with variations in the relative importance of dominant species and plant communities. Larsen (1980) divides the North American boreal forest 'zone' into 7 'regions', 5 of which occur in the western boreal forest. The 'ecological areas' devised jointly by the U.S. Environmental Protection Agency and Environment Canada (1994, manuscript on file, Corvallis, Oregon and Ottawa, Ontario) roughly parallel Larsen's forest regions. The Canadian Committee on Ecological Land Classification (CCELC 1987) delimits 7 'ecoclimatic regions' in western Canada, which attempt to reflect a combination of climate, physiography, and vegetation.

Other more regional classifications include the 'ecoregions' of Alberta (Strong and Leggat 1981), the Yukon (Oswald and Senyk 1977), British Columbia (Demarchi et al. 1990), and Alaska (Bailey 1994; Gallant et al. 1994); these all appear to be conceptually different. British Columbia also has 'biogeoclimatic zones' (Meidinger and Pojar 1991). Alaska has fairly detailed mapping of vegetation types (Viereck 1971) and more generalized 'forest formations and zones' (Zasada and Packee 1994).

At least two themes emerge from the welter of regional classifications. Broad-scale zonation of the western boreal forest reflects: 1) latitudinal climatic change–from south to north, forest-and-grass to boreal forest to subarctic woodland to forest-tundra ecotone; 2) major physiographic divisions and phytogeographic provinces–from west to east, Alaska-Beringia to Cordillera to Plains to Shield.

Physical Environment

Climate

Large regional variations in the climate within the boreal region cause major differences in the composition and growth of the forests. Low temperatures and short growing seasons dominate the northern half of the region (Boreal - Forest and Barren), with mean annual temperatures of -2° to -10°C. Here tree cover consists of slow-growing, sparse, open-canopied black or white spruce, with trees rooted in a thin, seasonally thawed layer over permafrost. Northward, the treed areas are smaller, occupying valleys where they are protected from exposure. Patches of tundra first appear on ridges, and finally dominate the landscape.

The main portion of the boreal forest (Boreal - Predominantly Forest) is characterized by close-canopied stands. The mean annual temperature ranges from +4.8°C in the most southern regions to -2.7°C in the north. The western boreal forest, west of 92° W, generally receives low annual precipitation (<500 mm). In the northern parts of the region, there is a trend of increasing precipitation from west (interior Alaska to the Yukon) to east (through northern Alberta to northern Manitoba) (Hogg 1994), and precipitation increases eastward from Ontario through the eastern boreal forest to the Atlantic coast.

Physiography, Geology and Soils

Three broad physiographic regions can be delimited in the boreal zone: the eastern Kazan Region, part of the granitic/gneissic Canadian Shield; the central Interior Plains; and the western Cordillera. The latter two are underlain by mostly sedimentary and metamorphic rock. Although till and outwash predominate, a large area in interior Alaska and the adjacent Yukon was ice-free in the Pleistocene, and has significant deposits of aeolian (wind-blown) material or loess.

Key factors in soil development include low temperatures, poor drainage, thick surface layers of moss/organic material, low levels of available nutrients and, in some areas, permafrost. Much of the western boreal forest is underlain by discontinuous permafrost, where patches of permanently frozen ground alternate with unfrozen ground during the summer months. In this region, permafrost is usually present in peatlands and in old, unburned stands where thick layers of insulating organic material have accumulated. Northwards, unfrozen ground becomes less prevalent, and eventually, near the limit of trees, all ground is underlain by permafrost.

Differences in soil type and drainage result in a mosaic of stand types in the boreal forest. Rapidly drained, coarse-textured soils are dominated by pines, as are the shallow soils over bedrock on the Canadian Shield. In areas with higher precipitation, however, the abundant moisture compensates for the coarse soil texture and mixed spruce-fir-birch stands are found, even on sandy soils. Hardwood-conifer mixtures generally occur on fine-textured soils that are well drained. Black spruce dominates the imperfectly drained sites, and stunted, open stands of black spruce, often mixed with tamarack, grow on poorly drained peatlands.

Soils on the Canadian Shield, derived from granites, are more acidic and more poorly developed than those derived from sedimentary materials, which dominate many cordilleran and lowland areas. Areas underlain by limestone are often associated with distinctive plant communities. Some moss, fern, and flowering plant species occur exclusively in association with limestone substrates. In the north, climate becomes the overwhelming factor limiting plant growth, so subtle distinctions among rock types become less important.

Time

The diversity and composition of plant communities in the western boreal forest have been profoundly influenced by the region's geologic history. Eighteen thousand years ago there was no boreal forest in what is now Canada and Alaska. Except for a few small pockets, the area now covered by this forest was largely blanketed by ice sheets several kilometres thick. After the ice sheets melted, plants colonized the resulting bare mineral substrates. The particular mix of plants, animals and soils that currently comprise the boreal forest has developed, for the most part, over the last 12,000 years. The plants that first colonized this area were the ones able to quickly take advantage of the vast areas of exposed land. These species were adapted to dispersal over long distances, and they could thrive in relatively simple communities. Today, many boreal species retain these pioneer characteristics, requiring a mineral substrate on which to reproduce, and many depend on wind for pollination and dispersal.

Vegetation

Forests

Trees can outcompete other forms of plants in the boreal region because trees live longer, grow taller, and can often tolerate a poor nutrient supply. Because some trees are shade-tolerant, they can persist under a dense forest canopy. Once established, trees can accommodate wide year-to-year fluctuations in climate better than most other kinds of plants. Conifers can photosynthesize at lower temperatures than deciduous trees, and so have a longer growing season in the relatively short, cool summers of the boreal forest. This is why conifers are so common worldwide at higher latitudes and elevations.

The vast mosaic of forest stands that forms the boreal forest reflects local and regional variations in temperature, moisture, soil development, and many other environmental factors. General patterns associated with changes in slope and elevation are presented schematically below. Countless combinations of environmental factors determine the success or failure of individual plants, and hence the composition of these plant communities. However, forest dynamics (disturbance and succession) must also be taken into consideration.

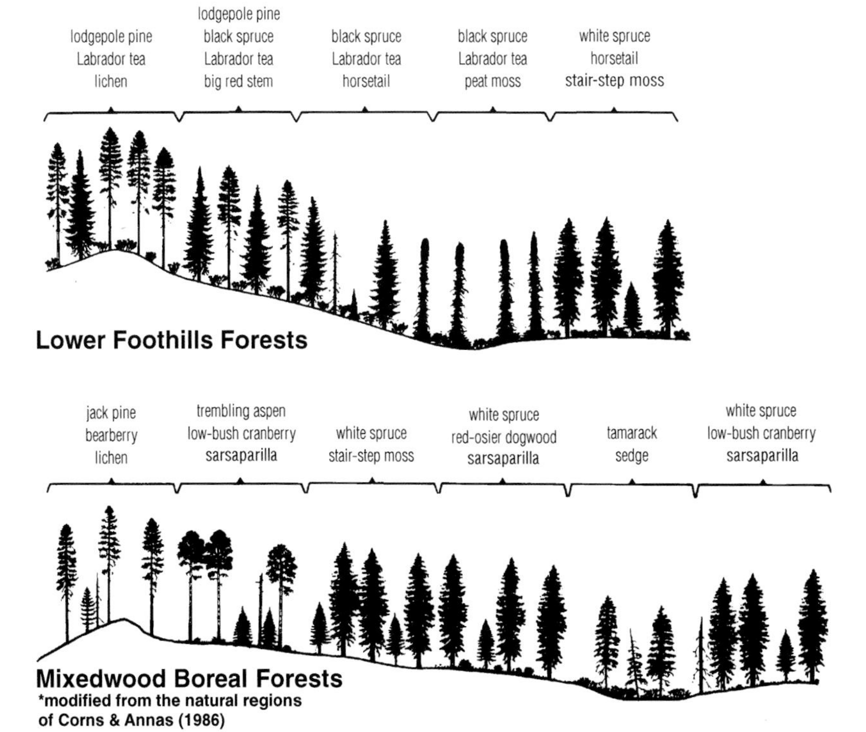

***modified from the natural regions of Corns & Annas (1986)**

Dynamics

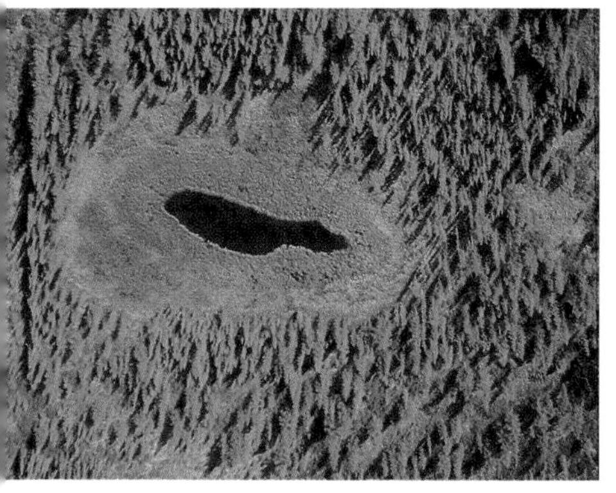

Plant communities develop through a process called 'succession', which involves change in community composition and structure over time. All communities are subject to natural disturbances of different kinds that can kill existing members of the community and 'reset' successional processes to varying degrees. Fires, insect and disease outbreaks, floods and windstorms all affect plant communities. However, in the boreal forest, fire is the dominant short-term influence, determining the distribution and growth of forest stands. Historic records show that areas burn every 50 to 150 years on average, depending on the local site conditions. Few boreal forest stands reach an age of more than 150-200 years. Because of frequent forest fires, the boreal forest is characterized by large areas of even-aged stands, composed mainly of pioneer species established after fire. Very dynamic disturbance regimes and complex succession have resulted in high ecosystem diversity, with a mosaic of habitats, vegetation types and successional stages over the landscape.

Many boreal tree species are adapted to surviving the effects of fire. Some trees (jack and lodgepole pine and black spruce) keep their cones for several years. The cones, containing viable seeds, are sealed until the cone scales are opened by the heat of a fire. The seeds are released after the fire has passed, and they germinate readily on the mineral soil exposed by the fire. Aspen has a different reproductive strategy: although its above-ground parts may be killed by the fire, most roots survive and produce sprouts. Paper birch usually produces suckers from the root collar, resulting in multiple-stemmed clumps. Understory plants can survive as seeds buried in the soil for decades, only to germinate when the overstory is removed.

Old-growth forests

As young forests grow into older ones, they pass through a series of characteristic developmental stages. Old growth is the final stage of forest development. Some changes that occur as a forest develops relate to the replacement of species characteristic of early successional stages by species of later stages. Fast-growing, shade-intolerant species such as aspen and pine are replaced with slower-growing but more shade-tolerant species such as white spruce and balsam fir as the canopy of the pioneer forest closes. If left undisturbed, these shade-tolerant species will dominate the forest and the pioneer species will decline in number. The next generation is composed of species that are able to regenerate under dense shade (balsam fir), with the other species occurring in occasional openings created by the death of old trees. However, this relatively stable (climax) community is rarely encountered in the boreal forest because this successional sequence is usually interrupted by fires.

There are many characteristics of old forests that are not found in younger forests, especially young forests managed for timber production. Structural attributes characteristic of older forests are a wide range of tree sizes and ages, and a patchy, open canopy punctuated by gaps beneath which the forest understory is especially well developed.

Various combinations of old-growth-like characteristics can certainly be found in some younger forests. This is especially true for stands regenerating without human interference after natural disturbances, or after logging in the early days of the industry, when harvesting operations were less efficient at removing all of the living trees, snags and logs. Modern forest plantations that are managed intensively for timber production on shorter rotations retain or develop few old-forest attributes. The challenges of redesigning forestry practices to mimic natural patterns of forest development and to retain or create old-forest-like characteristics in younger forests, are at the forefront of applied research in forest ecology and management.

Wetlands

The amount of land surface covered by wetlands in the boreal forest is significant: approximately 21% of Alberta, 16% of Saskatchewan, 38% of Manitoba and over 25% of the Northwest Territories is covered by wetlands. However, the classification of these wetlands is a complicated matter. Over time, different classification systems have created a profusion of terms. These terms have often been used interchangeably and this has resulted in considerable confusion about wetlands amongst the general public.

What is a wetland? A simplified definition of a wetland is an area of land where the water table is at or above the level of the mineral soil for the entire year. This is generally a workable definition except in the case of some marshes. Wetlands include mineral wetlands (areas of mineral soil that are influenced by excess water but which produce little [less than 40 cm] or no peat [the decaying remains of plants]; and organic wetlands or peatlands (areas with accumulations of more than 40 cm of peat above the mineral soil). The 40-cm limit was chosen because at this thickness the vast majority of wetland plants are rooted solely in peat.

We distinguish five classes of wetlands: bogs, fens, swamps, marshes, and shallow open water less than 2 m deep. The relative positions of these wetlands along gradients of water level fluctuations, water flow, nutrient availability, and rates of growth and decay are presented schematically below.

Types and Characteristics of Wetlands Found Within the Western Boreal Region*

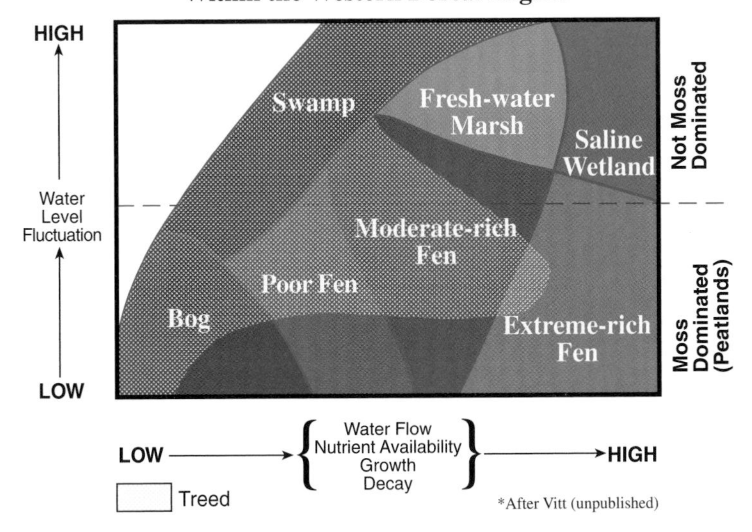

*After Vitt (unpublished)

Sphagnum bog

BOGS are peat-covered wetlands in which the vegetation shows the effects of a high water table and a general lack of nutrients. Bogs receive their nutrients only from rain water; the vegetation is not nourished by mineral-enriched groundwater. Calcium (Ca) and magnesium (Mg) levels in the groundwater are extremely low. The bog surface is virtually isolated from mineralized soil waters. The surface waters of bogs are strongly acidic (pH generally less than 4.6) and the upper peat layers are extremely low in nutrients. Cushion-forming *Sphagnum* mosses and heath shrubs are common. Trees may be present or absent; if present, they form open-canopied stands of low, stunted trees. *Sphagnum* mosses are the primary peat formers.

FENS are peatlands characterized by a high water table, but with very slow internal drainage by seepage down very gradual slopes. The slow-moving groundwater is enriched by nutrients (particularly Ca and Mg) from upslope materials and thus fens are more mineral-rich than bogs. The pH of the groundwater is generally 5.5 - 7.0. Fen vegetation reflects the quality and quantity of available water, resulting in three basic types: graminoid (usually sedge) fens, shrub fens, and treed fens. Sedges and "brown mosses" are the primary peat formers in fens.

Treed fen

Patterned fen or ribbed fen

SWAMPS are wetlands where standing or gently moving waters occur seasonally or persist for long periods, leaving the subsurface continuously waterlogged. The water table may drop seasonally, creating aerated conditions in the rooting zone of the vegetation. Swamp waters are almost neutral to moderately acid, and show little deficiency in oxygen or mineral nutrients. Swamps are nutritionally intermediate between bogs and fens. Their substrates consist of mixtures of mineral and organic materials, or woody, well-decomposed peat. The vegetation may be dense coniferous or deciduous forest, or tall shrub thickets. Most peat-forming mosses are absent, or present only in a subordinate role. Woody species are the primary peat formers in swamps, and in many swamps, peat formation is minimal.

Tall shrub swamp

Treed swamp

MARSHES are wetlands that are periodically inundated by standing or slow-moving water, and hence are rich in nutrients. They are mainly wet, mineral soil areas characterized by an emergent vegetation of reeds, rushes, sedges, or grasses. Peat formation is often minimal. Marshes are subject to water draw down, but water remains within the rooting zone for most of the growing season. Waters are usually almost neutral to slightly alkaline. The surface water levels of marshes may fluctuate seasonally, and the vegetation often has distinct zones reflecting water depth, frequency of drawdowns, and/or salinity.

All of these wetland classes reflect the nutrient status of the waters that feed them. The colloquial, and widely misunderstood, term '**muskeg**' is used to refer to the complex mosaic of fens, bogs, swamps, pools and scrubby forest that becomes increasingly common to the north.

Cattail marsh

Aspen Parkland

The aspen parkland forms a broad belt across the prairie provinces between the warm, dry prairies to the south and the cooler, moister boreal forest to the north. It consists of open grassland alternating with groves of trees (usually aspen), and it can be divided into three sections: eastern, central and western. In the eastern parkland, the grassland is tall-grass prairie, where big bluestem (*Andropogon gerardi*) and porcupine grass (*Stipa spartea*) predominate in the drier areas and prairie cord grass (*Spartina pectinata*) predominates in moister areas. Tree groves contain bur oak (*Quercus macrocarpa*), green ash (*Fraxinus pennsylvanica*), Manitoba maple (*Acer negundo*) and balsam poplar (*Populus balsamifera*), in addition to aspen. The central parklands have openings of mixed-grass prairie consisting of plains rough fescue (*Festuca hallii*), western porcupine grass (*Stipa curtiseta*) (replaced by *S. comata* on lighter soils), timber oat grass (*Danthonia intermedia*), and Hooker's oat grass (*Helictotrichon hookeri*). Tree groves consist predominantly of aspen, and often include some willows. In the western parkland, the grassland is often a fescue prairie, with plains rough fescue, needle-and-thread (*Stipa comata*) (replaced by *S. curtiseta* on heavier soils), and timber oat grass. Tree groves are predominantly aspen mixed with willows and balsam poplar.

Grasslands

Areas of grassland or shrub-grassland within the boreal forest are relatively uncommon and localized, but can be more extensive in drier, warmer parts of the region. Such communities typically occur on dry, often south-facing slopes above many of the major rivers. Many probably originated following fires. Some have been maintained by a combination of grazing and annual burning, but others are on sufficiently dry, extreme slopes that they persist with only occasional fires.

Disturbed Habitats

These include roadsides, railroad and powerline right-of-ways, trails, recently logged areas, agricultural areas, settled areas and other sites frequently disturbed by people. Disturbed habitats are often dominated by Eurasian weeds (particularly yellow-flowered composites) and introduced grasses, but native species also occur.

PLANTS AND PEOPLE

Aboriginal Peoples

Humans are thought to have moved into North America from Asia via a land bridge across the Bering Strait about 40,000 years ago. From Alaska and the Yukon, these nomadic tribes probably moved south along an ice-free corridor between glaciers, and dispersed south through North America. Then, as the glaciers receded, some tribes moved to inhabit northern regions, and eventually (7,000-10,000 years B.P.) groups settled in the subarctic region, in the boreal forest. The ancestors of some subarctic tribes may have been among the last to cross the shrinking Bering Land Bridge. A Chipewyan legend tells how an ancestral woman and her child waded across a shallow sea by walking continuously for two days. The woman was led by a wolf, and when she reached the other side she found a herd of caribou and killed one. A promising start to life in a new land.

Today, two major groups of native peoples inhabit the subarctic, separated on the basis of language - those speaking Athapascan and those speaking Algonkian. **Athapascan** speakers generally refer to themselves as **Dene** (Din-neh) or "people", in the sense of "our people." The Dene Nation extends from Alaska to northern Manitoba and includes over 2 dozen "tribes" or "bands," from the Ingalik and Tanana in the northwest to the Chipewyan and Beaver in the southeast. The northern **Algonkian** speakers inhabit the Precambrian Shield, from Saskatchewan to Labrador, from the Montagnais and Naskapi (Innu) of Labrador to the northern Ojibwa of Lake Superior and the **Cree** of the western boreal forest. There have also been periodic incursions of **Inuit Eskimo** people into the northern boreal forest region.

Population levels in the western boreal were always low (probably always less than 35,000 Athapaskans in precontact times), and most people lived in small, extended family groups. These bands readily helped one another, and people felt free to hunt outside their normal area in times of shortage. Tribal borders shifted through time, and bands were not always clearly delineated, as membership frequently shifted between groups within a culture but not between Cree and Dene.

Perhaps the most notable border lay between the Chipewyan and the Cree, where the two language groups met. The Cree were the focus of the second major phase of the fur trade, beginning with the incorporation of the Hudson's Bay Company in 1670. However, as the fur traders moved farther west, the Cree tried to prevent the Dene from dealing directly with the Europeans. In so doing, they forced the Chipewyan back to the north and west, and eventually established territory around Lake Athabasca.

Boreal Forest Tribes

These northern hunters shared many fundamental characteristics in their adaptation to the subarctic environment. Large game (caribou and moose in particular) provided the foundation for life. Big game was hunted with bows and arrows or spears, but because of the low density of animal populations in most areas, people often relied on traps and snares. Caribou were hunted along migration routes in some areas during the fall, when they were fat and their coats were in good condition. Sometimes several bands would gather to slaughter migrating herds from boats at river crossings, or to build corrals into which many animals could be herded and killed. Fish were also widely used for food, either trapped, netted, hooked, clubbed, or speared. During the summer, some bands congregated in fishing camps, but in winter they returned to their preferred hunting areas. Surplus game meat and fish were dried in thin strips and stored in caches. Smaller game, such as hare and ptarmigan, was usually snared or trapped by the women, and this added variety to the diet.

Animal-derived food (meat, organs, blood, fat, marrow, rumen contents, etc.) was estimated to comprise 95%-97% of the native diet. Although plant foods were limited and of low cultural status, some were eaten, especially in season. Berries were the main plant food consumed. However, a few roots, bulbs, and young shoots were used occasionally. Birch sap was considered a treat in spring, and spruce gum could be chewed year-round.

The typical winter home was made of a cone-shaped or rounded wooden frame covered with moose or caribou hides or with moss, spruce boughs and bark. This was then banked with snow for greater insulation. A distinctive feature of many camps was a raised food cache, either hung from a de-barked pole or laid out on a high platform. Winter clothing included caribou coats with hoods and moccasins on the feet for use in snowshoes.

Snowshoes were the primary means of transportation in the winter. Without snowshoes, these people could not have survived winter in the soft snow of the forest. Sleds were also used. Originally these were pulled by people, but with the introduction of rifles and commercial fishnets from the fur trade, it became possible to feed dogs as well as family members, so dog teams appeared. It was also necessary to have a dog team to run a trapline to supply the fur trade, which provided these "luxuries." During the summer, moosehide or birch-bark boats were used for transportation, and most human activity focused along the rivers, streams and lakes of the region.

Subarctic tribes shared many spiritual beliefs. Many hunters believed that success depended on the prey's willingness to support the life of the hunter and his family, and that this could be gained through rapport with the spirits. Young men often went alone to contact the spirit that would help them through life. Similarly, dreams and visions were thought to be messages from the spirit world. Those with frequent visions were called to be shamen or doctors, and were taught the medicinal value of plants, as well as skill in sleight-of-hand and ventriloquism. Their knowledge was passed from generation to generation by word of mouth.

Native peoples in the subarctic remained relatively unaffected by the presence of Europeans in North America for many years. Their land was not suitable for agriculture, and the people remained widely scattered in a vast wilderness. The advent of the fur trade in the 17th century brought some changes, with the introduction of guns, commercial fish nets and metal tools, and increased reliance on these items. It also brought alcohol and disease to many. However, people generally continued to live in much the same way as their ancestors had, until the mid-20th century.

In the past 70 years, subarctic natives have undergone drastic cultural and sociological changes. During that time, the governments assumed greater responsibility for the physical well-being of northerners, with increased social services and subsidies and the establishment of settlements. Family allowance, old age pensions and welfare allotments were introduced, along with day schools and hostels. With this government support came a general movement of the population away from the land and into towns and hamlets where they could access facilities. The advent of the electronic age, with radio, television, satellite disks and computer technology, bombarded homes with information and cultural values from other societies, further isolating young people from their traditional way of life.

Recently, many efforts have been made to reacquaint northern natives with their spiritual heritage and to teach young people the ways of the land that were practised by their ancestors. With this has come increased interest in subarctic ethnobotany and the recording of knowledge that was historically passed from one generation to the next by word of mouth.

The term "traditional" can be misleading, as it suggests a fixed way of doing things. All cultures are continually evolving, adapting to changes in their environment. When knowledge is based on an oral history, it too will change through time. The "traditional" uses discussed in this book were generally reported by elders in the past 5-15 years. Therefore, they probably reflect life among these people 50-100 years ago. Aboriginal use of plants in the subarctic prior to European contact is largely unknown.

Former Europeans

The impact of Europeans on the vast expanse of the western boreal forest has been minimal to date, compared to most other parts of the continent. Development remains scattered, usually in the form of small communities, mines, hydroelectric dams and scattered logging areas. Historically, Europeans considered the subarctic too marginal for agriculture and its trees too small and slow-growing to warrant harvest. But things have changed rapidly in the world economy, and the boreal forest is destined for change. In a paper-hungry world with decreasing forest cover, wood pulp has increased greatly in value. Vast areas of boreal forest have been designated for harvest to feed growing numbers of giant paper mills. Catastrophic events are not new to the boreal forest. Historically, fires (natural or man-made) have repeatedly burned large tracts of the subarctic, creating a mosaic of young and old stands. However, the impacts of extensive harvesting and road networks (with vastly increased access) remain to be seen.

Ethobotany

Ethnobotany, the study of the relationships between people and plants, has not been widely studied in the subarctic. Although northern natives were primarily nomadic hunters and fishers who relied on animal products for their food and clothing, plants were also important to many aspects of life. Trees provided wood, pitch, bark and roots, mosses provided absorbent diapers, and a wide range of plants provided medicines and food. This book, provides some details of the aboriginal uses of plants in the subarctic, in the hope that readers will gain a greater understanding and appreciation of the cultures and lifestyles of native peoples in this region.

The ethnobotanical information in this book was taken from a wide range of sources. Whenever possible, the uses of plants by aboriginal people in the subarctic was reported. Two important sources of such information were Marles (1984) and Leighton (1985). However, if a plant did not appear to have been used in this region (as many did not), but had a history of use in other places, some of this information is presented. Sources are listed in the "References" section.

Plants in aboriginal technology

Plant materials were used to make many items that were important to daily life, ranging from utensils, baskets and whistles to canoes, showshoes, and dwellings. Some articles (e.g. roasting sticks) could be produced with little alteration of the raw materials, but other items (e.g. birch bark canoes) required the collection of materials from several species, and the careful shaping and assembly of component parts. Almost all building materials were supplied by trees and shrubs, which provided wood, bark, boughs, pitch, poles and roots.

Fire was an important part of life in this northern region. A spark, generated by striking 2 rocks together, or heat generated with a drill, was used to ignite tinder. Then, a small piece was gently detached and used to ignite other fine plant materials such as dry grass, birch bark, or lichen-covered twigs. Fungal conks were most commonly used as tinder.

Several aromatic plants, alone or in combination, were used in trap lures. Although recipes varied, dried, chopped beaver castor glands were usually the principal ingredient in these lures. Plants also served as seasonal indicators of animal behaviour. For example, when fireweed was flowering, moose were feeding and fattening, prior to mating, but once the plants had gone to seed the mating season was in progress. This information could be used to determine the timing of a hunt.

Many plants, though principally mosses, were used for stuffing or padding, or as the absorbent material in dressings, diapers and menstrual pads. Sphagnum moss is not only soft, insulating and absorbent, it also has antibacterial properties. Winter supplies of moss had to be collected in the summer, dried and stored in a shelter, as they became inaccessible once the muskeg had frozen.

Plants as food

Plants represented only a small part of the traditional diet of aboriginal people in the subarctic (est. 3%-5%), and many northern residents still consider salads and vegetables "rabbit food." However, a variety of species was used for food. Rocktripe and cattail roots could be found year-round, but most plants had a limited season when they were collected. Birch sap and water parsnip roots were used in spring, and some fleshy fruits were collected in fall, but the greatest variety of plant foods was enjoyed during the summer.

Most plant foods were eaten fresh, but berries were often preserved by drying, freezing, or storing in a cool place (the muskeg). Wild berries are still collected in large quantities, but many traditional foods are no longer used. For example, crushed bearberries with raw fish eggs, or rocktripe in fish stew, may be eaten by elders, but these foods are unknown to most young adults.

Many plants were used to make herbal teas, in the days when water or broth were the only other beverages available. However, these have now been supplanted by store-bought teas. Wild mint is the exception, and it is still commonly mixed with tea for flavour. It is often difficult to draw the line between teas that are used as beverages, and those used as preventive medicines.

Plants as medicine

Relatively few plants were used traditionally for food, but a wide variety of species have provided medicine to treat or prevent ailments or illnesses. Because the use of medicinal plants was so closely linked with the aid of supernatural powers, information about herbal remedies is often complex and/or unclear. Many believed that the magical powers of healing plants diminished greatly following the adoption of Christianity, and there are no more "true" medicine men, those who manipulated magical powers for healing or harm. However, it is sometimes thought that the power of healing plants can be retained by following the proper procedures when collecting, preparing, administering and storing medicinal plants. For example, bark to be used as a laxative should be scraped from the trunk with downward (not upward) strokes, and the maximum strength of some barks will be gained by collecting from the side of the trunk facing the mid-day sun. Similarly, powdered medicine must swirl in the cup before sinking if it is to be effective.

The collection and preparation of plants for medicinal use required considerable effort, when compared to gathering plants for food. Sometimes it was necessary to travel several kilometres to find a certain species. Some plants were procured from people travelling from other regions. Often these were recognized by the smell and taste of the dried plant material, rather than by the characteristics of the living plant. Occasionally, plants that were difficult to find were cultivated. For example, the Cree sometimes grew cow parsnip (*Heracleum lanatum*), broad-leaved water-plantain (*Alisma plantago-aquatica*) and eastern mountain-ash (*Sorbus decora*), and the Chipewyan probably introduced sweet flag (*Acorus calamus*) to their region, well north of its usual range.

Traditionally, medicinal plants were collected, cleaned, air-dried and wrapped in separate containers for storage in a safe place. Some could be kept for long periods, but many had to be replaced each year as they lost their strength with age. Herbal remedies were administered in a variety of ways, including teas (steeped or boiled), ointments, salves, and poultices, depending on the ailment and the plant being used. They could also be inhaled as steam from a decoction or as smoke.

The benefit of many traditional medicines is undeniable. Many aboriginal people have reported cures or alleviation of symptoms following treatment of themselves or of other members of their family or community. Some traditional medicinal plants have been shown to have antibiotic or other pharmacological properties, but some are potentially toxic, especially if they are taken in the wrong dosage or improperly prepared.

PLEASE NOTE: This book records many historical uses of plants found in the boreal forest. Such information is presented to give the reader a better sense of the rich cultural and natural heritage of the region. **This guide is not meant to be a 'how-to' reference for consuming wild plants. We do not recommend experimentation by readers, and we caution that many plants in our region, including some traditional medicines, are poisonous or harmful.**

Trees

Trees are single-stemmed, woody plants that are usually taller than 10 m when mature. However, in the boreal forest, and especially in muskeg or near the northern limit of trees, tree species often have mature forms that are stunted or dwarfed, and less than 10 m tall.

Evergreen, needle-leaf conifers (spruces, pines and firs) dominate the boreal forest. Deciduous, needle-leaf conifers (tamarack) and the deciduous, broad-leaved trees (poplars and birches) are also widespread and often abundant. Although deciduous trees photosynthesize faster than evergreen trees, the evergreen trees are usually able to photosynthesize at lower temperatures, which gives them a longer growing season.

The two spruces found in our region are white spruce (*Picea glauca*) and black spruce (*Picea mariana*)—both transcontinental species that pretty well define the extent of the North American boreal forest. The pines and true firs of our region each consist of a pair of closely related species: jack pine (*Pinus banksiana*) and lodgepole pine (*Pinus contorta*); and balsam fir (*Abies balsamea*) and subalpine fir (*Abies lasiocarpa*). Jack pine and balsam fir are found east of the Rockies, and lodgepole pine and subalpine fir are found in the mountain regions. There are no pines or true firs in the Alaskan boreal forest. Tamarack (*Larix laricina*) is nearly as widespread as the spruces, but it has a patchier distribution, especially in the western mountainous part of our region. All three of our major hardwood trees—white birch (*Betula papyrifera*), aspen (*Populus tremuloides*) and balsam poplar (*Populus balsamifera*)—are widespread across the boreal forest.

Although they have huge ranges, all our boreal tree species are confined to North America. Similarly, the Eurasian boreal spruces, pines, larches, birches, and poplars are confined to Eurasia. The general principle seems to be that the smaller forest plants are more wide-ranging. Thus, there are no circumpolar trees, a few circumpolar shrubs, several circumpolar herbs, and very many circumpolar bryophytes and lichens. It is unclear why this is the case.

Key to the Trees

1a. Leaves needle-like, mostly evergreen; seeds in cones, not enclosed in fruit (conifers) 2

- **2a.** Leaves (needles) in clusters of 2 or more 3
  - **3a.** Needles in clusters of 10–20; deciduous .... ***Larix laricina***
  - **3b.** Needles in clusters of 2; evergreen 4
    - **4a.** Cones point straight out from stem or bend backwards; cone scales have prickles .... ***Pinus contorta***
    - **4b.** Cones point towards shoot tip; cone scales have very small (or no) prickles .... ***Pinus banksiana***
- **2b.** Leaves (needles) not in clusters, borne singly on twig 5
  - **5a.** Branches smooth where needles have fallen, with circular leaf scars; cones erect; needles flat; young bark smooth and covered with resin blisters 6
    - **6a.** Needles have many stomata on both upper and lower surfaces .... ***Abies lasiocarpa*** (see *A. balsamea*)
    - **6b.** Needles have few (or no) stomata on upper surface .... ***Abies balsamea***
  - **5b.** Branches rough where needles have fallen (due to presence of peg-like leaf bases); cones hang from stem; needles 4-sided in cross-section; bark thin and scaly 7
    - **7a.** Cones about 2.5 cm long, egg-shaped; branches often droop, but have upturned ends; tree-top knobby or club-shaped; young twigs densely hairy .... ***Picea mariana***
    - **7b.** Cones 2.5–5 cm long, cylindrical to narrowly oval; lower branches often droop, do not have upturned ends; tree-top rarely club-shaped; young twigs hairless .... ***Picea glauca***

1b. Leaves broad, fall off each year; seeds enclosed in fruit 8

- **8a.** Leaves opposite, compound (divided into leaflets) 9
  - **9a.** 3–5 leaflets, lobed, 6–7 cm long; fruits are V-shaped pairs of samaras .... ***Acer negundo***
  - **9b.** Usually 7 leaflets, not lobed, 7–12 cm long; fruits are single samaras .... ***Fraxinus pennsylvanica***
- **8b.** Leaves alternate, simple (not divided into leaflets, but may be lobed or toothed) 10

10a. Bark has many conspicuous horizontal markings (lenticels),
peels in large sheets .... ***Betula papyrifera***

10b. Bark has no conspicuous horizontal markings (lenticels), does not peel in large sheets 11

11a. Leaves deeply lobed; fruit an acorn .... ***Quercus macrocarpa***

11b. Leaves toothed but not lobed; fruit not an acorn 12

12a. Leaves coarsely double-toothed, lop-sided at base;
male and female flowers on same tree; seed a samara, in small clusters .. ***Ulmus americana***

12b. Leaves finely toothed, not lop-sided at base;
male and female flowers on separate trees; seed tiny,
with tuft of hair attached, in dense clusters of capsules (catkins) 13

13a. Leaf stalk flattened; mature bark smooth and waxy
but often becomes furrowed with age near base of trunk;
end bud about 0.6 cm long, not resinous;
leaves nearly circular in outline .... ***Populus tremuloides***

13b. Leaf stalk round; mature bark deeply furrowed;
end bud about 2 cm long, very resinous (gummy);
leaves broadly egg-shaped .... ***Populus balsamifera***

Picea glauca

Picea mariana

Pinus banksiana

Pinus contorta

Abies balsamea

Larix laricina

Quercus macrocarpa

Ulmus americana

Acer negundo

Fraxinus pensylvanicus

Populus balsmifera

Populus tremuloides

Betula papyrifera

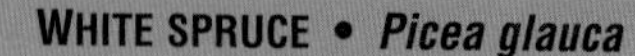

WHITE SPRUCE • *Picea glauca*

GENERAL: Stunted to erect evergreen tree, 7–20 m tall (sometimes to 40 m); **young twigs smooth and shiny**, not hairy; bark has loose ashy brown scales; inner bark has pinkish (sometimes reddish) tinge.

LEAVES: Needles 4-sided, 1–2.5 cm long, pointed, **stiff**, bluish green, tend to project from all sides of branches (like a bottlebrush), aromatic when crushed.

CONES: Pollen cones pale red; seed cones hang down, light brown to purplish, 2.5–5 cm long (usually 3.5 cm) at maturity; scales papery, smooth-edged.

WHERE FOUND: Grows well on well-drained, moist soils; widespread and common across the boreal forest.

NOTES: Black and white spruce are the most widespread trees in the boreal forest. • The Chipewyan mixed spruce pitch with a 'rich' fat (otter, bear or beaver fat), which they first burned in a frying pan, to make a salve to apply to sores, infections, insect bites, boils, chapped hands, scratches and cuts. Other tribes used the pitch, or a boiled decoction mixed with fat, to make a salve to treat burns (including sunburn), rashes, boils, scabs, and various skin diseases. Some people preferred to use black spruce for this. The Woods Cree used a decoction as an ingredient in an arthritis remedy. The Cree applied black or white spruce pitch to boils to draw out infection. They also chewed the gum at the first sign of a cough or sore throat. Some people took boiled pitch as a cough syrup, or they sucked on young cones or lumps of dried pitch and swallowed the juices to soothe a sore throat. • The Chipewyan boiled the inner bark of white spruce to make a medicinal wash or applied it directly as a poultice to treat skin sores and infected or decayed teeth. • In the winter, feet often went unwashed and were kept covered in moccasins for long periods of time. As a result, cracks would develop in the thick skin of the heel, and these were filled with thin spruce pitch or pitch salve to avoid infection and help healing. • The Chipewyan melted hardened spruce pitch or boiled it in water (sometimes with charcoal added to make it black), and used it to caulk the outside seams of canoes. These canoes had to be loaded and unloaded in the water, because scraping the bottom on land could break off pieces of the hardened caulking. Soft pitch provided a natural glue for sticking sheets of birch bark and twisted strands of willow bark twine together. It was also used to seal the inside seams of canoes and to waterproof and preserve babiche (strips of hide) in ropes and snares. • Some people chewed black or white spruce pitch like candy, though it is not as popular today. Chipewyan children would sometimes steal pitch from the caulked seams of canoes to chew, but parents kept an eye out for pale brown gum, the colour of boiled pitch. Gum straight from the tree is yellow. The chewing of spruce gum was credited with keeping the teeth of Chipewyan women so white. • Spruce branches have been used to make lean-to shelters, caribou-hide tipi poles, windbreaks, tent-base wind barriers, carpets, caches, meat-drying racks and dig shelters. • Once white spruce wood has been dried, it is light and splits easily and straightly, making it excellent for canoe frames, paddles, arrow shafts and snowshoes. • Dead standing white spruce trees were used to make a moose hide stretcher. • The Woods Cree collected the reddish, crumbly, rotted wood of white spruce from old stumps and burned it in a slow fire to smoke-tan hides. • Fresh or soaked black or white spruce roots were peeled, split, and used to make cord or rope for stitching the seams of birch-bark canoes or baskets, and for making fishnets. Watertight baskets woven from spruce roots could be used for cooking. Some women dyed small spruce root baskets different colours and used them to keep their sewing in. • Spruce beer was very important in preventing scurvy among early European travellers. The crews travelling with Captain James Cook received a ration of spruce beer every other day. • White spruce is the provincial tree of Manitoba.

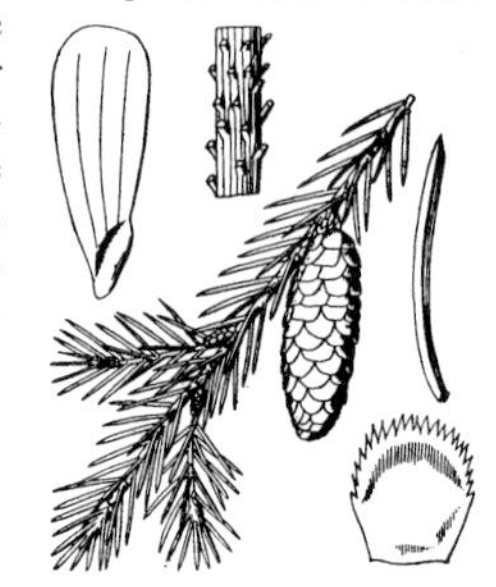

BLACK SPRUCE • *Picea mariana*

GENERAL: Small, often shrubby evergreen tree, 7–10 m tall (sometimes to 15 m); in older trees, lower branches drooping, lowest often root into ground ('layering'); short **uppermost branches clustered** to form 'crow's nest'; bark dark greyish to reddish brown, scaly; inner bark has greenish yellow tinge; **young twigs with tiny rusty hairs**.

LEAVES: Needles short, 1–2 cm long, 4-sided, stand out on all sides of branchlets or mostly pointed upwards; stomata on all sides.

CONES: Pollen cones dark red; **seed cones small**, 1.5–3 cm long (sometimes to 4 cm), broadly egg-shaped to nearly spherical, **purplish**, remain on tree for several years.

WHERE FOUND: Characteristic of cold, poorly drained, nutrient-poor sites; in bogs and swamps and upland sites near wetlands; also with jack or lodgepole pine and white spruce on upland sites; often defines the treeline; widespread and common across boreal forest.

NOTES: The Dene used black spruce for treating snow blindness or eye damage resulting from blows or scratches. The small twigs of saplings were split in 2 and slowly heated so that their resin could be collected. A bird's quill was used to gently coat the eye with this material. • The Chipewyan boiled young black spruce cones to make a mouth wash for treating infections, toothache, and sore throats, and for clearing phlegm from the throat. The Woods Cree gargled a similar preparation to soothe sore throats and drank it to relieve diarrhea. • Some tribes made a powder from spruce needles and dusted it on burns, blisters and cuts. • The Chipewyan wrapped bits of dry rotted spruce wood in a piece of caribou hide and pounded it to a fine powder. This made excellent baby powder that prevented chafing and acted as a deodorant. • Young spruce twigs and leaves can be steeped in boiling water to make a pleasant tea high in vitamin C, which has been used to treat colds. However, **do not use enamel or Pyrex pots** and do not boil the tea for long or it will become too strong. • Always use evergreen teas in **moderation**. When the tips of young branches are eaten by starving caribou cows, it causes **abortion** of their calves. • Spruce beer was very popular for many years. It is made from the young growing tips of the branches in the spring. • The Chipewyan boiled their fishnets with young black spruce cones to dye them brown and make them less visible to the fish. • Occasionally, the Slave and Dog Rib used spruce bark to make canoes when no large birch could be found. However, these canoes soon become dangerously brittle. The Chipewyan used the bark to make shingles for the roofs of log cabins. It could also be used as flooring. • Black spruce trees grow very slowly, and they are usually too small, twisted and knotty to provide lumber, but because of their long wood fibres they are important pulp trees, as their long wood fibres produce high quality paper. Their logs have been widely used in the past to build shelters and traps. The peeled poles were preferred for making tipi frames because they are generally straight and taper gradually. Spruce boughs could also be laid on the tipi hides or birch bark to hold them down and to add insulation. Similarly, a floor of spruce boughs was warmer and drier. • Black spruce was very important for trapping and the fur trade. Black spruce branches, sometimes mixed with alder twigs and bee's wax, were boiled with metal traps to remove the smell of people and to prevent rusting. Spruce branches or small trees were used as lures on traps. Spruce saplings were also used to make spring-pole, toss-pole, and dragging-pole snares for trapping porcupines, hares, beaver, lynx and even caribou. The babiche (rawhide) of some snares was rubbed with spruce pitch to prevent it from rotting. • Black spruce is widely used as fuel, and the standing dead trunks of trees that have been burned by forest fires are an important source of fuel for many people. • The cones are found almost exclusively at the top of the tree, giving the crown a distinctive, 'knobbily' appearance. Some naturalists say that the top bulb of branches—the 'crow's nest'—results from red squirrels clipping off cone-laden branches as close to the top of the crown as they can efficiently work.

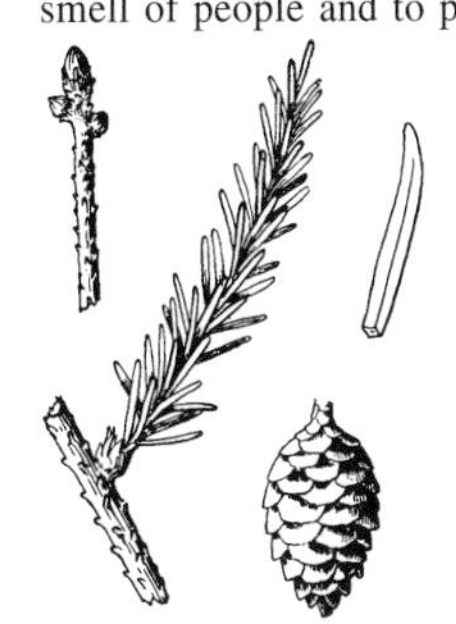

JACK PINE • *Pinus banksiana*

GENERAL: Variable evergreen tree, from short, limby and twisted when it grows in the open, to 20 m tall and straight when it grows in closed stands; lower branches often droop; twigs slender, yellow-green to dark grey-brown; older bark red-brown, flaky, becomes deeply grooved.

LEAVES: Yellowish green **needles in pairs, spreading, often twisted**, 2–5 cm long, sharp-pointed; needle clusters have papery sheaths at their bases.

CONES: Pollen cones small, about 1 cm long, in clusters at ends of branches; **seed cones** 3–7 cm long, tan-coloured, **usually curved and pointing towards end of branch**; **scales** thickened, **usually smooth**, occasionally with tiny prickles; seeds small, black, ribbed, winged.

WHERE FOUND: Well-drained sandy and gravelly areas; in pure stands or mixed with other trees; across boreal forest north and west Alberta, along Mackenzie valley to Great Bear Lake; particularly abundant in Saskatchewan and Manitoba.

NOTES: The Chipewyan dried and pounded pine needles to make a powder for treating frost sores. Other tribes sprinkled this powder on burns, blisters, or cuts, and then covered the wound. • The Panuanak Chipewyan powdered the crumbly, rotted wood of fallen jack pines to make a baby powder to prevent and heal diaper rash. • The Woods Cree used the soft inner bark as a poultice on deep cuts, though it was considered inferior to the inner bark of tamarack. • Pine bark tea, mixed with crowberries, was recommended for people with coughs. Also, pine bark is boiled to make a mild medicinal tea for people suffering from colds and flu. • Over the years, pine oil and pine tar have been used as disinfectants, antiseptics, insecticides, and deodorants. They have been used externally to kill parasites and internally to bring up phlegm from the lungs. The Cree melted lumps of pine pitch and used the smoke to fumigate the rooms of sick people and to kill germs that may have been left by visitors. • Pine needles were sometimes burned as incense. • The fresh, juicy inner bark of jack pine was eaten by the Chipewyan and Cree, but some other groups considered it inedible. Porcupines also enjoy eating pine bark and squirrels harvest the cones to eat the seeds. Boiling the male cones to remove excess resin makes them suitable for eating. • Pine wood is usually too knotted and full of pitch to produce good quality lumber, and it has too much pitch to be used for smoking hides or meat. However, pine cones were mixed with rotten white spruce wood and burned slowly to tan hides. Packed snow and pine branches were sometimes used to make windbreaks around tents, and pine wood has been used to make log cabins, sleds, toboggans, and floats for fishnets. Some tribes used the trunks of jack pine as canoe timbers, and pine pitch may have been used to caulk canoes, although some say it dries too slowly. • Today pine is used in general construction and for pulp, railway ties, mine timbers and pilings. • The pollen from the male cones is carried by the wind to female cones, where it sticks to one of the tiny drops exuded by the young, immature seeds in the female cones. The fertilized cones mature, but then development stops. Mature cones can remain high in the tree with their woody scales closed for many years, until a fire burns through the stand. Then the cones open, either as a result of the heat melting the resins (cone serotiny), or as a result of the death of the tree or its branches. Soon after the fire has passed, the seeds fall to the ground, which has been cleared of competing plants and fertilized with their ashes. Since pine forests are maintained by fire, they usually grow in dry, well-drained areas, and pine seedlings do best on mineral soil. These trees are very short-lived and shade-intolerant, and dense young stands soon undergo self-thinning. Jack pine was considered an evil tree by early settlers, probably because their crops failed to survive on the poor soils in which this tree often grows. • The species name, *banksiana*, is in honour of Sir Joseph Banks, a president of the Royal Society of London.

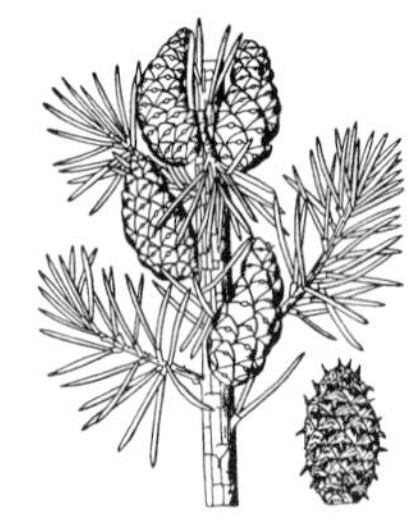

LODGEPOLE PINE • *Pinus contorta* var. *latifolia*

GENERAL: Straight evergreen tree, 20–30 m tall or more (or only 10 m); pyramidal crown; bark scaly, light reddish brown to greyish; twigs yellowish when young, becoming grey-brown and rough; branches generally curved upwards.

LEAVES: Needles in pairs, 2–4 cm long, often curved and twisted, deep green to yellowish green.

CONES: Pollen cones small, reddish green to yellow, in clusters on tips of branches in spring; seed cones egg-shaped, 2–5 cm long, usually slightly curved, spreading or bent backwards on branch; scales stiff, brown, **sharp prickle** at tip.

WHERE FOUND: Highly adaptable, tolerant of low-nutrient conditions; grows best on well-drained, coarse-textured soils; from western Alberta to southwestern N.W.T. and central Yukon.

NOTES: Lodgepole pine and jack pine can usually be distinguished from one another by their cones—lodgepole pine cones point straight out or backwards along the stem and have prickly scales, while jack pine cones point towards the branch tips and have very small prickles or none at all. However, where the ranges of lodgepole pine and jack pine overlap in central Alberta these two species often form hybrids. To further complicate matters, in some areas, such as northern British Columbia and the Yukon, lodgepole pine, which is the only pine in the region, is sometimes called 'jack pine.' • The Cree put gummy pine pitch in hollow teeth to relieve aching. Lumps of hardened pitch were chewed as a breath freshener and to soothe sore throats, or they were powdered and applied to sore throats with a swab. The pitch was taken internally to stimulate menstrual flow and to treat kidney problems, tuberculosis and sore throats. Pine pitch (often heated) was plastered over sore muscles, painful arthritic joints, swellings, insect bites, itching skin, boils and ulcers. European settlers preferred to dissolve it in alcoholic beverages for use as a wash on scalds, burns, and inflamed or itchy spots. • The bark of young trees was simmered and applied to burns and scalds to relieve inflammation and reduce infection. The inner bark was ground and hammered to a paste and applied to ulcers, carbuncles, wounds and sores. One tribe hung a piece of pine charcoal around the neck to treat laryngitis, and a decoction made from the roots was used to clean and heal wounds. • A tea made from pine pitch and common juniper berries was considered a good remedy for a cold or flu. Evergreen tea is very high in vitamin C, and it was used for many years in the north. European travellers soon learned its value for preventing and curing scurvy. A handful of young twigs, collected in the spring, makes a pleasant tea when steeped for 5 minutes in a teapot. It can be sweetened with sugar, honey, molasses or maple syrup, and spiced with cinnamon, nutmeg and orange peel. All evergreen teas should be used in **moderation**, as large amounts and high concentrations could be **toxic**. **Pregnant women** should not use this tea. • The juicy, white inner bark is rich in starches and sugars. It is cut off in strips or scraped off, taking care never to cut all the way around the trunk (this would kill the tree). The inner bark can be eaten raw, but it is difficult to digest when uncooked. Usually it is boiled alone or in stews and soups. It can also be dried, lightly roasted and ground to make a nutritious (though not very pleasant-tasting) flour. Young shoots (stripped of their needles) and young cones can be cooked in an emergency. The seeds of all pines can be eaten, but lodgepole and jack pine seeds are usually too small to make collection worthwhile. • The pitch provided a natural glue for small items such as baskets and headdresses, and it was a good waterproofing agent for moccasins. • The Cree used pine wood for making canoe frames. The common name, 'lodgepole pine,' comes from the historical use of these long, straight trunks in making the frames for tipis. They also make excellent travois poles and fences.

BALSAM FIR • *Abies balsamea*

GENERAL: Symmetrical evergreen tree to 18 m tall, with narrow pyramidal crown; **bark smooth, with resin blisters** when young, becomes thin and scaly with age.

LEAVES: Flat needles, 12–25 mm long, in 2 rows along branchlets, shiny green above with 2 silvery bands of stomata underneath, rounded or more often **slightly notched at tip**.

CONES: Pollen cones small, yellow-red or purple-tinged; **seed cones erect**, cylindrical, 5–10 cm long; bracts tipped with slender bristles, shorter than scales; **scales fall off at maturity**; seeds winged.

WHERE FOUND: Moist woods, often in association with aspen, birch and white spruce; in boreal forest of prairie provinces; becomes more common to east.

NOTES: Subalpine fir (*A. lasiocarpa*) is a closely related mountain species. It grows in our region primarily from northern British Columbia to the central Yukon and southwestern N.W.T. The needles of subalpine fir are narrower and paler than those of balsam fir, and have lines of stomata on both upper and lower surfaces. The cone-scale bracts have tips about 1/3 their length, compared to 1/5 their length in balsam fir. Although subalpine fir sometimes grows at lower elevations and in valley bottoms, it tends to be a tree of mountain slopes (especially on soils derived from acidic igneous rocks), ascending to the treeline and often above the upper limit of spruces. Subalpine fir and balsam fir often form hybrids in west-central Alberta, where their populations come into contact. • Balsam fir is one of the most distinctive trees in Canada. You can't mistake its symmetrical outline, and the narrow pyramidal crown is so spire-like that in some localities it is referred to as 'the church steeple.' • The smooth, greyish bark with its raised resin blisters is characteristic of the true firs, as is the smell of the resin when you burst one of the blisters. • The resin protects the tree by rapidly flowing from any breaks in the bark and hardening to seal out invading insects and fungus. The resin in the seed coats may deter squirrels and mice from eating them. • The small, winged seeds are carried on the wind as they fall from cones high in the tree. During storms some seeds have been carried as far as 7 km from the parent tree. • Although balsam fir usually appears late in the succession of a stand, balsam fir seedlings often form a dense understory in stands where windfall or insect damage has thinned the canopy. Balsam fir is shallowly rooted and does not stand up well to strong winds. • The Cree used the clear, fragrant resin to treat insect bites, boils, scabies, and other infections and skin sores. They applied it directly, or mixed it with grease or oil to make a salve or ointment. An ointment made by boiling fir pitch and sturgeon oil was said to cure tuberculosis. If someone had arthritis, a cloth covered with warmed pitch was wrapped around the sore joints and left on until it fell off. Then it could be re-heated and used again. The resin was even taken internally for colds, coughs and asthma. Some tribes chewed the resin as a gum to relieve heart and chest pains, or drank a beverage made from the gummy sap in an effort to cure gonorrhea. • Large doses of the resin may cause **nausea** and an emptying of the bowels. • The Cree also used the bark of the fir tree as a poultice to treat boils, infections and sores and in teas to purify the blood, to cure bowel troubles, gout, and worms, and to treat menstrual irregularity, coughs and colds. Similarly, they made tea from the inner bark to relieve chest pains, and they steeped twigs in water to make a laxative. The inner bark of the tree was dried, ground, mixed with flour or boiled with other ingredients and inhaled to speed childbirth. • The Cree sprinkled powdered needles on burns, blisters or cuts. Fir needles have also been used in poultices for treating fevers and chest colds, and a small piece of root was sometimes held in the mouth to relieve mouth sores. • Native peoples did not have much use for balsam fir wood, although they sometimes used it to make paddles. Today, it is important in the pulpwood industry and is sawed and sold as lumber. The yellow resin is marketed as Canada balsam and is used to seal microscope slides. Fir is also a popular Christmas tree, both for its fragrance and the fact that it holds its needles well. Many tree farms grow small fir trees for the Christmas market.

TAMARACK • *Larix laricina*
LARCH

GENERAL: Small deciduous, needle-leaved tree, 6–15 m tall (sometimes to 20 m); branches long, slender, pliable; bark thin, scaly, reddish brown.

LEAVES: Needles borne on woody projections, in small **clusters of 10–20**, 1–2 cm long, more or less flat, soft, pale green to blue-green, turn bright yellow in autumn and drop off (**deciduous**).

CONES: Pollen cones small, egg-shaped; seed cones erect, 10–20 mm long, broadly egg-shaped, dark red at flowering time, become leathery and brown with age; seeds winged.

WHERE FOUND: Fens, swamps and wet, mineral soils; widespread and common across boreal forest; north and west to Great Slave Lake, Mackenzie delta and western Alaska; circumpolar.

NOTES: Tamarack is the only native coniferous tree in our region to shed its needles in autumn. As a deciduous conifer, tamarack is somewhat of an ecological puzzle since it successfully lives in what is normally considered an evergreen world. It seems that these fast-growing trees can replace their needles each year at relatively low cost, and avoid the stress of keeping needles through the winter. • The Chipewyan call the soft inner bark *nidhe k'a* ('tamarack fat'), and they used it fresh as a poultice on burns and boils to draw out poison and speed healing. The Cree stripped the inner bark from tamaracks and used it to stop bleeding, and to treat hemorrhoids, earaches, inflamed eyes, jaundice, colic, and even melancholy. The Woods Cree boiled the inner bark before applying it to wounds. The Cree made a tea from the inner bark and used it to wash burns and running or gangrenous sores, and to stop itching. The Woods Cree used this tea as a wash and in poultices to heal frostbite and deep cuts. • Some tribes dried and powdered the leaves and inhaled the powder to relieve colds, bronchitis, and urinary tract problems. Tea made from the bark was also drunk as part of this treatment. • People starting to get a sore throat would chew tamarack gum. In the Yukon, tamarack pitch is used to keep deep cuts closed. • Tamarack needles are high in vitamin C and native peoples and European travellers made a tea from the needles and young twigs to prevent scurvy. Tea made from the needles, bark, and roots was recommended for treating sore muscles, arthritis and diabetes. Tamarack tea can be sweetened with sugar, honey, molasses, or maple syrup, and spiced with cloves, cinnamon, nutmeg, or grated orange peel. Evergreen teas should always be used in **moderation** and it is not wise to eat the foliage. • Although tamarack is not widely used as a food, the young buds are said to be quite sweet, either raw or cooked. **Caution** is required because some people are **allergic** to members of this family, and tamarack leaves can cause **cramps and paralysis** if consumed in large quantities. • The flexible roots were peeled and split, and used to stitch the seams of birch bark canoes. Rotted tamarack wood is burned for smoking fish and smoke-tanning hides, which gives the hides a yellow tinge. The Chipewyan and Woods Cree commonly used tamarack wood for making toboggans and it has also substituted for birch wood in the construction of snowshoes, drums and canoe paddles. • Europeans considered tamarack wood practically indestructible, even when exposed to the elements, and used it widely in ship-building, its twisting roots providing the natural bends and curves so precious to early craftsmen. • Tamarack has little economic importance today, though it was harvested in the past. An outbreak of an insect pest in the mid-1930s destroyed all tamaracks of any size, and in 1938 tamarack was no longer a commercial timber. One witness said, 'Their gaunt, bare skeletons, bare, grey and dry as tinder may still be seen standing in northern bogs and muskegs, a tribute to the species durability.' • 'Larch' comes from *larix*, the Latin name for a European larch, which may be derived from *lar*, 'fat,' since larches produce resin. 'Tamarack' is from the French Canadian *tamarac*, which was probably derived from an Algonquian name for this tree.

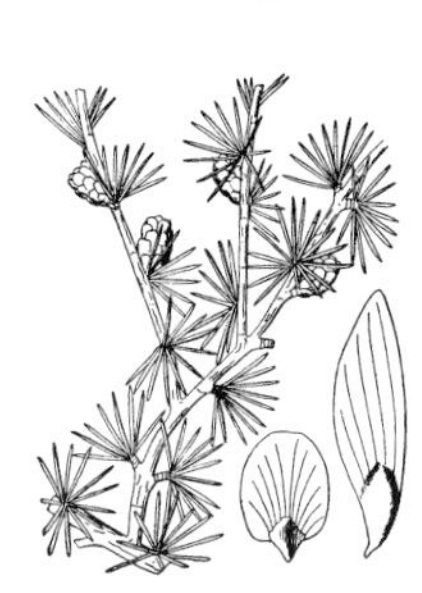

BUR OAK • *Quercus macrocarpa*

GENERAL: Deciduous tree to 15 m tall, trunk to 60 cm in diameter, but usually smaller, often shrubby in sandy or rocky sites; bark grey, flaky, becomes rough and deeply furrowed with age; twigs yellowish brown, somewhat hairy; buds hairy, blunt; wood pale brown, very hard.

LEAVES: About 17 cm long, 8 cm wide; bright, shiny green above, **greyish white and slightly woolly below**; highly variable in outline, most often with **broadly expanded upper portion** and irregular lobes that are gradually reduced towards base of leaf.

FLOWERS: Sexes in separate flowers on same tree, males in slender catkins, females solitary, appear after leaves unfold.

FRUITS: Acorns, 2–3 cm long, half or more rests in deep cup covered with large knobby scales and edged with noticeable **fringe**.

WHERE FOUND: Usually with other hardwoods on deep soils in rich bottomlands; occasional on coarser soils in upland situations; common in aspen-oak transition zone in southern Manitoba north to The Pas; in Saskatchewan only along Qu'Appelle River system where it is often shrub size.

NOTES: Bur oak is one of the 'white oaks.' It is the northernmost of the New World oaks, and one of the most drought-resistant. Its thick bark helps it to survive repeated burning, and once freed from competition by less fire-resistant species, these trees thrive. • Bur oaks flower in late spring, shortly after the leaves appear. The pollen of 1 tree germinates better on the stigmas of another, thus favouring cross-pollination. The acorns ripen in one season, and drop in autumn, as late as November. Trees 400 years old may still produce acorns, older than any other North American oak. Many acorns are distributed by squirrels, and a few may float away in water, but most simply fall to the ground. Bur oak acorns, with their deep, fringed cups, are the largest of all native North American oak acorns. • The acorns of bur oak are an important source of food for red and grey squirrels. Wood ducks, white-tailed deer, varying hares, pine chipmunks, 13-lined ground squirrels, grey squirrels and other rodents also eat them. • Acorns that mature in 1 season (rather than 2, as with black oak, red oak and pin oak) are generally sweet and edible. The inner white kernels of bur oak acorns are edible, though at times bitter. Their flavour improves with drying, and they were usually ground to make a nourishing flour. Often this was soaked in water for a day to remove the bitterness, then drained and molded into cakes or loaves. Sometimes it was mixed with other grains (e.g. corn meal) and baked in cakes. Acorns could also be roasted before they were ground, and this meal was used in cakes or as a substitute for coffee. Some native peoples gathered acorns in large quantities for winter use. • In the southeast of our region bur oak often invades prairie grassland, and it is often planted in shelterbelts and for shade on prairie homesteads. Bur oak is an excellent tree for city parks, because young trees transplant readily, and it is little affected by air pollution. • Oak bark is high in tannins and was universally used to tan leather. For this purpose it was stripped from the tree in the spring, when the sap was running. • The wood is valued for its combination of hardness and toughness, and is used in furniture, interior finishing, flooring, boat-building and barrels to hold liquids. Because of its elasticity and strength oak wood was considered particularly good for shipbuilding. • This species was named 'bur' oak because the cup of the acorn was thought to resemble the spiny bur of a chestnut. *Quercus* was the Latin name for an oak.

AMERICAN ELM • *Ulmus americana*
WHITE ELM

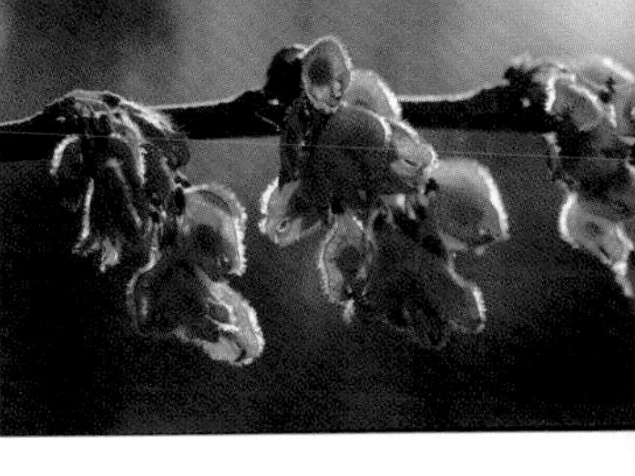

GENERAL: Deciduous tree, typically 20–25 m tall, 90–120 cm in diameter, larger on good sites; **crown umbrella-like on mature trees**; bark dark greyish brown with broad, deep, intersecting ridges, often scaly; **twigs markedly zigzag**, smooth, greyish brown, slightly hairy or hairless, often drooping; buds sharp-pointed, pale reddish brown with slightly hairy scales, somewhat flattened and lying close to twig; sapwood nearly white.

LEAVES: Oval, 10 cm long, 5–8 cm wide; dark green and slightly rough above, paler green below, with soft hairs; edges double-toothed; **leaf-base not symmetrical**; **leaf veins prominent**.

FLOWERS: In little bunches early in spring before leaves appear; both sexes in same flower.

FRUITS: Flat, oval, **1-seeded samaras**, about 1 cm wide; wing deeply notched at tip, hairless except fringe around edge.

WHERE FOUND: Common on river flats; does best on rich, moist, sandy or gravelly loams where water table is close to surface and drainage is good; eastern species that extends into southern Manitoba and southeastern Saskatchewan.

NOTES: The fruit ripens in May or early June in Canada. The flowers, which are largely self-sterile, are cross-pollinated by the wind. The small, light (about 7 mg), winged seeds are easily carried by the wind. Most fall within 100 m of the parent tree but some may be carried for 500 m or more on the wind and others may be carried farther by water. • The straight trunk, ascending limbs with drooping tips and the graceful, spreading, crown of American elm make this one of our most easily recognized trees. It is also one of our finest shade trees, widely used as an ornamental tree in cities and rural areas across much of Canada. Unfortunately, this familiar tree is no longer common in much of the eastern part of its range. Dutch elm disease, caused by the wilt fungus (*Ceratocystis ulmi*) and spread by elm bark beetles, was introduced accidentally to North America around 1930, and was first recorded in Manitoba in 1975. Since then, many trees have died. European elm bark beetle (*Scolytus multistriatus*) and the native elm bark beetle (*Hylurgopinus rufipes*), which tunnel into elm bark, carry the fungus from one tree to the next. A virus (*Morsus ulmi*), which kills the phloem of these trees, is also responsible for serious losses in the East and Midwest. Genetics research with elms has focused on finding varieties resistant to Dutch elm disease. American elm seldom hybridizes, and from about 20,000 controlled crosses between American elm and Siberian elm (*Ulmus pumila*), fewer than 100 seeds were obtained. Only a fraction of these germinated, but of the hybrids produced, one resisted repeated inoculations with the Dutch elm fungus. • Elms produce massive amounts of seed that can soon cover the ground with seedlings. This can be a problem when trees close to houses fill eave troughs with their fruits. In the natural environment, however, elm seed provides food for mice, squirrels, ruffed grouse, and other upland game birds. • Elm leaves decompose rapidly, and are high in potassium and calcium, making American elm a 'soil-improving' species. • American elm is also known as white elm, because of its 'white' wood. In many farm kitchens, the white, scrubbed, elm-topped table was a statement that the housewife insisted on cleanliness. Elm wood is used for dry-goods barrels, boxes, crates, furniture, flooring, panelling, caskets and boat-building. The unpolished veneer is used in fruit and vegetable containers and the fiber is used in roofing felt.

MANITOBA MAPLE • *Acer negundo*
BOX ELDER

GENERAL: Small deciduous tree to 12 m tall with widely spreading branches; bark grey-brown, smooth on young trees, becoming deeply grooved and darker on older trees; twigs light green, often have white bloom, sometimes hairy.

LEAVES: Pinnately compound, 6–7 cm long, **with 3–5 leaflets**; leaflets coarsely toothed to shallowly lobed, sharply pointed, light green above, grey-green below, occasionally hairy, **turn yellow in autumn**; **end leaflet stalked, often 3-lobed**.

FLOWERS: Appear with or before leaves, **sexes on separate trees**; male flowers red, in dense, drooping clusters on stalks that are 1–4 cm long; female flowers pale yellow-green, in open, drooping clusters; all flowers usually have 5 sepals, no petals.

FRUITS: V-shaped samaras (pairs of long-winged seeds joined near base), wrinkled, often with woolly hairs, 2–3 cm long; tips curve inwards; edges red-tinged when young; **hang in clusters from branches**; fall from tree in late winter or early spring.

WHERE FOUND: Moist soils along streams, rivers and lakes and in ravines and wooded valleys; primarily native to southern Manitoba and Saskatchewan; widely planted elsewhere.

NOTES: Manitoba maple has the paired winged fruits common to all maples, but it is easily distinguished from the other members of this genus by its compound, ash-like leaves. Manitoba maples in floodplain woods can produce long, straight trunks and high, open crowns. However, upland trees are usually about 8 m tall and have low, crooked branches that frequently sprout from near the trunk base. The branches are weak and easily broken. • Nowadays, Manitoba maple is often used for shelterbelts and as an ornamental because it is hardy and fast-growing, although short-lived. Female trees make better ornamentals than do male trees as they tend to be single-stemmed and to grow taller, whereas the male trees are often multi-stemmed and short. The only problem with the female trees is putting up with the annual flood of seeds. In eastern Canada, Manitoba maple is becoming a weed tree. • Although the sap of Manitoba maples is less sweet than that of the sugar maple (*Acer saccharum*), it is plentiful, and the Cree and other tribes used it for making syrup and sugar. In some parts of the United States, groves of these trees were planted for that purpose. • The Cree used the inner bark and the leaves to make a tea to treat liver and spleen problems and to calm the nerves. Tea from the inner bark was also used to induce vomiting. • The wood is soft, weak, and light, and the trees are usually too small to be of economic importance. However, it is used occasionally in construction and for making boxes. • *Acer* is the ancient Latin name for maple. The species name *negundo* is taken from *nurgundi*, a Sanskrit and Bengali term for the chaste tree of India, which has similar leaves. The common name 'box elder' is also widely used because the leaves of these trees are thought to resemble those of elders (*Sambucus*) and the whitish wood is similar to that of box (*Buxus sempervirens*).

GREEN ASH • *Fraxinus pennsylvanica*
RED ASH

GENERAL: Deciduous tree to 12 m tall; trunk to 45 cm in diameter, larger on good sites; bark greyish brown, becomes broken into firm, narrow, irregular, slightly raised ridges running into each other and giving somewhat **diamond-shaped pattern to bark**; twigs moderately stout, greyish brown, typically downy; buds reddish brown, hairy; wood hard, heavy, coarse-grained, moderately strong.

LEAVES: Pinnately compound, 25–30 cm long; typically **7 leaflets** (sometimes 5), **oval**, 7–12 cm long, 2–4 cm wide, tapered to each end; green above, paler and sparsely hairy below; edges toothed or smooth.

FLOWERS: Inconspicuous in compact, many-flowered clusters; appear at same time as leaves, with **sexes on separate trees**.

FRUITS: Samaras (seeds with long, membranous, yellowish green wing), to 4.5 cm long, 6 mm wide; **in hanging clusters**.

WHERE FOUND: Usually on low ground, riverbanks and lakeshores; occasionally in upland sites where competition is not too severe; southern Manitoba to southeastern Saskatchewan in our region.

NOTES: Also called *F. lanceolata* and *F. campestris*. • Moisture and winter temperature appear to be the most important factors affecting the distribution of this species. • Green ash flowers in the spring. Pollen from the male trees is spread by the wind, but travels relatively short distances, usually 60–90 m. Female flowers develop shortly after the male flowers. The winged seeds are also carried on the wind, but seldom travel more than 100 m from the parent tree. • Native peoples used tea made from the inner bark tea as a strong laxative to expel worms, as an afterbirth tonic, and as a medicine to induce vomiting, relieve stomach cramps, and remove bile from the intestine. It was also used as a wash for treating sores, itching, lice and even snakebites. The inner bark was chewed and applied as a poultice on sores, or made into a tea to relieve depression and fatigue. Some people said the seeds were an aphrodisiac. Ash was also used to stimulate sweating and urination. • Native peoples used green ash wood for making pipe stems, bows, arrows, tipi pins, pegs, drums and meat-drying racks. • Many superstitions have been associated with the ash tree. It had a reputation for magically curing warts. The wart was pricked with a new pin that had first been stuck into an ash tree. Then the pin was withdrawn and left in the tree, while the following charm was repeated: 'Ashen tree, ashen tree, pray, buy these warts from me.' According to another belief, if a live shrew was buried in a hole bored in an ash trunk and this was then plugged up, a sprig of this 'shrew ash' would cure paralysis caused by a shrew creeping over the sick person's limbs. • Green ash is a very successful hardwood in shelterbelts because it is hardy and fast-growing. It has also been planted on spoil banks of reclaimed strip mines and as an ornamental shade tree. • The wood is hard, heavy, moderately strong, and coarse-grained. It is sold as 'white ash' and used to make items where strength is required, such as baseball bats, handles, agricultural tools and furniture. • *Fraxinus* was the ancient Latin name used by Virgil for the ash tree. The derivation of the name is uncertain. It may be from the Greek *phraxis*, 'a separation,' because the wood may be easily split. The species name, *pennsylvanica*, is Latin for 'of Pennsylvania,' the place where this species was first collected.

BALSAM POPLAR • *Populus balsamifera*
BLACK POPLAR ssp. *balsamifera*

GENERAL: Deciduous tree to 25 m tall (usually 10–15 m); buds very sticky and **fragrant with resin**; old bark dark and **deeply furrowed**.

LEAVES: Thick, egg-shaped to nearly lance-shaped, 4–9 cm long, rounded to heart-shaped at base, sharp-pointed at tip; deep green above, pale or whitish below; edge finely round-toothed; **leaf stalk round**, 2–5 cm long, often with pair of glands at base of blade.

FLOWERS: Sexes on separate trees, both in catkins, appear before leaves; female flowers have 2 stigmas; male flowers have 20–30 stamens.

FRUITS: 2-valved, smooth capsules; seeds small, with **tuft of cottony hairs**.

WHERE FOUND: In moist depressions and on terraces along streams, rivers and floodplains; common and widespread across our region.

NOTES: The fragrant, resinous buds, collected early in the spring, were widely used in medicine. The Woods Cree put the sticky buds directly into their nostrils to stop bleeding. The Cree made a tea from the buds and used it as a gargle for sore throats. A mixture of the buds and inner bark was taken to prevent scurvy and applied to cuts to stop bleeding. Some tribes made a salve for treating wounds, sores, rashes, eczema, and frostbite, by heating (not boiling) the fragrant buds in grease and then straining and cooling the fat. Bear fat was said to be excellent for this, and deer fat was considered unsuitable by some. Today, similar salves are made using vegetable oil (with beeswax added to thicken it), Vaseline, cocoa butter, lard, lanolin and shortening. Balsam poplar salve was also put up the nose to cure head colds and bronchitis. • The Cree grated the bark and mixed it with warm water to make a paste that was applied to boils and left on overnight to draw out the infection. Northern tribes laid the inner bark on sore muscles to relieve the pain. The inner bark, leaves and buds were all used to relieve coughs, colds, and lung or urinary problems. Northern tribes boiled the bark to make a medicinal tea for treating chest colds, flu and tuberculosis. The Woods Cree placed fresh leaves directly on sores to draw out infection. • Young balsam poplar shoots contain bisabolol, which is active against the bacteria that cause tuberculosis. The buds contain salicin, a compound related to salicylic acid (the active ingredient in Aspirin), and would have some action against pain and fever. They also contain an antioxidant which has been extracted in alcohol and added to cosmetics to prevent them from becoming rancid. • Although there have been no reports of poisoning from poplar bark, buds or leaves, many people are **allergic** to the pollen of balsam poplar and develop rashes and respiratory problems in the spring when these trees are in flower. • Thick pieces of balsam poplar bark were carved into toy boats for children. The wood was also used for carving toys and for making spoons. Balsam poplar wood was used to make the hoops on cradle boards that protect the infant's face. Sometimes it was used to make paddles, but it was not preferred as the wood tended to become waterlogged and to break easily. In some northern regions, balsam poplar wood is the favourite fuel for smoking fish. • The Woods Cree mixed the aromatic, resinous buds with other ingredients to make lures to draw animals to their traps. • Moose and other ungulates browse on young balsam poplar trees. In the spring, bees collect the sticky, aromatic resin from the buds and young leaves and use it to cement and waterproof their hives. • The common name 'balsam poplar' refers to the aromatic, balsam-like smell of the resin of these trees.

TREMBLING ASPEN • *Populus tremuloides*
ASPEN, WHITE POPLAR

GENERAL: Small to medium-sized deciduous tree, up to 20 m tall (sometimes to 30 m); usually **forms stands by sending up suckers** from extensive shallow root system; **buds not resinous or fragrant**; bark greenish white, becomes blackish and roughened on lower trunk and around branch bases.

LEAVES: Oval to nearly circular, rounded to square-cut at base, sharp-pointed at tip, 3–7.5 cm long; **edges finely round-toothed**; **leaf stalks flattened**, allowing leaves to tremble in slightest breeze.

FLOWERS: In drooping catkins, appear in early spring, well before leaves; sexes on separate trees; male flowers with 5–12 stamens.

FRUITS: Capsules, 3–5 mm long; seeds tiny, with tuft of soft hairs.

WHERE FOUND: Dry ridges to rich, moist sites; grows best in well-drained, moist, loamy soils; common and widespread across our region.

NOTES: Aspen is one of the first trees to colonize areas that have been burned by forest fires. Female trees can produce millions of seeds each year. The seeds are carried on the wind for up to 30 km during storms. They remain viable for 3 days to 3 weeks, but they seldom find conditions suitable for germination. Most aspens reproduce without seeds. A parent tree sends out underground shoots and suckers grow up from these into new, genetically identical trees. A stand of trees that has grown from a single parent is called a 'clone.' Aspen clones can cover several hectares and they can best be distinguished in the spring and autumn, when all members of a clone produce or drop their leaves at the same time. • The Cree used the buds, inner bark and leaves as medicine. Aspen tea was considered a good spring tonic, especially for the elderly, and it was used in the treatment of rheumatism, diarrhea, liver and kidney problems. Aspen tea was made by boiling 3 or 4 spoonfuls of grated inner bark (fresh or dried) in 3 cups of water. This was cooled and taken in 1/2-cup doses every 2 hours with plenty of other liquids. Usually one 3-cup recipe was taken at the first sign of infection. However, a second, milder batch could be given to make sure the infection was gone completely. • The Woods Cree boiled the shredded inner bark to make a medicinal tea for treating coughs. They also boiled the outer bark and drank the tea as a cure for venereal disease, and they used the white powder from the bark to stop bleeding. • To treat an earache, the Cree boiled a spoonful of aspen bark and an equal amount of tree fungus in a 1/4 cup of water. This was cooled and used for eardrops. Chewed leaves are often plastered on insect stings to draw out the venom. Young seeds were chewed and swallowed by women trying to induce an abortion. This would have been strongly opposed by the Catholic Church, and the fact that some Chipewyans today believe that no part of the aspen has medicinal value may stem from earlier disinformation by the church. • The sweet pulpy material just under the outer bark was considered a treat for children. They scraped it off in strips with a knife during spring and early summer while the sap was running, and ate it raw. The Woods Cree considered it a mild purgative. The inner bark loses its sweetness and becomes bitter as the growing season progresses. Today, some consider it too strong to be edible. • Aspen wood has been used to make canoe paddles; knots have been made into bowls; and aspen poles were occasionally used to make tipi frames, and were also placed on the birch bark covering to hold it in place. In the spring, when the bark can be slid from a short section of branch in one piece, toy whistles can be made from fresh aspen branches. • The Chipewyan burned green aspen wood and then boiled the ash and strained the mixture to extract the lye, which is boiled with rendered caribou or dog fat to make lye soap. Once the mixture has cooled, it is cut in bars and dried on a wooden stage. The soap was used for rubbing into moose hides during tanning to make them soft, and the lye solution was also used to remove rust from traps by scrubbing them with grass. • The wood of trembling aspen has been ignored by Canadian timber interests until recently, when it has been harvested to make pulp, waferboard and chopsticks. The greenish bark of these trees is photosynthetic. • Aspen is the preferred food of beavers and many ungulates browse on the tender young twigs and leaves. In times of food shortage, some ungulates will eat aspen bark, leaving extensive scars on the lower trunks.

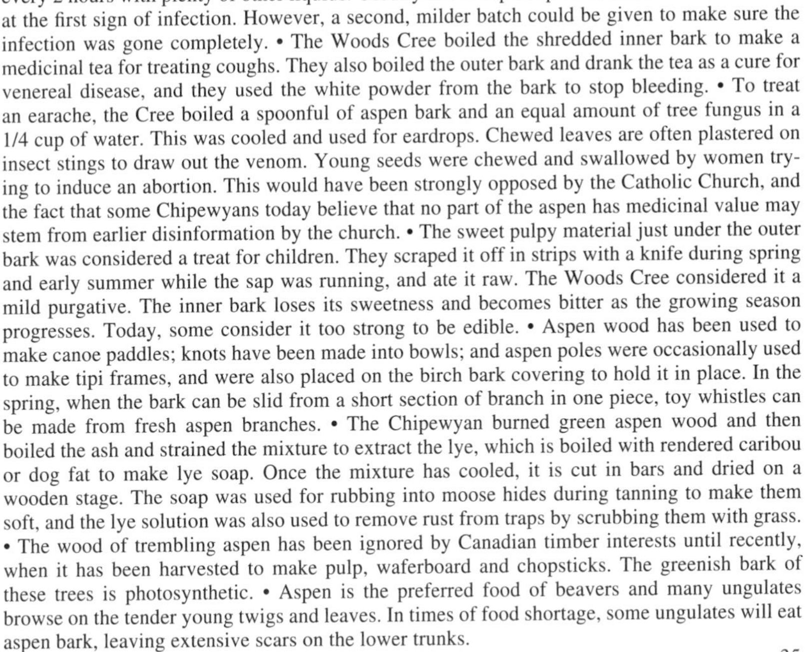

WHITE BIRCH • *Betula papyrifera* var. *papyrifera*
PAPER BIRCH, CANOE BIRCH

GENERAL: Small to medium-sized deciduous tree, usually to 15 m tall (sometimes to 30 m); **bark peels in papery strips**, usually **white to yellowish or copper brown**, smooth, marked with brown horizontal lines of raised pores; twigs brown, slender, hairy.

LEAVES: Egg- to diamond-shaped, 4–8.5 cm long (sometimes to 10 cm); **sharp-pointed**; greener above than below; lower surface with long hairs, usually with tufts in angles of larger veins; edges sharply double-toothed.

FLOWERS: Male and female flowers in separate catkins on same tree, 2–3 cm long (sometimes to 4 cm), appear with or before leaves; **catkins break up at maturity**.

FRUITS: Broadly oval nutlets, with wing broader than body.

WHERE FOUND: In open to dense woodland; grows best on well-drained but moist sites; widespread across boreal forest, occasional in parkland.

NOTES: The taxonomy of the paper birch complex can be confusing, especially if taken too seriously. **Alaska birch** (*B. papyrifera* var. *neoalaskana*, also called *B. neoalaskana* or *B. papyrifera* ssp. *humilis*) is another common variety in our region. It is a widespread, western tree, probably **the** white birch of Alaska, the Yukon and the District of Mackenzie. It has 3–5 cm long, hairless leaves with long, pointy tips, singly saw-toothed edges and fan-shaped bases, and its twigs have **yellow crystalline glands**. • Many northern native groups consider birch the most useful of all trees. They soaked sheets of bark in water to make them flexible and then made them into baskets, boxes, tubs, baby cradles, platters, bread pans, cups and other containers. A special birch-bark container held the beaver castor used in trap baits, and birch bark was also used to make documents for medicine men. Until the early 1930s, tipis and nearly all canoes were made of birch bark. The bark could be rolled and burned as a torch. Spear fishermen sometimes put such a torch in the bow of a canoe so they could see the fish at the bottom of the water at night. Birch bark could also be made into a cone-shaped trumpet and used to imitate the call of a cow or calf moose when hunting. The thin outer strips of birch bark make excellent kindling; a fire of birch wood is long-burning and sweet-smelling. • Apparently the Dene used tiny pieces of birch bark as scrapers to remove cataracts. The thin, papery, outer bark was used to bandage burns, and heavier bark, padded with sphagnum moss, was used to make casts or slings. The inner bark was used to treat rashes and sores. It was either boiled to make a wash, or dried, ground, and added to a mixture of spruce pitch and grease to make an ointment. • The leaves were chewed and applied to wasp stings to draw out the poison, and the buds were used to treat gonorrhea. • Birch wood is harder than that of most northern trees, and it was used to make sleds and snowshoes. These were usually constructed in the winter, when the wood was frozen, which made it easier to split in a straight line, and when there was less sap in it, which made it easier to steam and bend without splitting. Other items made from birch wood include: paddles, canoe ribs and carrying boards, wooden nails for canoe gunwales, bows, arrows, drums, axe handles, hammers, spoons, snowshoe webbing needles, dog whip handles (these were sometimes filled with pebbles or lead shot to make them rattle), and grease lamp bowls. To make a grease lamp, caribou fat was melted in the bowl and a wick, twisted from fine dry grass from a mouse's nest, was laid across the grease with one end sticking out to be lit. • Birch poles formed the frames of some tipis and were also used to make smoke-curing racks for tanning hides. • Dry, rotted birch wood also had many uses. It burns slowly, producing a smoky fire for curing meat and fish. Mixed with rotten white spruce wood and jack pine cones, it tans hides to a reddish colour. Dried and finely ground, it provided baby powder. Sometimes the rotted wood was first boiled with Labrador tea to improve the quality of the dusting powder. • Birch sap was collected in mid-May and boiled down to a syrup (often thickened with flour) that was used as a sauce with fish or bannock, much as maple syrup and corn syrup are used today. The sap has also been used to make beer, wine and soft drinks. This was common practice only 50 years ago and today, in Alaska, birch sap flows through a small cottage industry that produces birch syrup and various beverages. After the spring flow, birch sap becomes milky and bitter. • Strips of inner bark, taken fresh from the tree, were eaten as a snack.

Shrubs

Shrubs are woody plants that are less than 10 m tall when mature (usually less than 3 m in our region) and are usually multi-stemmed. Some dwarf shrubs—woody plants that are under 30 cm tall and sometimes do not look much like shrubs (e.g., crowberry)—are also included here.

The shrubs of our region represent a variety of plant families and genera; we have organized them by major plant families, and by similar species within those families.

We start with the catkin-bearers: birches, alders, hazelnut, sweet gale, and willows. Following the willows are two species often thought of as 'sort-of-willows': red-osier dogwood (also called red willow) and silverberry (also called wolf-willow). Then come buffaloberry and alder-leaved buckthorn, followed by the shrubs of the rose family. Next are shrubs with maple-like leaves: gooseberries and currants, mountain maple and bush cranberries, followed by the rest of the honeysuckle family shrubs and then the shrubs of the heather family. Finally, we describe crowberry, two shrubby species of juniper and sand heather.

Most of our shrubs are from one of five families: Salicaceae (willows), which is difficult to sort out; Grossulariaceae (gooseberries and currants), which can be tricky; Rosaceae (rose family), which has distinctive species; Caprifoliaceae (honeysuckle family), which also has distinctive species; and Ericaceae (heather family), which is easy to sort out.

Willows, birches, alders and shrubs in the heather family (Ericaceae) are particularly common in our region. The willows and birches dominate many open forests and non-forested habitats, while ericaceous shrubs seem to thrive in peatlands—the thick, wet, cold and acidic organic layers that cover the surface of much of our region. Ericaceous shrubs are also characteristic of the boreal forest ground cover and can be abundant in some open upland habitats.

Many of the boreal shrubs, especially those of the forest, have fleshy fruits—usually edible berries. This makes ecological sense, given the abundance of forests and the adaptive value of fruits that can be eaten and dispersed by mobile animals rather than by wind, which might not be particularly effective in the understory of a dense forest. Berries were, and still are, an important part of the diet of many native groups. However, a nearly equal number of shrubs (especially those of open forests and non-forested habitats, such as willows, alders and dust-seeded heathers) have light, sometimes winged or plumed seeds that are well adapted for wind dispersal. This mix of dispersal strategies reflects the complex dynamics of forest succession and the resulting shifts in forest structure and vegetation.

A key is provided to the various shrub genera, as well as a key to species for the willows (*Salix* species, p. 44) and the currants and gooseberries (*Ribes* species, p. 60).

Major Shrub Families

Birch
(pp. 39–41)

Willow
(pp. 43–53)

Oleaster
(p. 55)

Rose
(pp. 56–60)

Saxifrage
(pp. 60–63)

Honeysuckle
(pp. 64–68)

Heather
(pp. 68–73)

Juniper
(pp. 74–75)

Key to the Shrub Genera

1a. Leaves needle- or scale-like, evergreen 2
- **2a.** Leaves opposite or in whorls of 3; seeds in bluish, berry-like cone .... ***Juniperus* spp. (pp. 74–75)**
- **2b.** Leaves alternate or in whorls of 4; seeds in juicy, black, berry-like drupe .... ***Empetrum nigrum* (p. 74)**

1b. Leaves usually not needle- or scale-like (if scale-like, plants noticeably hairy) 3
- **3a.** Stems have spines or prickles 4
  - **4a.** Leaves simple, 3–5-lobed, shaped somewhat like maple leaf; fruits reddish to purplish black berries .... ***Ribes lacustre*** and ***R. oxyacanthoides* (p. 63)**
  - **4b.** Leaves compound (divided into leaflets); fruits red, not simple berries 5
    - **5a.** Flowers pink; fruits are 'hips' .... ***Rosa* spp. (p. 58)**
    - **5b.** Flowers white; fruits are raspberries .... ***Rubus idaeus* (p. 59)**
- **3b.** Stems have neither spines nor prickles 6
  - **6a.** Flowers and fruits in catkins (spikes of single-sexed flowers, somewhat resembling caterpillars) 7
    - **7a.** Fruits are nuts; leaves usually heart-shaped; young twigs, leaves and bud scales covered in long, white hairs .... ***Corylus cornuta* (p. 42)**
    - **7b.** Fruits are not nuts; leaves not heart-shaped; young twigs, leaves and bud scales not covered in long, white hairs 8
      - **8a.** Catkins and leaves yellow-waxy, fragrant .... ***Myrica gale* (p. 42)**
      - **8b.** Catkins and leaves neither waxy nor fragrant 9
        - **9a.** Buds enclosed in single bud scale .... ***Salix* spp. (p. 44–53)**
        - **9b.** Buds have 2 or more scales 10
          - **10a.** Bark has horizontal markings of raised pores (lenticels); leaves roundish; fruiting catkins neither woody nor persistent; pith usually flattened in cross-section; no bud at stem tip .... ***Betula* spp. (p. 39–40)**
          - **10b.** Bark lacks lenticel markings; leaves oval to elliptic; fruiting catkins woody and persistent ('cones'); pith usually 3-sided in cross-section; bud at stem tip .... ***Alnus* spp. (p. 41)**
  - **6b.** Flowers and fruits not in catkins 11
    - **11a.** Bark has lenticel markings; flowers white, fragrant; fruits red to purplish black cherries .... ***Prunus* spp. (p. 57)**
    - **11b.** Bark lacks lenticel markings; flowers and fruits various 12
      - **12a.** Leaves opposite 13
        - **13a.** Leaves leathery, evergreen; plants usually less than 40 cm tall; flowers showy, pink, saucer-shaped .... ***Kalmia polifolia* (p. 71)**
        - **13b.** Leaves deciduous 14
          - **14a.** Branches opposite; young stems usually bright red; leaves have 5–7 prominent, nearly parallel veins that converge at tip .... ***Cornus stolonifera* (p. 54)**
          - **14b.** Branches alternate; young stems not bright red; leaves have branching veins 15
            - **15a.** Leaves 3-lobed (except in *Viburnum lentago*, which has unlobed leaves) 16
              - **16a.** Large shrubs to 8 m tall; flowers pale yellowish green; fruits are samaras .... ***Acer spicatum* (p. 64)**
              - **16b.** Medium-sized shrubs to 4 m tall; flowers white; fruits red or orange berries (blue-black in *Viburnum lentago*) .... ***Viburnum* spp. (p. 64)**
            - **15b.** Leaves oval to elliptic, not lobed 17
              - **17a.** Flowers yellowish brown, single-sexed, inconspicuous (less than 4 mm across); branches and leaves covered with brown scabs or scales .... ***Shepherdia canadensis* (p. 55)**
              - **17b.** Flowers larger (more than 5 mm across), white, pink, yellow or orange, bell-shaped to tubular; branches and leaves neither scaly nor scabby 18
                - **18a.** Flowers usually less than 1 cm long, bell- or urn-shaped, white to pink; fruits white berries, remain through winter .... ***Symphoricarpos* spp. (p. 66)**
                - **18b.** Flowers usually more than 1 cm long, tubular, yellow to orange; fruits red, blue or purplish black berries (or capsules in *Diervilla*), do not remain through winter .... ***Lonicera* spp. (pp. 67–68)** and ***Diervilla lonicera*. (p. 65)**
      - **12b.** Leaves alternate 19
        - **19a.** Leaves evergreen, usually leathery 20
          - **20a.** Plants trailing or creeping, less than 20 cm tall 21
            - **21a.** Flowers yellow; fruits are capsules .... ***Hudsonia tomentosa* (p. 75)**
            - **21b.** Flowers white or pink; fruits are berries 22
              - **22a.** Petals bend backwards sharply (like miniature shootingstars); leaves widely spaced .... ***Oxycoccus* spp. (p. 72)**
              - **22b.** Flowers urn- to bell-shaped; leaves not widely spaced 23
                - **23a.** Fruits are white or red berries, with mild, wintergreen odour .... ***Gaultheria* spp. (p. 72)**
                - **23b.** Fruits are red berries 24

24a. Dark dots on leaf undersides; often on moist to wet sites .... *Vaccinium vitis-idaea* (p. 73)
24b. No dark dots on leaf undersides; usually on well-drained sites .... *Arctostaphylos* spp. (p. 73)
20b. Plants upright, usually more than 20 cm tall 25
25a. Leaves rusty below, with dense, woolly hairs .... *Ledum* spp. (p. 70)
25b. Leaves not rusty below 26
26a. Leaves have scales on lower surface; flowers white .... *Chamaedaphne calyculata* (p. 69)
26b. Leaves have waxy powder on lower surface; flowers pink .... *Andromeda* spp. (p. 71)
19b. Leaves deciduous, usually thin 27
27a. Leaves compound (divided into leaflets) 28
28a. 13–15 leaflets; flowers white, in dense clusters; fruits are red berries .... *Sorbus* spp. (p. 59)
28b. 3–7 leaflets; flowers yellow, single; fruits are achenes .... *Potentilla fruticosa* (p. 60)
27b. Leaves simple (not divided into leaflets) 29
29a. Leaves 3–7-lobed, somewhat maple-leaf-shaped; fruits reddish to black berries; leaves often sweet- or skunky-smelling .... **the currants** (*Ribes americanum, R. glandulosum, R. hudsonianum* and *R. triste*)
29b. Leaves not lobed 30
30a. Flowers white to pinkish, urn- to bell-shaped; fruits blue to blue-black berries . *Vaccinium* spp. (pp. 68–69)
30b. Flowers various colours, not urn- or bell-shaped; fruits various 31
31a. Flowers white 32
32a. Branches flat-sided or ridged, outer bark peels off in papery strips; fruits are papery pods .... *Spiraea* spp. (p. 56)
32b. Branches round, smooth; fruits are purple to black pomes (like miniature apples) .... *Amelanchier alnifolia* (p. 56)
31b. Flowers yellow or yellowish brown 33
33a. Plants usually more than 1 m tall; twigs densely covered with rusty brown scales; leaves silvery; fruits are silvery berries .... *Elaeagnus commutata* (p. 55)
33b. Plants usually less than 1 m tall; twigs smooth or finely grey-hairy; leaves green; fruits are bluish black 'berries' .... *Rhamnus alnifolia* (p. 54)

WATER BIRCH • *Betula occidentalis*
RIVER BIRCH

GENERAL: Tall shrub, with several spreading trunks, about 5–10 m tall; bark thin, glossy, dark reddish brown to nearly black on young trunks, **does not peel readily**, conspicuous horizontal markings of raised pores; twigs slender, with many glands, reddish brown.

LEAVES: Alternate, simple, broadly oval, 2–5 cm long, broadest below middle, with short taper to blunt or sharp tip; rounded or wedge-shaped at the base; deep green and shiny above, paler and dotted with small glands below; teeth sharp, of 2 sizes, none near stalk.

FLOWERS: Small, in cylindrical catkins; sexes in separate catkins on same plant, appear before leaves unfold; male catkins clustered, 4–5 cm long when mature; female catkins 2–3 cm long and 5–10 mm wide at maturity.

FRUITS: Small, winged nutlets in catkins that hang or spread slightly; scales hairy, 2 side lobes shorter than narrow, sharply pointed middle lobe, shed when ripe; seed wings at least as wide as hairy seed.

WHERE FOUND: Common on moist soils along streams, rivers and lakeshores and in moist depressions in sandhills; often associated with poplars, willows and alders; across most of our region.

NOTES: Also called *B. fontinalis*. • Hybrids between *B. papyrifera* and *B. glandulosa* are often included in this species. • This shrub can attain tree size. • If a storm prevents you from crossing a lake or bay, perhaps you'd like to try this solution. Make a tiny birchbark canoe and place a louse and a little tobacco in it. Push the canoe out into the water and watch carefully. If it upsets, the wind will end shortly. If nothing else, this project will occupy your time while you wait out the storm. • *Occidentalis* means 'western'; *fontinalis* means 'growing in or by springs'; both are appropriate descriptions for this species. Occasional hybrids with white birch contribute to a high degree of variability in the appearance of this species.

BOG BIRCH • *Betula glandulosa*
SCRUB BIRCH

GENERAL: Spreading to erect shrub, usually 0.3–2 m tall; twigs resinous, covered with **wart-like resin glands** that look like octopus suckers, and inconspicuous fine hairs.

LEAVES: Nearly circular, 1–2 cm long, with 2–3 side veins, and 6–10 rounded teeth per side, somewhat **thick and leathery**, dotted with glands, green on both sides, scarlet to red-brown in autumn.

FLOWERS: In small catkins, appear in spring at same time as leaves.

FRUITS: Small, round, slightly winged nutlets in upright catkins.

WHERE FOUND: In wetlands across boreal forest, also in open forest on coarse-textured, acidic soils; north to southern arctic islands; prefers more acidic habitats than dwarf birch (below).

NOTES: Also called *B. nana* var. *sibirica*. • It is easy to stumble over the branches of bog birch when you are hiking in the summer, and some people call them 'shin trippers' or 'shin tangle.' The branches spread out and sag down under the weight of the snow, often leaving only their tips exposed to the cold winter air. • The Dena'ina would sometimes put bog birch branches behind them on trails to keep away bad luck. • The Chipewyan chewed fresh twigs of bog birch and applied them to deep cuts on their hands or feet to stop bleeding. The leaves give a clear yellow dye, better than that of paper birch, and the branches produce a warm tan colour. A bundle of branches makes a good broom. They have also been used as matting under sleeping materials on the snow, and as padding on a hunter's back when he is packing out meat. • Although this shrub is not particularly palatable to moose and other large mammals, the buds are a favourite food of grouse and ptarmigan in the winter. • *Betula* is the ancient Latin name for birch; *glandulosa* means 'with glands,' referring to the twigs.

DWARF BIRCH • *Betula pumila* var. *glandulifera*
SWAMP BIRCH

GENERAL: Slender, many-branched shrub with erect or ascending stems, to 2 m tall; twigs finely hairy at first, gland-bearing; bark of older stems smooth, hairless, dark grey to reddish brown, with many scattered raised pores.

LEAVES: Alternate, simple, obovate to circular, 1–4 cm long, 1–2 cm wide, firm, leathery; **wedge-shaped or rounded at base**, rounded or blunt at tip; dark green above and paler below, hairless and dotted with glands on both surfaces; **edges coarsely toothed, 10–15 teeth on each side**; veins net-like, prominent below.

FLOWERS: In small catkins in May to early June; male catkins hanging, 12–20 mm long, fall off after flowering; female catkins erect, 12–25 mm long, about 6 mm thick.

FRUITS: Small, winged, nearly flat nutlets; in catkins, shed together with scales when ripe; **wings as broad as nutlets**.

WHERE FOUND: Shrubby and treed fens, swamps and moist depressions along lakeshores and riverbanks; widespread and common across boreal forest; north and west to Great Bear River.

NOTES: Also called *B. glandulifera*. • Dwarf birch cones were steeped in boiling water to make a tea to relieve menstrual cramps and to act as a tonic following childbirth. They were also roasted on the coals of a slow fire and the fumes were inhaled to treat chronic nasal infections. • The Cree mixed the chopped, dried roots of dwarf birch with Labrador tea to make a pleasant tea. • *Pumila* means 'dwarf,' although on average dwarf birch is taller than bog birch.

GREEN ALDER • *Alnus crispa*

GENERAL: Bushy shrub, 1–3 m tall; **buds pointed, stalkless**; branches more or less sticky and finely hairy when young, become smooth, reddish brown.

LEAVES: Oval to broadly elliptic, 4–8 cm long, yellowish green, usually shiny and hairless above, slightly hairy below; **edges not wavy-lobed**, finely double-toothed; young leaves hairy, sticky below; no ladder-like pattern between side veins.

FLOWERS: In male and female catkins on new twigs, **appear in May at same time as leaves**.

FRUITS: Seed cones 1–1.5 cm long, egg-shaped, stalk longer than cone; **nutlets with broad wings**.

WHERE FOUND: Open forest and slopes, sand hills and edges of wetlands and streams; prefers well-drained, coarse-textured soils; indicates seepage; widespread and common across boreal forest; north past treeline to Arctic coast.

NOTES: Also called *A. viridis*. • **Sitka alder** (*A. crispa* ssp. *sinuata*, also called *A. viridis* ssp. *sinuata* or *A. sinuata*), also called slide alder, has larger, thinner leaves with irregularly lobed or wavy, as well as doubly saw-toothed, edges. It usually forms thickets on coastal mountainsides, but also penetrates inland to southern Alaska, the southern Yukon, and northern British Columbia. • Green alder was used mainly as fuel for smoking meat and hides. The Chipewyan burned it with peat moss as part of the tanning process, and many Dena'ina preferred alder for smoking fish and meat because of the pleasant flavour it gave the food. Alder was widely used to make a red-brown dye for caribou hides, snowshoes and fishnets by boiling the inner bark or pieces of the bark and stems in water and soaking the items to be coloured. Fishnets were dyed because the darker colour made it harder for fish to see the net. However, dyed nets are weaker and don't last as long. • Dried cones were sometimes finely chopped and mixed with tobacco as an extender.

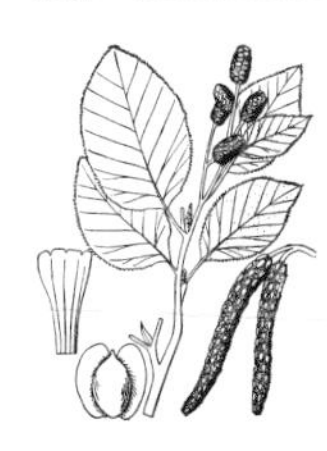

RIVER ALDER • *Alnus rugosa*
SPECKLED ALDER

GENERAL: Coarse shrub or small tree, 2–8 m tall, often in clumps; twigs hairless to woolly; **buds club-shaped, with short stalks**; bark yellow-brown with distinct raised pores.

LEAVES: Elliptic, 4–10 cm long, base rounded to somewhat heart-shaped; pale and hairy below, not sticky; green through much of autumn; **edges wavy-lobed**, double-toothed; ladder-like pattern between main side veins.

FLOWERS: Small, in catkins on previous year's twigs, **appear before leaves**.

FRUITS: Small nutlets, **wingless, with narrow ridge around edge**; in 1–2 cm long, short-stalked seed cones.

WHERE FOUND: Riverbanks and lakeshores; widespread in our region from Manitoba to Alaska.

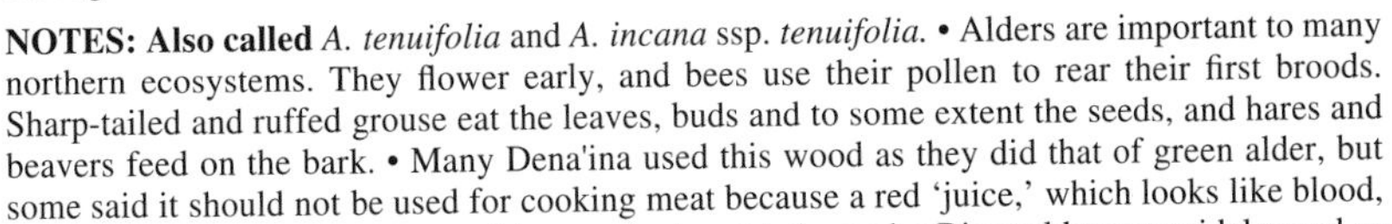

NOTES: Also called *A. tenuifolia* and *A. incana* ssp. *tenuifolia*. • Alders are important to many northern ecosystems. They flower early, and bees use their pollen to rear their first broods. Sharp-tailed and ruffed grouse eat the leaves, buds and to some extent the seeds, and hares and beavers feed on the bark. • Many Dena'ina used this wood as they did that of green alder, but some said it should not be used for cooking meat because a red 'juice,' which looks like blood, runs out of the wood when it is burned. • River alder was widely used as a dye. The Cree boiled the cones to dye quills yellow, but more often the bark was used to produce a red-brown dye for porcupine quills, birch bark baskets and hides. This process not only coloured the item, it also shortened the smoking time required for tanning hides and reconditioned the leather of old moccasins. The bark decoction was also used to soften toboggan boards for bending. The charcoal left by river alder wood was mixed with pitch and used to seal canoe seams.

BEAKED HAZELNUT • *Corylus cornuta*

GENERAL: Shrub, 1–3 m tall; young twigs, leaves, and **bud scales covered in long, white hairs**; twigs hairless after first season.

LEAVES: Elliptic to egg-shaped, 4–10 cm long, rounded to heart-shaped at base, sharp-pointed at tip; edges coarsely double-toothed; paler below than above, somewhat hairy below; **turn yellow in autumn**.

FLOWERS: In catkins, **appear before leaves** in April and early May; male catkins 4–7 cm long, hanging; female catkins tiny, few-flowered, with protruding red stigmas.

FRUITS: Thin-shelled, **spherical, edible nuts** enclosed in long tubular husks; husk light green, covered with stiff, prickly hairs, narrowly lobed at tip; in clusters of 2 or 3 at ends of branches.

WHERE FOUND: Moist but well-drained sites in thickets or woods; widespread across southern boreal forest and parkland.

NOTES: The small nuts are edible, once the shells have been removed, and the Woods Cree ate them as a trail snack or collected them in quantity for later use. The nuts can be roasted and eaten whole or ground into flour. Hazelnut flour makes delicious tortes (cakes). • Some Indian tribes used young, straight hazelnut shoots to make arrows. The roots and inner bark were used to make a blue dye. • Beaked hazelnuts are relished by jays, nutcrackers, crows, bears, squirrels, mice and voles. The dense, slender branches provide cover and nesting sites for birds. • A few Alberta gardeners grow beaked hazelnut as an ornamental. • The common name 'beaked hazelnut,' refers to the lobed bracts that form a sheath around the nut shell, and remain after it has ripened. The genus name *Corylus* comes from *korys*, meaning 'helmet,' and the species name *cornuta* means 'beaked' or 'horned.' Both refer to the bracts.

SWEET GALE • *Myrica gale*

GENERAL: Low shrub, 0.5–1.5 m tall, spreads by suckers; branches slender, ascending, brown.

LEAVES: Firm, **fragrant**, 2–6 cm long, **dotted above and below with bright yellow wax-glands**; whitish underneath; lance-shaped but broadest and rounded at tip, wedge-shaped at base; **edges toothed on upper third**.

FLOWERS: In greenish yellow, waxy catkins, appear before leaves; **sexes on separate plants**; male catkins to 2 cm long; female catkins to 5 cm long.

FRUITS: Small nutlets, egg-shaped, 2.5–3 mm long, coated with resinous wax, with 2 wing-like scales; **in erect, brown, cone-like catkins** that are 8–10 mm long.

WHERE FOUND: Swamps, shores and thickets; widespread across boreal forest.

NOTES: Sweet gale is dioecious—its plants bear flowers of 1 sex only. Unlike most other dioecious species, however, sweet gale plants can change sex from year to year, bearing male flowers 1 year and female flowers the next. • The Woods Cree gathered the female catkins in the fall, when they are largest and most fragrant, and dried and stored them for later use as an ingredient in lures. The Cree used the buds for dying porcupine quills, and the bark was made into a yellow dye used in tanning leather and dyeing wool. • Sweet gale was used to flavour English ale, long before hops were introduced, and gale ale was said to be 'fit to make a man quickly drunk.' The fruits, leaves, and nutlets are edible and are usually dried and used as a spice (similar to sage) in soups and stews. The dried leaves can be used to make tea. • Sweet gale is very aromatic and has been used in many households over the years to give a pleasant scent to linens and clothing, and to repel fleas. A safe, non-chemical flea collar can be made by stuffing a soft cloth collar with crushed sweet gale leaves. Smoking cigarettes made from these leaves is recommended as a sure mosquito repellent, but it would probably be best not to inhale too deeply. The Swedes boiled the plants to make a strong tea that was used to kill insects and vermin and to cure 'the itch.'

Willows

The willows are the largest group of shrub species included in this guide, but to most people only casually interested in plants a willow is a willow—why bother to include so many species in a field guide?

Willows are an interesting group of plants. Just because their individual flowers are small, are not particularly colourful and are present for only a short time each year, it does not mean they should be ignored. The willows, along with members of the birch family, are distinguished from most other shrubs in that their flowers are arranged in **catkins**. Willows are also the only shrubs in our area to have a single bud scale.

Willows are dioecious—a given plant will produce either male or female flowers, but not both. Many species are pollinated by wind in the spring, but usually willow pollen is carried from one shrub to the next by insects that are attracted to the bright red or yellow pollen and to the fragrant nectar produced by both male and female flowers.

Most willows in our region are browsed by herbivores, such as moose, deer, porcupines and hares. Willow thickets also provide bedding, cover and birthing areas for some of these species.

Native peoples in our area used willows for a variety of purposes: they weaved the stems and twigs into containers, and they used willows to start fires, to smoke meat and to make clothing.

As with most other sections of this guide, only the most common species found in our region are included. A key to species is provided, including pictures of mature leaves, as well as a list of the distinguishing characteritics of some of the willows. Readers are referred to Argus (1973), Brayshaw (1976) or Looman and Best (1979) for more complete treatments.

Learning to identify willows to species initially may be discouraging since their leaf shapes and growth forms are so variable, even within a species, but the reward of being able to recognize many species on sight is worth the effort. Once you get to know some of the willows, you will realize that they have some specific habitat preferences, and as you learn more about this diverse and important group of shrubs, you will discover that they are not as difficult to identify as you thought they were.

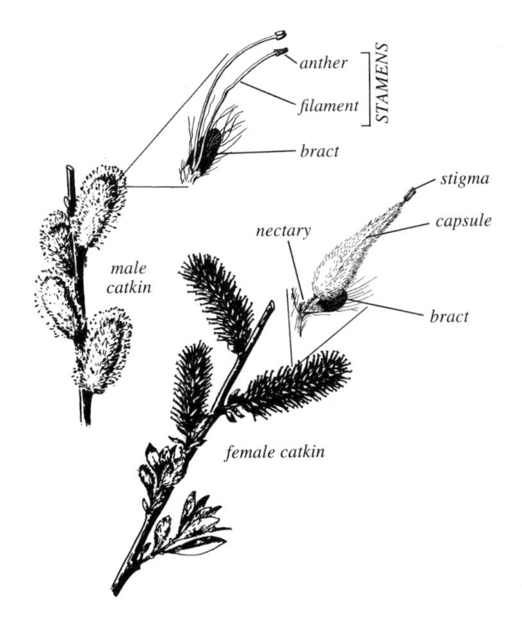

Willow Catkins

Key to the Willows (*Salix*)

1a. Leaf edges distinctly toothed 2

2a. Leaves hairy at maturity, at least below 3

3a. Leaf edge widely toothed with short, slender projections .... ***Salix exigua***

3b. Leaf edge closely saw-toothed or with rounded teeth 4

4a. Leaves broadly elliptic, base rounded to blunt .... ***S. bebbiana***

4b. Leaves narrowly elliptic, base sharp-pointed 5

5a. Leaves have long, silky hairs on both sides; edges closely saw-toothed .... ***S. petiolaris*** (see *S. bebbiana*)

5b. Leaves silky below with coating of short, flat-lying hairs; edges minutely saw-toothed .... ***S. arbusculoides***

2b. Leaves hairless, or inconspicuously hairy at maturity 6

6a. Leaves green below, without powdery bloom 7

7a. Capsules hairy .... ***S. maccalliana***

7b. Capsules not hairy 8

8a. Leaf stalks with glands at base of blade .... ***S. lucida*** (see *S. lasiandra*)

8b. Leaf stalks not glandular at base of blade .... ***S. myrtillifolia***

6b. Leaves have whitish bloom below 9

9a. Leaves lance-shaped, at least 3 times as long as broad 10

10a. Leaves about 3 times as long as broad, short-pointed at tip .... ***S. serissima***

10b. Leaves 4 times as long as broad (or longer), often drawn out into long point 11

11a. Leaves have whitish bloom below .... ***S. lasiandra***

11b. Leaves light green below .... ***S. lucida***

9b. Leaves broader, less than 3 times as long as broad 12

12a. Capsules hairy 13

13a. Leaves narrowly elliptic; catkins on short leafy shoots .... ***S. petiolaris*** (see *S. bebbiana*)

13b. Leaves broadly elliptic; catkins stalkless 14

14a. Leaves glossy above; capsule stalks 0.5 mm long; styles 0.6-1.8 mm long .... ***S. planifolia***

14b. Leaves dull above; capsule stalks 2-2.5 mm long; styles 0.3-0.7 mm long .... ***S. discolor***

12b. Capsules hairless 15

15a. Leaves lack stipules; unfolding leaves yellowish green, translucent .... ***S. pyrifolia***

15b. Leaves have prominent stipules; unfolding leaves reddish, opaque 16

16a. Catkins stalkless, no leafy bracts at base .... ***S. monticola***

16b. Catkins on leafy shoots, sometimes nearly stalkless, but have leafy bracts at base 17

17a. Leaves hairy above, especially on veins .... ***S. barclayi*** (see *S. monticola*)

17b. Leaves hairless on both sides 18

18a. Branches yellow to greyish yellow .... ***S. lutea***

18b. Branches reddish brown to pale brown ..... ***S. mackenzieana*** (see *S. lutea*)

1b. Leaves smooth-edged or indistinctly toothed 19

19a. Leaves smooth at maturity, not hairy 20

20a. Catkins at ends of short, leafy shoots; flowers in late spring .... ***S. pedicellaris***

20b. Catkins stalkless; stipules inconspicuous or absent; flowers in early spring 21

21a. Leaves glossy above; capsule stalks 0.5 mm long; styles 0.6-1.8 mm long .... ***S. planifolia***

21b. Leaves dull above; capsule stalks 2-2.5 mm long; styles 0.3-0.7 mm long .... ***S. discolor***

19b. Leaves distinctly hairy at maturity 22

22a. Leaves densely hairy below, leaf surface obscured 23

23a. Leaves long and narrow, more than 5 times as long as broad .... ***S. candida***

23b. Leaves usually less than 4 times as long as broad 24

24a. Leaves sparsely hairy below .... ***S. scouleriana***

24b. Leaves densely white-woolly below .... ***S. alaxensis*** (see *S. candida*)

22b. Leaves hairy below, leaf surface visible 25

25a. Leaves have some matted, rust-coloured hairs; leaf stalks velvety with short, erect hairs .... ***S. scouleriana***

25b. Leaf hairs white; leaf stalks not velvety; catkins at ends of short, leafy shoots 26

26a. Capsules on long stalks; catkins loosely-flowered .... ***S. bebbiana***

26b. Capsules stalkless or on short stalks; catkins compact 27

27a. Capsules stalkless or on very short stalks; catkins short-cylindrical to nearly round; leaf stalks less than 3 times length of buds .... ***S. brachycarpa*** (see *S. glauca*)

27b. Capsules on distinct stalks; catkins cylindrical; leaf stalks more than 3 times length of buds .... ***S. glauca***

SANDBAR WILLOW • *Salix exigua*
NARROW-LEAVED WILLOW

GENERAL: Shrub or small tree, 0.5–4 m tall, forms thickets by sending up shoots from roots; **branches greyish**, hairless; twigs brownish, silky when young, soon become hairless.

LEAVES: Long and narrow, 5–13 cm long, about 10 (between 7 and 20) times as long as wide; hairless or sparsely silky, no whitish bloom below; **edges usually have distinct teeth**; stalks short (0.8–5 mm long); stipules linear, up to 7 mm long.

FLOWERS: In stalkless or short-stalked catkins (1–7 cm long) on 3–6 cm long, leafy branches that are sometimes branched; appear with leaves or develop through summer; pistils hairless, long-beaked; bracts pale, fall off after flowering.

FRUITS: Hairless capsules, 4–7 mm long; stalk 0.3–0.6 mm long; style very short; stigmas fall off after flowering.

WHERE FOUND: Pioneer on sandy or gravelly floodplains and slough margins; found across our region.

NOTES: Also called *S. interior.* • The Woods Cree used the bark of sandbar willow, beaked willow and pussy willow to make nets for fishing or for straining the pitch that was used to seal birch bark canoes. The bark was also used as all-purpose cord to tie or fasten many things, including fish on roasting sticks, weights on fish nets, birch bark moose calls, snowshoe frames, and bundles of items for carrying. Several strips of bark could be twisted together to make rope. • The stems of these willows were also used to make many items: rims for birch bark baskets, bows and arrows, bead weaving looms, stoppers for sturgeon skin jars, and fish-roasting sticks. • The species name *exigua* means 'small' or 'very small,' though this is a fair-sized willow.

WESTERN SHINING WILLOW • *Salix lasiandra*
PACIFIC WILLOW

GENERAL: Tall shrub or tree, 2–9 m tall; branches light brown, usually hairless, brittle; **twigs glossy**, reddish brown, with **yellow, duckbill-shaped buds**, densely white pubescent, becoming hairless; bark fissured on older trees.

LEAVES: Lance-shaped, 6–15 cm long, **taper to long tip**; **edges have small gland-tipped teeth**; young leaves reddish, with dense, white and rust-coloured hairs; **older leaves hairless**, glossy above, **covered in white bloom below**; stipules prominent, glandular; stalks hairy, with **2 glands near leaf-base**.

FLOWERS: In catkins on long, leafy shoots; appear at same time as leaves; pistils hairless; bracts pale, fall off after flowering.

FRUITS: Hairless capsules, 5–7 mm long; style 0.5 mm long; stalk to 2 mm long.

WHERE FOUND: Riverbanks, floodplains, lakeshores and wet meadows; often in quiet river backwaters; widespread across our region, from Saskatchewan to interior Alaska.

NOTES: Also called *S. lucida* ssp. *lasiandra.* • Western shining willow is one of our largest native willows. • **Shining willow** (*S. lucida*) is similar to western shining willow, but it is generally smaller (less than 3 m tall), has smaller female catkins (less than 5 cm long). Some taxonomists classify *S. lasiandra* as part of *S. lucida.* • The Chipewyan burned and powdered willow bark and placed it on 'green' wounds and ulcers and places that would trigger perspiration. • Salicin, which gives willow bark its bitter taste, is related to acetylsalicylic acid, the active ingredient in Aspirin. Both of these chemicals were named for the genus *Salix*, since they were first derived from willow plants. They have been used for many years to treat rheumatism, arthritis, aches and pains, and fever. The earliest records of the use of willows to relieve pain come from Greece more than 2,400 years ago.

AUTUMN WILLOW • *Salix serissima*

GENERAL: Shrub, 1–4 m tall; branches yellowish brown to greenish brown, shiny, hairless; bark of older stems grey-brown.

LEAVES: Alternate, simple, elliptic to oblong-lance-shaped, 5–10 cm long and 1–3 cm wide; **reddish when unfolding**, become firm and somewhat leathery; tip sharp-pointed or quickly tapered, base sharp-pointed to rounded; dark green, hairless, shiny above; paler, slightly whitened below; edges have small gland-tipped teeth; **fine network of veins**; stipules absent or minute.

FLOWERS: In catkins on leafy branchlets; appearing from late May to mid-July; male catkins 1–3.5 cm long, usually 5 stamens; female catkins 1.5–4 cm long, up to 2 cm thick; **pistils reddish**, hairless; bracts oblong to obovate, pale yellow, hairy, fall off after flowering.

FRUITS: Hairless capsules; style short; stalk 0.5–2 mm long (about twice as long as nectary).

WHERE FOUND: Lakeshores, streambanks, slough edges, fens, and seepage areas; favours calcium-rich soils; across much of the boreal forest, north and west to Great Bear River.

NOTES: According to native folklore, if a woman sees a bear following her she should cut a willow branch, twist it into a circle and spit on it. Then she must throw the ring in the direction she wants the bear to go. • The species name *serissima* means 'very late.' Of all the willows in this book, autumn willow is the last to flower. The catkins, which mature in late summer, often last well into the fall, hence the common name autumn willow.

YELLOW WILLOW • *Salix lutea*

GENERAL: Shrub, 1.5–4 m tall; **small branches yellowish to greyish yellow**, hairless.

LEAVES: Alternate, simple, narrowly elliptic, 5–10 cm long, 1.5–3 cm wide; dark green and smooth above, paler below; **young leaves reddish**; tip abruptly tapered or sharp-pointed; base round or sharp-pointed; edges shallowly toothed or almost smooth; **stipules separate, egg-shaped, lobed at base**.

FLOWERS: In catkins that are stalkless or on short, leafy branchlets; male catkins 2–4 cm long; 2 stamens; female catkins 4–5 cm long; pistils hairless; bracts narrow, tawny to brown, hairy.

FRUITS: Hairless capsules; style 0.2–0.5 mm long; stalk 1.2–4 mm long (as long as or longer than bract).

WHERE FOUND: Forms thickets on riverbanks and slough edges; occasional in coulees and on active sand dunes; in prairie and parkland, north into boreal forest along Slave and Mackenzie rivers to 65°N.

NOTES: Mackenzie's willow (*S. mackenzieana*, also called *S. prolixa* or *S. rigida*) is closely related to yellow willow, from which it can be distinguished by its reddish brown branches. Mackenzie's willow also forms thickets on moist sandy riverbanks, primarily along the Mackenzie River and its tributaries (the Peace, Athabasca and Liard rivers). • Willow bark can be collected in strips and cooked like spaghetti, or dried and ground into flour. • Try picking willow leaf buds in the spring, just as they start to turn green. Serve them in a bowl with milk and sugar, like cereal, for a healthy breakfast with lots of vitamin C. Willow is 7–10 times richer in vitamin C than an orange. Buds and young leaves make a good salad addition or nibble, and they can also be cooked as a vegetable. • *Lutea* means 'yellow,' in reference to the colour of the branches.

S. mackenzieana

BALSAM WILLOW • *Salix pyrifolia*

GENERAL: Shrub, 1–3 m tall; twigs and branches widely spreading, hairless, dark reddish brown or purplish; **buds and leaves have balsam-like fragrance**.

LEAVES: Elliptic or egg-shaped, 3–10 cm long; base rounded or heart-shaped; edges have rounded, gland-tipped teeth; **young leaves translucent**, yellow-green; **older leaves hairless, net-veined, bluish white bloom below**; stalks 9–20 mm long, with glands at base of blade.

FLOWERS: In catkins on leafy shoots; appear at same time as leaves; female catkins 2.5–5 cm long; bracts pale, with silky hairs.

FRUITS: Hairless capsules, 4.5–8 mm long; style 0.3–0.5 mm long; stalk 1.6–2.8 mm long.

WHERE FOUND: Poplar and aspen woods, fens and swamps, streambanks, lakeshores, and moist clearings; widespread across boreal forest from southern Manitoba to southern N.W.T. and eastern B.C.

NOTES: The balsam-like fragrance is similar to that of balsam poplar, and lasts in dried specimens. • The Chipewyan used this willow, as well as beaked willow and myrtle-leaved willow, to make ptarmigan snares. To do this, they took a willow branch with its leaves still on and bent it into a ring. This was set about 20 cm above the ground on small legs made by bending willow branches together and lashing them in place with willow bark twine. A ptarmigan snare of sinew thread was then suspended between the legs. Sometimes willow bark strips were used to protect rawhide snares from moisture and deterioration. • Chipewyan women used willow branches to curl their hair. Forked twigs were stripped of their bark and heated in a bed of hot sand. Then the women would roll a lock of hair around each twig. Were these the first heated rollers?

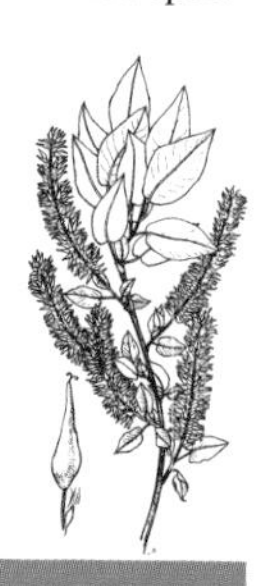

MOUNTAIN WILLOW • *Salix monticola*

GENERAL: Erect shrub, 1–4 m tall; **branches dark reddish brown to blackish**, glossy; twigs lighter, usually hairless.

LEAVES: Narrowly to broadly elliptic to obovate, 4–8 cm long, pointed at the tip, rounded to heart-shaped at the base; edges with rounded, gland-tipped teeth; **young leaves reddish**; reddish leaf stalks; older leaves hairless, **bluish white bloom below**; stipules egg-shaped.

FLOWERS: In catkins that are **stalkless** or on very short, leafy branches; appear before or with leaves, female catkins usually 4–8 cm long, hairless; bracts dark brown.

FRUITS: Hairless capsules, about 4.5 mm long; style 0.7–1.2 mm long; stalk 0.9–2 mm long.

WHERE FOUND: Wetlands, moist forest edges and openings, riverbanks and lakeshores; widespread through our region.

NOTES: Also called *S. pseudomonticola* and *S. padophylla*. • Mountain willow forms large thickets in some valley bottoms. • **Barclay's willow** (*S. barclayi*) is a similar species, but it has greenish (rather than reddish) young leaves; mature leaves that are green on both surfaces; dark, hairy twigs; and catkins on leafy stalks. Barclay's willow is a 1-3 m tall, thicket-forming, northwestern willow. It occurs in and west of the Rockies in our region, and grows in forest openings, mountain fens and avalanche tracks. • Willow suckers can grow more than 1 m in a single season, producing slender, straight shoots that are ideal for weaving into baskets and mats. Peeled willow branches have been used to make bows and arrows. The bark was used to make cooking baskets because it swells when it is wet, thereby sealing the seams of the container. If the bark was slid off smaller branches as a tube it could be used to make drinking straws and whistles. Willow bark contains large amounts of tannin, and it was sometimes used to tan hides. Skins boiled with willow twigs took on a yellow colour. • Willows are easy to propagate and they grow quickly and produce extensive root systems. They are often used in revegetation projects to hold soil in place and prevent erosion in wet areas.

HOARY WILLOW • *Salix candida*

GENERAL: Short shrub, 0.3–1 m tall (sometimes to 1.5 m); branches brown, sparsely hairy; twigs short, with **white, woolly hairs**.

LEAVES: Narrowly elliptic or oblong, 2–10 cm long, 0.5–1.5 cm wide; **edges somewhat rolled under**, smooth or nearly so; dark green, sparsely woolly above; **dense, dull white, woolly hairs below**; stalks 3–9 mm long (sometimes to 12 mm); stipules narrowly elliptic.

FLOWERS: In catkins on short, leafy shoots; appear with leaves; female catkins 2–5 cm long, with **dense, white, woolly hairs**; bracts brownish.

FRUITS: Hairy capsules; style 0.8–1.6 mm long, bright red at first; stalk 0.5–1.2 mm long; nectary red at first.

WHERE FOUND: Fens, lakeshores and riverbank thickets, usually where soil is calcium-rich or alkaline; widespread across our region north and west to eastern Alaska.

NOTES: Hoary willow is an uncommon but locally abundant species. Its narrow, woolly leaves are unmistakable. • Across the Northwest Territories and through Alaska, another common willow with white woolly hairs on its lower leaf surface is **Alaska willow** (*S. alaxensis*) (inset photo), also called felt-leaved willow. Alaska willow is easily distinguished from hoary willow by its broader (1.5–3 cm wide), less linear leaves. Also, the catkins of Alaska willow appear before the leaves, whereas those of hoary willow are produced later, with or after the leaves. Alaska willow is very common on streambanks and lakeshores in the northwest of our region, where it can reach the size of a small tree. It is often a favourite food of moose and grouse, especially during winter when many other shrubs are buried in snow.

MYRTLE-LEAVED WILLOW • *Salix myrtillifolia*
LOW BLUEBERRY WILLOW

GENERAL: Short shrub, 10–100 cm tall; branches greyish, soon hairless, **usually spreading on ground and rooting**; twigs greenish to reddish brown, with sparse, short, curved hairs.

LEAVES: Narrowly lance- to egg-shaped (usually widest above middle), 2–6 cm long; **hairless, green on both sides**; tip broad and pointed; **edges round-toothed**; stipules 0.2–2 mm long, narrowly elliptic.

FLOWERS: In catkins on short, leafy branchlets; appear at same time as leaves; female catkins 1.5–4 cm long (usually 2–3 cm); pistils hairless, pale green; bracts light brown, with dark tip; male catkins sweetly scented.

FRUITS: Hairless capsules, 4.5–7 mm long; style 0.3–0.5 mm long; stalk 0.6–2 mm long.

WHERE FOUND: In deep moss in peatlands and on lakeshores and riverbanks, floodplains and moist coniferous forests; widespread through western boreal forest.

NOTES: Also called *S. novae-angliae*. • **Tall blueberry willow** (*S. myrtillifolia* var. *pseudomyrsinites*, sometimes called *S. myrtillifolia* var. *cordata*) is 1–4 m tall. Its leaves are hairy (at least on the upper midvein) and paler below; its stipules are 1–5 mm long; its styles are 0.5–1 mm long; and it is generally found in drier habitats than myrtle-leaved willow. • The Chipewyan used myrtle-leaved willow, beaked willow, and balsam willow to make a variety of items. Slender branches were split, hollowed and tied back together with spruce roots to make pipe stems. Flutes and whistles were made by burning a hole through the length of a stick and adding 4–5 finger holes. Willow wood was used to make bows, canoe ribs, make-shift snowshoes, and wooden fasteners (nails). Flexible willow shoots were woven into baskets.

LITTLE-TREE WILLOW • *Salix arbusculoides*
SHRUBBY WILLOW

GENERAL: Shrub to small tree, 1–7 m tall; branches slender, reddish brown, shiny; twigs sparsely covered in velvety hairs.

LEAVES: Narrowly elliptic to lance-shaped, 2–6 cm long; edges **finely toothed**; **glossy and hairless above**; **underside covered with whitish bloom and short stiff hairs** pointing towards tip of leaf; stalks 5–8 mm long; stipules linear.

FLOWERS: In catkins that are 2–7.5 cm long and stalkless or on short, leafy shoots; appear at same time as leaves; pistils have dense, short, silky hairs (some rust-coloured).

FRUITS: Sparsely silky capsules, 4–6 mm long; style 0.3–0.5 mm long; stalk 0.6–0.9 mm long.

WHERE FOUND: Wetlands, floodplains, streambanks, lakeshores and open spruce woods; widespread in our region; especially common in northwestern parts.

NOTES: Alaskan peoples used pieces of these shiny red branches as chewing sticks to treat mouth sores. They also chewed the leaves to treat sore throats. • Willow bark was taken off the stems in strips and the fine fibres were twisted and spliced together by rolling them on the leg. The resulting thread or string was used to make nets and lines for fishing. Willow string and cord becomes very brittle when it dries, so nets had to be stored in water and cord had to be soaked before it could be used. Willow roots were also used for sewing birch bark baskets. • *Arbusculoides* is from *arbuscula*, which means 'a low shrub,' but little tree willows are sometimes more than 5 m tall.

FLAT-LEAVED WILLOW • *Salix planifolia*
PLANE-LEAVED WILLOW

GENERAL: Erect shrub, 0.5–4 m tall; branches dark brown to reddish brown, hairless, sometimes with bluish white bloom; twigs hairless, widely spreading.

LEAVES: Elliptic to narrowly elliptic, 3–5 cm long; edges sometimes rolled under, usually almost smooth but sometimes with a few small teeth; young leaves sparsely hairy with short, silky, white and rust-coloured hairs; **older leaves hairless and shiny above, sparse silky hairs and bluish white bloom below**; many primary veins, closely spaced, parallel; stipules to 2.5 mm long, soon fall off.

FLOWERS: In **stalkless** catkins on previous year's branches; appear before leaves; female catkins 2–6 cm long, with long, dense, silky hairs, compactly fruited; bracts dark brown to black, with long hairs.

FRUITS: Sparsely silky-hairy capsules, 5–6 mm long (sometimes to 8 mm); style 0.8–1.3 mm long; stalk 0.5–0.9 mm long.

WHERE FOUND: Fens, moist coniferous woods, forest openings, lakeshores and streambanks; widespread across our region and north to Arctic coast.

NOTES: Also called *S. phylicifolia*. • Two subspecies occur in our region—ssp. *planifolia* and ssp. *pulchra*. These two subspecies look very similar, but ssp. *pulchra* is readily distinguished by the many dead leaves and longer stipules that often remain on its twigs for over a year. Also, it is generally found only north of 60°N or at high elevations, whereas ssp. *planifolia* is widespread across our region. • Some Inuit chewed the leaves of flat-leaved willow to cure mouth sores. They also used the bark and leaves to make medicinal teas for aches and pains. In the Arctic, this was the willow species most highly regarded as spring food. Young shoots, catkins and inner bark were all used as food, either raw or cooked. Often they were mixed with seal oil. Flat-leaved willow also seems to be preferred by moose.

Velvet-fruited Willow • *Salix maccalliana*
Maccall's Willow

GENERAL: Shrub, 1–5 m tall, rarely over 1 m at northern limit of range; branches dark, reddish brown, glossy; twigs brown to yellowish, hairless to sparsely hairy.

LEAVES: Leathery, narrowly elliptic to oblong, 5–7 cm long (or smaller); young leaves have matted, white and rust coloured hairs; **older leaves glossy and green on both sides**; edges have small, **rounded, gland-tipped teeth**; stipules are tiny glandular lobes, up to 0.5 mm long.

FLOWERS: In catkins that are 2–6 cm long and on short, leafy shoots; appear as leaves expand; bracts pale.

FRUITS: Hairy capsules, 6–8 mm long; style 0.8–1.2 mm long; stalk 1–2 mm long.

WHERE FOUND: Wetlands and riverbanks; widespread across boreal forest; occasional in parkland; north and west to southern N.W.T. and eastern B.C.

NOTES: Velvet-fruited willow stands out in early spring with the bright red bark of the previous year's growth. • The Dene used the inner bark of willows as tobacco, preferring it to the more commonly used kinnikinnick. The leaves and young twigs of many willows have been boiled in water to make a hair rinse for dandruff. • All parts of the willow can be used for dyeing: catkins give a soft green; leaves a bright yellow; branches a soft pinkish yellow; and roots a deep, warm brown or red-brown.

Bog Willow • *Salix pedicellaris*

GENERAL: Small erect shrub, 20–150 cm tall (rarely over 1 m in the north); branches dark grey-brown, hairless; twigs reddish, hairless or with a few short, erect hairs.

LEAVES: Leathery, narrowly egg-shaped or elliptic, 2–5 cm long, often widest above middle, **tip usually rounded**; **always hairless**; bluish green, often **whitish bloom on both sides**; **smooth-edged**; stipules narrowly elliptic, blunt-tipped, 0.2–2 mm long.

FLOWERS: In catkins on leafy shoots; appear at same time as leaves; female catkins 1.5–3.5 cm long, hairless, loosely flowered; bracts pale.

FRUITS: Hairless, slender capsules, 5–7 mm long, **dark reddish brown when young**, tawny when mature; style very short; stalk 2–4 mm long.

WHERE FOUND: Fens and swamps; widespread across boreal forest; north and west to Great Bear Lake and eastern Yukon.

NOTES: Hybrids between bog willow and grey-leaved willow (p. 52) have patches of hairs on the pistils, and hairy leaves. • If you want to be sure a shrub is a willow, take a close look at the buds on its stems. Willow buds are covered by a single, smooth scale that curves around and encloses the immature leaf or branch. The buds of other shrubs have 2 or more overlapping scales.

GREY-LEAVED WILLOW • *Salix glauca*
BLUE-GREEN WILLOW, SMOOTH WILLOW

GENERAL: Erect shrub, 0.5-1 m tall (sometimes to 2 m); branches dull greyish to reddish brown, usually hairy.

LEAVES: Elliptic to narrowly egg-shaped, usually widest above middle, 2–5 cm long (sometimes to 10 cm), smooth-edged, usually **somewhat hairy** (especially below); smooth, green above, **greyish bloom below**; stalks 4–15 mm long, usually over 3 times as long as buds; stipules less than 1 mm to 8 mm long, with glands along edges.

FLOWERS: In **dense catkins** that are **2–7 cm long** and **on leafy branchlets**; **appear at same time as leaves**; **female catkins** 2–5 cm long (sometimes to 7.5 cm), with **dense silky hairs** on pistils; male catkins 1.5–2.5 cm long; bracts yellowish to light brown, hairy.

FRUITS: Grey-hairy capsules, 3–5 mm long (sometimes to 8 mm), green or yellowish while living; stalks 0.5–2 mm long, longer than nectary.

WHERE FOUND: Along rivers, in thickets and in wetlands across northern boreal forest; very common from northwestern Saskatchewan to Alaska; circumpolar.

NOTES: INCLUDES Athabasca willow (*S. athabascensis*). • Grey-leaved willow is an extremely variable species that has been divided into 4 or more subspecies or varieties by various taxonomists. It is treated in the broad sense here. When growing in exposed areas, grey-leaved willow could be confused with a similar but usually smaller species, **short-capsuled willow** (*S. brachycarpa*). These two willows are most easily separated on the basis of their leaf stalks. Grey-leaved willow has relatively long leaf stalks (4–15 mm), whereas short-capsuled willow has stalkless leaves or leaf stalks that are 1–2 mm long. Short-capsuled willow has short, fat catkins and it grows in open forests, gravelly floodplains, seepage areas and sedge fens, in limestone country.

BEAKED WILLOW • *Salix bebbiana*
BEBB'S WILLOW

GENERAL: Shrub or small tree, 0.5–5 m tall; branches usually widely divergent, reddish brown; twigs have dense to straggly hairs.

LEAVES: Elliptic to obovate, 2–6 cm long; edges smooth to scalloped; dull green above; **grey below, covered with whitish bloom, prominent venation**; **sparsely hairy on both sides**; stalks 2–9 mm long; stipules small, fall off early.

FLOWERS: In catkins on short leafy shoots; appear **just before or with leaves**; female catkins 2–5 cm long, loosely flowered; pistils finely silky; **bracts pale**.

FRUITS: Sparsely hairy capsules, 6–9 mm long; stalk 2–5 mm long.

WHERE FOUND: Woods, thickets, riverbanks, wetlands and disturbed sites; one of the most common willows; widespread across our region.

NOTES: Basket willow (*S. petiolaris*, also called *S. gracilis*) is similar to beaked willow (and pussy willow, p. 53) but has longer, narrower, long-pointed, closely toothed leaves with a whitish bloom below. It is a species of slough edges, sedge meadows, lakeshores and streambanks. • The Woods Cree chewed the inner bark from the roots of beaked willow to stop bleeding, reduce infection and promote healing of deep cuts, such as axe cuts. • Beaked willow is one of several species of willow commonly called 'diamond willow.' The trunks and large branches can become infected with 1 or more species of fungi that darken some of the wood to a reddish brown. These dark sections form diamond-shaped patterns around the bases of branches and scars. Sections of the trunks or branches that have been stripped of their bark take on attractive patterns and native peoples used diamond willow to make ceremonial staffs and sticks. They are also used to make decorative canes, lamp posts, furniture and candle holders.

PUSSY WILLOW • *Salix discolor*

GENERAL: Large shrub or small tree, 2–6 m tall (sometimes to 11 m); bark greyish brown; branches spreading, yellowish brown to reddish, hairless with age.

LEAVES: Elliptic to obovate, 3–10 cm long, usually pointed at tip and base, smooth-edged and wavy to irregularly round-toothed, **dark green above**, **greyish below with white bloom** and widely spaced veins; dense white or reddish silky when young, slightly hairy to hairless when mature.

FLOWERS: In **essentially stalkless catkins**, appear long before leaves on branches of previous year; male catkins 2–4 cm long; 2 stamens; female catkins 4–7 cm long (sometimes only 2 cm), pistils sparsely silky; bracts dark brown to black.

FRUITS: Long-beaked capsules, 8–10 mm long (sometimes only 6 mm), **short-hairy**; scales black, silky hairy; style 0.3–0.7 mm long; stalk 2–2.5 mm long, shorter than or equal to scales.

WHERE FOUND: Riverside thickets, swamps, shrub fens, edges of marshes and lakes, open forest and moist clearings; common and widespread across southern boreal forest and parkland, north to southern N.W.T.

NOTES: Flat-leaved willow (p. 50) is very similar to pussy willow, but flat-leaved willow can be distinguished by its longer styles (0.8-1.3 mm), shorter capsule stalks (0.5–0.9 mm) and relatively shiny upper leaf surfaces. Identification is made more difficult by the fact that pussy willow and flat-leaved willow often form hybrids. • The Woods Cree steeped the inner bark of pussy willow in boiling water to make a medicinal tea to relieve diarrhea. A brew to relieve pain and fever was made by steeping the twigs. • Native peoples used the wood for furniture, tipi pegs and sweat-lodges, the spring buds for a red dye and the tannin in the bark to tan hides. • The species name *discolor* is Latin for 'non-uniform in colour.' It refers to the leaves, which are dark green above and light greyish below.

SCOULER'S WILLOW • *Salix scouleriana*

GENERAL: Tall spindly shrub or multi-stemmed small tree, 2–7 m tall (sometimes to 10 m); branches dark brown to yellowish brown, often velvety; twigs densely velvety to woolly.

LEAVES: Obovate or oblanceolate to narrowly elliptic, wedge-shaped at base, 5–8 cm long; edges smooth or with small, gland-tipped teeth, often rolled under; **young leaves densely velvety**; **older leaves firm, dark green above, sparsely hairy** with short, stiff, flat-lying hairs (**some rust-coloured**), or densely hairy with wavy hairs, **bluish white bloom below**.

FLOWERS: In catkins on previous year's branches; usually appear **well before** leaves; catkins are stalkless or on short branchlets with small leaves and bracts; female catkins 2.5–6 cm long; pistils densely silky; **bracts dark brown to black**, with silky hairs.

FRUITS: Silky capsules, 5–8 mm long; style 0.2–0.6 mm long; stigma 1–2 times as long as style; **stalk 0.8–2 mm long**.

WHERE FOUND: Well-drained coniferous forests, floodplains, sand beaches and sand dunes; widespread across boreal forest from central Manitoba to Alaska.

NOTES: Willows are fast-growing and they produce large amounts of nutritious forage for many wild animals, including snowshoe hares, porcupines, beavers, lemmings, grizzly bears, deer, caribou and moose. Because these shrubs are usually tall enough to stick up through the snow, their buds are an important winter food for grouse and ptarmigan. • This common species is named for Dr. John Scouler, an associate of David Douglas, an explorer-botanist of the Northwest.

RED-OSIER DOGWOOD • *Cornus stolonifera*
RED WILLOW

GENERAL: Erect to spreading shrub, 1–3 m tall; branches opposite; lower branches often root in ground; young stems, **usually bright red**, but sometimes greenish to purplish.

LEAVES: Opposite, oval to egg- or lance–shaped, 2–8 cm long; usually rounded at base, pointed at tip; **5–7 prominent parallel veins** converge towards tip; filmy white threads running through veins can be seen if leaf is split crosswise and gently pulled apart.

FLOWERS: In many-flowered, **dense**, **flat-topped clusters** (2–5 cm across), **at branch tips**; **white to greenish**, small; petals 2–3 mm long; appear in late May to July.

FRUITS: Berry-like drupes, **white** (occasionally bluish), 5–6 mm across, juicy but bitter; stones somewhat flattened.

WHERE FOUND: Moist woods, thickets, clearings and riverbanks; widespread across our region, north and west to interior Alaska.

NOTES: Also called *C. alba* and *C. sericea*. • Many tribes have scraped or shaved off the soft, green, inner bark, dried the pulp, and used it for smoking, either alone or mixed with tobacco. Northern groups are said to have learned this use from the plains cultures to the south. It does not add any taste but is said to give a pleasant aroma. Some elders still use it today. • The Chipewyan boiled the roots of red-osier dogwood to make a medicinal tea to treat dizziness. They also used the stems to make a tea for people with chest trouble or difficulty urinating. The Cree used the fruit or the stem pith (when berries were unavailable) to make a wash to treat snow blindness. They also treated cataracts with the pith. They used the bark of young branches to induce vomiting, and bark tea was taken as a treatment for coughs, fevers, and pain. • This species is extremely important moose winter browse, and bears like the bitter white berries. • The Chipewyan do not eat the berries, but many other native groups do, usually mixing them with other sweeter berries to mask their bitter flavour. Occasionally the 'nuts' inside the fruits were cracked open and eaten. The inner bark was also used to make tea. • The flexible, red branches were often used to weave baskets or make rims for birch bark containers.

ALDER-LEAVED BUCKTHORN • *Rhamnus alnifolia*

GENERAL: Low, erect or spreading shrub, usually under 1 m tall; stems have fine grey hairs; become smooth, dark grey-brown with age.

LEAVES: Simple, alternate; 3–10 cm long, lance-shaped to egg-shaped, thin, **6–8 prominent veins**; glands along finely scalloped edges; stalks short; stipules 3–6 mm long, soon fall off.

FLOWERS: Clustered in the axils of lower leaves; usually **single-sexed, with sexes on separate plants**; usually 5 sepals that are **greenish yellow**, 4–5 mm long; no petals; flowers mostly in June.

FRUITS: Reddish to blue-black **berry-like drupes**, 6–10 mm long, with 3 seeds; **poisonous**.

WHERE FOUND: Moist soil along streams or in shady woods; scattered across boreal forest from Manitoba to northeastern Alberta.

NOTES: The bark, leaves and fruits of alder-leaved buckthorn all contain glycosides, which cause **vomiting and diarrhea**. The inner bark of this shrub has been used as a laxative, like that of cascara (*R. purshiana*), which is found south from southwestern British Columbia. However, alder-leaved buckthorn is much faster acting and more insistent than cascara. • 'Buckthorn' is from a translation of *cervi spina*, 'stag's thorn,' the modern Latin name for a European species. The genus name *Rhamnus* is from *rhamnos*, the Greek name for a European species, or possibly the Celtic *rham*, meaning 'a tuft of branches.' *Alnifolia* means 'alder-like leaves,' and is from the Latin *alnus*, 'an alder,' and *folium*, 'a leaf.'

SILVERBERRY • *Elaeagnus commutata*
WOLF-WILLOW

GENERAL: Upright shrub or small tree, 1–4 m tall; grows from spreading rhizomes; **twigs densely covered with rusty brown scales**.

LEAVES: Alternate, oval to egg- or lance–shaped, 2–6 cm long (sometimes to 10 cm), **densely silvery with tiny star-shaped hairs on both surfaces**; paler, sometimes has brown scales below.

FLOWERS: Many clusters of 1–3 flowers in leaf axils; **yellowish**, funnel-shaped, 12–15 mm long, strongly sweet-scented; appear in June to July.

FRUITS: Oval, **silvery**, dry, mealy berries, about 1 cm long, contain 1 large nutlet.

WHERE FOUND: Streambanks, lakeshores, floodplains and open slopes; widespread across our region.

NOTES: The flowers can be detected from many meters away by their heavy perfume. Many people enjoy this fragrance but some find it too overwhelming. The wood also has a strong fragrance when it is burned. If green wood from silverberry is put in the fire it gives off a strong smell of human excrement. Some practical jokers enjoy sneaking branches into the fire and watching the reactions of fellow campers. • The Cree and Blood ate silverberries, but many groups considered the dry, mealy berries good only as famine food. The berries were mixed with blood and cooked, mixed with lard and eaten raw or frozen like ice cream, or they were fried in moose fat. Apparently silverberries make good jam. • The Cree used the bark to make cord, and several tribes used the seeds as decorative beads. The fruits were boiled to remove the flesh, and while the seeds were still soft, a hole was made through each. They were then threaded, dried, oiled and polished, and used to make jewellery and to decorate clothing.

CANADA BUFFALOBERRY • *Shepherdia canadensis*
SOOPOLALLIE; SOAPBERRY

GENERAL: Spreading shrub, 0.5–3 m tall; **branches brownish, covered with small, bran-like scabs**.

LEAVES: Opposite, elliptic to narrowly oval; dark greenish above; **whitish, silvery, with felt of star-shaped hairs and rusty, brown scales below**.

FLOWERS: Single or in small clusters; **yellowish brown**, inconspicuous, about 4 mm wide; male and female flowers on separate plants; appear in late April to early May, just before leaves.

FRUITS: Bright red to yellowish, **translucent**, spherical to **oval, juicy berries**, 3–6 mm long (usually 4-5 mm), extremely bitter, soapy to touch.

WHERE FOUND: Open woods, thickets, and riverbanks; widespread across our region, north to Arctic coast.

NOTES: Northern natives used Canada buffaloberry to make medicinal teas for treating stomachaches, constipation, heart problems, arthritis, tuberculosis and gallstones • Buffaloberry is also known as 'soapberry' because of the foaming properties of its small scarlet berries, and because of their taste. Although these berries were widely used as food by many native peoples, the Chipewyan considered them inedible. The berries, which are rich in vitamin C and iron, stay on the shrubs for a long time and are easy to find, so they make a good emergency food. Although they are bitter, the flavour improves a bit after the first frost. The bitter, soapy substance (saponin) in these berries is not dangerous in small amounts, but if large amounts are eaten, it can cause **diarrhea, vomiting and abdominal pain**. • The berries were often made into a foamy dessert, much like whipped cream or meringue, widely known as 'Indian ice cream.' They were also added to stews, boiled into a sauce, canned or made into jelly. • After giving its bark for strings and its berries for food, the buffaloberry refused to give its wood for firewood. Some tribes called it 'stinkwood.' If you try burning it you will understand why. • The name 'soopolallie' is Chinook jargon meaning 'soap' (*soop*) 'berry' (*olallie*).

NARROW-LEAVED MEADOWSWEET • *Spiraea alba*

GENERAL: Slender, erect shrub with few branches (in our area), to 1.5 m tall; branches not round, but more or less **angled or ridged**; twigs yellowish to reddish brown, hairless or minutely hairy; **outer bark peels off in narrow, papery strips**.

LEAVES: Simple, alternate, 4–6 cm long, narrowly lance-shaped to elliptic, pointed at both ends; edges finely toothed; dark green above, paler below; hairless or occasionally with fine hairs on veins below.

FLOWERS: In dense, long, finely hairy, many-flowered clusters at branch tips; 5 sepals, broadly triangular; 5 petals, white, 2–3 mm long; flowers from June to September.

FRUITS: Clusters of small, papery 'pods' (follicles) that open along 1 side; seeds few, tapered at both ends.

WHERE FOUND: Moist meadows, roadsides, shores and edges of fens and marshes; widespread across northern parkland and southern boreal forest in prairie provinces.

NOTES: Alaska spiraea (*S. beauverdiana*, also called *S. stevenii*) (inset photo) also has many small white flowers, but they are in a flat-topped cluster. Alaska spiraea is found on both sides of the Bering Strait (extending east to the Mackenzie River valley) and grows along streambanks and in wet meadows, muskeg, shrub thickets and open forest. • *Spiraea* was one of the first sources of salicylic acid. Apparently, the drug Aspirin takes its name from *a*, for 'acetyl' (it is composed of acetylsalicylic acid), and *spir*, for 'spiraea,' since it was found in *Spiraea* blossoms. This suggests that teas made from these flowers should be good for reducing fever and relieving headaches.

S. beauverdiana

SASKATOON • *Amelanchier alnifolia*
SERVICEBERRY; JUNEBERRY

GENERAL: Shrub or small tree, 1–5 m tall, often spreading by rhizomes or stolons and forming dense thickets; branches smooth, dark grey to reddish brown.

LEAVES: Alternate; thin, elliptic to obovate, 1–5 cm long; rounded at tip, squared to rounded at base; **coarsely sharp-toothed** on top half.

FLOWERS: 3–20 **in short, leafy clusters**; white, **showy**, about 8–12 mm across, drooping to erect; appear in spring.

FRUITS: Berry-like pomes (like miniature apples), 6–10 mm across, purple to nearly black, with whitish bloom, juicy and sweet.

WHERE FOUND: Dry to moist forests, thickets and open hillsides on well-drained soils; widespread across our region, north and west to interior Alaska.

NOTES: Most native peoples harvested these delicious fruits and in some regions this was the only fruit available in large quantities. Traditionally they were spread out and dried separately, like raisins, or mashed and formed into blocks for drying. Once dried they were eaten for a snack, mixed with dried, powdered meat and fat to make pemmican, or rehydrated by boiling in water, often in soups or stews. Because they are so sweet, saskatoons were often mixed with less palatable berries. Saskatoons are widely collected for use in pies, pancakes, muffins and syrups, or fresh on cereals and desserts—much like blueberries. • The Cree boiled 4–5 barked, split sticks, each about 10 cm long, in sturgeon oil for about 10 minutes to keep the oil fresh during storage. • Saskatoon has hard, strong, straight stems, and the wood has been used to make arrows, canes, pipestems, digging sticks, spears, canoe cross-pieces, and rims for birch bark baskets. • Saskatoon is an excellent ornamental shrub for northern gardeners. It is hardy and easily propagated by seeds or suckers. In early spring, it is covered with beautiful white blossoms, in late summer it produces delicious fruit, and in fall its leaves often turn pale to deep scarlet.

PIN CHERRY • *Prunus pensylvanica*

GENERAL: Shrub or small tree, 1–5 m tall; bark reddish brown, peeling in horizontal strips, with **prominent, raised pores**.

LEAVES: Oval to lance-shaped, 3–10 cm long, gradually taper to long point at tip, rounded at base; edges have small rounded teeth; 2 small glands on stalk near base of blade.

FLOWERS: In **flat-topped clusters**; white, 6–10 mm across; appear at same time as leaves.

FRUITS: Bright-red cherries, 5–8 mm across, sour.

WHERE FOUND: Forest clearings, hillsides and riverbanks; usually on well-drained sites; widespread across our region north and west to southern N.W.T. and northeastern B.C.

NOTES: Warning: Pin cherry leaves, bark, wood, and seeds (stones), like other *Prunus* species, contain hydrocyanic acid and therefore can cause **cyanide poisoning**. Crushed leaves or thin strips of bark will kill insects in an enclosed space. The flesh of a cherry is the only edible part—the stone should always be discarded. • Pin cherries can be eaten raw from the bush as a tart nibble, but usually they are cooked, strained and made into jelly or wine. The fruit seldom contains enough natural pectin to produce a firm jelly, so additional pectin must be added. Pin cherries are still collected in large quantities today. • The Cree used the inner bark to make a medicinal tea for treating sore eyes. The bark was collected in short strips in winter or spring and dried for later use.

CHOKE CHERRY • *Prunus virginiana*

GENERAL: Shrub or small tree, 1–6 m tall; bark smooth, reddish brown to grey-brown, becomes dark with age, does not peel readily, with **inconspicuous, raised pores**.

LEAVES: Thin, **elliptic to obovate**, 2–10 cm long, sharp-pointed to rounded at tip, blunt at base; bright green and hairless above, paler below; edges have fine, sharp teeth; stalks have 2 or 3 prominent glands.

FLOWERS: In many-flowered, **bottlebrush-like clusters (5–15 cm long) at ends of branches**; flowers are white, 10–12 mm across; appear in May to June.

FRUITS: Shiny, red, purple or black cherries, about 8 mm across, edible but astringent.

WHERE FOUND: Woods, clearings, hillsides and river terraces; often on dry and exposed sites; widespread across our region, north and west to southern N.W.T. and northern B.C.

NOTES: All parts of the choke cherry (except the flesh of the fruit) contain **hydrocyanic acid**. There are reports of children **dying of cyanide poisoning** after eating large amounts of choke cherries without removing the stones. Dried choke cherries were also added to pemmican, or were cooked with meat or stew. The Cree believed that large quantities would cause constipation, so they always mixed grease with choke cherries, and sometimes they also added fish eggs. • Choke cherries are very popular for making jellies, syrups, sauces and wine, and large quantities are still collected in the wild. • The Cree made medicinal teas from dried chokecherry bark to treat coughs, colds and pneumonia. This was also given to singers and speakers to clear their throats. The Woods Cree boiled fresh bark or roots in water to make a drink for relieving diarrhea. • Choke cherry bark was listed in the U.S. Dispensatory from about 1800–1975. Wild cherry cough drops and syrups have been made for many years using this bark. In folk medicine the bark was boiled to make a strong medicinal tea. This was taken to expel worms, or it was used to wash ulcers or abscesses. • Choke cherry wood is very hard and does not burn easily, and forked sticks of these shrubs were often used to move or carry hot items such as heated rocks for sweat lodges and hot coals for smudges. The hard wood was also useful to make digging sticks, tent pegs, back rests and household tools.

PRICKLY ROSE • *Rosa acicularis*

GENERAL: Bushy shrub, 0.3–1.5 m tall; stems stout, usually **densely covered with many straight, weak bristles** and straight, slender thorns.

LEAVES: Compound, **3–9 oblong leaflets** (usually 5), each 2–5 cm long (usually 3–4 cm), sharply double-toothed, usually somewhat hairy beneath.

FLOWERS: Single, on short side branches; **pink**, showy, 5–7 cm across; appear in late May to July.

FRUITS: Scarlet, spherical to pear-shaped, fleshy '**hips**,' about 1.5 cm long.

WHERE FOUND: Open forest, thickets, riverbanks, and clearings; widespread and common across our region; nearly circumpolar.

R. woodsii

NOTES: Prickly rose could be confused with **common wild rose** (*R. woodsii*), also called wood rose, but the upper stems of that species lack the many small bristles and prickles that cover the stems of prickly rose. Instead, common wild rose has a few scattered thorns, usually at stem nodes. It is found infrequently across the boreal forest from southern Manitoba to eastern Alaska. Despite its name, common wild rose is by far the less common of the two species in our region. Native peoples used common wild rose in much the same way as prickly rose. • The beautiful, common flower of prickly rose is the floral emblem of Alberta. • The scarlet hips remain on the shrubs all winter, standing out against the snow, and they can be an important survival food in the winter months. Most parts of these shrubs are edible. Three rose hips are said to contain as much vitamin C as a whole orange. Although ripe rose hips are a delicious snack, rich in vitamins A, B, C, E, and K, they should be eaten in **moderation**. Too many petals or hips can cause diarrhea. Also, the small seeds inside the hips should be discarded as they are covered with tiny sliver-like hairs that **irritate** the digestive tract on their way through, and cause an 'itchy bum' on their way out. Rose hips are often boiled to make tea, jam, jelly, or syrup. They have also been used to make wine. • Rose hips can be dried, ground to a fine powder, and used as a flavouring and sugar-substitute. The flower petals can be candied, added to salad or fruit punch as a garnish, or used in sandwiches. Although large amounts of mature petals must be gathered, many people say it is worth the trouble to produce the delicately scented teas and jellies that can be made from rose flowers. Young, green shoots can be cooked as a potherb, or they can be peeled and eaten raw. • Many northern groups poured boiling water over the petals and used the cooled liquid as a gentle eye wash. The Cree steeped the lower stems and roots to make eye drops, while the Chipewyan used just the roots. They steeped them until the water was tea-coloured, then cooled it and dripped it into the eyes to treat snow-blindness and other irritations. The Cree and Slave boiled the roots to make a tea for treating coughs. • Necklaces were sometimes made by stringing together firm rose hips, and the Chipewyan used green rose hips to make toy pipes for children. The end of the hip was cut off and the seeds were scraped out; then a hollow piece of grass, fireweed or a willow twig was stuck into the bottom as a pipe stem. Some children actually stuffed their small pipes with tobacco and smoked them. Some native peoples used the inner bark of rose shrubs like tobacco. • The rose flower produces one of our most popular fragrances. Rose water and rose oil extracted directly from rose flowers are very expensive. They are used in special nonallergenic cosmetics. It takes 60,000 roses to produce 28 ml of pure essential oil. However, you can produce rose-scented oil much more easily by soaking rose petals in vegetable oil. To make rose water, add bruised rose petals to warm water, heat gently and let the mixture sit for a few hours. Rose water does more than just give a lovely scent—it is slightly astringent and has a mild cleansing action, so it will help to keep your skin healthy. • In the 16th century, the rose was the sign of secrecy and silence, and in the 1500s and 1600s roses were often carved or painted on the ceilings of council chambers to accentuate the stealthy mood.

WESTERN MOUNTAIN-ASH • *Sorbus scopulina*

GENERAL: Erect, several-stemmed shrub, 1–4 m tall; bark reddish grey to yellowish; winter buds and young growth slightly white-hairy and **sticky**.

LEAVES: Alternate, pinnately compound; 9–17 leaflets (usually 11–13); leaflets are 2–9 cm long, usually **sharp-pointed**, **base rounded**, **edge sharply toothed almost to base**.

FLOWERS: Many (up to 200) in flat-topped to rounded clusters that are 9–15 cm across; flowers are white, small.

FRUITS: Orange to scarlet berries, about 5–8 mm across.

WHERE FOUND: Woods and thickets; most vigorous in open sites but also in closed coniferous stands; widespread in boreal forest from northern Saskatchewan to west-central Alaska.

NOTES: Eastern mountain-ash (*S. decora*) has 4–8 cm long leaflets with **shorter pointy tips** than those of western mountain-ash, and its fruits are larger (8–10 mm thick). It grows in open woods and thickets, entering our region in southern Manitoba and east-central Saskatchewan. • The Woods Cree peeled the branches of western mountain-ash and boiled them to make a medicinal tea for back pain. Inner bark, removed from the base of stems that were not producing fruit, was also boiled to make a tea to relieve back pain and rheumatism. • The Chipewyan considered the bitter mountain-ash berries to be inedible, but their name for this shrub means 'medicine stick,' reflecting its importance in their traditional medicines. They sometimes chewed the stems, or boiled them and inhaled the steam, to relieve the pain of headaches and sore chests. They also made a tea from the branches for colds, headaches and sore chests, and to treat 'blood in the body,' which might be the bleeding associated with tuberculosis. The tea was also given to women to make their labour easier. • Another medicine for making birth easier was a drink make by boiling an otter's nose. It was said that this medicine would make the baby slip out of the womb as quickly and as smoothly as an otter slips through water.

WILD RED RASPBERRY • *Rubus idaeus*

GENERAL: Erect, perennial shrub, 1–2 m tall, stems (canes) upright, biennial, **prickly**, often with **gland-tipped hairs**; **bark shredding**, yellow to cinnamon brown; similar to cultivated raspberry.

LEAVES: Compound; **3–5 leaflets** per leaf on first-year canes, **egg-shaped**, sharply pointed, doubly saw-toothed; usually 3 leaflets per leaf on second-year (flowering) canes, end leaflet largest.

FLOWERS: Single or in small clusters; drooping, white, 8–12 mm across.

FRUITS: Red drupelets; in dense clusters (**raspberries**), about 1 cm across, that fall intact; smaller but tastier than domestic raspberries.

WHERE FOUND: Thickets, clearings and open woods; widespread across our region; circumpolar.

NOTES: Also called *R. strigosus*. • Wild red raspberries are a favourite fruit of many people, and they are often used to make jams, jellies, pies and preserves. The Woods Cree ate red raspberries with dried fish and fish oil. • The Cree also ate the tender young shoots after peeling off the outer layer. The fresh or dried leaves make excellent tea, but wilted leaves can be **toxic**. Extended use of raspberry tea can irritate the stomach and bowels. Use in **moderation**. • Cree women drank raspberry leaf tea during childbirth, and boiled the stem and upper roots to make a medicinal tea when recovering from childbirth or experiencing excessive menstrual bleeding. Pharmacologists have validated the use of raspberry leaves as an anti-spasmodic for treating painful menstruation. • The active ingredient, fragarine, acts as both a relaxant and a stimulant on the uterine muscles. Modern-day 'Motherese' tablets contain dry, aqueous extracts of raspberry leaf. Raspberry tea was also given to children who were suffering from sickness associated with teething. Raspberry leaves and willow bark were mixed together and used to make a tea for cholera infantum and dysentery.

SHRUBBY CINQUEFOIL • *Potentilla fruticosa*

GENERAL: Spreading to erect, freely branching shrub, 20–100 cm tall (sometimes to 1.5 m); silky hairs on young branches; **bark reddish brown, shredding**.

LEAVES: Compound, **3–7 closely crowded leaflets** (usually 5); lightly hairy, **greyish green**; edges often rolled under.

FLOWERS: Single or in small clusters near branch tips; **golden yellow, buttercup-like**, 2–3 cm across; appear from June to September.

FRUITS: Densely hairy achenes, in compact clusters.

WHERE FOUND: Open to partly wooded plains and wetlands; widespread across our region, north to Arctic coast; circumpolar with large gaps.

NOTES: Shrubby cinquefoil leaves, dried or fresh, can be steeped for 5–10 minutes to make a refreshing, golden-coloured tea that is high in calcium. Shrubby cinquefoil is mildly astringent, and the roots, stems and leaves have been boiled to make medicinal teas to treat various forms of internal bleeding and congestion, including tuberculosis. • Small, dry, papery strips of bark from the branches of shrubby cinquefoil were used as tinder in the days when fires were started by twirling sticks. • This hardy shrub is widely used as an ornamental in gardens and public places, where its buttercup-like, yellow flowers provide colour from spring to fall. It is also used to some extent for erosion control, especially along highways. The cuttings root readily in sand and the mature shrubs require little maintenance. • 'Cinquefoil' is from the Latin *quinque*, meaning 'five,' and *folium*, meaning 'leaf,' and refers to the fact that many *Potentilla* species have 5 leaflets. *Potentilla* is from the Latin *potens*, 'powerful' or 'strong,' and refers to the medicinal properties of some cinquefoils; *fruticosa* means 'shrubby.'

Key to the Gooseberries and Currants (*Ribes*)

1a. Stems more or less spiny, usually bristly (the gooseberries) 2

2a. Flowers saucer-shaped, 7–15 in spreading or drooping clusters; flower stalks jointed just below ovary (or berry); berries dark purple or black, with bristly glands; leaves without glands, usually not hairy below, heart-shaped at base .... ***R. lacustre***

2b. 1-3 flowers, bell-shaped; flower stalks not jointed near top; berries reddish to bluish purple, smooth; leaves usually have gland-tipped hairs below, squared-off to slightly heart-shaped at base .... ***R. oxyacanthoides***

1b. Stems not spiny or bristly (the currants) 3

3a. Leaves sprinkled with yellow resin glands below; mature fruits black, with whitish bloom 4

4a. Flower stalks much longer than tiny bracts at their base; sepals densely hairy outside; flowers white, in erect clusters, have unpleasant odour .... ***R. hudsonianum***

4b. Flower stalks much shorter than conspicuous bracts surrounding them; sepals smooth or slightly hairy outside; flowers creamy yellow, in drooping clusters, smell somewhat of currants (not unpleasant) .... ***R. americanum***

3b. Leaves have no resin dots; fruits reddish when mature 5

5a. Ovary and fruit smooth; leaves primarily 3-lobed, lobes rounded .... ***R. triste***

5b. Ovary and fruit bristly with stalked glands; leaves 5–7-lobed, lobes sharp-pointed .... ***R. glandulosum***

WILD BLACK CURRANT • *Ribes americanum*

GENERAL: Low shrub, usually 90–120 cm tall, sometimes taller, stems erect, without spines; **outer layer of bark often peels off** to expose inner yellowish red layer; branchlets greyish, finely hairy, more or less angled (ridged); older branches hairless, reddish to black.

LEAVES: Alternate, simple, nearly circular in outline, 3–8 cm long, slightly wider, sharply 3–5-lobed; broadly wedge-shaped to shallowly heart-shaped at base; dark green above with scattered resinous dots; paler, somewhat hairy, **many resinous dots below**; edges coarsely double-toothed.

FLOWERS: In **drooping, many-flowered clusters**; perfect, **creamy white to yellowish**, bell-shaped, about 9 mm long; bracts are longer than flower stalks, remain on plant; flowers appear in May and early June.

FRUITS: Black berries, without glands or bristles, 6–10 mm across, in drooping clusters, edible.

WHERE FOUND: Moist soil along streams and in woods; across prairie provinces in parkland and southern boreal forest, north to about 56°N.

NOTES: The fruits of all currant and gooseberry species are used by northern tribes. They can be eaten raw or dried, but they are usually made into jam, jelly, preserves, chutney, relish, syrup, and wine. These berries are also good in pies, cakes, muffins, breads, rolls, and dessert dishes. The young leaves and flowers were also eaten. • Native peoples used the bark to make a medicinal tea for colds, flu and tuberculosis. This tea was also cooled and used as an eyewash. Some native groups made a medicinal tea from the bark of the roots and used it to treat kidney ailments and expel intestinal parasites. The root bark was also used as a poultice to relieve swelling. • 'Currant,' an abbreviation of 'raisins of Corauntz'—from the French *raisins de Corinthe*, 'grapes of Corinth'—was originally used for a dwarf seedless variety of grape grown in the eastern Mediterranean region and used in cookery. The name 'currant' was later applied to the fruit of *Ribes* species.

SKUNK CURRANT • *Ribes glandulosum*

GENERAL: Ascending to trailing shrub, to 1 m tall; **strong skunky odour** when bruised; branches have no prickles or bristles.

LEAVES: Maple-leaf-shaped, 5–7 lobes, 2–7 cm across; lobes pointed and sharp-toothed; fresh green, hairless or with sparse gland-tipped hairs on veins below.

FLOWERS: 6–15 in erect or ascending clusters; **whitish to pink**, with gland-tipped hairs; stalks jointed, bear gland-tipped hairs; sepals 2–2.5 mm long, without hairs on outside; petals longer than broad; flowers appear in May to June.

FRUITS: Dark red berries, about 6 mm across, **bristly, with stalked glands**, disagreeable odour and flavour.

WHERE FOUND: Moist woods, thickets, rocky slopes and clearings; across our region.

NOTES: The Woods Cree boiled the stems, sometimes with the addition of red raspberry, to make a bitter tea that was given to women after birth to prevent clotting. • Some native children used green (unripe) currants or gooseberries to play a game of stoicism. Two teams would face each other and begin chewing the sour, unripe fruits. The team with the most members who could finish eating without making faces, won. • Skunk currant berries are often collected in quantity and eaten fresh, or they are cooked to make jelly. • This plant is ill-smelling (skunky) when bruised, hence its name 'skunk currant' or 'fetid currant.' The species name *glandulosum* refers to the characteristic gland-tipped hairs on the fruit.

NORTHERN BLACK CURRANT • *Ribes hudsonianum*

GENERAL: Erect to ascending shrub, 0.5–2 m tall, with sweet, **'tomcat' odour**, unpleasant to some people; branches have no prickles or bristles, covered in yellow glands (resin dots).

LEAVES: Often wider than long, 3–5-lobed, 5–7 cm across, sharply toothed, maple–leaf shaped; covered with **yellow resin dots** below, usually hairy on veins.

FLOWERS: 6–12 in spreading to **erect clusters**; **white**, small; ovary has a few resin dots; flower stalks jointed, with no glands or hairs; flowers appear in late May and early June.

FRUITS: Black berries, 5–10 mm across, with whitish, waxy bloom, usually speckled with a few resin dots; bitter.

WHERE FOUND: Moist to wet woods and streambanks; widespread across western boreal forest and northern parkland.

NOTES: The Cree boiled dried mint and black currant leaves (or root) to make a medicinal tea. This was cooled and given to women, along with plenty of black currants if possible, to prevent miscarriage. Bed rest was also required. A tea of the chopped roots, stems and bark of skunk currant and northern black currant was given to women to help them conceive. The Woods Cree boiled sections of black currant and northern gooseberry stems to make a medicinal tea for women suffering from sickness after childbirth. • Black currant jam was enjoyed with fish, meat, or bannock. According to Hearne (1802), large quantities of northern black currants can be strongly purgative for some people, and can even cause vomiting at the same time; but not if they are mixed with cranberries.

WILD RED CURRANT • *Ribes triste*
SWAMP RED CURRANT

GENERAL: Reclining to ascending shrub, to 1 m tall; branches do not have prickles, often root at lower nodes.

LEAVES: Maple–leaf shaped, **3-lobed** (rarely 5-lobed), shallowly heart-shaped or (rarely) squared at base; lobes broadly triangular, toothed, may be hairy below but **lack resin dots**.

FLOWERS: 6–15 in **drooping clusters**; **reddish or greenish purple**; petals about 1 mm long; ovary smooth; flower stalks jointed, usually bear gland-tipped hairs; flowers appear in late May and early June.

FRUITS: Bright red, smooth berries, about 6 mm across; sour but enjoyed by some, flavour similar to garden red currant.

WHERE FOUND: Moist woods and swamps; widespread across our region, north to Arctic coast.

NOTES: Also called *R. propinquum*. • Wild red currant is very common and is one of the favourite currants of the Dena'ina. Many native peoples collect wild red currants and use them to make pies, jams, and jellies. Alaskan native peoples mixed them with salmon roe and stored them for winter use. The Woods Cree, however, did not eat the berries. • Some tribes 'skinned' the berries, put them in boiling water and then wrapped them as a poultice for sore eyes. Similar preparations were also used to treat the eyes of puppies. The stems and inner bark of wild red currant were boiled (sometimes together) to make a medicinal tea to treat colds, flu, and tuberculosis. • The Cree name for wild red currant, *athikmin*, means 'frog berry.'

Black Gooseberry • *Ribes lacustre*
Bristly Black Currant

GENERAL: Erect to spreading shrub, 0.5–1 m tall; branches covered with **small, sharp prickles, with larger thorns at nodes**; **older bark cinnamon-coloured**.

LEAVES: Maple-like, 3–4 cm across, heart-shaped base; with **3–5 deeply cut lobes**; smooth to sparsely hairy; toothed.

FLOWERS: 7–15 in **drooping clusters**; **reddish, shallowly saucer-shaped**; petals about 1.3 mm long; ovary has gland-tipped bristles; stalks jointed; flowers appear in May to June.

FRUITS: Dark purple or black berries, 5–6 mm across, bristly, with **stalked glands**.

WHERE FOUND: Moist woods and streambanks to open slopes; widespread across western boreal forest.

NOTES: The spines of these plants can cause a serious allergic reaction in some people. Some native groups considered them highly poisonous to touch and believed that they would cause violent swelling. In some cases this belief was extended to encompass the fruit, which was said to be poisonous. However, the berries are definitely edible. But if you are stuck by a spine while collecting the fruit, you could try a chant from the 1800s: 'Christ was of a virgin born / And he was pricked with a thorn / And it did neither bell nor swell / And I trust in Jesus this never will.' • In the early 1900s, 10 million pine seedlings were brought to North America from Europe. With them came the blister rust, *Cronartium ribicola*, which soon spread to infect the native 5-needled pines. In order to reproduce, rust species must spend part of their life cycle on an alternate host. Shrubs of the genus *Ribes* are the intermediate host for blister rust. Therefore, a program to eradicate currants and gooseberries was introduced in an attempt to control the spread of this parasite; it has not worked.

Northern Gooseberry • *Ribes oxyacanthoides*

GENERAL: Erect to sprawling shrub, 30–150 cm tall; branches bristly, often with 1–3 spines up to 1 cm long at nodes; **older bark whitish grey**.

LEAVES: Somewhat maple-like, 3–4 cm across, 3–5-lobed, squared to slightly heart-shaped at base, usually has **gland-tipped hairs below**; edges irregularly round-toothed.

FLOWERS: 1–3 in drooping clusters; **greenish yellow to whitish**, **bell-shaped**; petals 2–2.5 mm long; ovary smooth; stalks not jointed; flowers appear in May.

FRUITS: Smooth berries, reddish to bluish purple, 10–15 mm across; edible.

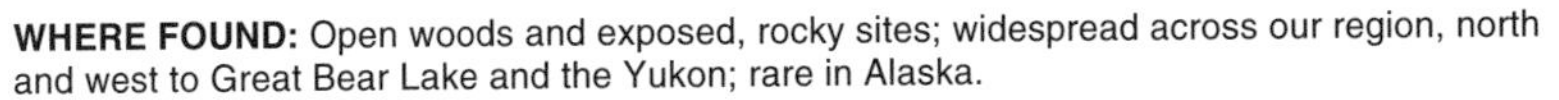

WHERE FOUND: Open woods and exposed, rocky sites; widespread across our region, north and west to Great Bear Lake and the Yukon; rare in Alaska.

NOTES: Also called *R. hirtellum* and *R. setosum*. • The Cree gathered the stems and roots of northern gooseberry in the summer and stored them for later use. For bladder problems, the roots of northern gooseberry and wild black currant were boiled together and the decoction was drunk in large quantities until the patient felt relief. The Chipewyan boiled northern gooseberry roots with a water weed to make a medicinal tea for women whose menstrual period was delayed. The Woods Cree boiled the stems of northern gooseberry and northern black currant to make a tea for women who were sick after childbirth. • Gooseberry thorns were often used as needles for probing boils, removing splinters and tattooing. • Northern gooseberries are often eaten fresh, and they make excellent jams and pies.

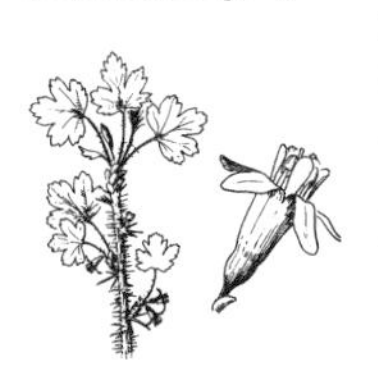

MOUNTAIN MAPLE • *Acer spicatum*
WHITE MAPLE

GENERAL: Tall, coarse shrub, 3–8 m tall; bark thin, dull, reddish to greyish brown, smooth or slightly grooved; twigs slender, yellowish green to reddish brown, coated with very short grey hairs; buds slender, stalked, covered with grey hairs.

LEAVES: Opposite, simple, 6–8 cm wide, **3-lobed**, **end lobe triangular**; coarsely and irregularly single-toothed; yellowish green above, with soft, whitish hairs below; stalk slender, reddish, usually longer than blade.

FLOWERS: In dense upright clusters at branchlet tips; sexes in separate flowers usually in same flower cluster; flowers are small, pale yellowish green; appear after leaves in late May to early June.

FRUITS: Samaras, **often brilliant red**; wings 2–3 cm long, with acute angle (less than 90°) between them; seed portion indented on 1 side; mature in late summer.

WHERE FOUND: Well-drained, moist soils along streams, in ravines and on moist hillsides in eastern forests; common on recently cut-over forest land; from lake country of Manitoba to eastern Saskatchewan in our region.

NOTES: Mountain maple is a hardy shrub, adapted to growing in partial shade. Although it is the smallest of the eastern maples, usually growing as a coarse shrub, mountain maple can reach 8 m in height with trunks 7–18 cm in diameter. • The wood is moderately light and soft, with little strength and no economic importance. However, these large shrubs or small trees are sometimes planted as ornamentals, especially varieties that have strikingly red fruits. The leaves also turn various shades of red and yellow in the autumn. • It is browsed to some extent by hoofed mammals, and some birds eat the seeds. Rabbits, beavers, deer, and moose feed on the bark of mountain maple and ruffed grouse eat the buds.

LOW BUSH-CRANBERRY • *Viburnum edule*
SQUASHBERRY; MOOSEBERRY

GENERAL: Straggly to erect shrub, 0.5–2 m tall; bark smooth, reddish to dark grey.

LEAVES: Opposite; round to obovate, 4–10 cm wide, shallowly **3-lobed**, **sharply toothed**, hairy below.

FLOWERS: 3–30 in flat-topped to rounded **clusters** (**1–3 cm across**) on short stems from leaf axils; flowers are small, **white**; appear in late May to early June.

FRUITS: Light **red or orange, berry-like** drupes, 8–10 mm long, with single, large, flattened stone; edible, juicy, acidic, strong smelling.

WHERE FOUND: Moist forests, thickets, margins of wetlands and streambanks; widespread across our region.

NOTES: *V. edule* is sometimes called **highbush-cranberry**, especially in northern regions, which can lead to confusion between it and *V. opulus* (p. 65). • Cree medicine used the whole shrub in medicinal teas for treating the sick. The barks of both high bush-cranberry and low bush-cranberry were widely used to relieve pain from menstrual cramps. Cree women boiled fresh or dried bark in water and then took 1 glass a day during their menstrual period. The Woods Cree chewed the unopened flower buds and applied them to sores on the lips. For sore throats they chewed the twigs and swallowed the juice or made a tea from the leaves and stems. The roots were boiled to make a medicinal tea to treat sickness associated with teething. Northern native peoples drank the leaf tea to relieve kidney troubles, colds in the back and backaches. • Low bush-cranberry is an excellent source of vitamin C. The fruits are picked in the fall and boiled to make juice or wine. Some liken the smell to dirty socks, but the flavour is good. The fruits are sometimes eaten raw, but they are very sour and can cause **vomiting and cramps** when eaten in large quantities. However, they make excellent jams, jellies and pies, and require little or no thickening as they contain their own pectin. Low bush-cranberry sauce is delicious with wild game.

AMERICAN BUSH-CRANBERRY • *Viburnum opulus*
HIGH BUSH-CRANBERRY; PEMBINA

GENERAL: Upright shrub, to 4 m tall; bark smooth, grey.

LEAVES: Opposite; simple, **with 3 long, pointed, spreading lobes**; smooth, 6–12 cm long, irregularly toothed, deep green above, paler below, **red in fall**; stalks reddish; stipules to 6 mm long, narrow, often gland-tipped.

FLOWERS: In flat-topped clusters that are 5–15 cm across; white; **outer flowers large**, 1–2 cm across, **sterile**; **inner flowers smaller**, 3–4 mm across, **fertile**; appear from late May to July.

FRUITS: Orange to red, 1-seeded, berry-like drupes, 8–10 mm across, in drooping clusters at branch tips; edible, juicy but acidic.

WHERE FOUND: Poplar groves, river valleys and moist open woods across northern parkland and southern boreal forest of prairie provinces; more common to east.

NOTES: Also called *V. trilobum*. **Nannyberry** (*V. lentago*) is a tall shrub or small tree (to 6 m tall) that, unlike high-bush cranberry, has unlobed, pinnately veined leaves, flower clusters with edge flowers like the central flowers, and blue-black fruits with a whitish bloom. It is a species of forests, openings and roadsides in the southeastern part of our region. • The Cree used the leaves, stems and roots of American bush-cranberry to treat a variety of sicknesses. They collected the inner bark in the spring, as the sap began to flow, and dried it in small pieces for later use. A small piece boiled in a cup of water was used as a pain-reliever. The bark tea is still used by many women for relief from menstrual cramps, and the leaves are still boiled to make teas to soothe sore throats. • The fruit of American bush-cranberry makes excellent jams, jellies and pies. The fruits should not be eaten raw in quantity, as uncooked fruit can cause **vomiting and severe cramps**.

V. lentago

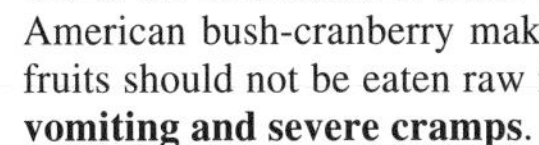

BUSH HONEYSUCKLE • *Diervilla lonicera*

GENERAL: Low shrub, 0.5–1 m tall; branchlets green or reddish; **often has 2 lines of minute hairs along length of stems**; older stems brownish to grey.

LEAVES: Simple, opposite, short-stalked, oblong-egg-shaped to lance-shaped, 5–12 cm long, 1.5–6 cm wide, dark green above, paler below; tip long, tapering, sometimes curved; base rounded or wedge-shaped, often asymmetrical; edges **sharply toothed, with fringe of short hairs**.

FLOWERS: 1–6 (usually 3) in short-stalked clusters in leaf axils and at branch tips; yellow (orange or brownish red with age); narrow, funnel-shaped, 5-parted, about 2 cm long; appear in June and July.

FRUITS: Slender, long-beaked, brown **capsules, with bristle-like calyx lobes at end**.

WHERE FOUND: Dry woods, clearings and rocky thickets across southern Manitoba to east-central Saskatchewan.

NOTES: The Woods Cree steeped or boiled the roots or stem of bush honeysuckle and used the cooled liquid as an eyewash for soothing sore eyes. Tea made from the roots (alone or mixed with other herbs) was given to women to ease childbirth and was also taken later to ensure a good supply of breast milk. European herbalists used various bush honeysuckles internally to stimulate urination and externally to relieve itching. • This genus was named for French surgeon Dr. N. Diereville, who travelled in eastern Canada about 1700 and later introduced the shrub to France.

COMMON SNOWBERRY • *Symphoricarpos albus*
FEW-FLOWERED SNOWBERRY

GENERAL: Erect shrub, 0.5–1 m tall; **often forms thickets from suckers**; many branches, slender, with reddish brown, shreddy bark.

LEAVES: Opposite, **thin**, oval, usually 2–4 cm long, may be slightly lobed on young stems.

FLOWERS: In small, dense, clusters at branch tips (or in axils of upper leaves); **pink to white**, bell-shaped, 4–7 mm long, hairy within; stamens and **non-hairy** style do not protrude from flower; appear in June to July.

FRUITS: White, waxy, oval to round, **berry-like** drupes, about 6 mm long (sometimes to 12 mm), with 2 seeds; lasts through winter; inedible, considered **poisonous** by many.

WHERE FOUND: Open woods, thickets, and valley slopes; most abundant in dry habitats but also on moist sites; widespread across southern boreal forest and parkland, north and west to southern N.W.T. and Alaska's panhandle.

NOTES: Snowberries are **toxic** when eaten in quantity, and other parts of the plant can produce mild symptoms of illness, vomiting and diarrhea. Many native peoples considered them **poisonous**. Some native groups believed snowberries were the ghosts of saskatoons and part of the spirit world. These 'saskatoons of the dead' were not to be eaten by the living. • The Woods Cree steeped the fruits in boiling water to make a wash for sore eyes, and they used the whole plant to make a medicinal tea and a wash for rashes. The roots and stems were boiled with other ingredients to make a tea for venereal disease and fevers associated with teething. Fresh leaves, fruit or bark were placed on burns, sores, cuts, scrapes and wounds as a healing poultice. Sometimes the fruits were rubbed into the hair as soap. • The species name, *albus*, means 'dead-white,' in reference to the white fruits. It is called 'snowberry' either because the fruit is white or because it stays on the plant through the winter.

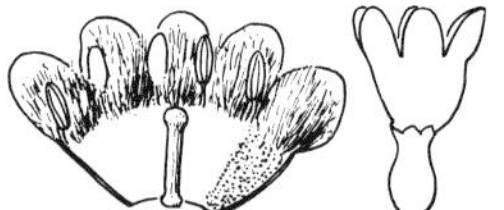

WESTERN SNOWBERRY • *Symphoricarpos occidentalis*
BUCKBRUSH

GENERAL: Erect, much-branched shrub, up to 1 m tall; young twigs pale green to light red-brown; older bark grey-brown, shredding; **readily produces suckers to form dense thickets**.

LEAVES: Simple, **opposite**, oblong, 3–6 cm long, **thick**, grey-green, paler below; edges smooth, wavy-toothed or occasionally lobed.

FLOWERS: In small clusters at branch tips or in leaf axils; pinkish white, urn-shaped, 4–10 mm long, white-hairy inside; **stamens usually protrude from flower**; **style hairy**, 4–5 mm long; appear in late June to August.

FRUITS: Greenish white, berry-like drupes, 6–10 mm long, **in dense clusters**; turn purple in autumn, **often last through winter**; believed **poisonous**.

WHERE FOUND: Dry open woodland, river valleys, hillsides, ravines and overgrazed prairie; widespread across our region, north and west to southern N.W.T. and eastern B.C.

NOTES: The Blackfoot and Cree steeped the leaves and inner bark in water to make a wash for sore eyes. They also boiled the berries and gave the liquid to horses to stimulate urination. • Western snowberry branches were used to make arrow shafts and pipe stems, and the smaller twigs were tied together to make brooms. • Depending on where you are, 'buckbrush' can be the local, colloquial name for this species as well as for bog birch, willows and other low, branching shrubs. The genus name *Symphoricarpos* comes from the Greek *syn*, 'together,' *phorein*, 'to bear,' and *karpos*, 'fruit,' in reference to the dense clusters of fruit. *Occidentalis* is Latin for 'western.'

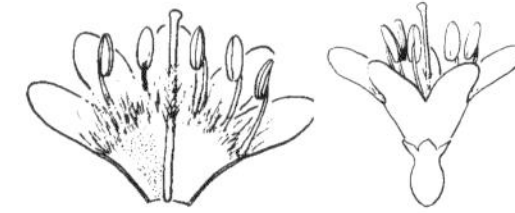

TWINING HONEYSUCKLE • *Lonicera dioica* var. *glaucescens*
RED HONEYSUCKLE

GENERAL: Semi-erect shrub or **twining woody vine**; light-coloured shreddy bark; branchlets smooth, green or purplish; older stems turn brown to grey.

LEAVES: Opposite, simple, **obovate or oval**, 5–8 cm long; rounded or blunt at tip; dark green, smooth above; pale, often hairy below, particularly on veins; smooth edged; **uppermost leaves** joined at base, **form cup around stem**.

FLOWERS: In **clusters** at branch tips; **yellow to orange** (often turn reddish with age), tubular to funnel-shaped, 2–2.5 cm long, with 2 spreading lips; appear in late May to early July.

FRUITS: Red berries, 8–12 mm long, in clusters **surrounded by leafy cups**; inedible, extremely bitter.

WHERE FOUND: Dry woods, thickets and rocky slopes; common and widespread across our region, north to southern N.W.T.

NOTES: This twining shrub can climb to heights of 5 m or more on the trunks of deciduous trees. • The Woods Cree used tea made from the inner bark, or from peeled sections of stems cut from between the branches, to stimulate urination. They also used the stem tea to treat flu. Honeysuckle stems were an ingredient in stronger teas to control blood clotting after childbirth and to treat venereal disease. • **Nausea and vomiting** have been reported by people who have eaten honeysuckle. • Sections of honeysuckle stem between nodes are hollow and they were used as pipe stems for corn cob pipes and for toy rose-hip pipes. They were also used as straws for children. • The species name *dioica* means 'dioecious' or 'in 2 households,' which is hardly appropriate for a species with bisexual (perfect) flowers.

BRACTED HONEYSUCKLE • *Lonicera involucrata*
BLACK TWINBERRY

GENERAL: Erect to ascending shrub, 1–3 m tall (usually less than 1.5 m); stems somewhat downy; **young twigs 4-sided**, greenish.

LEAVES: Opposite, somewhat **elliptic to broadly lance-shaped**, 5–10 cm long, often broadest towards tip; hairy, dotted with glands below.

FLOWERS: In opposite pairs in leaf axils, cupped by 1–2 cm long, green to purplish bracts; **yellow, tubular**, 10–13 cm long, with gland-tipped hairs; appear in May to July.

FRUITS: Shiny, **purple-black** berries, about 8 mm across, in pairs **cupped by pairs of deep purplish maroon bracts**; unpleasant taste.

WHERE FOUND: Moist or wet soil in forests, clearings, riverbanks, swamps and thickets; widespread across boreal forest, north and west to north-central Alberta and the northern part of Alaska's panhandle.

NOTES: Some sources say these berries are pleasant-tasting, but most consider them inedible; some even say they are poisonous to humans. However, birds, bears and other animals eat them in large quantities. • Many native groups used the berry juice as a purple dye and for painting faces on dolls. Bracted honeysuckle was often associated with the crow, perhaps because of the colour of its berries. • *Involucrata* means 'with an involucre,' referring to the pair of bracts surrounding the flower. The bracts later become enlarged in fruit.

FLY HONEYSUCKLE • *Lonicera villosa*
MOUNTAIN FLY HONEYSUCKLE

GENERAL: Low, erect or ascending shrub, usually less than 1 m tall; **young stems finely hairy**; bark of older twigs reddish brown to grey, **outer papery layers soon split and peel** to expose reddish brown inner layers.

LEAVES: Simple, **opposite**, **oblong, widest above middle**, 2–6 cm long, short-stalked, round-tipped, with **fringe of hairs**, **hairy below**.

FLOWERS: In pairs on short stalks **from leaf axils**, enclosed at base by narrow green bracts; petals **yellow**, united to form a 5-lobed, often hairy, tube, 8–15 mm long; appear in late May to June.

FRUITS: Many-seeded, **blue-black berries**, up to 1 cm long, covered with blue bloom, **edible**.

WHERE FOUND: Swamps, treed fens and streambanks across boreal forest of prairie provinces.

NOTES: Also called *L. caerulea* var. *villosa*. • **Swamp fly honeysuckle** (*L. oblongifolia*) (inset photo) differs in having leaves that are not fringed with hairs and are widest below the middle, flower stalks that are longer than the flowers, and orange-yellow to deep red berries. It occurs in wet woods and treed fens in the boreal forest east of central Saskatchewan. • The genus name *Lonicera* honours German botanist Adam Lonitzer (1528–86). *Caerulea* is Latin for 'sky blue,' and refers to the bluish bloom on the berries.

VELVET-LEAVED BLUEBERRY • *Vaccinium myrtilloides*
COMMON BLUEBERRY

GENERAL: Low shrub, 10–50 cm tall; often forms dense colonies; **twigs have velvety hairs**.

LEAVES: Thin, **soft-hairy**, elliptic- to oblong-lance-shaped, 1–4 cm long; edges smooth.

FLOWERS: Single or in small **clusters at branch tips**, **greenish white or pinkish**, cylindrical bells, 3–5 mm long; appear in late May to June.

FRUITS: Blue berries, 4–8 mm across, with **pale blue bloom**, sweet.

WHERE FOUND: Gravelly or sandy soils in open forests (usually coniferous stands) and clearings; widespread and common across boreal forest, north and west to southern N.W.T. and northeastern B.C.

NOTES: Another blueberry, **low sweet blueberry** (*V. angustifolium*), grows in similar habitats in the southeastern part of our region. It resembles velvet-leaved blueberry, but it has toothed, hairless leaves, its twigs are not velvety, and its fruits are not covered with a bloom. • In many parts of the boreal forest, blueberry and bog bilberry (p. 69) were the most important fruits for local native peoples. The Chipewyan preserved blueberries for winter storage by cooking them in lard and allowing the fat to solidify, or by drying them in the sun or in birch bark baskets or gunny sacks over a low fire. The dried berries were later reconstituted by boiling, or were pounded into pemmican. The Cree ate blueberry jam with fish or bannock. • The Cree used blueberry plants to treat cancer. The Chipewyan boiled the roots to make a medicinal tea for headaches. The Woods Cree used the plants or leafy stems in 'woman's medicine' teas to prevent miscarriage, increase bleeding after childbirth, regulate menstruation, and stimulate sweating. Tea made by boiling the stems was said to prevent pregnancy.

V. angustifolium

DWARF BILBERRY • *Vaccinium caespitosum*
DWARF BLUEBERRY

GENERAL: Densely branched, **matted, dwarf shrub**, 5–30 cm tall; twigs rounded, yellowish green to reddish, hairless or with sparse, tiny hairs.

LEAVES: Oblong to lance-shaped, widest above middle, pointed or blunt at tip, tapers to base, 1–3 cm long; **distinctly toothed** on upper half; **bright green and shiny on both sides**, with **pronounced network of veins below**.

FLOWERS: 1–4 in leaf axils, whitish to pink, **narrowly urn-shaped with 5 lobes**, 4–5 mm long; appear in late May to early July.

FRUITS: Light **blue berries, with pale grey bloom**, about 6 mm across; sweet.

WHERE FOUND: Woods (especially pine woods) to open slopes; widespread across southern boreal forest, north and west to central Yukon and southern Alaska.

NOTES: Bog bilberry (*V. uliginosum*) resembles dwarf bilberry, but it has elliptic to oval, smooth-edged leaves that are green above and pale beneath, and its flowers have 4 rather than 5 lobes. It is a circumpolar species of acidic soils, and tends to be more northern than dwarf bilberry. • Dwarf bilberries were extremely popular and were often traded. They were eaten fresh or were dried in cakes for winter use. • All northern species of blueberries are edible and provide vitamins A, B and C, as well as traces of calcium, phosphorus and iron. Flavours vary from fruity to insipid and from tart to sweet, depending on the species and the time of year. • Northern native peoples recommend blueberry tea as a refreshing drink and as a remedy for diarrhea. • Many wild birds and mammals feed on *Vaccinium* species. The berries are an important part of the diet of black bears, chipmunks, tanagers, and grouse, and the leaves and twigs are browsed by deer, elk and hares.

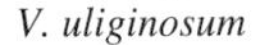

V. uliginosum

LEATHERLEAF • *Chamaedaphne calyculata*

GENERAL: Low, branching, **evergreen** shrub, usually 20–60 cm tall; twigs hairy or scaly.

LEAVES: Alternate; oblong, elliptic or lance-shaped, 1–4 cm long; **leathery, with white or brown scales**, especially on lower surface; stalks short; edges minutely round-toothed or smooth.

FLOWERS: In **1-sided, long, leafy clusters at branch tips**; white, small, urn-shaped to tubular; petals 5-7 mm long; appear in May to early June.

FRUITS: 5-chambered, round, many-seeded capsules, 3-5 mm wide, split open along lengthwise slits.

WHERE FOUND: Usually with sphagnum moss in bogs and poor fens; often forms dense thickets; characteristic of boreal and subarctic peatlands across our region; circumpolar.

NOTES: Although some native groups reportedly made tea from leatherleaf, this beverage could be **dangerous**. Any boiling would extract andromedotoxin from the leaves, making the drink **poisonous**. • These shrubs require 1.5 growing seasons to produce seed. In the first year, leafy shoots produce flower buds, and the flower parts mature. Although they are ready to open, the buds remain closed through the fall and winter. The following spring, they blossom early and by mid-summer they have produced seed. If the shrub is exposed to the air all winter, only 1–2 buds closest to the stem survive, but if it is completely covered by snow, almost all of the flowers will bloom. The seed capsule forms a kind of rattlebox, and wind, rain, or animals shake the seeds out through slits that open along the seams.

COMMON LABRADOR TEA • *Ledum groenlandicum*

GENERAL: Erect evergreen shrub, 30–80 cm tall; spicy fragrance; many branches, with dense rusty hairs.

LEAVES: Oblong to narrowly oblong, 1–5 cm long, often drooping; leathery; deep green above; **rusty below, with dense woolly hairs** (hairs on young leaves may not be rusty); edges rolled under.

FLOWERS: In loose, umbrella-like clusters at branch tips; **white**, 5–8 mm long, with protruding stamens; stalks white-hairy, long.

FRUITS: Drooping, 5-parted, dry, finely hairy capsules, 5–7 mm long; in clusters at branch tips.

WHERE FOUND: Bogs, swamps and moist woods; indicator of acidic, nutrient-poor soils; widespread across boreal forest, north to Arctic coast.

L. groenlandicum

L. decumbens

NOTES: Also called *L. palustre* var. *latifolium.* • **Northern Labrador tea** (*L. decumbens* or *L. palustre* var. *decumbens*) (lower inset photo) is essentially a dwarfed version of Labrador tea. It has shorter, narrower, pointed leaves, 10 rather than 8 stamens and flower stalks with reddish (rather than white) hairs. Northern Labrador tea grows on rocky, peaty heaths, most commonly in the northern and western portions of our region, although it is circumpolar. • **Warning:** Do not confuse Labrador tea with other heath family shrubs such as northern bog-laurel or dwarf bog-rosemary (both p. 71), which contain **toxic** alkaloids. The reddish brown fuzz on the lower side of Labrador tea leaves is distinctive. • Although most people find this tea safe for regular use, it can be **toxic** if large amounts are taken. Therefore, it is probably best to use Labrador tea in **moderation** and to avoid strong tea. Like many other plants in the heath family, Labrador tea contains andromedotoxin, which can cause headaches, cramps, indigestion and other problems. It also contains narcotic substances and an active oil, 'ledol,' that can cause cramps and paralysis in large doses. • Most tribes have used the leaves of Labrador tea to make a beverage tea that is rich in vitamin C. Occasionally the flowers are used as well. In western Canada, many of the native peoples' names for Labrador tea are derived from the English word 'tea.' This suggests that the Europeans, probably officials of the Hudson's Bay Company, introduced the idea of using this plant for making tea. Although the leaves are usually used to make tea, either alone or added to black tea, they have also been used to put a head on beers. Some recommend this tea as a 'pick-me-up,' similar to caffeine, while others suggest using it to get to sleep. • The Cree used the tea as a sedative, especially if someone had trouble sleeping, and it has also been used to relieve stomachaches, headaches, colds, and fevers. Occasionally they added bits of stems and roots to make a more bitter tea. The Chipewyan chewed the leaves or drank the tea to relieve stomach flu and diarrhea. The leaves were dried and powdered and dusted on burns, wet eczema and broken blisters. The Woods Cree mixed the leaves with grease (sometimes with pitch added) to make a salve for burns. They also made a medicinal tea that they used to wash burns, itchy skin, bites, sores on the hands, chapped skin and cracked nipples. The same medicinal tea was used for pneumonia, and was mixed with sweet flag (p. 217) for whooping cough. A little fish oil on the lower side of a Labrador tea leaf was applied to the umbilical scab to speed healing. Dried, powdered leaves were used for diaper powder, and fresh, chewed leaves were applied to wounds as a dressing. • The leaves of Labrador tea are very aromatic and have often been used to repel and kill insects. Europeans used them to repel fleas, clothes moths and other insects, and they even put them in corn cribs to keep out mice and rats. A strong tea made by boiling the leaves was used as a wash to kill lice.

NORTHERN BOG-LAUREL • *Kalmia polifolia*
PALE BOG-LAUREL; SWAMP-LAUREL

GENERAL: Erect or matted, evergreen shrubs, usually 30–40 cm tall; spreads by rooting branches and short rhizomes; branches slender, 2-edged.

LEAVES: Opposite; narrowly lance-shaped, 2–4 cm long, dark green; leathery above; dense, short, **white hairs below**; **edges rolled under**.

FLOWERS: In loose clusters at branch tips; showy, long stalked, **rose-pink**, **saucer-shaped**, 15–20 mm across; 10 stamens, tip of each tucked into small pouch in petal and held under tension (like a catapult); appear in May to June.

FRUITS: 5-valved, egg-shaped to round capsules, about 6 mm long.

WHERE FOUND: Bogs and acidic lakeshores; widespread across boreal forest north and west to central Yukon.

NOTES: This species includes 2 varieties: var. *polifolia* is described above; var. *microphylla* is small (to 20 cm tall), and has leaves less than 2 cm long and flowers less than 1 cm across. It is found from western Alberta north and west to the Yukon. • Northern bog-laurel resembles Labrador tea, but the 2 species are easily distinguished. Labrador tea has white rather than pink flowers, and the undersides of its leaves are covered with fuzzy, rust-coloured hairs not found on northern bog-laurel. • Northern bog-laurel contains andromedotoxin, and has caused the **poisoning** and death of cattle, sheep, goats and horses. Children have been poisoned by sucking on the flowers and honey made from northern laurel nectar can also be toxic. Grieve (1931) recommends whiskey as an antidote, but it is better to induce vomiting and call a physician. • Each stamen is like a tiny catapult, with 2 minute cups of pollen ready to fire at the slightest touch and dust visiting insects. If an anther remains anchored, it is not yet ready to release its pollen. A single capsule can produce 500–700 tiny seeds that fall out through slits in the seed pod and are carried away by the wind.

var. *microphylla*

var. *polifolia*

DWARF BOG-ROSEMARY • *Andromeda polifolia*

GENERAL: Erect to spreading, **evergreen** shrub, 10–40 cm tall, spreads by creeping rhizomes.

LEAVES: Leathery, 1–5 cm long (usually 2–3 cm), very narrow to narrowly oblong, sharp-pointed; dull green above, with sunken veins; white below, with **fine, waxy powder**; edges rolled under.

FLOWERS: 2–6 in drooping clusters at branch tips; **Pinkish**, 5–7 mm long, **urn-shaped;** appear in late May to June.

FRUITS: Round, erect, 5-valved capsules.

WHERE FOUND: Fens and occasionally in bogs; widespread across boreal forest, north past treeline to Arctic coast; circumpolar.

NOTES: Bog-rosemary (*A. glaucophylla*) is an eastern species that reaches eastern Saskatchewan in our region. It is a taller shrub (to 60 cm), and its leaves are larger and have a layer of short, white hairs below. Also, its flowers tend to be in denser clusters and on shorter stalks than those of dwarf bog-rosemary. • Andromedotoxin, a common toxin of the heath family, was first isolated from bog rosemary and later discovered in many other species. It causes low blood pressure, breathing difficulties and intestinal upsets. • Linnaeus named the genus after the mythological Ethiopian princess, Andromeda, the daughter of Cassiopeia and Cepheus in Greek mythology, who was rescued from a sea monster and then married by Perseus. As Linnaeus wrote, this plant 'is always fixed on some turfy hillock in the midst of the swamps, as Andromeda herself was chained to a rock in the sea, which bathed her feet as the fresh water does the roots of the plant.'

SMALL BOG CRANBERRY • *Oxycoccus microcarpus*

GENERAL: Tiny, vine-like, **evergreen** shrub; stems thread-like, slender, 10–50 cm long, creeping and rooting.

LEAVES: Widely spaced, leathery, elliptic to egg-shaped, 2–6 mm long, 1.5–2 mm wide, sharp-pointed; dark green above; grey-waxy below; **edges rolled under**.

FLOWERS: 1–3 on slender, 1–4 cm long, hairless stalks; deep pink, drooping; 4 petals, 5–7 mm long, **sharply bent backwards** like miniature shooting stars; 8 protruding stamens; appear from late May to July.

FRUITS: Pale-pink to dark red (often spotted) berries, 5–10 mm across; look too big for plant.

WHERE FOUND: Bogs, swamps and fens, usually on sphagnum moss; widespread across boreal forest, north to Arctic coast; circumpolar.

NOTES: Also called *Vaccinium oxycoccos*. • Some taxonomists distinguish another species, *O. quadripetalus*, on the basis of its thicker, longer, more branched stems, larger leaves (6–11 mm long and 2–4 mm wide), finely haired flower stalks and larger fruits (8–14 mm across). *O. quadripetalus* is much less widespread but is circumpolar in peatlands. It is found as far north as Great Bear Lake, but in our region it is mostly more southern than small bog cranberry. • The Cree boiled the stems, leaves and roots to make a medicinal tea to treat bladder problems. Many doctors todayrecommend that people who are predisposed to bladder infections should drink cranberry juice to prevent and help in the treatment of infections. • Bog cranberries are widely used to make jams and jellies. They are also delicious in bannock, stuffing, muffins and pancakes. The Woods Cree ate stewed bog cranberries with smoked fish. Berries were usually collected in autumn after the first frost and stored outside. Large amounts of sugar are usually added to cranberries to make a sauce or jelly. However, the addition of 5 ml of salt, prior to adding sugar, can counteract much of the acidity and half the amount of sugar needed for 1 litre of berries.

CREEPING-SNOWBERRY • *Gaultheria hispidula*

GENERAL: Delicate, creeping, **evergreen** shrub, often forms mats; stems slender, 10–14 cm long, bristling with reddish hairs (especially when young).

LEAVES: Alternate, **closely spaced**, leathery, oval to round, 3–6 mm long; bristly beneath with **brown scale-like hairs**; edges rolled under.

FLOWERS: Usually single in leaf axils; drooping, **bell-shaped**, pinkish to white, about 2 mm long; stalks 1 mm long.

FRUITS: White berries, 5–10 mm across, hairy, mealy; mild wintergreen flavour and smell.

WHERE FOUND: Bogs, swamps, and damp coniferous woods in southern boreal forest, seldom north of 56°N.

NOTES: Teaberry (*G. procumbens*) has erect leafy stems 10-20 cm high, hairless leaves, barrel-shaped flowers with 5 lobes, and red fruits. It grows on acidic soils in dry or moist forests in the southeastern part of our region. • Wintergreen oil, used as a flavouring, was first extracted from teaberry plants, and teaberries and creeping-snowberries have a delicious wintergreen flavour. • These plants contain methyl salicylate, which is closely related to the active ingredient in Aspirin and helps relieve pain. Native peoples and Europeans used teaberry both internally and externally to treat rheumatism. Teaberry tea has also been used to treat colds, headaches, stomachaches, fevers, kidney ailments, diarrhea, and gas in infants. Wintergreen oil has been shown to relieve pain, expel gas, reduce inflammation, and prevent infection; it is sometimes added to baths or saunas, often to relieve painful or swollen joints. **Caution**: wintergreen oil is considered **highly toxic**; it can be **absorbed through skin**, can cause vomiting and will harm the liver and kidney. • Today wintergreen oil has been replaced by synthetics, and the genuine article is enjoyed mainly by deer, partridge, ruffed grouse, and other animals during the winter.

LINGONBERRY • *Vaccinium vitis-idaea*
BOG CRANBERRY

GENERAL: Dwarf, mat-forming, **evergreen** shrub, 10–20 cm tall; many branches, creeping or trailing.

LEAVES: Leathery, narrowly elliptic to egg-shaped, 6–15 mm long, rounded at tip, shiny, dark green above, pale, with **dark dots below**; **edges smooth**, rolled under.

FLOWERS: Few, in short clusters at branch tips; drooping, pinkish, cup-shaped, with **4 short lobes**, about 5 mm long; appear from late May to July.

FRUITS: Red berries, 5–10 mm across, edible but acidic.

WHERE FOUND: Raised areas in bogs, moist forests, rocky barrens, open slopes and dry woods; very common and widespread across boreal forest, north past treeline to southern arctic islands; circumpolar.

NOTES: Commonly called **lowbush cranberry** in the Yukon and Alaska. Lingonberry is a good indicator of boreal climates. To distinguish it from kinnikinnick look for the dark glands (like bristles) on the lower side of its leaves. • In the northern boreal forest, lingonberry was traditionally the third most important fruit, after blueberries and cloudberries. It is the only fruit used in pemmican by some tribes. The Chipewyan cooked the berries with grease and ate them like ketchup with cooked or dried meat. The Woods Cree ate stewed lingonberries with fish or meat. They also mixed the berries with whitefish eggs, livers, and air bladders, sucker livers and fat, and pickerel fat. • Northern native peoples used lingonberries raw to relieve fevers, sore throats or upset stomachs. The berries were also used in hot packs to treat swellings, aches, pains and headaches. • The Chipewyan used dried lingonberry leaves as a tobacco stretcher. The berries were strung to make necklaces or they were used to dye porcupine quills.

COMMON BEARBERRY • *Arctostaphylos uva-ursi*
KINNIKINNICK

GENERAL: Trailing, evergreen shrub, 7.5–10 cm tall; often forms mats with 50–100 cm long flexible rooting branches; bark brownish red to dark grey, peeling.

LEAVES: Leathery, oval to spoon-shaped, widest at rounded tip, tapered to base, 1–3 cm long, dark green, somewhat shiny above, paler below, smooth-edged.

FLOWERS: In small, drooping clusters at branch tips; pinkish to white, 4–6 mm long, urn-shaped; appear in May to June.

FRUITS: Dull red drupes, 6–10 mm across, look like miniature apples; edible but mealy, tasteless.

WHERE FOUND: Sandy and well-drained sites in woodlands and open areas; widespread across our region, north to Arctic coast; circumpolar.

NOTES: Alpine bearberry (*A. rubra*), also called red bearberry, has toothed deciduous leaves that are wrinkled and deeply veined. It is a matted, trailing shrub of wet mossy forest, muskeg, and rocky tundra, and produces scarlet leaves and juicy red berries in the fall. This circumpolar species is most abundant in the western part of our region. Alpine bearberry can be confused with **net-veined willow** (*Salix reticulata*), in the northwest part of our region, but that species has catkins and single bud scales, and its leaves are pale on their undersides. • The Cree and Chipewyan cooked bearberries in lard, pounded them, and mixed them with jackfish or whitefish eggs, which moistened the dry fruit. This mixture was often sweetened with birch syrup or sugar. Yukon Athapaskans boiled the fruits and then fried them in grease and sugar. • Although common bearberry is widely used, there are many **warnings** against extended use of the leaf tea. It can lead to **stomach and liver problems**, especially in children, and the stimulant effects of a strong tea can produce **uterine contractions**. The fruits, used as a remedy for diarrhea, can cause constipation if too many are eaten. • The common name 'kinnikinnick' is from an Eastern Algonquian word meaning 'admixture,' in reference to its use as a tobacco stretcher.

CROWBERRY • *Empetrum nigrum*

GENERAL: Creeping, matted **evergreen** shrub, to 15 cm tall; branches to 40 cm long.

LEAVES: Needle-like, 3–8 mm long, spreading, **deeply grooved beneath**; edges rolled under.

FLOWERS: Solitary **in leaf axils**; **purplish crimson, inconspicuous**; male, female or with both sexes; appear in May to June.

FRUITS: Juicy, **black, berry-like drupes**, with 6–9 white nutlets; edible but insipid.

WHERE FOUND: Acid peatlands, cold coniferous forest, and acidic rocky slopes; widespread across northern boreal forest, north through arctic islands; circumpolar.

NOTES: Crowberries are a favourite food of bears, and most northern peoples ate them as well. They are usually eaten fresh but sometimes they are mixed with grease, or they are cooked and strained to make syrup. They can also be made into jams, jellies, pies and even wine, and they can be added to muffins, pancakes, ice cream, or custard. • The Woods Cree boiled the leafy branches to make a medicinal tea, or chewed them and swallowed the juice, to increase urination and to relieve fevers in children. The tea was also used for external application, or the branches were ground and mixed with grease. In the Yukon, pine bark tea mixed with crowberries is taken to relieve colds. • Crowberry is thought to be named for the black colour of the fruit, because crows feed on it, or because it is only good for crows, not humans. It is also called 'curlew berry' and 'crakeberry' (*crake* is an Old Norse word for crow). *Empetrum* is from the Greek *en petros*, or 'on rock,' referring to the common arctic, alpine and rocky slope habitat of this species, and *nigrum* means 'black.'

COMMON JUNIPER • *Juniperus communis*
GROUND JUNIPER

GENERAL: Evergreen, **prostrate or spreading** shrub to 1 m tall; bark very thin, reddish brown, shredding, scaly.

LEAVES: Needle-like to narrowly lance-shaped, 5–12 mm long (sometimes to 15 mm), jointed at base, **very prickly**; whitish above, dark green below; in 3s.

FRUITS: Female cones **berry-like**, 6–10 mm in diameter, **bluish** with white-grey bloom, fleshy, maturing in the second season; male cones smaller, catkin-like; sexes on separate plants.

WHERE FOUND: Dry open woods, gravelly ridges, outcrops, and open rocky slopes; throughout our region, north to Arctic coast; circumpolar.

NOTES: The Chipewyan considered juniper inedible, but ate single cones as a cure-all medicine or smoked them to relieve asthma. They boiled green cones to make a tea to treat back pain associated with kidney trouble. Juniper berries have also been used to help digestion and increase appetite, to stimulate sweating and to increase mucous secretion. The boiled tea was applied to injuries to relieve swelling and inflammation. • The Woods Cree boiled the bark to make a medicinal tea to treat diarrhea and chest pains due to lung infections. • The needles were dried and powdered for dusting on skin diseases such as psoriasis and eczema. Fresh needles were often burned like incense on top of a wood stove. • The bluish 'berries' (fleshy cones) of common juniper have a bitter taste, and people with **kidney problems** and **pregnant women** should never take any part of this plant internally. One of the old country names for juniper was 'bastard killer' because it was used to produce abortions, sometimes with fatal results. • The species name *communis* means 'common,' which this species is (over much of the globe). Common juniper is the only circumpolar conifer of the northern hemisphere.

CREEPING JUNIPER • *Juniperus horizontalis*

GENERAL: Prostrate, matted, evergreen shrub, less than 30 cm tall; long trailing branches, with many short side branches.

LEAVES: Scale-like, overlapping, about 1.5 mm long, closely pressed to stem, not prickly, green to grey-green or bluish.

FRUITS: Male and female cones on separate plants; female cones **berry-like**, about 6 mm long, fleshy, bluish purple with grey bloom, mature in 1 year.

WHERE FOUND: Dry, rocky or sandy, open slopes and forests; widespread in our region, north and west to Mackenzie delta and southern Alaska.

NOTES: The Cree used the bitter tasting 'berries' of creeping juniper to treat kidney ailments and sore chests. Traditionally, creeping juniper was applied externally to treat facial blemishes, sores, warts, bleeding, swelling, and headaches. In Iceland in the 1400s, it was used to stimulate **contractions** when women had miscarried. Creeping juniper should be used with **caution**. • Creeping juniper's low growth form and spreading branches make it an excellent ground-cover shrub. • The name 'juniper' is from the Latin *juniperus*, which appears to be derived from a word that means 'something used for binding'—in reference to the tough juniper boughs (of other species) used to bind things together in ancient times. The species name *horizontalis* refers to the creeping habit of this shrub.

SAND HEATHER • *Hudsonia tomentosa*

GENERAL: Low, branched shrub, usually under 20 cm tall, forms mats or low mounds; **branchlets greyish green, with whitish hairs**, almost hidden by close-pressed, scale-like leaves; older stems grey to reddish brown.

LEAVES: Alternate, simple, **evergreen**, egg- to lance-shaped, **overlap like shingles, almost scale-like**, 2–4 mm long, densely grey-hairy.

FLOWERS: Many, crowded near branch tips, 5-parted; 5 bright yellow petals surrounded by hairy calyx; outer 2 sepals are much shorter than inner 3; appear in late June and July.

FRUITS: Smooth, egg-shaped, few-seeded capsules, surrounded by calyx, which remains after flowering.

WHERE FOUND: Sand dunes, sandy pine woods and clearings; occurring in boreal forest from southern Manitoba across Saskatchewan and Alberta, north to Lake Athabasca and Great Slave Lake.

NOTES: Sand heather is well adapted for growing in sand and on shifting dunes because of its long wiry taproot. Its woody base lives for several years but its branches die back after flowering. Although the leaves become quite brown in the fall, they remain for several seasons. Sand heather belongs to the rockrose family (Cistaceae), which is usually tropical or subtropical, with a few species extending into the warmer areas of the temperate zone. Sand heather, however, has adapted to survive the severe winters of the North. • The genus *Hudsonia* is named in honour of William Hudson, an English botanist and apothecary. The species name, *tomentosa*, is in reference to the entangled, cottony hairs that cover the stems.

Wildflowers

Flowering plants are generally divided into two subclasses: Dicotyledoneae (the 'dicots') and Monocotyledoneae (the 'monocots').

Dicots have two seed leaves, usually have net-veined leaves, and their flower parts are in fours or fives or multiples of four or five. This is by far the larger of the two groups, and includes some of the simplest flowers (such as buttercups and pinks) and some of the most complex flowers (such as asters) in our region.

Monocots have only one seed leaf (cotyledon), usually have relatively long, narrow, parallel-veined leaves and their flower parts in threes or multiples of three. They include some of the showiest flowers (such as orchids and lilies) and some of the least conspicuous flowers (such as grasses and sedges) in the boreal forest.

In addition to the lilies and orchids, other monocots in this guide include common blue-eyed-grass (*Sisyrinchium montanum*) from the iris family, the sedges, rushes and grasses, and several species and families of mostly aquatic plants.

This section includes all non-woody flowering plants except the aquatic, saprophytic, insectivorous and parasitic plants, and the grasses, sedges and rushes. It is divided into 18 subsections: one for each of the 17 largest flowering plant families of our region, and a final subsection that includes flowers from smaller families.

The wildflowers have been organized by family so that similar species can be compared easily. Within families we have ordered entries to allow easy comparison of similar genera and species. The order of the major families generally follows that set out in the Flora of Canada (Scoggan 1978, 1979), and reflects apparent increasing evolutionary complexity and specialization in flower structure.

Identification to family is usually the first step in identification, especially for less-common species that are not illustrated and cannot be picture-keyed. Representative plants for the 17 major plant families are illustrated below. If the leaves, flowers or fruits of your 'unknown plant' resemble one of those shown below, then you have probably found its family and you can turn to that subsection. If not, your plant may be found in the 'Other Families,' 'Aquatics' or 'Eating for a Living' section.

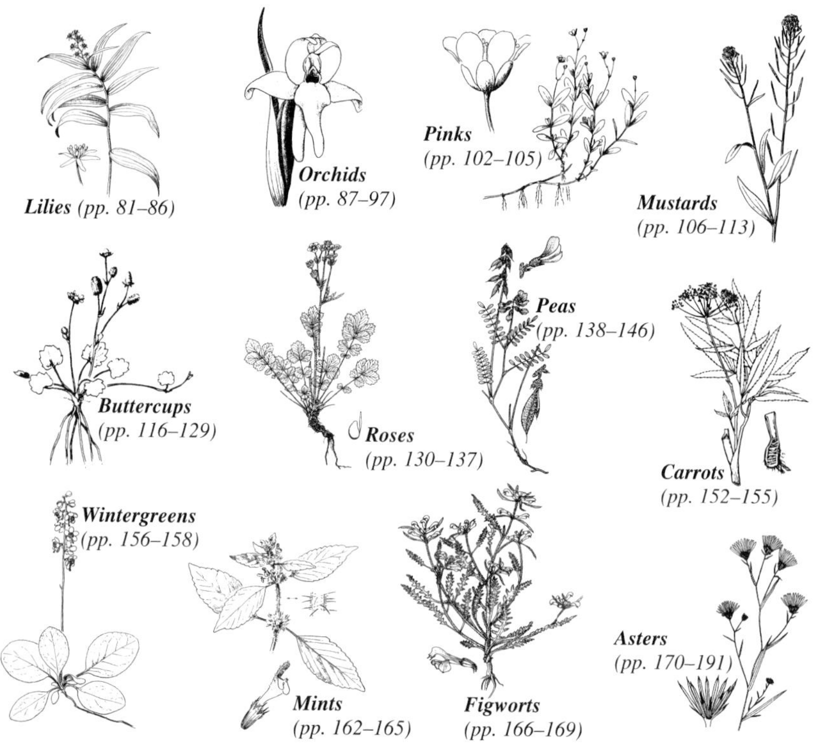

Wildflower Colour Key

Boreal wildflowers come in all sizes, shapes and colours. The wildflowers of this guide are organized by family, with flowers of similar shapes and sizes placed as close to one another as possible. However, one striking flower characteristic spans most families and major groups—colour. The following 4 pages provide a general colour key to the wildflowers section. All of our wildflowers are not shown, but representatives of the major groups are included to guide you to the section of the book where flowers of each type are found.

p. 93 p. 120 p. 176 p. 115 p. 153
p. 100 p. 154 p. 128 p. 157 p. 158
p.1 56 p. 85 p. 83 p. 90 p. 114
p.1 97 p. 115 p. 161 p. 132 p. 134
p. 133 p. 125 p. 120 p. 121 p. 102

p. 103 p. 104 p. 110 p. 112 p. 129
p. 126 p. 84 p. 95 p. 98 p.1 41
p. 145 p. 155 p. 202 p. 178 p. 176
p. 184 p. 186 p. 160 p. 204 p. 113
p. 148 p. 92 p. 196 p. 164 p. 99
p. 96 p. 101 p. 193 p. 194 p. 185

p. 177 p. 85 p. 159 p. 194 p. 150
p. 149 p. 158 p. 179 p. 132 p. 159
p. 140 p. 167 p. 197 p. 91 p. 92
p. 167 p. 144 p. 141 p. 143 p. 144
p. 163 p. 165 p. 162 p. 180 p. 181
p. 119 p. 147 p. 169 p. 86 p. 126

p. 201 p. 204 p. 146 p. 127 p. 127

p. 168 p. 82 p. 97 p. 90 p. 195

p. 160 p. 151 p. 121 p. 123 p. 122

p. 134 p. 136 p. 135 p. 108 p. 146

p. 145 p. 166 p. 168 p. 161 p. 152

p. 189 p. 190 p. 187 p. 175 p. 187

Liliaceae (Lily Family)

The Liliaceae is a large and varied family, with about 3,500 species worldwide, most of which are perennial herbs that grow from rhizomes, bulbs or fleshy roots. The parallel-veined leaves are either all at the stem base, or they are alternate or in whorls along the stem. Lily flowers, which are often large and showy, have radially symmetrical flower parts in threes (except for *Maianthemum* species, which have 4-parted flowers), and have a single, superior ovary. The fruit is a 3-chambered capsule or a berry.

Onions, garlic, leeks and asparagus are all edible members of the lily family, but some members, such as white death-camas (*Zygadenus elegans*) are very poisonous. Beautiful garden ornamentals include tulips, lilies, day lilies (*Hemerocallis*), hyacinths, scillas, and hostas.

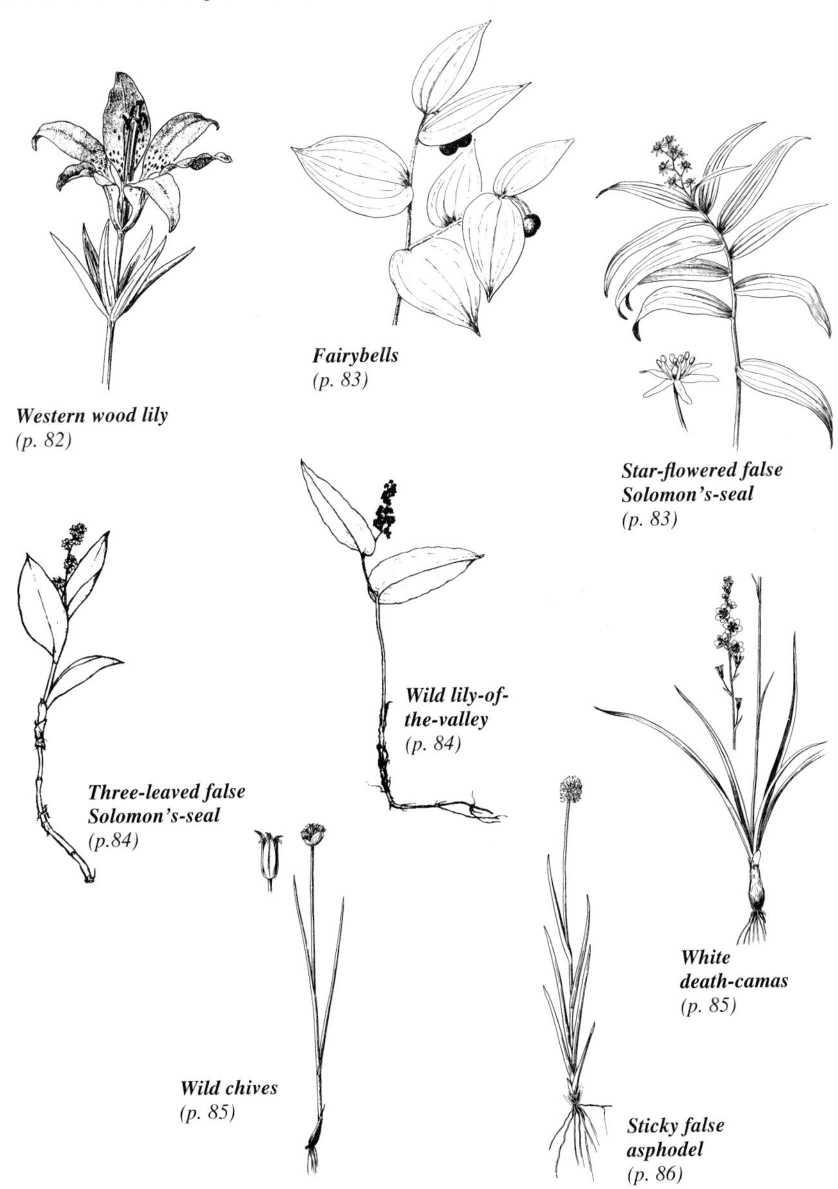

Western wood lily *(p. 82)*

Fairybells *(p. 83)*

Star-flowered false Solomon's-seal *(p. 83)*

Three-leaved false Solomon's-seal *(p.84)*

Wild lily-of-the-valley *(p. 84)*

White death-camas *(p. 85)*

Wild chives *(p. 85)*

Sticky false asphodel *(p. 86)*

Key to the Lily Family

1a. Leaves grass-like, linear 2

2a. Flowers pink to purplish; entire plant smells strongly of onion .... ***Allium schoenoprasum***

2b. Flowers greenish or white; plant does not smell of onion 3

3a. Grows from bulb; stem not sticky; common in prairies, grassy meadows and on sunny, gravelly slopes, also found in dry open forests; flowers smell foul .... ***Zygadenus elegans***

3b. Grows from rhizome; stem sticky; usually in wetland habitats; flowers have no particular smell .... ***Tofieldia glutinosa***

1b. Leaves not grass-like 4

4a. Leaves in whorls of 6–9 on stem (sometimes alternate beneath whorl); flowers large, reddish orange to brick-red .... ***Lilium philadelphicum***

4b. Leaves not in whorls; flowers not reddish orange to brick-red 5

5a. Stems branched; 1–3 hanging, bell-shaped flowers .... ***Disporum trachycarpum***

5b. Stems unbranched; 5 or more upright flowers (or as few as 2 in *Smilacina trifolia*) 6

6a. 4 petals; 4 sepals; 1-3 leaves, short-stalked to stalkless, heart-shaped .... ***Maianthemum canadense***

6b. 6 petals; 6 sepals; leaves stalkless 7

7a. Less than 20 cm tall; 2-4 leaves (usually 3), oval; grows in bogs and fens .... ***Smilacina trifolia***

7b. More than 20 cm tall; more than 4 leaves, lance-shaped; usually not in bogs and fens .... ***Smilacina stellata***

WESTERN WOOD LILY • *Lilium philadelphicum* var. *andinum*

GENERAL: Perennial from **whitish, scaly bulb**, 30–60 cm tall; stems single, hairless.

LEAVES: Lower leaves alternate, **very narrow to lance-shaped**, 5–10 cm long, 0.3–1 cm wide, hairless; **uppermost leaves whorled** (usually in a cluster of 6).

FLOWERS: 1–5 at stem tips; large, showy; sepals and petals 5–8 cm long, **orange to brick-red, yellowish at base with purplish black spots**.

FRUITS: Cylindrical capsules, 2–4 cm long, broad at top, taper slightly at base.

WHERE FOUND: Moist meadows, ditches, thickets and open deciduous woods; common and widespread in our region in prairie provinces.

NOTES: Eastern wood lily (*L. philadelphicum* var. *philadelphicum*) is the dominant wood lily in the eastern boreal forest. It occurs in our region only in southeastern Manitoba, and it differs from western wood lily in that all or most of its leaves are in whorls, as opposed to only the uppermost. • Wood lily roots were used to make medicinal teas for stomach disorders, coughs, tuberculosis, and fever. It was also given to new mothers to help expel the placenta, and was applied externally to relieve swelling and heal bruises, wounds, and sores. The flowers were used as a poultice on spider bites. • The flower, seeds and underground bulbs were all used as food. The Woods Cree and other tribes ate the bulbs fresh, or dried them for later use. Wood lily bulbs are strong-tasting, peppery and bitter—definitely an acquired taste!—and they were usually boiled and eaten with other foods. • Western wood lily is the floral emblem of Saskatchewan. • When these flowers are picked or mowed, the food-producing leaves are removed and the plant dies. Overpicking and digging have led to a decline in this species in the settled southern parts of the boreal forest.

FAIRYBELLS • *Disporum trachycarpum*
ROUGH-FRUITED FAIRYBELLS

GENERAL: Perennial from creeping rhizomes, 30–60 cm tall; stems leafy, with few branches, slightly hairy when young.

LEAVES: Alternate, egg- to lance-shaped, 3–8 cm long, somewhat pointed tips, rounded to heart-shaped at base, clasp stem.

FLOWERS: 1–3 at branch tips; hanging, creamy white or tinged green, narrowly **bell-shaped**; style hairless.

FRUITS: Berries, 8–10 mm long; become lemon yellow, then orange to red; **warty, roughened surface** appears velvety; 4–18 seeds (usually 6–12).

WHERE FOUND: Moist woods, thickets and clearings; widespread and common across southern boreal forest, aspen parkland and prairie; extends north to central B.C. and central prairie provinces.

NOTES: The Blackfoot occasionally ate fairybells berries raw, but they have very little flavour. • The Blackfoot name for fairybells was 'dog feet,' perhaps because the velvety, fleshy fruits resembled the pads on the foot of a dog. One glance at the delicate, hanging flowers explains the common name fairybells. The scientific name *Disporum* is derived from the Greek *dis*, 'double,' and *sporo*, 'seed.' It refers to the pairs of seeds that often develop from each ovary cell, producing pairs of bright scarlet berries. The species name, *trachycarpum*, is derived from the Greek *trachys*, 'rough,' and *karpos*, 'fruit,' and refers to the hundreds of minuscule bumps that cover the surface of the fruits and give them their velvety appearance.

STAR-FLOWERED FALSE SOLOMON'S-SEAL • *Smilacina stellata*

GENERAL: Perennial from extensive branching rhizomes; stems erect, unbranched, slightly arching, **15–60 cm tall** (usually 20–50 cm), finely hairy.

LEAVES: Alternate, 6–12 in 2 rows, narrowly **lance-shaped**, 3–12 cm long (usually 5–10 cm); flat, sometimes folded down centre when young; prominent veins; clasp stem somewhat; minute hairs below but edges hairless.

FLOWERS: 5–10 in short (1.5–5 cm long), **unbranched clusters at stem tips**; 3–5 mm long, creamy white, star-like.

FRUITS: Berries, **greenish yellow** with 6 blue-purple stripes, **dark blue or reddish black at maturity**.

WHERE FOUND: Moist to dry, open woodlands, riverbanks and lakeshores; widespread across our region, north and west to central Yukon and southern Alaska.

NOTES: Some people consider star-flowered false Solomon's-seal to be inedible. It is not classified as poisonous, but it can be **purgative**. The starchy rhizomes were occasionally used as food, cooked overnight in lye (made with white ashes) to remove the bitterness and then rinsed several times to remove the lye. Apparently, these rhizomes have also been pickled. The young leaves and shoots were cooked like asparagus as a potherb or cooked vegetable. The bitter-sweet berries are high in vitamin C and can be eaten raw. Too many will cause **diarrhea**, but cooking reduces the purgative effect and improves their taste. • The species name, *stellata*, means 'star-like,' and refers to the radiant white flowers. The name 'Solomon's-seal' (originally given to the species *Polygonatum multiflorum*) is traditionally thought to refer to the rhizomes, which bear surface scars or show markings when freshly cut that resemble the seal of Solomon—a 6-pointed star. However, Grigson (1974) claims the original medieval Latin referred to one of the flowers hanging like a seal on a document.

THREE-LEAVED FALSE SOLOMON'S-SEAL • *Smilacina trifolia*

GENERAL: Perennial from very slender, branching, whitish rhizomes; stems erect, slender, weak, **5–20 cm tall** (sometimes to 30 cm).

LEAVES: Alternate, 3–13 cm long (usually 4–10 cm), oval to oblong-lance-shaped, hairless; **2–4 (usually 3) on stem**; stalkless, **clasp stem somewhat**.

FLOWERS: In long-stalked, **unbranched clusters at stem tips**; white to greenish white.

FRUITS: Dark red berries, up to 6 mm across.

WHERE FOUND: Bogs, swamps, fens and moist woods and openings; often with sphagnum moss; widespread across boreal forest, north and west to Great Bear Lake and northern B.C.

NOTES: Three-leaved false Solomon's-seal might be mistaken for wild lily-of-the-valley (below), but wild lily-of-the-valley has 4-parted flowers and does not grow in bogs or fens. • Three-leaved false Solomon's-seal differs from star-flowered false Solomon's-seal in that it has 2–4 rather than 6–12 leaves, and it grows in wetland rather than upland habitats. • The berries of three-leaved false Solomon's-seal are not classified as poisonous, but the Chipewyan considered them inedible. • The genus name *Smilacina* is the diminutive form of *Smilax*, another genus of the lily family. *Trifolia* means 'three-leaved,' and refers to the 3 stem leaves commonly found on these small plants.

WILD LILY-OF-THE-VALLEY • *Maianthemum canadense* TWO-LEAVED SOLOMON'S-SEAL

GENERAL: Perennial from slender, branched, creeping rhizomes; stems single, erect, 5–25 cm tall, hairy at least above.

LEAVES: Alternate, 1–3 on flowering stem; broadly heart-shaped to oval, 2–8 cm long, pointed at tip, somewhat **hairy**; **short stalked or stalkless**; 1 long-stalked leaf at stem base.

FLOWERS: In clusters at stem tips; white, 4–6 mm wide, **4-parted** (unlike most of the lily family).

FRUITS: Berries, 3–5 mm wide; hard, green, mottled with brown; become red and soft; 2-seeded.

WHERE FOUND: Moist, rich to dry, often sandy woods and clearings; widespread across much of our region, from Manitoba to southwestern N.W.T. and northeastern B.C.

NOTES: Native peoples used wild lily-of-the-valley to make a medicinal tea for headaches, to keep the kidneys 'open' during pregnancy, and to soothe sore throats (as a gargle). However, there is evidence that the berries are a **strong purgative**, and they may contain **glycosides** (heart stimulants). • The tiny flowers have a wonderfully sweet, strong fragrance, which is easily detected within 50 cm of a patch of flowers. • The roots produce small tuberous swellings, like tiny potatoes, that store nutrients for the growth of new shoots in the spring. They were sometimes carried as good-luck charms for winning games. • The genus name is derived from the Latin *maius*, 'May,' and *anthemon*, 'flower,' referring to the flowering time of these plants (though northern plants may bloom much later). The species name, *canadense*, indicates that this species was first identified in Canada. It is called 'wild lily-of-the-valley' because its leaves resemble those of lily-of-the-valley (*Convallaria majalis*), a European species.

WHITE DEATH-CAMAS • *Zygadenus elegans*
MOUNTAIN DEATH-CAMAS

GENERAL: Perennial from long, oval bulbs **covered with blackish scales**; stems 20–60 cm tall, slender, often pinkish with whitish bloom.

LEAVES: Mainly at stem base, pale green, grass-like, with slight ridge along length, 8–30 cm long, reduced to bracts upwards.

FLOWERS: Lily-like, foul smelling; few to several in loose, spreading clusters at stem tips; tepals 7–10 mm long, greenish to yellowish white, with **dark green, heart-shaped gland at base**.

FRUITS: 3-lobed capsules, 1.5–2 cm long, egg-shaped to oblong; many ridged seeds, 5–6 mm long.

WHERE FOUND: Open woods and damp meadows; widespread across our region, north to Arctic coast.

NOTES: All parts of white death-camas contain the **poisonous alkaloid** zygadenine. Humans and livestock have died from eating these plants. Some claim that the toxic effect is twice as potent as that from strychnine. Two bulbs, raw or cooked can be fatal. If someone has eaten this plant, induce vomiting and get medical help. Most cases of human poisoning have resulted from the confusion of these bulbs with those of edible species such as wild onion. Onions can be easily identified by their smell. If there is any doubt about a plant's identity, do not eat it. • The Blackfoot mashed white death-camas roots to make a paste to apply to swollen knees and aching legs, to cure boils and to relieve the pain of rheumatism, bruises and sprains.

WILD CHIVES • *Allium schoenoprasum*

GENERAL: Onion-smelling perennial from 1–2 oblong-egg-shaped bulbs; outer bulb coats membranous; flowering stems 20–60 cm tall, usually longer than leaves.

LEAVES: At stem base, **round** or nearly so, **hollow**, up to 3 mm thick.

FLOWERS: In compact cluster that is 2–5 cm across; **short-stalked**; tepals bright rose-pink to purplish with dark midveins, 8–14 mm long, lance-shaped, slender-pointed.

FRUITS: Small, egg-shaped, 3-lobed capsule; seeds black, with honey-combed surface, 1–2 per cavity.

WHERE FOUND: Wet meadows, banks and shores; scattered across our region; circumpolar.

NOTES: Care must be taken not to confuse wild chives with its **poisonous** relative white death-camas (above). Wild chives are easily distinguished by their onion-like smell. Never eat a wild onion that doesn't smell like an onion. • Probably the most common medicinal use of wild chives in the past has been in treating coughs and colds. The juice was either boiled down to a thick syrup or a sliced bulb was placed in sugar and the resulting syrup was taken. Alternatively, a smudge of the bulb was used to fumigate the patient and snuff made by grinding the dried bulb was used to open the sinuses. Wild chives were also said to stimulate appetite and aid digestion, though water in which they had been crushed and soaked for 12 hours was swallowed on an empty stomach to rid the system of worms. The juice is somewhat antiseptic, and it was used by native peoples and sourdoughs to moisten sphagnum moss for use as a dressing on wounds and festering sores. Crushed bulbs were also used to treat insect bites and stings, hives, burns, scalds, sores, blemishes, and even snakebites. • Today wild chives are seldom used for food as they taste like green onions, which are easily bought at a store. Some people find them too strong by themselves, but others pickle them or use them as a cooked vegetable, served hot with butter. • The flowering stems dry well and make a beautiful addition to dried flower arrangements.

STICKY FALSE ASPHODEL • *Tofieldia glutinosa*

GENERAL: Perennial from vertical rhizomes covered with fibrous remains of old leaf bases; stems 10–50 cm tall; upper part of stem **densely covered with gland-tipped hairs, sticky**.

LEAVES: Alternate; 2–4 at base, sheathing, **iris-like**, 5–20 cm long; often 1–2 much smaller leaves on stem.

FLOWERS: White or greenish white, about 4 mm long; in **dense, 2–5 cm clusters** at stem tips; stalks sticky; anthers often purplish and conspicuous.

FRUITS: Erect, yellowish to **reddish purple capsules**, 5–6 mm long.

WHERE FOUND: Calcium-rich fens, lakeshores and riverbanks; widespread across our region, north and west to lower Mackenzie River and southern Alaska.

NOTES: Native peoples of northwestern North America boiled or baked the edible rhizomes of sticky false asphodel. • The origin of the name 'asphodel' is ancient and obscure. Homer named it as the flower of the Elysian Fields—*And rest at last, where souls unbodied dwell/In ever-flow'ring meads of asphodel* (from *The Odyssey*: xxiv.1.19 of Pope's translation). This would fit with 1 possible origin—the Greek *a*, 'not,' and *spodos*, 'ashes.' Another source says it is from the Greek *a*, 'not,' and *sphallo*, 'I surpass,' to indicate a stately plant of 'unsurpassed' beauty. In medieval England the name was somehow corrupted to the now widely used 'daffodil' (*Narcissus* sp.). The *Tofieldia* genus is named for English botanist Thomas Tofield (1730–79). *Glutinosa*, refers to the sticky (glutinous) flower stalks and upper stems.

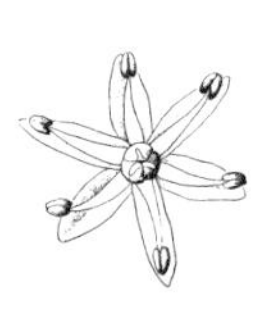

COMMON BLUE-EYED-GRASS • *Sisyrinchium montanum*

GENERAL: Slender, tufted perennial from short rhizomes and fibrous roots; **stems flattened and ridged**, erect, 10–50 cm tall (usually 20–30 cm), 1–4 mm wide, stiff.

LEAVES: At stem base, very narrow, 1–3.5 mm wide, about half as long as flower stalk.

FLOWERS: 3–6 in clusters at stem tips, from 2 erect, unequal, pale green or purplish, leaf-like bracts; delicate, blue-violet; 6 tepals, 8–10 mm long (sometimes to 15 mm), sharp-pointed.

FRUITS: Round capsules, 3–6 mm long, split open at tip; seeds round to egg-shaped, black.

WHERE FOUND: Moist slopes and meadows; widespread across our region, north and west to Great Bear River and southwestern Yukon.

NOTES: Also called *S. angustifolium* (in part). • Native peoples used the roots of common blue-eyed-grass to make a tea for diarrhea, especially in children. A tea made from the entire plant was taken to cure stomachaches and to expel intestinal worms. Herbalists have used common blue-eyed-grass to make medicinal teas for treating menstrual disorders, and regular doses of the leaf tea were taken for birth control. Several species of blue-eyed-grass have been used as laxatives. • *Sisyrinchium* was the name given to an iris-like plant by the Greek philosopher and scientist Theophrastus (c. 372–c. 287 B.C.). The species name, *montanum*, means 'of the mountains.' The common name of this species is apt, for the delicate, blue flowers with their yellow centres look like little eyes peeking out from the grass-like leaves.

Orchidaceae (Orchid Family)

The orchid family is very large—it vies with the aster family (Asteraceae) for top spot—but most orchids are found in the tropics. Our region has a fair number of orchids, but they are not particularly showy, except for *Cypripedium*, *Calypso* and *Arethusa* species.

Orchid flowers are intricate, complex and often wonderfully scented. They have bilateral symmetry, with 3 sepals and 3 petals. One of the sepals is usually specially modified, and the lower petal is usually modified into a **lip**, or sometimes inflated into a **pouch**, and it may have a hollow **spur** extending from it. The ovary is inferior, and it and its stalk are twisted through 180°. The fruit is usually a 1-chambered capsule containing sometimes thousands of tiny seeds.

Economically the orchid family is important primarily for the many beautiful, ornamental plants and flowers it contributes to the florist and horticultural trades. Natural vanilla extract comes from capsules of the tropical genus *Vanilla*.

It is important to remember that many of our orchids grow with an intimate relationship between their roots and soil fungi. This makes them very difficult to transplant and they are best left in the wild to be enjoyed.

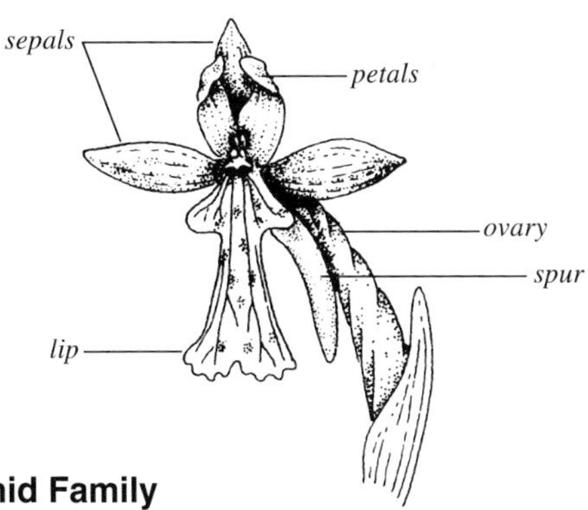

Key to the Orchid Family

1a. Plants saprophytic, without green leaves; roots coral-like (the coralroots) 2

2a. Plants often less than 20 cm tall; flowers less than 1 cm long, yellowish to nearly white, sometimes with a few red spots ..... ***Corallorhiza trifida***

2b. Plants frequently more than 20 cm tall; flowers more than 1 cm long, with either purplish spots or stripes 3

3a. Flowers yellowish pink to reddish purple, lip white with wine red spots, 3-lobed .... ***Corallorhiza maculata***

3b. Flowers pink to purple, with purplish stripes, lip unlobed .... ***Corallorhiza striata*** (see *C. maculata*)

1b. Plants with green leaves; roots not coral-like 4

4a. Flowers large and showy, single at end of stem; lower lip inflated, pouch-like 5

5a. 1 leaf, at stem base; lower lip appears slipper-like but not true pouch .... ***Calypso bulbosa***

5b. 2 or more leaves; lower lip inflated into nearly closed pouch (the lady's-slippers) 6

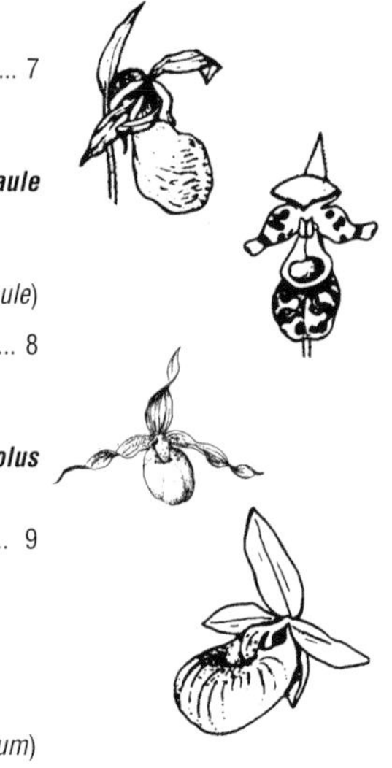

6a. 2 leaves at or near stem base 7

7a. Leaves at stem base, opposite, broadly elliptic; lip 3–6 cm long, pink with reddish veins .... ***Cypripedium acaule***

7b. Leaves low on stem; lip 2 cm long, white with purple blotches ..... ***Cypripedium guttatum*** (see *C. acaule*)

6b. 3 or more leaves, well distributed along stem 8

8a. Lip yellow, 1.5–4 cm long, often veined or spotted with purple inside .... ***Cypripedium calceolus***

8b. Lip white, or white with pinkish veins and blotches 9

9a. Lip large, white with pinkish veins and blotches, up to 5 cm long; leaves large (up to 20 cm long, 15 cm wide); plants up to 80 cm tall .... ***Cypripedium reginae*** (see *C. passerinum*)

9b. Lip small, white, 1.5–2 cm long; leaves smaller (to about 5 cm wide); plants up to 40 cm tall .... ***Cypripedium passerinum***

4b. Flowers generally smaller; lip variously shaped, but not truly pouch-like 10

10a. Flowers single, spurless, lip bearded or crested 11

11a. Flowering stem leafless; single leaf at stem base, oval to rounded; flower pink to white, lip shaped like sugar scoop, crested with yellow hairs .... ***Calypso bulbosa***

11b. Single, stalkless, linear leaf on lower half of flowering stem; flower pink, lip shallowly notched, crested with white hairs .... ***Arethusa bulbosa***

10b. 2 or more flowers, in spikes 12

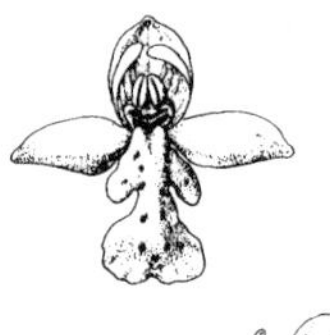

12a. Flowers with distinct, long, slender spur at base of lip 13

13a. Flowers white to pinkish with purple spots on lip; 1 leaf, oval, at stem base .... ***Orchis rotundifolia***

13b. Flowers white to greenish, not spotted (the bog-orchids) 14

14a. 1 or 2 leaves at stem base 15

15a. 1 leaf, lance-egg-shaped; plants usually less than 20 cm tall .... ***Habenaria obtusata***

15b. 2 leaves, oval to circular; plants usually more than 20 cm tall .... ***Habenaria orbiculata*** (see *H. obtusata*)

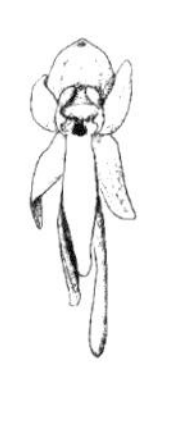

14b. More than 2 leaves, alternate along stem 16

16a. Flowers white, sweet-scented; lip widens abruptly at base .... ***Habenaria dilatata*** (see *H. hyperborea*)

16b. Flowers greenish, not sweet-scented 17

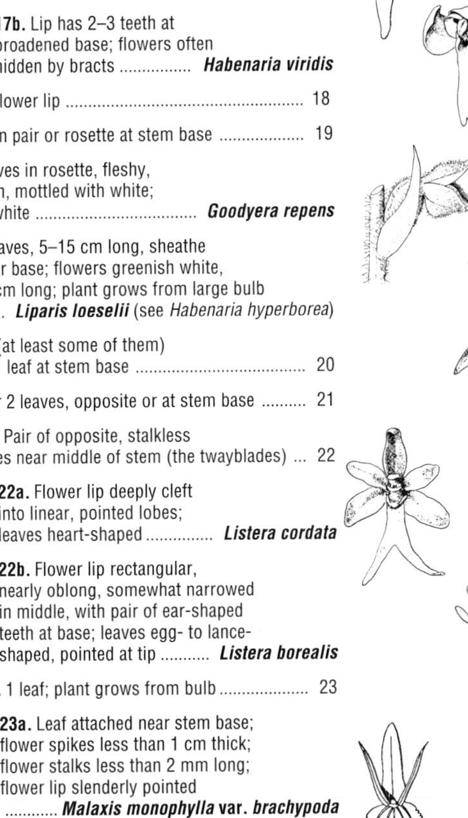

17a. Lip tapers towards rounded base; bracts much shorter than flowers .... ***Habenaria hyperborea***

17b. Lip has 2–3 teeth at broadened base; flowers often hidden by bracts .... ***Habenaria viridis***

12b. No spur on flower lip 18

18a. Leaves in pair or rosette at stem base 19

19a. Leaves in rosette, fleshy, evergreen, mottled with white; flowers white .... ***Goodyera repens***

19b. 2 leaves, 5–15 cm long, sheathe stem near base; flowers greenish white, about 1 cm long; plant grows from large bulb .... ***Liparis loeselii*** (see *Habenaria hyperborea*)

18b. Leaves (at least some of them) on stem, or 1 leaf at stem base 20

20a. 1 or 2 leaves, opposite or at stem base 21

21a. Pair of opposite, stalkless leaves near middle of stem (the twayblades) ... 22

22a. Flower lip deeply cleft into linear, pointed lobes; leaves heart-shaped .... ***Listera cordata***

22b. Flower lip rectangular, nearly oblong, somewhat narrowed in middle, with pair of ear-shaped teeth at base; leaves egg- to lance-shaped, pointed at tip .... ***Listera borealis***

21b. 1 leaf; plant grows from bulb 23

23a. Leaf attached near stem base; flower spikes less than 1 cm thick; flower stalks less than 2 mm long; flower lip slenderly pointed .... ***Malaxis monophylla* var. *brachypoda***

23b. Leaf attached near middle of stem; flower spikes more than 1 cm thick; flower stalks to 8 mm long; flower lip cleft into 2 narrow lobes .... ***Malaxis unifolia*** (see *M. monophylla* var. *brachypoda*)

20b. 2 or more leaves, alternate, stalkless, clasp stem 24

24a. Plant grows from solid bulb; flowers yellowish green, spike not spirally twisted .... ***Malaxis paludosa*** (see *M. monophylla*)

24b. Plant grows from tuberous-thickened roots; flowers white, in spirally twisted spike .... ***Spiranthes romanzoffiana***

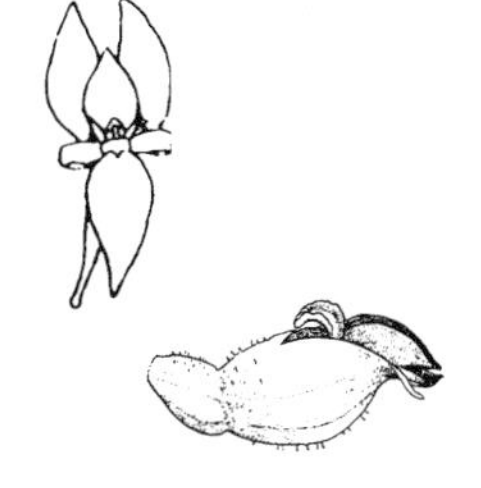

YELLOW LADY'S-SLIPPER • *Cypripedium calceolus*

GENERAL: Perennial, 20–40 cm tall; stems simple, leafy, sparsely hairy, with few to several glands.

LEAVES: Alternate, simple, **lance- to egg-shaped**, 5–15 cm long, up to 7 cm wide; **more or less sheathe stem**; sparsely hairy, **usually gland-bearing**.

FLOWERS: 1 or rarely 2 at stem tip; **yellow, often marked with purple near mouth**, fragrant; lip inflated into showy pouch, 2–6 cm long; **sepals** egg- to lance-shaped, yellowish or greenish, **striped with purple**, 3–8 cm long; **petals** narrower than sepals, greenish yellow to purplish brown, **usually shallowly wavy and spirally twisted**.

FRUITS: Erect ellipsoid, glandular-hairy capsules, with thousands of tiny seeds.

WHERE FOUND: Moist woods, ditches and banks; rarely in bogs; our most common lady's-slipper orchid; widespread across our region, north and west to eastern Alaska; circumpolar.

NOTES: Yellow lady's-slipper comes in 2 varieties: **var. *parviflorum*** has smaller flowers, with 3–5 cm long sepals and a 2–3.5 cm long lip; **var. *pubescens*** has 5–8 cm long sepals and a 3.5–6 cm long lip. The former is widespread across our region; the latter can be found in Manitoba and eastern Saskatchewan. • Lady's-slipper orchids tend to occur on lime-rich soil. • The Cree used the whole plant to treat female disorders and used the roots in tonics. A boiled medicinal tea of these roots was once widely used to calm the nerves of patients suffering from irritability, insomnia, nervous headache, and similar disorders. It was even prescribed for hysteria, delirium and epilepsy. **Warning:** Large doses can be **dangerous**, and the gland-tipped hairs of these plants may **irritate** sensitive skin. • Most orchids grow in close association with fungi in the soil, which makes them difficult, if not impossible, to transplant. They should never be moved from their natural habitat. However, lady's-slippers are available from many nurseries. Yellow lady's-slipper will bloom after 2–3 years. Unfortunately, these beautiful orchids are declining in abundance near settled areas because they are overpicked and dug up for gardens. • *Calceolus* means 'small shoe,' in reference to the pouch.

SPARROW'S-EGG LADY'S-SLIPPER • *Cypripedium passerinum*
FRANKLIN'S LADY'S-SLIPPER

GENERAL: Perennial, 10–40 cm tall; stems simple, leafy, **often hairy with downwards-pointing hairs**.

LEAVES: Alternate, simple, **lance-shaped to oval**, 5–15 cm long, stalkless, **with few to several sticky hairs**.

FLOWERS: 1–3 at stem tip; **white with purple spots inside**, fragrant; lip inflated into small pouch, 15–20 mm long; other sepals and petals green, oblong, blunt, 10–15 mm long.

FRUITS: Erect ellipsoid, pubescent, 2–3 cm long, capsules, with 10,000 to over 25,000 tiny seeds.

WHERE FOUND: Moist woods, ditches and streambanks; usually on calcium-rich mineral soils; widespread but not common across our region.

NOTES: Showy lady's-slipper (*C. reginae*)(inset photo) is a splendid but uncommon species of fens, swamps and wet forests in the southeastern part of our region. It is large (to 80 cm tall), has very leafy stems, and has a 3-4-cm long, white or pinkish, reddish-striped lip. • The gland-tipped hairs on lady's-slipper stems and leaves may **irritate** sensitive skin. • Never pick the flowers; not only do they wilt quickly, but picking them kills the whole plant. • The slippers of these beautiful flowers are very smooth, inside and out, with the exception of a small section at the 'heel,' where there is a small, dense strip of long hairs. Bees and flies can enter the slipper easily through the hole at the top, but once inside, the only way out is to climb up the hairy strip at the back. Semi-transparent sections in the slipper wall let light in and help guide insects to the exit. When they finally depart, the insects are forced to pass first the stigma, where they may deposit any pollen they are carrying, and then one of the anthers, where they pick up a sticky mass of pollen to deliver to the next flower they visit. • The inflated lip of the flower looks like a sparrow's egg, which inspired its discoverer, John Richardson, to name the species *passerinum* after the sparrow genus name *Passer*.

STEMLESS LADY'S-SLIPPER • *Cypripedium acaule*
MOCCASIN FLOWER

GENERAL: Perennial from thickened roots, 10–30 cm tall; stems single, erect, hairy.

LEAVES: Simple, **2 at stem base, nearly opposite, narrowly elliptic to obovate**, 10–20 cm long, 5–8 cm wide, sparsely hairy.

FLOWERS: Pink with reddish veins, fragrant, **1** at stem tip; **lip inflated into showy pouch**, 3–6 cm long, **drooping, divided down front**; other sepals and petals lance-shaped, yellow-green to greenish brown, 3–5 cm long.

FRUITS: Ascending capsules, with thousands of tiny seeds.

WHERE FOUND: Adapted to many habitats; bogs, open woods and sand dunes; infrequent in boreal forest from Manitoba north and west to Lake Athabasca.

NOTES: Speckled lady's-slipper (*C. guttatum*) occurs in open mossy forests from the District of Mackenzie through Alaska. It also has 2, nearly paired leaves low on the stem and has a 2-cm-long lip, which is white with purple blotches. • Be careful around lady's-slippers if you have sensitive skin. Some people develop a **rash** after touching the gland-tipped hairs on its stem and leaves. • Some native peoples used the roots in a medicinal tea for headaches or simply chewed them as a sedative. • The genus name comes from the Greek *kypris*, 'Venus,' and *pedilon*, 'a slipper,' in reference to the slipper-like shape of the flowers.

SWAMP-PINK • *Arethusa bulbosa*
DRAGON'S-MOUTH

GENERAL: Perennial from solid bulb, 10–30 cm tall; stems single, erect, hairless, **bear 1–3 loose, sheathing bracts**.

LEAVES: Single, on lower half of stem, very narrow, 2–4 mm wide, hairless, **develops after flowers**.

FLOWERS: Magenta-pink, 1 on leafless stalk; **pinkish white lip shallowly notched, usually spotted and streaked with purple, with 3 or 5 yellow- or white-fringed crests**.

FRUITS: Elliptic, strongly ribbed capsules, with innumerable tiny seeds.

WHERE FOUND: Bogs and wet meadows; often in sphagnum moss; uncommon to rare in boreal forest of Manitoba and Saskatchewan north to Lake Athabasca.

NOTES: Swamp-pink is widely scattered over much of North America, but it is rare throughout its range. It has been classified as rare, threatened, endangered or extirpated in 3 provinces and 12 states. Populations can fluctuate greatly from year to year, especially if late frosts kill the flowers. This suggests that swamp-pink is short-lived and relies on heavy seed production to maintain its populations. • Transplanting usually kills this orchid. • This genus is named after Arethusa, a Greek nymph who was changed into a spring to escape the river god Alpheus, perhaps because of the wet habitats preferred by this beautiful orchid.

VENUS'-SLIPPER • *Calypso bulbosa*
FAIRYSLIPPER

GENERAL: Perennial from solid, **bulb-like corm** with fleshy, beansprout-like roots and sometimes a fleshy, coral-like rhizome; stems delicate, purplish, covered in membranous, sheathing bracts, 8–15 cm tall (sometimes only 5 cm).

LEAVES: Single, dark-green, 2–4 cm long, **broadly egg-shaped**; stalked, at stem base; develop in autumn, remain through winter and wither in following summer.

FLOWERS: Single, 15–20 mm long, **rose purple, showy**; 3 sepals (pale, purple) and 2 petals (narrow, pointed, twisted) sit erect above lid-like anther; lower lip large, **slipper-like**, yellow to whitish, streaked and spotted with purple, cluster of golden hairs within; spotted double spur below the lip; sweet smelling.

FRUITS: Erect ellipsoid capsules, 2–3 cm long, with 10,000 to over 20,000 tiny seeds.

WHERE FOUND: Dry to moist coniferous forests; scattered across boreal forest; circumpolar.

NOTES: Although widespread, this beautiful little orchid is being exterminated in populated areas due to trampling and picking. Its delicate roots are easily damaged by even the slightest tug, so picking the flower usually kills the plant. Transplanting rarely succeeds because orchids need to grow in association with the soil fungi of their natural habitat—fungi that are not found in gardens. • Venus'-slipper is a good example of pollination by deception. The flowers contain no nectar and their pollen is inaccessible to foraging insects, but their colour and perfume mimic insect food flowers. Because Venus'-slipper is among the first plants to flower in the spring, it can take advantage of naive worker and queen bumblebees on their early foraging flights. The colour (pink to yellow and finally white), patterning and scent of the flowers change with time and vary greatly from 1 plant to the next. This reduces the chance that a 'wronged' bumblebee will avoid the next Venus'-slipper it encounters. After 2 or 3 fruitless visits, the bees may learn to avoid Venus'-slipper, but by then they have cross-pollinated 1 or more of these beautiful orchids.

ROUND-LEAVED ORCHID • *Orchis rotundifolia*

GENERAL: Perennial with fleshy fibrous roots, 10–20 cm tall, no leafy stems.

LEAVES: Single, **at stem base**, elliptic to circular, 3–7 cm long, 2–5 cm wide.

FLOWERS: 2–8 clustered at stem tips; pink to white, 12–15 mm long; **lip** 6–9 mm long, **white with purple spots, 3-lobed, middle lobe large and notched at tip**; spur slender, curved, much shorter than lip.

FRUITS: Erect, ellipsoid, to 15 mm long capsules; over 1,000 tiny seeds.

WHERE FOUND: Moist woods, swamps, fens and well-drained streambanks; often on calcium-rich soils; common across our region.

NOTES: Also called *Amerorchis rotundifolia*. • In Greek mythology, Orchis was the offspring of a nymph and a satyr, which meant he was a creature of unbounded passion. He attacked a priestess at a festival of Bacchus and the crowd fell upon him and tore him limb from limb. The gods refused to put him back together again. Instead, they decided that since he had been such a nuisance in life, he should be a satisfaction in death and they changed him into a beautiful flower—the orchid. • The roots of orchids were once considered a strong aphrodisiac. Full firm roots were said to provoke intense lust, while dry or withered roots would restrain unwanted passion. A drink made from these roots, called 'salep' or 'saloop,' was imported to England from Italy and Turkey in the 1700–1800s. In 1855 it was reported to be 'a favourite repast of porters, coal-heavers, and other hard-working men.' Old beliefs die hard, and the crowds frequenting the salep-houses were probably hoping for more than a relaxing nightcap. • Round-leaved orchids reproduce mainly by seed. They do not transplant well, so they are best left in their natural habitat. • The common name 'orchid' comes from the Greek *orchis*, 'testicle,' in reference to the shape of the swollen tubers of some species of the genus. Similarly, 'salep,' the name of the orchid root drink, comes from the Arabic *sah.lab*, a contraction of *khus.a ath-tha'lab*, meaning 'fox's testicles.'

WHITE ADDER'S-MOUTH • *Malaxis monophylla* var. *brachypoda*

GENERAL: Perennial from solid bulb, 5–20 cm tall; single stem.

LEAVES: Single, at or near stem base, clasps stem, elliptic to egg-shaped, 3–6 cm long.

FLOWERS: In slender, tapering cluster at stem tip; greenish to yellowish, very small; **lip** egg-shaped, **drooping, slenderly pointed**; **sepals egg- to lance-shaped**, 2–2.5 mm long; **petals very narrow**, about 1 mm long; sepals and petals eventually bend backwards and flatten against capsule.

FRUITS: Erect, ovoid, 3–6 mm long capsules, with many tiny seeds.

WHERE FOUND: Damp woods, banks and fens; rare across southern boreal forest to northern B.C. and south-central Alaska.

NOTES: Also called *M. brachypoda*. • **Bog adder's-mouth** (*M. paludosa* or *Hammarbya paludosa*) (inset photo) is similar to white adder's-mouth but has 2–5 small, broadly lance-shaped leaves at the stem base that enclose a small daughter bulb, and its flower lip is erect. Although bog adder's-mouth is much smaller than other adder's-mouths, its yellowish green flowers are as large or larger. It is rare and local with large gaps in its distribution across the boreal forest into southern Alaska. • **Green adder's-mouth** (*M. unifolia*) has only 1 stem leaf, but unlike white adder's-mouth it has a broad, deeply lobed (not narrowly pointed) flower lip. It is a rare resident of damp woods, fens and bogs in the boreal forest of southern Manitoba. • The genus name *Malaxis* is from the Greek word for 'a softening,' perhaps alluding to the tender nature of these small orchids.

BLUNT-LEAVED BOG-ORCHID • *Habenaria obtusata*
ONE-LEAVED REIN-ORCHID

GENERAL: Perennial from fleshy roots; stems 6–30 cm tall (usually 10–20 cm), leafless; sheathing bracts rare.

LEAVES: 1 (rarely 2) at stem base, **4–12 cm long, egg- to lance-shaped**, broadest above middle; blunt-tipped, tapers to base.

FLOWERS: 3–12 in loose, slender cluster at stem tip; greenish white to yellowish green; upper sepal broad, rounded; other sepals and petals lance-shaped; lip narrow, 5–8 mm long; spur slender, tapering, same length or shorter than lip.

FRUITS: Erect, many-seeded, ellipsoid, up to 1 cm long capsules.

WHERE FOUND: Bogs, fens, swamps and moist to wet woods; widespread across boreal forest, north to tundra; more or less circumpolar.

H. obtusata

H. orbiculata

NOTES: Also called *Platanthera obtusata*. • **Round-leaved bog-orchid** (*H. orbiculata*) (inset photo) has 2 stem-base leaves that are oval or round and look fat and shiny. It usually grows in moist poplar or coniferous forest (not bogs), from northern British Columbia and the southern Northwest Territories east through the more southern parts of our region. • Blunt-leaved bog-orchid flowers throughout its June to September growing season. It is the smallest native bog-orchid and mosquitoes (*Aedes* sp.) are its most important pollinators. It is not unusual to see a mosquito flying by with club-shaped pollen clusters stuck to its head like tiny yellow 'horns.' Blunt-leaved bog-orchids store their nectar in a slender spur that projects down from the base of the flower. To reach the nectar a mosquito must enter through the mouth of the flower. This triggers the pollinia, which spring forwards and cement themselves to the mosquito's head by their sticky bases. When the mosquito visits its next flower, its head brushes past the stigma, and the pollen is successfully transferred to that flower. To see this pollination mechanism in action simply poke a small twig under the hood of a flower and watch the pollinia spring down and cement themselves to the twig.

NORTHERN GREEN BOG-ORCHID • *Habenaria hyperborea*
GREEN-FLOWERED BOG-ORCHID

GENERAL: Perennial from fleshy, thickened rhizome, 20–60 cm tall; stems slender to stout, leafy, hairless.

LEAVES: Alternate, simple, **lance-shaped to oblanceolate**, 4–10 cm long, 1–3 cm wide, tip blunt to slenderly pointed, **become bracts higher on stem, sheathe stem**.

FLOWERS: In loose to dense, many-flowered, 2–10 cm long cluster at stem tip; green or yellowish green; lower lip lance-shaped, 4–7 mm long, **gradually and uniformly widened towards base; spur slender, about as long as lip**.

FRUITS: Erect, ellipsoid, up to 15 mm long capsules, with many tiny seeds.

WHERE FOUND: Riverbanks, lakeshores, fens, wet meadows and moist coniferous woods; often on calcium-rich soil; tolerant of deep shade; common and widespread across our region.

H. hyperborea

H. dilatata

Liparis loeselii

NOTES: Also called *Platanthera hyperborea.* • **Tall white bog-orchid** (*H. dilatata* or *Platanthera dilatata*), also called **scent candle**, has wonderfully perfumed white flowers with a lip that is flared at the base, but otherwise it is similar to northern green bog-orchid. It grows in wet meadows, fens and swamps scattered across the southern half of our region; it is most common in British Columbia and the southern Yukon. • Another small but rare orchid of the boreal forest is **twayblade** (*Liparis loeselii*). Its small clusters of greenish flowers could be mistaken for those of a bog-orchid (*Habenaria* sp.), but closer examination will show that its tiny flowers do not have a sac-like spur at their base. Twayblade has 2 lance-shaped stem base leaves, hence its common name. It has been found on wet organic soils in fens and woods in southern Manitoba and Saskatchewan, and at a single location in the southwestern Northwest Territories. • Orchids are usually difficult to propagate because their seeds germinate only if a specific fungus is present in the soil. However, experimenters have had some success germinating northern green bog-orchid seeds in the absence of its symbiotic fungus, with up to 54% germinating after 8 months. • The genus name comes from the Latin *habena*, 'a thong' or 'a strap,' which the lip of the flower somewhat resembles. *Hyperborea* means 'of the far north,' appropriate if you consider most of Canada to be 'the far north'!

BRACTED BOG-ORCHID • *Habenaria viridis*
FROG ORCHID; LONG-BRACTED ORCHID

GENERAL: Stout perennial from fleshy, thickened rhizomes; **stems leafy**, 15–30 cm tall (sometimes to 60 cm).

LEAVES: Alternate, oblong to lance-shaped, 4–12 cm long, sheathe stem, smaller and narrower higher on stem.

FLOWERS: In loose to dense, many-flowered cluster at stem tip; each green flower is **almost hidden by large, lance-shaped, leafy bract; lower lip** oblong, 6–8 mm long, with **2–3 teeth at tip; spur sac-like**, 2–4 mm long.

FRUITS: Many-seeded, ellipsoid, up to 1 cm long capsules.

WHERE FOUND: Moist meadows and open woods; scattered across our region; circumpolar.

NOTES: Also called *Platanthera viridis* and *Coeloglossum viride.* • The tuber-like roots of bog-orchids are edible, raw or cooked. However, because many species are rare, they should not be harvested. • In the 1700s, bog-orchids were highly valued in Norway, because they were thought to protect cattle from illness and to cure sickness when it did occur. • Bracted bog-orchid is often called 'long-bracted orchid' because of the long bracts that stick out between the flowers. It may have acquired the name 'frog orchid' because it grows in places often inhabited by frogs, or perhaps because its flowers are green, like some frogs. The species name, *viridis*, means 'green.' Bracted bog-orchid is sometimes classified in a genus named *Coeloglossum*, meaning 'hollow tongue.'

HOODED LADIES'-TRESSES • *Spiranthes romanzoffiana*

GENERAL: Rather stout perennial from fleshy, tuberous roots, 10–40 cm tall.

LEAVES: Alternate, long and narrow, 5–15 cm long, **largest at base**; upper leaves much smaller, bract-like.

FLOWERS: In dense spike-like cluster at stem tip, in 3 **spiralling rows**; **white or creamy to greenish white**, 7–10 mm long, petals form **upwards-arching hood**; small, lower lip 9–12 mm long, strongly bent downwards.

FRUITS: Dry, many-seeded capsules, to 10 mm long.

WHERE FOUND: Fens, wet meadows and wet, open woods across our region.

NOTES: Although hooded ladies'-tresses was not widely used in the north, some southern groups used it to treat venereal disease and urinary disorders. The plants were also added to babies' baths to promote general health and make children stronger. • Bend down and smell the flowers when you find this orchid. They have a strong, vanilla-like fragrance both day and night. The 2 pollinia are each 2-parted and split into delicate plates of granular pollen connected by elastic threads. • The spiralling spike of flowers was thought to resemble a neat braid of hair, hence the common name, 'ladies'-tresses.' The species name, *romanzoffiana*, honours Nikolei Rumliantzev, Count Romanoff (1754–1826), a Russian patron of science and the person who sent Kotzebue to explore Alaska.

LESSER RATTLESNAKE-PLANTAIN • *Goodyera repens*

GENERAL: Evergreen perennial from slender creeping rhizomes; stems leafless, slender, 10–20 cm tall (sometimes to 30 cm), with tiny gland-tipped hairs.

LEAVES: In **rosette at stem base**, broadly egg-shaped to oval-lance-shaped, **1.5–3 cm long**; dark green with darker veins that are usually conspicuously white-margined.

FLOWERS: Several in 1-sided cluster at stem tip; white or pale green, 3.5–5 mm long, hooded; lower lip deeply pouched.

FRUITS: Many-seeded egg-shaped capsules, 5–9 mm long, ascending to spreading, tipped with withered flowers.

WHERE FOUND: Dry to moist forest, common in many areas; widespread across boreal forest; circumpolar.

NOTES: Lesser rattlesnake plantain reportedly grows for 7 years before it blooms. The sweetly scented flowers are fragrant both day and night, and attract as many moths and butterflies as possible to carry their pollen. The flowers mature in mid-summer. After they are fertilized they produce from 200 to 400 seeds in each small pod. Lesser rattlesnake-plantain does not rely entirely on its seeds to reproduce. It also forms extensive cloned colonies by sending up new plants from its spreading rhizomes. • Lesser rattlesnake-plantain was not widely used by native peoples in our region. Farther east, however, several groups used this orchid in blood tonics, appetite stimulants, toothache remedies, and in medicines for treating colds, sore eyes, thrush, female disorders, bladder problems, stomach diseases, and even snakebite. Fresh leaves or whole plants, dried and powdered, were used as a soothing poultice on scratches, insect bites, burns and similar ailments. European settlers used the leaves to soothe mucous membranes, and to treat tuberculosis of the lymph glands and eye diseases. • The species name *repens*, from the Latin *repere*, 'to creep,' refers to the creeping habit of the buried stems.

SPOTTED CORALROOT • *Corallorhiza maculata*

GENERAL: Perennial saprophyte from branched, coral-like rhizomes; stems **20–50 cm tall, purplish to reddish**, or light yellow to tan.

LEAVES: Reduced to thin, semi-transparent sheaths.

FLOWERS: 10–40 in loose cluster at stem tip, 12–20 mm across; **reddish purple to whitish, with red to purple spots**; lip white, spotted with purple, 6–8 mm long.

FRUITS: Oval capsules, about 2 cm long, drooping to hanging, many-seeded.

WHERE FOUND: Moist to dry forests; scattered across southern part of our region, north to 55°N.

NOTES: Striped coralroot (*C. striata*) has purplish to yellowish pink flowers with purplish stripes, and could be confused with spotted coralroots that do not have spots—as sometimes happens. However, the tongue-shaped lip of striped coralroot is not lobed, whereas the lip of spotted coralroot has 2 flange-like lobes at its base. Striped coralroot is scattered across the southern part of our region. • Spotted coralroot is not green because it has no chlorophyll. It is saprophytic and takes its nutrients from dead organic matter, through a symbiotic relationship with the fungi that grow among the needles on the forest floor in coniferous stands. • Coralroots do not have roots. Instead, they have fleshy, underground rhizomes with several, short, thickened branches that make them look like small pieces of coral, hence the name 'coralroot.' • Northern native peoples did not use spotted coralroot as a medicinal herb. However, native groups in the southwest used the stems to make a blood tonic for patients with pneumonia, and used the roots to make medicinal tea to induce sweating, to calm the patient and to expel intestinal worms. • This species is called 'spotted' coralroot because of the spots on the flowers. *Corallorhiza* means 'coral-like root' and *maculata* means 'spotted.'

C. maculata

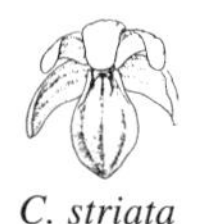

C. striata

PALE CORALROOT • *Corallorhiza trifida*
YELLOW CORALROOT

GENERAL: Perennial **saprophyte** from branched, coral-like rhizomes; **stems slender, pale yellowish**, 10–30 cm tall.

LEAVES: Reduced to thin, semi-transparent sheaths.

FLOWERS: 3–12 in loose cluster at stem tip; **yellowish white to greenish** or occasionally purplish, less than 1 cm across; lip nearly white, sometimes red- or purple-spotted.

FRUITS: Capsules, drooping to hanging, elliptic, with 1,200–1,500 tiny seeds.

WHERE FOUND: Moist to dry forests, thickets, fens, swamps and streambanks; scattered across our region; circumpolar.

NOTES: Pale coralroot is the most common saprophytic orchid in North America. Unlike the other coralroots, pale coralroot has a slight greenish tinge, especially in its seed capsules, which suggests it has some chlorophyll and is capable of manufacturing a small part of its food. • Coralroots have no developed leaves and little or no chlorophyll, so they cannot manufacture food as most plants do. Instead of parasitizing other plants, coralroots have developed a means of taking up the by-products left by fungi and other organisms that break down rotting plants and wood in the soil. All coralroots grow in close symbiotic association with soil fungi, and as with most saprophytes, they cannot be cultivated. • In Europe and the eastern United States, pale coralroot was considered an excellent medicinal herb for stimulating perspiration and sedating patients. However, it was not used by western native peoples. • The species name, *trifida*, from the Latin *tri*, 'three,' and *findere*, 'to divide,' refers to the 3-lobed lip of the flower, similar to that of spotted coralroot but quite different from the unlobed lip of striped coralroot.

NORTHERN TWAYBLADE • *Listera borealis*

GENERAL: Perennial from fleshy fibrous roots; stems 5–25 cm tall, 4-sided, glandular hairy towards top.

LEAVES: One pair, nearly opposite, just above middle of stem, stalkless, **egg-shaped to elliptic**, 1–5 cm long.

FLOWERS: 3–15 in open to dense cluster at stem tip; **pale green to yellowish green** with deep green veins; sepals and petals very narrow or lance-shaped, small; lip 7–12 mm long, rectangular to oblong, broadest and **shallowly 2-lobed** at tip, tiny ear-like lobes at base, with gland-tipped hairs above.

FRUITS: Many-seeded, oval capsules, to 5 mm long.

WHERE FOUND: Moist woods, thickets, meadows and wetlands; scattered across most of boreal forest; rare in northern Saskatchewan.

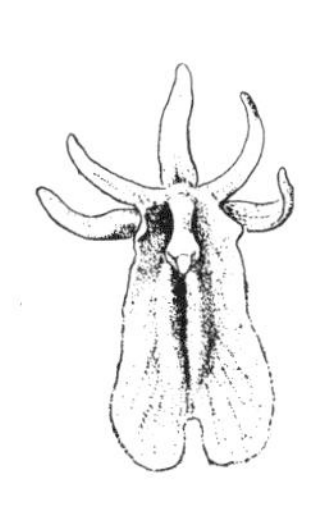

NOTES: The flowers of northern twayblade produce nectar in a narrow furrow down the centre of the lower lip. When small bees, wasps, or beetles land on the lip, they lick the nectar from the furrow. As it feeds, the insect follows the furrow to the top of the petal, where a tiny drop of sticky fluid is put onto the insect's forehead. A pollen bundle (pollinium) is stuck to the drop, which soon hardens to cement the pollen in place. The insect then flies off to find another flower, carrying the bundle of pollen on its forehead. • The name 'twayblade' comes from the archaic word *tway*, meaning 'two.' Plants of this genus earned the name because of the single pair of leaves found on the flowering stem. They are also sometimes called 'big ears'—perhaps because the 2 leaves resemble a pair of large ears. The species name *borealis*, which means 'northern,' comes from *Boreas*, the north wind in Greek mythology.

HEART-LEAVED TWAYBLADE • *Listera cordata*

GENERAL: Perennial from slender creeping rhizomes; stems 6–20 cm tall, smooth or with a few gland-tipped hairs towards top.

LEAVES: 1 opposite pair, at or below middle of stem, broad, **heart-shaped**, 1–4 cm long.

FLOWERS: 5–16 in long cluster at stem tip; **pale green to purplish brown**; sepals and petals 2–3 mm long; lip 3–6 mm long, tip split into **2 very narrow or lance-shaped lobes**, 2 tiny horn-like teeth at base.

FRUITS: Many-seeded, egg-shaped capsules, 4–6 mm long.

WHERE FOUND: Dry to wet mossy woods, thickets and bogs; scattered across boreal forest north to Churchill, Man., central Yukon and southern Alaska; circumpolar.

NOTES: Heart-leaved twayblade is our most common *Listera* but nonetheless it is an uncommon wildflower, small and easily overlooked. • The intricate pollination mechanisms of *Listera* species fascinated Charles Darwin, who studied them intensively. The pollen is blown out explosively within a drop of viscous fluid that glues the pollinia to unsuspecting insects (or to your finger if you touch the top of the column!). • This species is also called 'mannikin twayblade.' 'Mannikin' is from the Dutch *manneken*, which means 'little man' or 'dwarf.' The genus *Listera* is named in honour of Dr. Martin Lister, an English naturalist who lived from 1638–1711. The species name *cordata* means heart-shaped, in reference to the heart-shaped leaves.

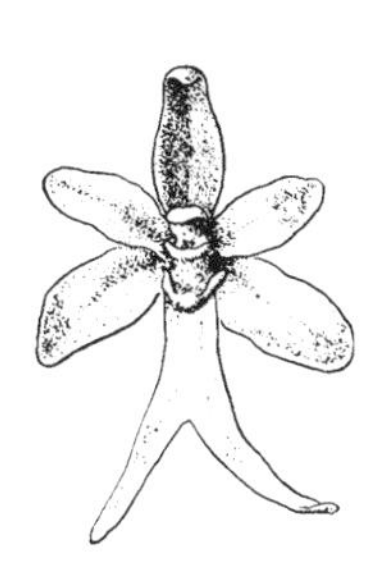

Polygonaceae (Buckwheat Family)

The buckwheat family is medium-sized, with about 750 species worldwide. Most buckwheats are found in north temperate regions, although *Polygonum* species are found on all continents.

The buckwheats are annual or perennial herbs that sometimes grow in twining, vine-like forms. Many species (particularly the *Polygonums*) have swollen stem nodes. They have alternate leaves and sheathing stipules at the base of the leaf stalks. The small, sometimes single-sexed flowers are usually in many-flowered clusters at the stem tips or from the leaf axils.

The flowers are radially symmetrical and usually have 6 scale-like sepals, which are in 2 whorls of 3 each, and no petals. Sometimes there are only 3, 4 or 5 sepals. The inner sepals are sometimes enlarged or modified with hooks, spines, wings or bumps. In *Rumex* species the inner sepals remain after fertilization and become enlarged and membranous in the fruit. Buckwheat family flowers usually have 6 to 9 stamens, and 1 single-chambered, superior ovary. The fruit is a flat, ridged or winged, nut-like achene with 1 seed.

This family is of minor economic importance, although buckwheat (*Fagopyrum*) and rhubarb (*Rheum*) are significant foods. Some species, especially some of our *Polygonum* and *Rumex* species, are widespread weeds.

Note that water smartweed (*Polygonum amphibium*) and long-spiked water smartweed (*Polygonum coccineum*) are treated in the 'Aquatics' section.

ALPINE BISTORT • *Polygonum viviparum*

GENERAL: Small, slender perennial from short, twisted, starchy rootstock; stems single, erect, 5–20 cm tall (sometimes to 30 cm).

LEAVES: Simple, alternate, lance-shaped to elongate-egg-shaped; slender stalked at stem base, dark green, often shiny; upper leaves nearly linear, small, firm; sheath membranous, brownish, translucent, open down 1 side.

FLOWERS: In narrow, cylindrical, spike-like clusters 2–6 cm long (sometimes to 10 cm), usually **with egg-shaped bulblets below and sterile flowers above**; no petals; sepals pink to white; stamens noticeable, protruding.

FRUITS: Dull brown, 3-sided achenes, occasionally with granular surface.

WHERE FOUND: Moist woods and meadows; extremely common across northern boreal forest and throughout Arctic but uncommon to south; circumpolar.

NOTES: Alpine bistort is one of the most interesting members of the genus *Polygonum*. It is very common and widespread in northern regions, and part of its success is due to the fact that the lower flowers in the spike are replaced by vegetative bulblets. These bulblets fall off the plant and start a new plant immediately without going through the seed stage. Often, the bulblets produce roots while they are still on the parent plant, drawing nutrients from the established adult plant to increase their chances of success. This type of reproduction occurs frequently in arctic plants. Many small mammals and birds eat the bulblets and disperse them like seeds. They are often found in the gizzards of ptarmigan (*Lagopus* species), from which some are regurgitated undamaged. • The leaves of alpine bistort are high in vitamins A and C, and can be eaten raw or cooked. The small bulblets are also eaten raw as a trail snack or as an addition to salads. Some northern native peoples gathered the roots of alpine bistort and ate them raw or roasted, as we might eat nuts or raisins. They are said to taste like almonds or small potatoes. Older roots and leaves become tough and acidic, so they are usually boiled.

PALE PERSICARIA • *Polygonum lapathifolium*
DOCKLEAF SMARTWEED

GENERAL: Annual, 20–60 cm tall; stems usually branched, erect or occasionally prostrate, hairless to rather hairy.

LEAVES: Simple, alternate, lance-shaped to oblong-lance-shaped, 5–15 cm long, short-stalked, hairless to densely hairy, especially below; hairless **gland-dotted, membranous sheath** 5–20 mm long surrounds stem at leaf base; edge sometimes fringed with short hairs.

FLOWERS: Many, in erect or spreading, cylindrical spikes, 1–6 cm long, white, greenish or pink, 2–3 mm long, **strongly 3-veined**; **yellow glands on flower stalks**.

FRUITS: Smooth, shiny achenes 2–3 mm long, broadly oblong or egg-shaped, dark brown to black.

WHERE FOUND: Streambanks, slough edges, low wet areas and occasionally in cultivated fields; widespread across our region, scattered north and west to interior Alaska.

NOTES: The introduced, weedy, **common knotweed** (*P. arenastrum*, *P. aviculare* in part) grows near settlements and along roadsides, and is scattered throughout our region. It is a much-branched, prostrate to ascending annual, with elliptic-oblong, stalkless, blue-green, 1–2-cm-long leaves, and tiny, greenish, stalkless flowers. • **Striate knotweed** (*P. erectum*, also called *P. achoreum*) looks like common knotweed, but has fluted stems and larger, oval to egg-shaped leaves that commonly are crowded on the stems. It too is an introduced weed of roadsides and waste places near settlements. • Many *Polygonum* species make good eating, but they often have the tart flavour of rhubarb. Usually these plants are steamed or boiled like spinach. With added sugar they can be made into jam or condiments. The roots are also edible. • The *Polygonums* are related to rhubarb, and like rhubarb they should be eaten in moderation. Large amounts can have **laxative** effects and the greens also contain some hydrocyanic acid. These plants are suspected of causing **skin irritation** in people with sensitive skin.

SHEEP SORREL • *Rumex acetosella*
SOUR WEED

GENERAL: Hairless annual or perennial with slender, creeping **rhizomes**; 1-several stems, thin, simple or branched from base, **10–30 cm tall** (sometimes to 50 cm).

LEAVES: Many, variable, narrowly **arrowhead-shaped** with spreading lobes at leaf-base, 1–8 cm long (sometimes to 10 cm); stalks long at base, shorter or absent upwards; membranous sheath surrounds stem leaf base.

FLOWERS: Many in **loose**, narrow, **leafless** clusters, **reddish or yellowish**, small; no petals; 6 sepals, scale-like, 0.5–1.8 mm long; **sexes on separate plants**; male flowers have long stamens hanging like red chandeliers; female flowers have 3 inner sepals covering achene at first; stalks jointed immediately under flower.

FRUITS: Achenes, nut-like, glossy golden brown, 1–2 mm long, 3-sided; enveloped in 3 smooth, 1–2 mm long scales.

WHERE FOUND: Introduced aggressive weed of roadsides, cultivated land and waste places; widespread across temperate North America; north and west to Lake Athabasca and southern Alaska; circumpolar.

NOTES: Green sorrel (*R. acetosa* ssp. *alpestris*) is a similar species found in the northwestern part of this region. It is distinguished by its showier, larger fruits (3-4 mm long), which have valves that are much larger than the achene. Both species have arrowhead-shaped leaves, but the lobes at the base of sheep sorrel leaves usually flare out to the sides whereas those of green sorrel curve downwards.• *Rumex* flowers are wind-pollinated and the large amounts of pollen produced each year are the bane of many allergy sufferers. A single plant produces approximately 4,000,000 pollen grains. The seeds are carried to new locations by birds such as crows, sparrows, wood pigeons and starlings. However, once the fruits have been shelled in the beak and ground in the gizzard, only 4–5% are likely to survive, and of these, only 25% may germinate. • These plants should always be used in moderation, as they contain oxalic acid.

GOLDEN DOCK • *Rumex maritimus* var. *fueginus*

GENERAL: Annual or biennial from fleshy root, 20–60 cm tall; stems thick, **hollow**, usually **diffusely branched**, covered with very short hairs.

LEAVES: Simple, alternate, lance-shaped to narrowly oblong, 4–10 cm long, 1–3 cm wide, smaller upwards; stalk shorter than blade; base rounded, straight or occasionally heart-shaped; **edges more or less wavy**.

FLOWERS: In numerous large, **tight, leafy clusters that become continuous towards top;** greenish brown, 1–2 mm long.

FRUITS: Smooth, shiny achenes, 3-sided, 1–4 mm long.

WHERE FOUND: Wet places, salty shores and disturbed areas; often forming large patches; common and widespread across the southern part of our region north to eastern Alaska.

NOTES: Wild dock species should be used with **caution**. Some cause skin irritation, though this varies with the plant and the sensitivity of the person. Dock roots are high in tannins, which irritate the bowels and kidneys, so they should not be eaten. The leaves contain oxalic acid (like rhubarb leaves), which is poisonous if large amounts are eaten. The oxalic acid content can be reduced by parboiling, and the hazard can be further reduced by serving a calcium-rich food (a cream sauce, for example) with dock greens. However, small children, elderly people, and people with gout, rheumatism, or kidney ailments are advised to avoid these plants. Large doses of 'medicinal' teas can also be poisonous. • All northern docks are edible and nutritious. They can contain more vitamin C then oranges and more vitamin A than carrots, as well as calcium, iron, potassium, phosphorus, thiamin, niacin and riboflavin. Tender young dock leaves are harvested in the spring or late fall for use as a potherb, a cooked vegetable, or a flavouring for cream sauces or soups. Boiling in 2 changes of water helps to get rid of bitterness. The plentiful seeds can be dried, ground into meal, and added to bread, pancakes, and muffins.

NARROW-LEAVED DOCK • *Rumex salicifolius*
WILLOW DOCK

GENERAL: Perennial from branching taproot, 30–60 cm tall; stems hairless, **reclining to erect, often tufted, freely branched in lower leaf axils**.

LEAVES: Simple, alternate, mostly along stem, slightly smaller upwards, narrowly lance-shaped to oblong-elliptic, 5–10 cm long, 1–3 cm wide; lower leaves have short stalks, upper leaves nearly stalkless, pale green, **flat**, smooth-edged, taper at both ends, **often folded**.

FLOWERS: Many, in clusters along upper stem and branches, greenish brown to reddish purple; 3 inner sepals have grain-like swelling near base.

FRUITS: Smooth, shiny, dark red achenes, faces slightly concave, 2 mm long.

WHERE FOUND: Moist open ground, riverbanks, lakeshores and roadside ditches; widespread across our region, north and west to the Yukon.

NOTES: Also called *R. mexicanus* and *R. triangulivalvis*. • **Arctic dock** (*R. arcticus*) (inset photo) has erect stems from a large, fleshy rhizome. Its dark green to reddish, somewhat fleshy, oblong to lance-shaped leaves are mostly at the stem base. The plant is purplish-tinged at the base. Arctic dock is very common in wet turfy places and shores of ponds and lakes, from the Mackenzie River valley westwards. • Arctic dock is the most common species of *Rumex* in the northwest. Northern native peoples use arctic dock leaves as a cooked vegetable or serve them for dessert with sugar added to taste. The leaves are also used as poultices to remove slivers. • The Woods Cree boiled willow dock to make a wash for aching joints. • In Europe, different species of wild dock have been used for many years to make poultices, laxatives, and tonics. Research has shown that several *Rumex* species contain mildly antiseptic compounds and that these plants are also laxative. • The flower clusters of willow dock are not nearly as dense as those of most other docks.

WESTERN DOCK • *Rumex occidentalis*

GENERAL: Perennial from taproot, 0.5–1.5 m tall; **stems simple**, stout, erect, **often red-tinged**, hairless.

LEAVES: Simple, alternate, smaller and fewer upwards; lower leaves oblong-lance-shaped, 5–20 cm long, long-stalked, **usually heart-shaped at base**, taper to narrow tip; **edges somewhat irregularly curled and wavy**.

FLOWERS: Many, in dense clusters on upper 20–50 cm of stem, greenish to reddish, small, **nodding or bent backwards; flower stalks not jointed**.

FRUITS: Smooth, shiny, short-pointed, chestnut-brown achenes 2–4 mm long; **reddish brown, net-veined wings enclose fruit**.

WHERE FOUND: Marshes and wet meadows; common across our region north to south-interior Alaska.

NOTES: Also called *R. fenestratus*. • **Curled dock** (*R. crispus*) differs from western dock in that its flower stalks are jointed near the middle; its leaves are narrowly oblong and blunt-tipped, and have fan-shaped bases and strongly curled and wavy edges; and the 3 wings enclosing the fruit have a grain-like swelling. Curled dock is an introduced, weedy species that is scattered across our region but it is most common in the prairies and parkland in the eastern part of our region. • **Water dock** (*R. orbiculatus*, also called *R. britannica*) is a tall (to 1.5 m), native dock with leaves shaped like those of curled dock but with flat edges. Water dock also has fruiting wings, each with a seed-like swelling. It grows in marshes, lakeshores and ditches across our region north to the southern Northwest Territories. • The Cree dried, grated and boiled the large yellow roots of western dock to make a yellow dye. They also collected the roots for medicinal use. The roots were washed, split into eighths and dried for storage. The dried, chipped root was boiled in water to make a medicinal tea to relieve a tickling throat. This was sweetened to taste and swallowed slowly. • Several western native groups used western dock for food. The stems were cooked and eaten like rhubarb is today, and the leaves were used like spinach. Young stems and leaves were also eaten raw, and seeds were ground into meal for addition to other foods. • The seeds of curled dock were often used as a tobacco stretcher or substitute, and this plant was often called 'Indian tobacco.' It is still recommended by some as a reasonably good substitute. • The roots of curled dock were dried, ground and used to make an abrasive tooth powder. They were also used in Europe as a laxative and tonic. • The Woods Cree boiled water dock plants to make a wash for the ache of painful joints.

R. occidentalis

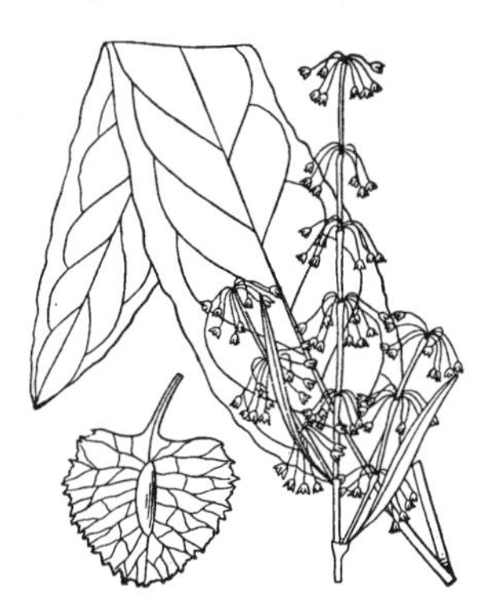

R. orbiculatus

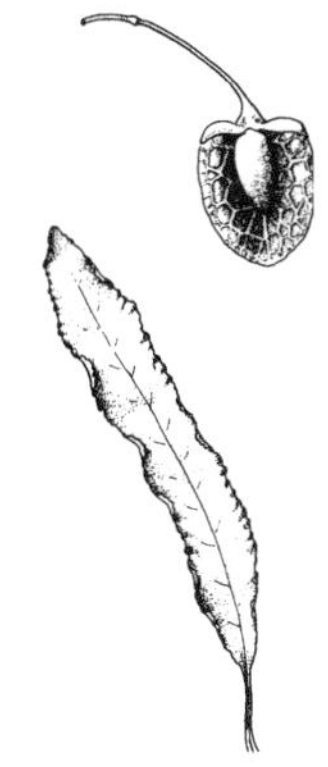

R. crispus

Caryophyllaceae (Pink Family)

The pinks are a family of annual or perennial herbs that characteristically have swollen stem nodes; opposite, mostly narrow leaves; and radially symmetrical flowers with 4 or 5 sepals (sometimes fused into a tube), 4 or 5 petals, 5–10 stamens, and a single, superior ovary. The fruits are usually 1-chambered capsules that open at the top by valves, teeth or lids. The capsules usually contain many seeds.

The pink family is fairly large—about 2,000 species worldwide—and most of its members are found in north temperate regions, with a centre of distribution in the Mediterranean basin.

This family includes a large numer of commercially important ornamentals, notably carnation (*Dianthus caryophyllus*) and many other *Dianthus* species, baby's-breath (*Gypsophila*), catchfly (*Silene*), maltese cross (*Lychnis*), sandwort (*Arenaria*) and the mouse-ear chickweeds (*Cerastium*). Several species are widespread weeds, among them the chickweeds (*Cerastium* and *Stellaria*) and spurry (*Spergula*).

Arenaria
p. 102

Cerastium
p. 103

Stellaria
pp. 103–105

BLUNT-LEAVED SANDWORT • *Arenaria lateriflora*

GENERAL: Perennial from thread-like rhizomes or stolons; stems thin, ascending to erect, single or in small clumps, 5–20 cm tall, covered in minute hairs.

LEAVES: Opposite, in 2–5 (sometimes 7) pairs on stem, thin, **egg-shaped to oblong or lance-shaped**, 0.3–4 cm long (usually 1–3 cm); middle leaves largest, minutely hairy.

FLOWERS: Usually single, but can have 2–5 in clusters at stem tips; erect, white; 5 petals, **3–8 mm long, not lobed**; 5 sepals, 2–2.5 mm long; stalks thread-like, 5–15 mm long.

FRUITS: Capsules, 3–6 mm long; tip opens by 6 'teeth'; seeds black, shiny, 0.8–1.3 mm wide.

WHERE FOUND: Moist meadows, shores, thickets, and woods; widespread across our region; circumpolar.

NOTES: Also called *Moehringia lateriflora*. • **Large-leaved sandwort** (*Arenaria macrophylla*, also called *Moehringia macrophylla*) has flat-sided stems, sepals that are pointed rather than blunt, and larger (to 7 cm long), narrower, pointier leaves than those of blunt-leaved sandwort. Large-leaved sandwort is a rare species of dry, open or wooded boreal habitats, northwest to Great Slave and Athabasca lakes. Unlike its common and widespread cousin (blunt-leaved sandwort), large-leaved sandwort has very specific habitat requirements. It always grows on soils derived from magnesium or ultrabasic rock, and therefore its distribution is very spotty. • The species name is from the Latin *lateralis*, 'lateral' (of, at, from or towards the side), and *flos*, 'a flower.'

NODDING CHICKWEED • *Cerastium nutans*
LONG-STALKED MOUSE-EAR CHICKWEED

GENERAL: Annual, 5–25 cm tall; stems tufted, erect or often curved at base, with **gland-tipped hairs**.

LEAVES: Simple, opposite; lower leaves spoon-shaped to oblong-lance-shaped, 7–25 mm long, short-stalked; upper stem leaves narrowly lance-shaped to egg-shaped, stalkless.

FLOWERS: White, small, **nodding**; in loose, leafy terminal groups on stalks 5–20 mm long; **5 petals** (rarely absent), deeply 2-lobed, 4–6 mm long, **as much as 1 1/2 times as long as** broadly lance-shaped **sepals**.

FRUITS: Cylindrical capsules, 2–3 times as long as sepals.

WHERE FOUND: Moist ground and open woods; infrequent across southern boreal forest north to southern Northwest Territories.

C. vulgatum

NOTES: Common mouse-ear chickweed (*C. vulgatum*) is a cosmopolitan weed of roadsides and other disturbed areas, mostly in the southern part of our region. It is a hairy (sometimes also with glands), short-lived perennial with firm but trailing or reclining, much-branched stems. **Field mouse-ear chickweed** (*C. arvense*, photo), an abundant species throughout the prairies and parkland, grows in meadows, sandy or gravelly places, and on rocky slopes scattered across the boreal forest. It is a hairy, matted perennial with narrow leaves and fairly large showy flowers. Its conspicuous white petals are at least 1 1/2 times as long as the sepals. Field mouse-ear chickweed often forms large patches on favourable sites. • The name 'chickweed' comes from the fact that these plants and common chickweed, which infests your garden, were fed to chickens, goslings, and caged birds, especially if they were ill. • The genus name, *Cerastium*, is derived from the Greek *ceras*, 'a horn,' in reference to the long and curved capsules of some species. *Nutans*, which means 'nodding,' refers to the nodding flowers and mature capsules of long-stalked mouse-ear chickweed.

COMMON CHICKWEED • *Stellaria media*
COMMON STARWORT

GENERAL: Delicate **annual** from taproot; stems trailing to ascending, 10–50 cm long, rooting at nodes, **often matted**, branched, leafy, with lines of fine, white hairs.

LEAVES: Broadly egg-shaped, 5–30 mm long, often with tiny blisters; lower leaves **stalked**, become stalkless upwards.

FLOWERS: Single or in few-flowered, leafy clusters; **petals short or absent**; sepals 3–6 mm long, more or less hairy and blistered.

FRUITS: Many-seeded, straw-coloured capsules, 4–8 mm long, split open into 6 sections; seeds 0.8–1.2 mm long, reddish brown, uniformly warty.

WHERE FOUND: Introduced weed of cultivated and disturbed ground; widespread across our region and most of North America; circumpolar.

NOTES: This aggressive weed can produce as many as 5 generations of offspring in 1 year. At times a gardener may seem to be fighting a losing battle, but perhaps it would be better to harvest, rather than compost chickweed. The plants remain tender through the summer and are excellent as a potherb or as an addition to salads or soups. However, chickweed is a favourite host plant for many insects, so it should be picked early to avoid having to pick off large numbers of insect eggs. • Herbal chickweed leaf tea is sold commercially. Because of its nutrients, chickweed was said to build up patients who had lost strength from extended illness. On the other hand, chickweed tea (a laxative) was said to slim the obese. Medicinally, chickweed is a very controversial plant. Some herbalists praise it as a remedy for skin diseases, burns, inflammations, colds, coughs, tumors, haemorrhoids, sore eyes, and rheumatism while others consider it nearly worthless. • Chickweed was also called 'hen's inheritance' because the succulent greens are a favourite food of chickens. It is also recommended as a food for caged birds.

LONG-STALKED CHICKWEED • *Stellaria longipes*
LONG-STALKED STARWORT

GENERAL: Perennial from slender, branched rhizomes; forms small to large tufts or mats; stems slender, hairless, 5–20 cm tall (sometimes to 30 cm).

LEAVES: Opposite, stalkless, **stiff**, usually hairless and **shiny**, linear to narrowly lance-shaped, 10–25 mm long, **often tinged bluish grey**.

FLOWERS: White on **slender, erect stalks**; 1 to several in loose clusters; 5 petals, 3–8 mm long, deeply 2-lobed and sometimes looks like 10.

FRUITS: Dark brown to purplish black, shiny capsules, 4–6 mm long; tip opens into 6 'teeth'; seeds 0.6–0.9 mm long, roughened.

WHERE FOUND: Dry to moist open areas and woodlands; widespread across our region, northwards and upwards (elevationally) into tundra; circumpolar.

NOTES: Recent taxonomic work (Chinnappa & Morton, 1991, and Emery & Chinnappa, 1994) has concluded that *S. longipes* should be treated as a single, variable species with no subspecies or varieties except for ***S. longipes* ssp. *arenicola***, a population of the Lake Athabasca sand dunes with straw-coloured capsules, and ***S. longipes* var. *monantha***, the bluish green, arctic-alpine form. • All chickweeds are good sources of vitamin C and minerals. The young leaves and stems of all make an excellent salad green, potherb or cooked vegetable, similar to spinach. Plants with hairy stems should always be cooked before eating. Chopped chickweed can also be used to make a flavourful creamed soup or puree, and it has even been added to pancakes. A refreshing tea can be made from fresh or dried plants.

LONG-LEAVED CHICKWEED • *Stellaria longifolia*
LONG-LEAVED STARWORT

GENERAL: Perennial, 10–40 cm tall; stems slender, weak, reclining to erect, **sharply 4-sided**, hairless but **minutely roughened towards top**.

LEAVES: Simple, opposite, stalkless, **linear or narrowly lance-shaped**, 2–5 cm long, up to 4 mm wide, pointed at both ends, hairless or **sometimes with hairy fringe at base**; edges minutely roughened.

FLOWERS: In many-flowered, open groups at stem tips, white, small; **stalks spreading or bent backwards**; bracts papery; **petals deeply 2-lobed**, 3–5 mm long, **as long as or slightly longer than sepals**.

FRUITS: Straw-coloured capsules, longer than flowers, tip opens by 6 'teeth.'

WHERE FOUND: Moist woods, meadows and shores; common and widespread across our region.

NOTES: There is evidence that *S. longifolia* and another species, ***S. porsildii***, hybridized to produce long-stalked chickweed (*S. longipes*). • Long-leaved chickweed is often a pioneering species on rich, moist, disturbed soils. It can form extensive mats, sometimes to the short term exclusion of almost all other species. • The name *longifolia*, meaning 'long-leaved,' is appropriate as this species has the longest and narrowest leaves of any of the chickweeds.

FLESHY STITCHWORT • *Stellaria crassifolia*

GENERAL: Perennial, **freely branching and often matted**; stems weak, **reclining to erect**, 5–15 cm long, hairless.

LEAVES: Simple, opposite, many, **somewhat fleshy**, lance-shaped to egg-lance-shaped, 5–15 mm long, 1–4 mm wide, hairless, narrow at base, sharp-pointed to blunt at tip, stalkless.

FLOWERS: Few; **usually single in leaf axils** but sometimes in few-flowered, open, groups at stem tips, white, small, **petals deeply 2-lobed**, 3–5 mm long, **usually longer than sepals**.

FRUITS: Straw-coloured capsules, longer than flower; tip opens by 6 'teeth.'

WHERE FOUND: Wet ground and shaded woods; often forms mats around bases of grasses and sedges; widespread across boreal forest north to low Arctic.

NOTES: Fleshy stitchwort was boiled and used as a poultice for treating sores and swellings. Similarly, bruised leaves were mixed with fat (lard or Vaseline) to make an ointment for bruises, irritations and other skin problems. Fleshy stitchwort can also be boiled to make a laxative tea to treat serious constipation. In the 19th century, fleshy stitchwort juice was prescribed for scurvy and eye inflammation. In China, fleshy stitchwort was boiled and applied externally as a treatment for colds, pimples, snakebite and traumatic injuries. • Near its northern limit fleshy stitchwort does not regularly produce viable seeds. Instead, it reproduces vegetatively by forming over-wintering buds on its stems and at the tips of fragile thread-like runners that grow from the lower leaf axils. • The *Stellaria* genus takes its name from the Latin *stella*, 'a star,' in reference to the star-like shape of the flowers. *Crassifolia* means 'thick-leaved,' in reference to the slightly fleshy texture of the leaves.

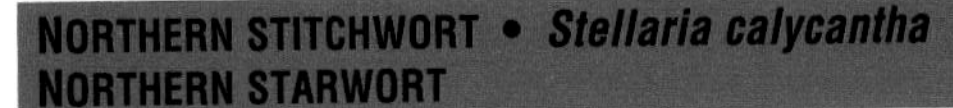

NORTHERN STITCHWORT • *Stellaria calycantha*
NORTHERN STARWORT

GENERAL: Low, often sprawling or matted perennial from long rhizomes; stems slender, prostrate to ascending or erect, 5–50 cm or more long, hairless to sparsely hairy.

LEAVES: Opposite, stalkless, **elliptic to narrowly lance-shaped**, 0.5–5 cm long, thin; hairless except for a few short hairs at leaf-base.

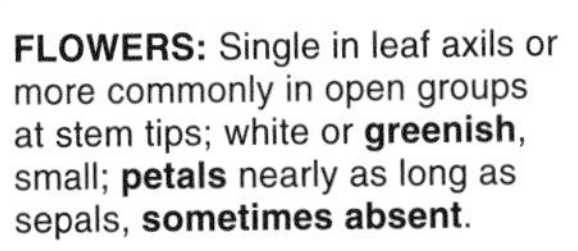

FLOWERS: Single in leaf axils or more commonly in open groups at stem tips; white or **greenish**, small; **petals** nearly as long as sepals, **sometimes absent**.

FRUITS: Straw-coloured to purplish capsules, much longer than flowers, open by 6 'teeth'; seeds 0.5–1 mm long, reddish brown.

WHERE FOUND: Locally common in wet meadows, thickets, streambanks, open moist forests, clearings, and roadsides, across northern parkland and boreal forest.

NOTES: Northern stitchwort resembles a long-leaved chickweed, but its flowers are in the axils of normal green leaves. • The species name *calycantha* is from the Latin *calyc*, 'calyx,' and *antho*, 'flower,' because the petals are much reduced, leaving the calyx to form the flower.

Brassicaceae (Mustard Family)

The mustard family (*Brassicaceae*, also known as *Cruciferae*) includes about 3,000 species, most of which are found in the cooler regions of the northern hemisphere. It consists mainly of annual, biennial or perennial herbs with a watery sap, and often with forked or star-shaped hairs. The leaves are alternate and simple to finely divided. The flowers are radially symmetrical, usually in clusters that lengthen with age, and they have 4 sepals, 4 petals, 6 stamens, and 1 superior, 2-chambered ovary. The fruits are usually pod-like, with 2 chambers separated by a membranous partition, and they open by 2 valves. If the fruits are at least 3 times longer than wide they are called '**siliques**'; if they are shorter they are called **'silicles.'**

This family has considerable economic importance as food crops, weeds, and ornamentals. Important food crops include cabbage, cauliflower, broccoli, kohlrabi, Brussel sprouts (all from 1 species—*Brassica oleracea*), rape (*Brassica rapa* and *B. napus*), rutabaga (*Brassica napus*), radish (*Raphanus*), and watercress (*Nasturtium*). The family produces 2 condiments, mustard (*Brassica*) and horseradish (*Armoracia rusticana*). Troublesome weeds like shepherd's purse (*Capsella bursa-pastoris*), stinkweed (*Thlaspi arvense*), mustards (*Brassica* and *Barbarea*), flixweeds (*Descurainia*), and pepper-grasses (*Lepidium*) are widespread.

Key to the Mustard Family

1a. Fruits composed of 2 sections, with a large beak, one-third to half as long as body .... ***Brassica kaber*** **(see *Barbarea orthoceras*)**

1b. Fruits composed of only 1 segment 3

2a. Pods are silicles, oval, elliptic, oblong or heart-shaped (scarcely longer than wide) 4

3a. Silicles flattened in same plane as partition, not rounded or circular in cross section .... ***Draba* spp.** (see *D. nemorosa*)

3b. Silicles flattened at right angles to partition between chambers of pod 5

4a. Silicles heart-shaped to triangular .... ***Capsella bursa-pastoris***

4b. Silicles not heart-shaped or triangular 6

5a. More than 1 seed per chamber; silicles more than 8 mm long .... ***Thlaspi arvense***

5b. 1 seed per chamber; silicles usually less than 5 mm long (the pepper-grasses) 7

6a. Silicles elliptic to egg-shaped, less than 3 mm long and 2 mm wide, with pointed teeth at tip; flowers have tiny white petals, in many flower clusters at stem tips and in leaf axils .... ***Lepidium ramosissimum*** (see *L. densiflorum*)

6b. Silicles round to obovate, often more than 3 mm long and 2 mm wide, with rounded teeth at tip; petals usually absent from flowers; 1 to few flower clusters .... ***Lepidium densiflorum***

2b. Pods are linear siliques, at least 3 times as long as wide 8

7a. Leaves (at least those at or near stem base) pinnately lobed or divided 9

8a. Flowers white .... ***Cardamine pensylvanica***

8b. Flowers yellow 10

9a. Plants hairy with branched or star-shaped hairs (the tansy mustards or flixweeds) 11

10a. Siliques generally less than 12 mm long; upper leaves once-pinnate 12

11a. Leaflets green; siliques somewhat club-shaped, rounded at top; at least some seeds in 2 rows .... ***Descuriana pinnata*** (see *D. sophia*)

11b. Leaflets grey-green; siliques linear, pointed at tip; seeds in 1 row .... ***Descurainia richardsonii*** (see *D. sophia*)

10b. Siliques generally more than 12 mm long; upper leaves 2–3 times pinnate 13

12a. Silique stalks 1–9 mm long; stems have stalked glands .... ***Descurainia sophioides*** (see *D. sophia*)

12b. Silique stalks 7–12 mm long; stems without stalked glands .... ***Descurainia sophia***

9b. Plants hairless, or hairy with unbranched hairs 14

13a. Ear-like lobes at base of stem leaves clasp stem; siliques to 5 cm long .... ***Barbarea orthoceras***

13b. Stem leaves not clasping; siliques to 7 mm long .... ***Rorippa islandica***

7b. Leaves not pinnately lobed or divided, sometimes toothed, lobed, or lyrate 15

14a. Flowers yellow or orange (the wallflowers) 16

15a. Petals less than 6 mm long; siliques generally 1-3 cm long . ***Erysimum cheiranthoides***

15b. Petals more than 6 mm long; siliques generally 2-5 cm long .... ***Erysimum inconspicuum*** (see *E. cheiranthoides*)

14b. Flowers white to pink or purple 17

16a. Siliques cylindrical, with constrictions at regular intervals .... ***Braya humilis*** (see *Arabis hirsuta*)

16b. Siliques somewhat flattened, without regular constrictions (the rock cresses) 18

17a. Siliques angle downwards when ripe .... ***Arabis holboellii*** (see *A. divaricarpa*)

17b. Siliques point upwards or towards branch tip 19

18a. Stem leaves do not clasp stem; leaves smooth or only slightly hairy; leaves at stem base lyrate .... ***Arabis lyrata***

18b. Stem leaves stalkless, clasp stem; leaves at stem base (and often those low on stem) covered with star-shaped hairs 20

19a. Siliques spread away from stem .... ***Arabis divaricarpa***

19b. Siliques tightly pressed against stem 21

20a. Siliques 2-3 mm wide; seeds in 2 rows; flowers pink or purple .... ***Arabis drummondii*** (see *A. divaricarpa*)

20b. Siliques less than 1.5 mm wide; seeds in 1 row; flowers white or creamy .... ***Arabis hirsuta***

WORMSEED MUSTARD • *Erysimum cheiranthoides*

GENERAL: Slightly hairy, annual from taproot; stems **20–60 cm tall**, slender, green to purplish, sparsely covered with short-stalked, parallel, T-shaped hairs.

LEAVES: Alternate, **narrowly lance-shaped**, to 8 cm long, **with or without teeth**.

FLOWERS: At stem tips in dense clusters 2–3 cm across which lengthen in fruit; 4 petals, **pale yellow, 2–4 mm long**; stalks thread-like, spreading to ascending, 6–12 mm long.

FRUITS: Nearly erect, linear pods (siliques), 1–3 cm long, **round to square** in cross-section.

WHERE FOUND: Introduced and native weed of cultivated and waste land, but also on lakeshores and streambanks; widespread across our region; essentially circumpolar.

NOTES: Small-flowered rocket (*E. inconspicuum*) is also widespread in the boreal forest. It is a native of dry, sandy or gravelly, open areas, and is most common in our region in the Yukon and Alberta and along the southern boundary of our region in Saskatchewan and Manitoba. It is a greyish green perennial with larger yellow petals (7–10 mm long), longer pods (2–5 cm long) on stout curved stalks, and a rosette of leaves at the stem base. • The seeds of wormseed mustard contain mustard oil (allyl isothiocyanate). Large quantities of seed can be **dangerous**. Swine fed wormseed mustard containing 1.7% seed by weight, were killed. • The seeds are extremely bitter and have been used to rid the system of intestinal worms, hence the name wormseed. Some native groups boiled the roots of 1–4 plants in 1 litre of water to make a wash for pimples. • The Latin *Erysimum* is from *erysio* meaning 'to draw,' as in drawing out pain or causing blisters, because *Erysimum* spp. were often used as poultices.

AMERICAN WINTER CRESS • *Barbarea orthoceras*

GENERAL: Biennial herb from woody base and weak taproot; stems erect, **angled, usually hairless**, 20–50 cm tall.

LEAVES: Lower leaves 4–15 cm long with a large lobe/leaflet at tip; stem leaves alternate, smaller and simpler upwards.

FLOWERS: Many, in small clusters at stem tips; 4 petals, **yellow, 4–5.5 mm long**; sepals yellowish, hairless, 2.5–3.5 mm long.

FRUITS: Pods (siliques), strongly ascending, usually 20–40 mm long, **somewhat 4-sided**; stalks thicken towards tip; seeds brownish, with tiny wrinkles.

WHERE FOUND: Moist woods, streambanks and sheltered lakeshores; widespread across northern and central boreal forest; circumpolar.

NOTES: Another yellow-flowered mustard, **wild mustard** (*Brassica kaber*) is easily identified by the large beak at the top of its pods. Wild mustard is an introduced annual weed of cultivated and disturbed ground, scattered across our region. The young plants and buds of American winter cress are eaten in Alaska and northeastern Asia and are said to be high in vitamin C. They are used raw, as a radish-like addition to salads, or they are boiled. • The genus *Barbarea* was named for St. Barbara, a martyr of the 4th century A.D. who refused to renounce her belief in God. The protection of St. Barbara was usually invoked against lightning and fire, so she became the patron saint of military architects, artillery men, and miners (people who work with gunpowder and flames). A German interpretation is that winter cresses are eaten by barbels—a freshwater carp.

Brassica kaber pod

MARSH YELLOW CRESS • *Rorippa islandica*

GENERAL: Annual or biennial; stems **erect**, 20–60 cm tall, usually much branched towards top.

LEAVES: Alternate, **6–15 cm long**, oblong-lance-shaped in outline, deeply cut into leaflets, often with large lobe/leaflet at tip, variously toothed, become smaller, simpler and stalkless upwards, with lobed leaf-base.

FLOWERS: In long clusters at stem tips and in leaf axils, **yellow**; 4 petals, about **2 mm long**.

FRUITS: Sausage-shaped to nearly spherical pods, 3–7 mm long; stalks spreading, 4–8 mm long; seeds in 2 rows.

WHERE FOUND: Marshy ground across our region, north to treeline.

NOTES: Also called *R. palustris*. • Marsh yellow cress is edible and can be used in the same way as European watercress (*R. nasturtium-aquaticum*, also known as *N. officinale*). It is usually added to soups, meat dishes and salads as a flavouring, but it can also be used as a potherb. Some suggest mixing it with parsley and chives to flavour herb butter. Dried leaves can be used to make tea, or they can be ground and used as a seasoning. • The Romans used a mixture of watercress and vinegar to treat people with mental illnesses. From this came the Greek proverb, 'Eat cress and learn more wit.' • *Rorippa* is from *rorippen*, the Old Saxon name for this plant. The species name *palustris* describes its habitat (marshy places).

FLIXWEED • *Descurainia sophia*
TANSY MUSTARD

GENERAL: Annual weed; stems erect, 20–100 cm tall, usually many-branched, greyish green with tiny, star-shaped hairs.

LEAVES: Mainly on stem, 2–9 cm long, greyish green with fine hairs (some star-shaped), **2–3 times pinnate into narrow segments**, smaller and simpler upwards.

FLOWERS: In clusters much elongated in fruit at stem tips; 4 petals, yellow or whitish, 0.8–1.5 mm long.

FRUITS: Narrowly linear, slightly curved **pods** (siliques), **15–30 mm long**, 0.5–1 mm wide, erect ascending; **stalks 7–12 mm long**, spreading to ascending; several to many seeds, in 1 row.

WHERE FOUND: Common, **introduced** weed of cultivated and waste ground; widespread across our region, north and west to interior Alaska; circumpolar.

NOTES: There are 4 tansy mustards in our region. **Green tansy mustard** (*D. pinnata*), primarily a weed of the prairies, has once-pinnate leaves, is green and has gland-tipped hairs. **Grey tansy mustard** (*D. richardsonii*) is widespread across the boreal forest, north to Great Bear Lake and interior Alaska. Like green tansy mustard it has once-pinnate leaves, but its leaflets are grey-green and toothed, and it has non-glandular hairs. Grey tansy mustard has smaller pods than flixweed (1 cm long or less) on short, strongly ascending stalks (see illustration). **Northern flixweed** (*D. sophioides*, inset photo) is much like flixweed, but it is bright green with gland-tipped hairs, and it is found mostly west of Great Slave Lake. • Flixweed is a widespread, introduced weed, while green tansy mustard, grey tansy mustard, and northern flixweed all appear to be native species that behave like weeds. •Because of its abundance, it is a good survival food. Some native groups used young plants of grey tansy mustard as a potherb, either fresh or dried for future use. Often it was baked in a fire pit, alternating layers of greens with layers of hot stones. The pit was then covered and the plants were steamed for half an hour. Changing the water twice during the cooking time will remove some of the unpleasant odour and bitter taste.

D.richardsonii

PENNSYLVANIAN BITTER CRESS • *Cardamine pensylvanica*

GENERAL: Biennial or short-lived perennial, 10–50 cm tall; stems simple, freely branching towards top, erect or spreading, sometimes reclining with ascending tip, often bristly-hairy at base, otherwise hairless.

LEAVES: Stem leaves deeply pinnately lobed with 2–8 pairs of side lobes, linear-oblanceolate to oval; larger end lobe; bases of side lobes extend down midvein; leaves at stem base soon wither.

FLOWERS: Several in clusters at stem tips which lengthen considerably in fruit, white, small; 4 petals, about 4 mm long.

FRUITS: Very narrow pods (siliques), 1–3 cm long, spreading or ascending.

WHERE FOUND: Streambanks, lakeshores and wet soil; occasionally in shallow, standing water; widespread, but uncommon, across our region north to southern Yukon.

NOTES: All *Cardamine*s can be eaten raw in salads, and bitter cress is one of the best species for this. However, these plants are usually better cooked, either as a potherb or added to soups, stews and other dishes. • There is some disagreement regarding the origin of the genus name *Cardamine*. Some say it is derived from the Greek *kardamon*, a name the Greek physician Dioscorides (40–70 A.D.) used for a plant in the mustard family that was later applied to the bitter cresses. Others say it comes from the Greek *kardia*, 'heart,' and *damao*, 'to overpower,' suggesting that this plant was a heart medicine or a heart poison. Still others suggest that it is derived from *kardia*, 'heart,' and *damao*, 'to calm,' because these plants were used as sedatives. The species name, *pensylvanica*, is for Pennsylvania, the place where the first specimens were identified.

LYRE-LEAVED ROCK CRESS • *Arabis lyrata*

GENERAL: Biennial or short-lived perennial, 10–40 cm tall; stems slender, **often irregularly twisted**, simple to more often freely branched.

LEAVES: Mostly in rosette at stem base, spoon-shaped in outline, **usually lyrate** (see notes), sometimes merely toothed, 2–4 cm long, 5–17 mm wide, hairless or sometimes with simple or forked hairs; stem leaves narrow, tapered at base, **do not clasp stem**, hairless; edges smooth to few-toothed.

FLOWERS: 3–20 in clusters at stem tips which lengthen with age, white to pinkish, small; 4 petals, 3–8 mm long.

FRUITS: Long (15–40 mm), narrow (1-1.5 mm) pods (siliques); on stalks 10–14 mm long, **erect-ascending**.

WHERE FOUND: River flats and floodplains, open sandy woods and dunes; scattered across our region into Alaska, but more common to west.

NOTES: All rock cresses are edible, but lyre-leaved rock cress is said to be particularly tasty. In the summer the Inuit of Alaska ate the young leaves of these plants raw or cooked. They also liked to ferment them for use in the winter. • Lyre-leaved rock cress has been classified as rare in Manitoba, the continental Northwest Territories, Ontario, Iowa and Virginia, and threatened in North Carolina, so it's best to find other salad greens. • Lyre-leaved rock cress gets its name from the shape of its leaves, which are 'lyrate-pinnatifid.' That means they have an enlarged lobe at the tip, and then they taper to much smaller leaflets towards the base.

HAIRY ROCK CRESS • *Arabis hirsuta*

GENERAL: Biennial or short-lived perennial herb from taproot; 1 to several stems, slender, unbranched, 20–70 cm tall, usually hairy.

LEAVES: In rosette at stem base, short-stalked, 1.5–8 cm long, **hairy, often purplish beneath**; 5–15 stem leaves, alternate, oblong to **lance-shaped**, 1–5 cm long (sometimes to 7 cm), stalkless, clasp stem.

FLOWERS: Several to many in simple or branched clusters at stem tips; **white to somewhat greenish or pinkish**, 3–5 mm long (sometimes to 8 mm).

FRUITS: Erect, flat, smooth pods (siliques), 2–3 cm long (sometimes to 5 cm); seeds in 1 row, winged.

WHERE FOUND: Disturbed sites; dry, rocky banks and ledges, open woods and fields; widespread across our region; circumpolar.

NOTES: Leafy braya (*Braya humilis*) also grows in open, disturbed habitats, especially on silty riverflats in limestone country mostly west of Great Slave Lake in our region. All of its leaves are lance-shaped, toothed or smooth-edged and its white to purplish-tinged flowers are in a cluster that is head-like at first but much longer later. Leafy braya has distinctive siliques that are cylindrical with constrictions at regular intervals. • Although all rock cresses are edible, woolly or very hairy species, like hairy rock cress, are usually good only after cooking, and even then they are considered inferior to the smooth-leaved species. • The leaves of this plant are hirsute—shaggy, hairy and bristly—hence the species name *hirsuta*. *Arabis* means 'from Arabia.' The name 'cress' derives from an old Indo-European word meaning 'to nibble or eat,' because the leaves of cresses were (and are) widely used for salads. *Arabis* plants are closely related to cresses, but grow in dry, rocky places—hence, 'rock cress.'

PURPLE ROCK CRESS • *Arabis divaricarpa*

GENERAL: Biennial from weak taproot, 30–80 cm tall; stem simple or branched towards top, hairless towards top, sparsely hairy below.

LEAVES: In rosette at stem base, narrowly spoon-shaped, 3–6 cm long, 3–8 mm wide, pointed tip, narrows to slender stalk, **covered with star-shaped, white hairs**; edges smooth to sparsely toothed; stem leaves smaller, overlapping, stalkless, **clasp stem**, narrowly oblong to lance-shaped.

FLOWERS: Several to many in cluster at stem tip which lengthens with age, pale pink to purple, small; 4 petals, 6–10 mm long.

FRUITS: Long, narrow pods (siliques); **erect to spreading** stalks, 7–15 mm long.

WHERE FOUND: Dry slopes and sandy areas; common and widespread across our region, north and west to interior Alaska.

A. drummondii

NOTES: Reflexed or **Holboell's rock cress** (*A. holboellii*) (inset photo) resembles purple rock cress, but its leaves are hairier below and its mature siliques are at first spreading, but ultimately bent backwards and drooping. These 2 rock cresses also have similar habitat preferences and distribution. • **Drummond's rock cress** (*A. drummondii*) is another widespread biennial species found on dry sites across our region. It is identified by its broader (3 mm wide), erect pods with seeds in 2 rows, rather than 1. • All rock cresses are edible, with the typical sharp flavour of plants of the mustard family. The tender leaves and flowers are usually added to salads and sandwiches for flavour, but some people like their hot, horseradish flavour and enjoy them alone. They are also cooked as a potherb, often mixed with other, milder greens. Rock cresses are considered especially good in combination with eggs or cheese and as an addition to potato soup.

ANNUAL WHITLOW-GRASS • *Draba nemorosa*
WOODS DRABA

GENERAL: Delicate, often weedy annual from slender taproot; stems erect, 3–25 cm tall (sometimes to 35 cm), with simple, forked and star-shaped hairs.

LEAVES: Alternate, mainly on lower third of stem, egg- to lance-shaped, smooth or finely toothed, 2–30 mm long (sometimes to 45 mm), with simple and long-stalked, branched hairs.

FLOWERS: Simple open clusters of few to many flowers, yellow (occasionally white), 1.2–4 mm long.

FRUITS: Elliptic pods (silicles), 3–10 mm long, 1–3 mm wide; stalks 5–25 mm long, spreading-ascending; seeds in 2 rows.

WHERE FOUND: Open, usually cultivated or waste ground; widespread across our region, north and west to interior Alaska; more or less circumpolar.

NOTES: Annual whitlow-grass is one of several species of *Draba* that range over the boreal region. Although *Draba* has over 30 species in North America, most of these are low, tufted, arctic-alpine perennials. • However, **northern whitlow-grass** (*D. borealis*), **slender whitlow-grass** or **Alaska draba** (*D. stenoloba*), **thick-leaved whitlow-grass** (*D. crassifolia*, also called *D. albertina*), **lance-leaved whitlow-grass** (*D. cana*, also called *D. lanceolata*) and **tall whitlow-grass** (*D. praealta*) are also found in our region. • The identification of different species in the genus *Draba* usually requires close examination of the tiny hairs on both sides of the leaves. • Many *Draba*s grow in exposed, northern sites, and they have developed strategies to ensure seed production in cold, windy areas. The flower buds often open while they are still nestled close to the rosette of leaves at the stem base. This holds them in the warmest layer of air, close to the ground, where cross-pollinating insects are most likely to visit, and where higher temperatures will help the early development of the seeds. As the summer progresses, the stalk of the flower cluster lengthens, raising the seed pods from the ground and increasing the chances of seed dispersal either by wind or by animals.

SHEPHERD'S-PURSE • *Capsella bursa-pastoris*

GENERAL: Finely hairy annual; stems 10–50 cm tall, usually branched.

LEAVES: Mainly in rosette at stem base, 3–8 cm long, broadly lance-shaped, **toothed to pinnately divided**; stem leaves alternate, lance-shaped to oblong, stalkless; base **clasps** stem.

FLOWERS: Usually many, densely clustered at first, later spread out; petals **white, 2–4 mm long**.

FRUITS: Strongly flattened, triangular to heart-shaped pods (silicles), 4–8 mm long, on long, slender stalks; about 20 seeds per pod, sticky when wet.

WHERE FOUND: Common, **introduced** weed of cultivated and waste land; widespread across much of our region; circumpolar.

NOTES: The entire plant is edible, though it has a biting taste. It is usually used as a potherb, but young plants are eaten raw in salads. The pods and seeds, dried or fresh, can be used as a substitute for mustard in soups and stews. Fresh or dried roots have been used as a substitute for ginger, and have even been candied in sugar syrup. • Many small birds are fond of shepherd's-purse fruits. A single plant can produce 40,000–64,000 seeds. When the seeds come into contact with water they produce a sticky mucilage that traps small insects and other invertebrates. The tiny seeds have only small reserves of food, and they often find themselves on poor ground. The gelatine-like substance they produce contains the enzyme protease, which probably digests the trapped invertebrates and it is possible that the nutritious products of this digestion are used for the early growth of the seedlings. • Shepherd's-purse was once considered one of the most important drug plants in the mustard family. Made into a tea, salve, or poultice, it was used primarily as a haemostatic to stop internal or external bleeding, as an astringent to treat diarrhea, dysentery, or hemorrhoids, and as a diuretic to treat kidney or bladder problems. Shepherd's-purse is high in vitamin C and it was used to cure or prevent scurvy. It is also high in vitamin K (the blood-clotting vitamin), calcium, sulphur and sodium. The tea was also applied externally to relieve rheumatism and poison ivy.

STINKWEED • *Thlaspi arvense*
PENNYCRESS

GENERAL: Hairless, yellow-green annual from taproot; stems 20–50 cm tall.

LEAVES: Few at stem base, stalked, 1.5–7 cm long, spoon-shaped, **wavy-edged to toothed**, soon fall off; stem leaves stalkless, oblong to **arrowhead-shaped**, base clasps stem.

FLOWERS: In long clusters at stem tips, white; 4 petals, 3–4 mm long; stalks 3–15 mm long, ascending to spreading.

FRUITS: Flat, broadly heart-shaped pods (silicles), 9–18 mm long, broadly **winged all around**, notched at tip; 4–16 seeds per pod, flattened, nearly round, blackish with concentric rings (like fingerprint).

WHERE FOUND: Introduced weed of cultivated and waste ground; widespread across our region, north and west to interior Alaska; circumpolar.

NOTES: Stinkweed was one of the first weeds introduced to North America by Europeans. Its seeds contain mustard-oil (isothiocyanate) and cattle have been poisoned and even killed by eating hay contaminated with 25% or more stinkweed. Dairy products and meat can be tainted if cattle graze it. • Some people are said to have used the seeds for flavouring, but this is probably unwise. The leaves have a characteristic mustard 'bite,' but young tender shoots have been used in salads or cooked like spinach. Stinkweed is usually mixed with other, blander plants to mask its bitterness, and even then it is not to the taste of most people. • This species is called 'stinkweed' because the leaves have a rank odour when crushed. The round, flat pods look like shiny, silver pennies and give rise to the common name 'pennycress.' • *Thlaspi* was the name the Greek physician Dioscorides (40–70 A.D.) used for 'the cress of corruption and ruination,' and was a medieval name for a poisonous buttercup. The word appears to be from the Greek *thlao*, which means 'to compress,' and presumably describes the flattened fruits.

COMMON PEPPER-GRASS • *Lepidium densiflorum*
PRAIRIE PEPPER-GRASS

GENERAL: Grey-green **annual** from slender taproot; stems erect, single, **freely branched**, 15–60 cm tall, with **dense, tiny hairs**.

LEAVES: In rosette at stem base, 1.5–8 cm long, stalked, **toothed to pinnately lobed**, usually absent at flowering time; stem leaves alternate, narrowly lance-shaped, smaller, simpler and stalkless upwards.

FLOWERS: Tiny, **white**; **in many-flowered, dense clusters**, 5–15 cm long; sepals about **1 mm long**; **petals absent** or thread-like, shorter than sepals.

FRUITS: Oblong pods, 2–3.5 mm long, broadly rounded at notched, winged tip; on short stalks; 2 seeds per pod.

WHERE FOUND: Weedy, in dry, open areas and disturbed sites such as cultivated and waste ground; widely scattered across our region (usually near settlements), north and west to interior Alaska.

NOTES: Common pepper-grass could be confused with **branched pepper-grass** (*L. ramosissimum*) in southern parts of the boreal forest. The flowers of branched pepper-grass have tiny white petals; common pepper-grass flowers usually do not have petals. Also, branched pepper-grass seed pods are widest at or below the middle, with pointed teeth at the tips and tiny hairs along the edges; the seed pods of common pepper-grass are widest above the middle, with rounded teeth and no hairs. • All pepper-grasses are edible. In spring, their tender shoots provide a zesty addition to salads, sandwiches and hors-d'oeuvres similar to watercress. Most people prefer the plant raw, but it can also be added to soups, sauces, and casseroles. The pods and seeds of pepper-grass have a stronger taste, but they have also been used. They make a peppery addition to meats and salads, usually mixed with salt and vinegar. The seeds can be collected and dried for year-round use. Pepper-grass is a good source of vitamins and minerals. In the past, it was used by seamen as a source of vitamin C to prevent scurvy. • The name 'pepper-grass' appears to have derived from 'pepper-wort,' a name given to broad-leaved pepper-grass (*L. latifolium*) because of its burning taste.

L. densiflorum

L. ramosissimum

THREE-TOOTHED SAXIFRAGE • *Saxifraga tricuspidata*

GENERAL: Low, loose to densely matted perennial, **often forms large cushions**; stems densely leafy, **often with withered leaves**.

LEAVES: Evergreen, stiff, **leathery**, alternate, overlapping, often reddish tinged, oblong, broadest at tip, wedge-shaped at base, 7–14 mm long, **3 prickly teeth at tip**.

FLOWERS: Few to several in rounded to flat-topped clusters at stem tips; petals creamy **white**, usually maroon- or **orange-dotted**, 4–7 mm long, much longer than sepals.

FRUITS: 2-beaked capsules, 4–8 mm long.

WHERE FOUND: Dry, open, sandy, gravelly, and rocky places; widespread across northern boreal forest, north through arctic islands to Greenland.

NOTES: With over 30 North American species, *Saxifraga* is another of the major circumpolar genera. But, as with *Draba*, most of the species are arctic-alpine, not residents of the boreal forest. However, **yellow mountain saxifrage** (*S. aizoides*) is common at low elevations in the western Northwest Territories and eastern Yukon, typically on the moist silt and gravel of riverflats in limestone terrain. • The leaves of three-toothed saxifrage live for 3 years. After that time the dead leaves remain on the stem for several more years. They provide some protection for the developing winter buds, which helps these hardy plants survive on the exposed rocky or gravelly ridges where they usually grow. • According to the Doctrine of Signatures, saxifrages should be used to cure stones in the bladder, because they grow in rocky fissures. • Three-toothed saxifrage can be grown from seed, and it is an excellent addition to a rock garden. It does best in sunny, well-drained sites, but can tolerate some shade and moisture.

RICHARDSON'S ALUMROOT • *Heuchera richardsonii*

GENERAL: Perennial from stout base and scaly rhizomes; stems bristly, 30–40 cm tall, with gland-tipped hairs towards top.

LEAVES: At stem base, coarse, long-stalked, broadly heart-shaped, 2–6 cm across, nearly hairless above, bristly below; 5–9 shallow lobes edged with broadly egg-shaped teeth.

FLOWERS: In long, narrow clusters on 30–50 cm high leafless stalks, pale purplish; bilaterally symmetrical, about 1 cm long, dotted with glands near base; 5 spoon-shaped petals edged with gland-tipped hairs.

FRUITS: 2-beaked capsules, split in half lengthwise; seeds egg-shaped, flattened on 1 or 2 sides, about 0.7 mm long.

WHERE FOUND: Open woods and meadows; widespread across our region, north to central Manitoba, Great Slave Lake and northeastern British Columbia.

NOTES: To treat diarrhea, the Woods Cree chewed raw alumroot roots or boiled them to make a medicinal tea. A milder tea was used by the Cree as a wash for sore eyes, and the chewed roots were applied to wounds and sores to stop bleeding and speed healing. The Blackfoot mixed the dried root with buffalo fat to make a salve for treating saddle sores. Europeans used alumroot tea to treat fevers and general debility, and as a gargle or injection for sore throats, mouth and throat ulcers and piles. • Alumroot is astringent, so it stops bleeding of the nose, mouth and skin. It should not be used frequently as it can dry the mucous membranes. • The young stems can be used as a cooked vegetable or potherb. • This genus was named after Johann van Heucher (1677–1747), a German medical professor.

BISHOP'S-CAP • *Mitella nuda*
COMMON MITREWORT

GENERAL: Small perennial, with long, slender, creeping rhizomes (often stolon-like); stems erect, **3–20 cm tall**, with fine gland-tipped hairs, usually leafless.

LEAVES: At stem base, **few**, long-stalked, heart–shaped to kidney–shaped, **2–5 cm across**, round-toothed, with scattered, stiffly erect hairs above.

FLOWERS: In **few-flowered clusters** at stem tips, greenish yellow, small, **saucer–shaped**, inconspicuous; petals divided into 4 pairs of thread-like lobes (like television antenna); **10 stamens**.

FRUITS: Capsules, 2–3 mm long, open widely into shallow cups; seeds shiny, black.

WHERE FOUND: Moist forests, thickets and streambanks; widespread across our region, north to southern N.W.T. and southern Yukon.

NOTES: The Woods Cree used bishop's-cap to treat earaches. The leaves were crushed, wrapped in a cloth, and inserted into the ear. • These small plants are easily overlooked. Take a close look at their tiny, unusual flowers. The delicate petals have been reduced to 5 branched, hair-like structures. • Seed dispersal is at least partly by a 'splash-cup' mechanism, where falling drops of water are caught by the cups and 'splash' the small seeds out of the capsule. • The common and Latin names come from the diminutive of *mitra*, which means 'cap' or 'mitre.' Presumably the seed capsule was thought to resemble a bishop's mitre, though one reference suggests that it looks more like 'a tattered French-Canadian toque'! The species name, *nuda*, means 'naked,' in reference to the bare stem.

NORTHERN GRASS-OF-PARNASSUS • *Parnassia palustris*

GENERAL: Hairless perennial from short rhizomes; 1 to several flowering stems, 10–30 cm tall, with 1 heart-shaped, **clasping leaf, usually on lower half**.

LEAVES: Mainly at stem base, **more or less heart-shaped**, 5–20 mm wide; stalks 0.5–10 cm long.

FLOWERS: Single; petals white, showy, up to twice as long as sepals, 8–15 mm long, with 7–9 faint greenish or yellow veins; 5 fertile stamens alternate with broad, sterile stamens with 7–15 slender, gland-tipped segments.

FRUITS: Ovoid, capsules, 8–12 mm long, with many seeds.

WHERE FOUND: Wet, shady places across our region, less common in south; circumpolar.

NOTES: Small grass-of-Parnassus (*P. kotzebuei*) is only 5–20 cm tall, lacks a stem leaf (or has 1 attached near the stem base), and has small flowers in which the white, 3-veined petals are no longer than the 5–7 mm sepals. Small grass-of-Parnassus ranges across the Arctic, and it is also fairly common (but often overlooked) in the northern boreal region on gravelly-sandy shores and banks, and in wet meadows, seeps and thickets. • The flowers of northern grass-of-Parnassus are designed to attract flies. Each flower has five 7–15-pronged sterile stamens, and each prong appears to be tipped with a glistening drop of nectar. However, there is no liquid in these 'drops'; they are false nectaries. The anthers mature first to increase the chance of cross-pollination. Later, when the 2-pronged stigma opens, most of that flower's own pollen is gone. • The faint perfume of these flowers can be smelled on warm days when the plant is in bright sunshine.

GREEN SAXIFRAGE • *Chrysosplenium tetrandrum*
NORTHERN WATER-CARPET

GENERAL: Delicate, **yellowish green**, perennial, 2–10 cm tall (sometimes to 15 cm) from creeping stems; stems erect, often branching from near middle; stolons leafy.

LEAVES: Simple, alternate; lower leaves **round to kidney-shaped**, 3–15 mm wide, shallowly 3–7 lobed, **inconspicuously veined above**, essentially hairless, leaf stalks 1–3 cm long; 1–4 stem leaves, almost kidney-shaped to obovate, lobed, hairless.

FLOWERS: Usually 3–10 in compact, top-shaped cluster, inconspicuous, 2–3 mm across; no petals; 4 sepals, **green**, often purple-dotted, faintly 1-veined, 1 mm wide and long; **4 stamens**, opposite sepals.

FRUITS: Top- to bell-shaped capsules, split open along upper edge to form cups; seeds light reddish brown, 0.6–0.7 mm long, shiny.

WHERE FOUND: Moist, shady sites, especially on rich soil and in wetlands; widespread across northern boreal forest and north to the Arctic, less common in south; circumpolar.

C. tetrandrum

NOTES: Also called *C. alternifolium* var. *tetrandrum*. • **Golden saxifrage** (*C. iowense*) (inset photo) is a similar plant that has larger central flowers (3–5 mm across), golden-yellow sepals with the outer pair wider (rather than green sepals of equal size), 2–8 stamens (rather than 4) and leaves that are conspicuously veined above. It grows in similar habitats in the boreal forest, but is more common than green saxifrage southwards. • The tiny seeds of golden saxifrage are dispersed by the 'splash cup' mechanism. When drops of rain fall into the cup-shaped capsules they splash out the seeds. This mechanism works best when large drops fall vertically from great heights. In the open northern fens and thicket swamps where golden saxifrage is usually found, the rain usually falls obliquely and in small drops. Under these conditions, it is unlikely that the seeds are thrown more then 15 cm from the cup. Larger drops may accumulate on and fall from tall grasses and sedges.

C. iowense

Ranunculaceae (Buttercup Family)

The buttercup family is fairly large—about 1,800 species worldwide—and its members are mainly found in cooler, northern, temperate regions.

The buttercups are mostly annual or perennial herbs, though some *Clematis* species are shrubs or vines. Their leaves are mostly alternate (opposite in *Clematis*), often compound or deeply divided (except *Caltha* species and some *Ranunculus*), and most lack stipules (compare with Rosaceae).

Buttercup flowers are radially symmetrical, except in the *Aconitum* and *Delphinium* species. They have both sepals and petals, or undifferentiated, usually showy, petal-like segments. These segments are distinct from one another and variable in number. The flowers usually have many stamens arranged in a spiral, and several to many superior ovaries, which are also distinct and spirally arranged.

The fruits of some species are 1-chambered follicles or berries, containing several to many seeds, but most members of this family produce single-seeded achenes.

The buttercup family has many important ornamentals, with several species from each of the following genera: *Anemone*, *Delphinium*, *Aconitum*, *Aquilegia*, *Helleborus*, *Thalictrum*, *Paeonia*, *Ranunculus* and *Trollius*. *Aconitum* species are also a source of anti-fever compounds important in internal medicine. Several members of this family are strongly **poisonous**, most notably some *Delphinium* and *Aconitum* species.

See the 'Aquatics' section (pp. 212–226) for descriptions of the aquatic *Ranunculus* species (the water-crowfoots).

Key to the Buttercup Family

1a. Flowers bilaterally symmetrical, deep blue to purple 2

2a. Upper sepal hooded but not spurred; 2 petals, covered by hood .... ***Aconitum delphinifolium*** (see *Delphinium glaucum*)

2b. Upper sepal spurred but not hooded; 4 petals, not hidden by sepals .... ***Delphinium glaucum***

1b. Flowers radially symmetrical, various colours 3

3a. Petals prominently spurred; flowers blue or red 4

4a. Flowers red .... ***Aquilegia canadensis***

4b. Flowers blue .... ***Aquilegia brevistyla***

3b. Petals not prominently spurred; flowers mostly white or yellow 5

5a. 1 ovary; fruits red or white berries .... ***Actaea rubra***

5b. Usually 2 or more ovaries; fruits achenes or follicles 6

6a. Fruits 2- to many-seeded follicles, split when mature 7

7a. Leaves divided into 3s, leathery, evergreen; rhizomes bright yellow .... ***Coptis trifolia***

7b. Leaves simple, not divided into 3s, not leathery or evergreen; rhizomes not bright yellow (the marsh-marigolds) 8

8a. Flowers yellow; plants usually ascending or lying on ground in mud or peat .... ***Caltha palustris***

8b. Flowers white or pinkish; plants usually floating, or creeping on mud .... ***Caltha natans*** (see *C. palustris*)

6b. Fruits 1-seeded achenes, do not split when mature 9

9a. Flowers have both sepals and petals; petals showier than sepals and mostly yellow (the buttercups) 10

10a. Plants aquatic; leaves both floating and submerged, submerged leaves finely dissected .... **aquatic *Ranunculus* spp.** (p. 225)

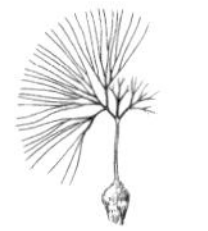

10b. Plants terrestrial; leaves of 1 type, not finely dissected 11

11a. Leaves at stem base toothless to saw-toothed, sometimes shallowly lobed 12

12a. Leaves at stem base linear to spoon-shaped .... ***Ranunculus reptans***

12b. Leaves at stem base rounded, with scalloped edges 13

13a. Plants with creeping stolons .... ***Ranunculus cymbalaria***

13b. Plants upright .... ***Ranunculus abortivus***

11b. Leaves at stem base (at least some of them) deeply lobed or compound 14

14a. Leaves at stem base simple (merely lobed, not compound) 15

15a. 1 leaf (or none) and 1 flower per stem; flower has 3 sepals .... ***Ranunculus lapponicus***

15b. Stems have more than 1 leaf and more than 1 flower; flower has 5 sepals 16

16a. Plants very hairy, especially leaf stalks .... ***Ranunculus acris***

16b. Plants smooth to inconspicuously hairy .... ***Ranunculus sceleratus***

14b. Leaves at stem base compound (divided into leaflets) 17

17a. Flowers 6-8 mm across; petals shorter than sepals; fruiting heads oblong-cylindrical; stems erect .... ***Ranunculus pensylvanicus*** (see *R. macounii*)

17b. Flowers more than 8 mm across; petals longer than sepals; fruiting heads spherical; stems erect to prostrate, root at nodes .... ***Ranunculus macounii***

9b. Flowers have sepals only, or have sepals and less showy petals; sepals often petal-like but seldom yellow 18

18a. Leaves at stem base often absent at flowering time; stem leaves alternate, 2–3 times compound (the meadow rues) 19

19a. Plants have male and female parts in each flower (perfect flowers); achenes flattened .... ***Thalictrum sparsiflorum***

19b. Plants have male and female parts in separate flowers on separate plants; achenes inflated 20

20a. Leaflets 3-lobed, lobes toothless, mostly dark green above, hairy below .... ***Thalictrum dasycarpum***

20b. Leaflets 3–7-lobed, if 3-lobed then lobes themselves usually toothed or lobed, pale green above, prominently veined, not hairy below .... ***Thalictrum venulosum***

18b. All leaves at stem base, palmate, compound or deeply lobed; whorl of bracts below flower cluster (the anemones) 21

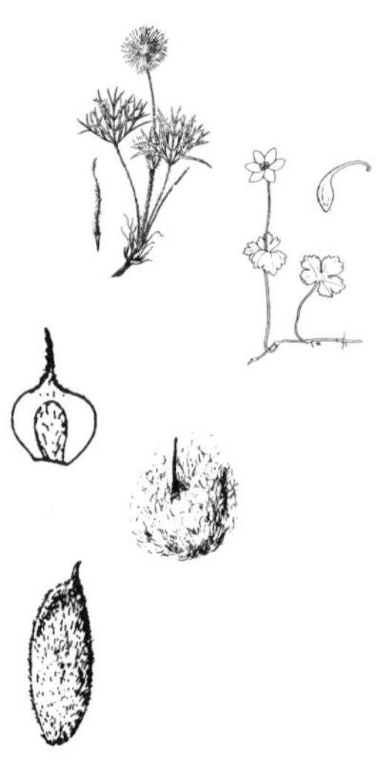

21a. Styles long (1.5–3.5 cm) and feathery at maturity; sepals blue to purplish .... ***Anemone patens***

21b. Styles less than 1.5 cm long, not feathery; sepals creamy to white, or mauve 22

22a. 1 flower, yellow .... ***Anemone richardsonii*** (see *Ranunculus lapponicus*)

22b. 1 to several flowers, creamy to white 23

23a. Achenes hairless to slightly hairy .... ***Anemone canadensis***

23b. Achenes densely hairy or woolly 24

24a. Involucral (stem) leaves on long stalks; flowers white or greenish white 25

25a. Plants 10-20 cm tall, grow from rhizomes; achenes hairy .... ***Anemone quinquefolia***

25b. Plants 20-80 cm tall, no rhizomes; achenes densely woolly 26

26a. Individual flower stems with their own involucres .... ***Anemone riparia***

26b. Individual flower stems primarily without their own involucres, the involucres only below the whole inflorescence .... ***Anemone cylindrica*** (see *A. multifida*)

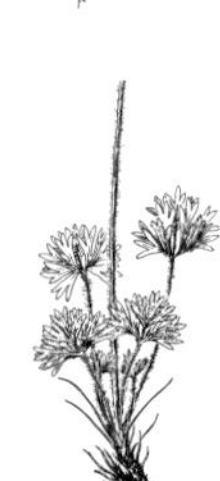

24b. Involucral leaves with short (or no) leaf stalks; flowers various colours (purple, pink, white or yellowish green) 27

27a. Leaves with 3 wedge-shaped leaflets .... ***Anemone parviflora*** (see *A. multifida*)

27b. Leaves 2–4 times narrowly divided into 3s .... ***Anemone multifida***

PRAIRIE CROCUS • *Anemone patens*
PASQUE FLOWER

GENERAL: Tufted perennial from stout branched rhizome, 10–40 cm tall, with silky hairs throughout.

LEAVES: Several at stem base, with **silky hairs**, **appear after flowers**, long-stalked, divided in 3s; divisions further divided into linear or lance-shaped segments; leaves on flowering stem similar but smaller and stalkless.

FLOWERS: Blue, purple or nearly white; 5–7 sepals, 2–4 cm long, **hairy on back**; no petals; **stems lengthen considerably in fruit**.

FRUITS: Achenes, 2–4 cm long, **with feathery style**, in heads.

WHERE FOUND: Prairies, hillsides and dry open woods; across our region, but most common in prairie and parkland; north to Arctic coast and Banks Island, N.W.T.

NOTES: Also called *Pulsatilla ludoviciana* and *P. patens*. • All parts of the prairie crocus contain compounds that are **poisonous** if taken internally and **irritating** externally. These plants can raise blisters on sensitive mucous membranes. • The wind carries the achene by its parachute-like, feathery style. When the achene lands, the style twists and untwists with changes in humidity, and these movements may help the achene work its way into the ground. • This beautiful spring flower is the floral emblem of both Manitoba and South Dakota. It is seldom abundant in the boreal forest, but in the prairie regions to the south it can form vast sheets of mauve early in the growing season. Dense stands of prairie crocus are an indicator of overgrazing. • The Blood apparently used prairie crocuses to play jokes on people. If it is used for toilet paper it can burn or sting, and may even cause the skin to peel. Some native peoples are said to have stuffed the sepals up the nose to stop bleeding or bound them over boils and sores to draw out infection, but this must have been very irritating. Prairie crocus was also crushed and applied externally to relieve rheumatism, as a counter-irritant.

CUT-LEAVED ANEMONE • *Anemone multifida*

GENERAL: Hairy, tufted perennial from thickened stem bases; 1–7 stems **15–50 cm tall**.

LEAVES: Most at stem base, long-stalked, **2 or 4 times narrowly divided into 3s**; stem leaves in whorl below flower cluster, short-stalked but otherwise similar to basal leaves.

FLOWERS: Single or less commonly in clusters of 2–4 long-stalked, fairly showy, **creamy white, yellowish, or pinkish** and often tinged with red, blue, or purple on the outer surface; sepals 5–10 mm long; no petals.

FRUITS: Silky-woolly achenes in egg-shaped or spherical heads 1 cm wide.

WHERE FOUND: Dry to moist open woods and meadows; widespread across our region; occasionally north past treeline.

A. multifida

A. parviflora

NOTES: Long-fruited anemone (*A. cylindrica*) (inset photo) is similar to cut-leaved anemone, but has long-stalked, less finely divided leaves below the flower cluster, 2–6 flowers, and cylindrical heads of achenes. It grows in grassy openings and open forest along the southern edge of our region. • **Small wood anemone** (*A. parviflora*), also known as northern anemone, has leaves with 3 broad wedge-shaped segments, much less divided than the other anemones with woolly fruits. A wide-ranging subarctic species, it is found in open spruce woods and along river flats from Newfoundland to Alaska, north to the arctic islands. • The Blackfoot burned the 'cotton' of the ripe seed heads on hot coals and inhaled the smoke to relieve headaches. • Anemones contain ranunculin, a harmless glycoside that produces protoanemonin, a volatile, strongly irritant, unstable oil. The leaves of most species are irritating, and have been boiled to make a strong tea used to kill fleas and lice.

TALL ANEMONE • *Anemone riparia*
RIVERBANK ANEMONE

GENERAL: Perennial, 30–80 cm tall; **stems coarse**, stiffly erect, sparsely hairy.

LEAVES: Several at stem base, long-stalked, broader than long, **deeply 3–5-parted**, each part deeply notched and sharply toothed; 3 stem leaves below flowers, similar to leaves at stem base, but smaller.

FLOWERS: 1–3 on long stalks from stem tips, greenish white or white; 4–6 sepals, 7–20 mm long; no petals.

FRUITS: Densely woolly achenes in slenderly egg-shaped to nearly cylindrical **heads**; style remains at tip.

WHERE FOUND: Moist woods and shaded thickets; fairly common in southern boreal forest and northern parkland west into British Columbia.

NOTES: Anemone was planted over graves in China, and it was widely associated with grief and suffering and even regarded as dangerous. • Because it is related to the poisonous delphinium and the acrid buttercups, anemones are **potentially dangerous** and should probably be avoided. • Some say the name 'anemone' comes from the Semitic word *Na'aman*, 'handsome,' an epithet for Adonis, the lover of Aphrodite. The Greek poet Bion wrote that anemones first sprang from the tears of Aphrodite as she wept over the body of her slain lover. Other sources report it comes from the Greek, *anemo*, 'wind,' because it was supposed that the flowers did not open until beaten by the wind. 'Wind-flower' is another common name for this group of plants. • The species name *riparia*, meaning 'growing beside rivers,' is somewhat inappropriate because in our region tall anemone is most often found in wooded areas rather than along watercourses.

CANADA ANEMONE • *Anemone canadensis*

GENERAL: Tufted perennial from short, slender rhizome, 20–50 cm tall; stems erect, hairy.

LEAVES: Several at stem base, long-stalked, broader than long, **strongly veined, deeply 3–5-parted**; divisions mostly 3-cleft and sharply toothed; stem leaves below flower similar but stalkless and more deeply divided.

FLOWERS: Single on long stalk from set of bracts, white; 5 sepals, 1–2 cm long; no petals; secondary sets of bracts form in extra-healthy plants, each bearing 1 flower.

FRUITS: Achenes, beaked, hairy, flat, in spherical heads.

WHERE FOUND: Damp meadows, thickets, shores and ditches; across our region, north and west to Fort Simpson, N.W.T.

NOTES: Canada anemone is one of the most common anemones in our region, east of the Rockies. It often forms large patches at the edges of woodlands and in low moist areas. • The roots and leaves of Canada anemone have been used as poultices or made into teas for washes to stop nosebleeds and to treat wounds and sores. The tea was even used as an eyewash to cure the twitching of crossed eyes. In some native groups, singers would chew the roots to clear their throats before performing. Some Plains Indians held Canada anemone roots in such high esteem as an external medicine for many ailments, that they attributed mystical powers to them. However, it is not wise to use anemones medicinally. All of our anemones contain some of the **caustic irritants** so prevalent in the buttercup family, and therefore they could be more harmful than helpful.

WOOD ANEMONE • *Anemone quinquefolia*

GENERAL: Delicate perennial from slender, horizontal rhizome, 10–20 cm tall; stems nearly hairless.

LEAVES: Long-stalked at stem base, with **3–5 leaflets**, coarsely and unevenly toothed or deeply notched; **appear after flowering stem**; 3 leaves below flower, **stalked**, each with 3–5 leaflets, deeply notched or divided.

FLOWERS: White; 1 on slender hairy stalk; 4–9 sepals (commonly 5), 1–2 cm long; no petals.

FRUITS: **Achenes**, with **short hairs**; stiff, hooked beak at tip, 1–2 mm long; in spherical heads.

WHERE FOUND: Moist woods; boreal forest and parkland of southern Manitoba and eastern Saskatchewan; rare in Alberta.

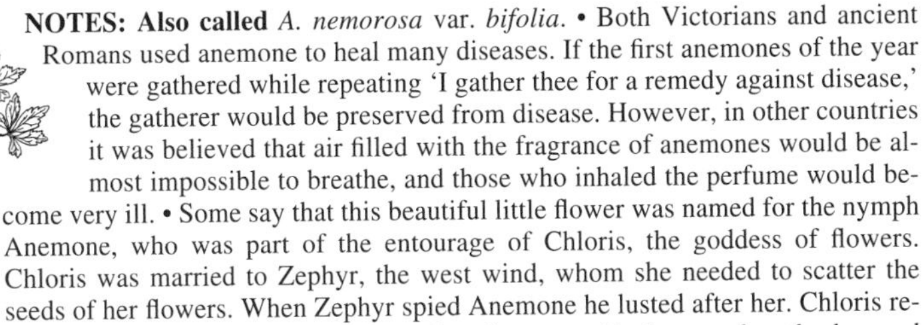

NOTES: Also called *A. nemorosa* var. *bifolia*. • Both Victorians and ancient Romans used anemone to heal many diseases. If the first anemones of the year were gathered while repeating 'I gather thee for a remedy against disease,' the gatherer would be preserved from disease. However, in other countries it was believed that air filled with the fragrance of anemones would be almost impossible to breathe, and those who inhaled the perfume would become very ill. • Some say that this beautiful little flower was named for the nymph Anemone, who was part of the entourage of Chloris, the goddess of flowers. Chloris was married to Zephyr, the west wind, whom she needed to scatter the seeds of her flowers. When Zephyr spied Anemone he lusted after her. Chloris realized what was happening and banished Anemone from her court. Zephyr was heartbroken and asked Aphrodite to change the nymph into a flower, which she did. Alas, the gods are fickle, and Zephyr soon tired of waiting for Anemone to bloom each year, but Boreas, the north wind, fell in love with her; he forced her blossoms open and she faded immediately.

YELLOW MARSH-MARIGOLD • *Caltha palustris*

GENERAL: Perennial from coarse, fleshy roots; stems ascending or prostrate with erect tips, smooth, **hollow**, branched towards top, 20–60 cm long.

LEAVES: Long-stalked at stem base, **kidney-shaped to circular**, broadly heart-shaped at base, the 2 lobes touch or overlap; edges toothed to nearly smooth; upper leaves similar but stalkless or nearly so.

FLOWERS: Solitary on short or lengthened stalks from axils of stem leaves, **bright yellow**, 1.5–4 cm across; 5–9 sepals, broadly spoon-shaped or elliptic; no petals.

FRUITS: Cluster of 6–12 many-seeded, curved pods (follicles), (each) 10–15 mm long with **backwards-curving beak**.

WHERE FOUND: Wet meadows, marshes, fens, swamps, ditches and shallow water; often among grasses or sedges; widespread across our region; circumpolar.

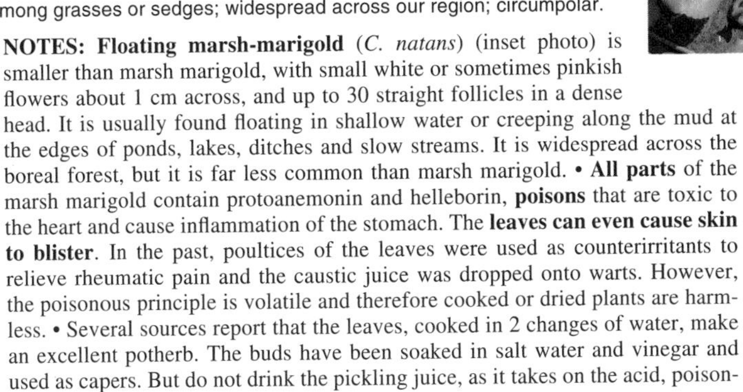

C. palustris

C. natans

NOTES: Floating marsh-marigold (*C. natans*) (inset photo) is smaller than marsh marigold, with small white or sometimes pinkish flowers about 1 cm across, and up to 30 straight follicles in a dense head. It is usually found floating in shallow water or creeping along the mud at the edges of ponds, lakes, ditches and slow streams. It is widespread across the boreal forest, but it is far less common than marsh marigold. • **All parts** of the marsh marigold contain protoanemonin and helleborin, **poisons** that are toxic to the heart and cause inflammation of the stomach. The **leaves can even cause skin to blister**. In the past, poultices of the leaves were used as counterirritants to relieve rheumatic pain and the caustic juice was dropped onto warts. However, the poisonous principle is volatile and therefore cooked or dried plants are harmless. • Several sources report that the leaves, cooked in 2 changes of water, make an excellent potherb. The buds have been soaked in salt water and vinegar and used as capers. But do not drink the pickling juice, as it takes on the acid, poisonous character of the buds.

SMALL-FLOWERED BUTTERCUP • *Ranunculus abortivus*

GENERAL: Biennial or short-lived perennial with slender fibrous roots, 15–50 cm tall; stems usually single, erect, fleshy, smooth, simple or freely branching towards top.

LEAVES: Basal, usually simple, long-stalked, rounded egg-shaped to kidney-shaped, 1–4 cm long, heart-shaped at base; edges toothed to shallowly lobed; stem leaves short-stalked or stalkless, commonly divided into 3 narrowly lance-shaped, smooth-edged segments.

FLOWERS: On stalks to 10 cm long from axils of stem leaves, yellow; 5 sepals, 2.5–5 mm long, greenish yellow, **usually purple-tinged**, spreading or turned downwards, **soon fall off**; 5 petals, **shorter than sepals**.

FRUITS: Hairless achenes about 1.5 mm long, with minute beak, 20–40 in spherical or egg-shaped heads.

WHERE FOUND: Open woods, wet meadows, streambanks and other moist places; fairly common across prairie and parkland; less common in boreal forest north to central Yukon.

NOTES: Small-flowered buttercups growing along the edges of streams or sloughs and in wet spots have large, branched flower clusters with many small flowers. In drier, less favourable spots they are more erect and have fewer flowers in less-branched clusters. These 2 phases can be mistaken for different species. • Buttercups can **blister or inflame skin**. If eaten, they will cause **severe problems** in the digestive tract. Young plants contain only about 1/6 as much protoanemonin (the poisonous compound found in buttercups) as mature plants. Boiling or drying may destroy this, but it is safest not to eat these plants at all. Buttercup poisoning of animals is rare, because the plants are strongly distasteful. • Small-flowered buttercup has the smallest flowers of any of the buttercups.

CELERY-LEAVED BUTTERCUP • *Ranunculus sceleratus*

GENERAL: Annual or short-lived perennial from slender, fleshy roots, 20–60 cm tall; stems stout, erect, **hollow**, usually hairless, freely branched towards top.

LEAVES: Long-stalked at stem base, fleshy, kidney-shaped in outline, 2.5–4 cm long, deeply 3-parted, divisions again divided or lobed; stem leaves alternate, lower leaves more deeply notched or divided than at stem base; upper stem leaves smaller, nearly stalkless, in 3 narrow segments or undivided.

FLOWERS: Many, on short stalks from upper leaf axils, pale yellow; 5 sepals, turned downwards, soon fall off; 5 petals, 2–4 mm long, **shorter than or as long as sepals**.

FRUITS: Achenes, hairless, about 1 mm long, with minute beak and narrow ridges, 100–250 in cylindrical heads.

WHERE FOUND: Ditches, marshes, ponds and lakeshores; often semi-aquatic; common and widespread across our region.

NOTES: Often a dominant plant along the edges of drying sloughs. Sometimes called 'cursed crowfoot' because of its **burning, poisonous juice**, which blisters the skin and produces intestinal inflammation if eaten. Celery-leaved buttercup has high concentrations of protoanemonin, especially at flowering time, and is considered one of our **most toxic** buttercups. • Some buttercups have been used as emergency food in different parts of the world. Young tops and roots have been boiled and then the poison-bearing water poured off. Young flowers have been preserved in vinegar as pickles (but do not drink the poison-containing vinegar). Also, the seeds have been parched and ground to make meal to add to bread. However, these plants should not be eaten except in extreme emergencies.

TALL BUTTERCUP • *Ranunculus acris*
MEADOW BUTTERCUP

GENERAL: Tall, **hairy** perennial from fleshy, fibrous roots; **1 to several stems, erect, branched, hollow**, 30–80 cm tall.

LEAVES: Long-stalked at stem base, blades broadly 5-sided to heart-shaped in outline, palmately **3–5-lobed**; each lobe divided 2–3 times into narrow, sharp-pointed segments, giving leaves ragged appearance; stem leaves mostly short-stalked and 3-lobed, reduced to 3–5-lobed bracts near stem tip.

FLOWERS: Several in loose clusters at stem tips, glossy yellow, **15–35 mm across**; sepals greenish, hairy, soon fall off; 5 petals; stalks long, slender, hairy.

FRUITS: Smooth achenes, 2–3 mm long, with tiny, **flattened, curved beaks**, 25–40 in spherical clusters.

WHERE FOUND: Damp meadows, clearings and roadsides; **introduced and naturalized** across much of North America; north to southernmost N.W.T. in our region.

NOTES: Fresh leaves of tall buttercup were used to irritate and redden the skin when treating rheumatism, arthritis, and neuralgia (severe pain along a nerve). Native peoples used the roots as a poultice on boils and abscesses, and they often added it to salves. **Warning:** The sap of these plants can cause intense pain and burning of the mouth and mucous membranes and can raise blisters on the skin. The poisonous material is volatile and dried plants are said to be harmless. • The species name *acris* refers to the acrid or burning nature of these plants.

MACOUN'S BUTTERCUP • *Ranunculus macounii*

GENERAL: Annual to short-lived perennial from thick, somewhat fleshy roots, 30–70 cm tall; stems coarse, branched, leafy, with long, bristly hairs, erect or sometimes lying on ground and rooting at nodes.

LEAVES: Long-stalked at stem base, divided into 3 stalked and deeply toothed segments, usually stiff-hairy; usually only 1 or 2 stem leaves, reduced to much simpler bracts below flowers.

FLOWERS: Single, on stalks to 6 cm long at stem tips, yellow; 5 sepals, yellowish green, turned downwards, soon fall off; 5 petals, 4–6 mm long, **as long as or slightly longer than sepals**.

FRUITS: Achenes, smooth, hairless, about 3 mm long with sharp, stout, nearly straight beak about 1/4 length of body, 30–60 in spherical or egg-shaped heads.

WHERE FOUND: Moist to wet thickets, woods and meadows; common across our region north and west to interior Alaska.

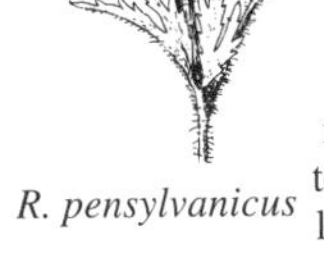

R. pensylvanicus

NOTES: Bristly buttercup (*R. pensylvanicus*) is similar to Macoun's buttercup, but it tends to be a more upright plant; its petals are shorter than the sepals, and its head of achenes is cylindrical rather than spherical. Bristly buttercup is far less common, occurring in wet places in the northern parkland and boreal forest north and west to the southern Northwest Territories and interior Alaska. • The same compounds that make buttercup plants poisonous could be useful if applied externally. Anemonin and protoanemonin are said to have antibiotic action against a broad spectrum of bacteria. Therefore, dried plants could be used to treat cuts and abrasions. They have also been applied to aching teeth and to muscles or joints affected by rheumatism. • If you want to find out if someone likes butter, just hold a buttercup under his or her chin. If the chin appears yellow, you have a butter-lover; if not, you may as well buy margarine. • This species is named in honour of John Macoun, Canada's foremost botanist in the late 19th century.

SHORE BUTTERCUP • *Ranunculus cymbalaria* SEASIDE BUTTERCUP

GENERAL: Low, tufted, somewhat fleshy, perennial with **strawberry-like runners** that root at nodes and produce new plants.

LEAVES: At stem base, 5–15 mm long (sometimes to 25 mm); egg-, heart- or kidney–shaped; **edges scalloped or shallowly lobed**; stalks relatively long.

FLOWERS: Few, on 5–20 cm long stalks, yellow, **5–10 mm across**; 5 sepals, greenish, hairless, soon fall off; 5 petals, 3–5 mm long, slightly longer than sepals.

FRUITS: Achenes, 2 mm long, hairless, with **length-wise grooves** and **short, straight beak**, numerous in cylindrical clusters.

WHERE FOUND: Shores of ponds, lakes, rivers and streams and in wet meadows; widespread across our region.

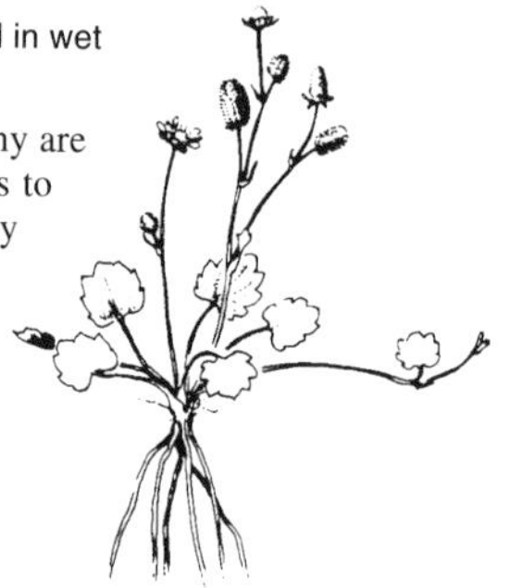

NOTES: All buttercups have **poisonous, burning sap** and many are very irritating to the skin. • In the past beggars used buttercups to create and maintain wounds, as a means of gaining sympathy and, hopefully, a few coins as well.• Pliny, a Roman naturalist who lived from 23–79 A.D., reported that buttercups had merit, in that they would 'stir the eater into such a gale of laughter that he scarce contains himself; in fact unless he drinks pine-apple kernels and pepper in date wine, he may guffaw his way into the next world in a most unseemly manner.'

LAPLAND BUTTERCUP • *Ranunculus lapponicus*

GENERAL: Perennial from slender, creeping roots; stems single, 10–20 cm tall.

LEAVES: 1 (rarely 2) at stem base, long-stalked, **kidney-shaped**, 2–4 cm wide, **deeply 3-parted**; segments wedge-shaped to broadly spoon-shaped; edges coarsely toothed to shallowly lobed; 1 stem leaf, small, sometimes absent.

FLOWERS: Solitary on long stalks, yellow or whitish, about 10 mm across, sweet-scented; 3 sepals; 6–10 petals.

FRUITS: Achenes, about 3 mm long, with long hooked beak almost as long as achene, 3–15 in spherical head.

WHERE FOUND: Wet mossy woods and sphagnum bogs; widespread across boreal forest.

NOTES: Yellow anemone (*Anemone richardsonii*) has deeply 5-cleft leaves at the stem base that look like those of Lapland buttercup, and it has solitary flowers. But yellow anemone also has sharply-toothed leaves below the flower cluster, and its flowers are bigger—15–25 mm across. Yellow anemone is widespread in wet meadows, moist thickets and swamps across the northern boreal forest. • The habitat of Lapland buttercup is unlike that of most of the other buttercup species. Its roots are frequently deeply buried in moss, and its stems root readily from the nodes. • Some Lapland buttercups bloom very early in the spring. The flowers have been known to start opening while still under 10 cm or more of snow. • The name *Ranunculus* is derived from the Greek word *rana*, frog, probably in reference to the marshy home of many buttercups.

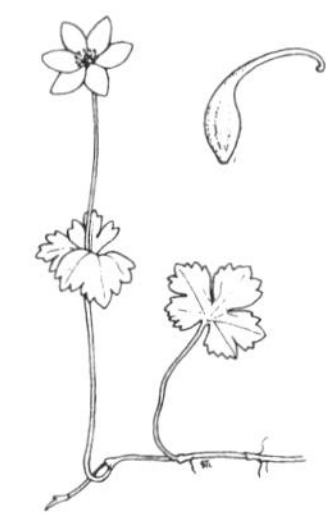

A. richardsonii

R. lapponicus

CREEPING SPEARWORT • *Ranunculus reptans*
LESSER SPEARWORT

GENERAL: Creeping amphibious perennial; stems slender, lie flat on ground, **root freely at nodes**, up to 50 cm long.

LEAVES: Simple, smooth-edged, 1–3 cm long, **highly variable in shape** from linear to narrowly spoon-shaped; stalked at stem base, short-stalked or stalkless higher on stem.

FLOWERS: Single, on short bract-bearing stalks from leaf axils, yellow; usually 5 petals, 3–5 mm long; 5 sepals, **hairy, about as long as petals**.

FRUITS: Achenes, hairless, 2–3 mm long, with very short curved beak; 5–20 in nearly spherical heads.

WHERE FOUND: Edges of rivers, ponds and lakes; widespread but uncommon across our region; circumpolar.

NOTES: Also called *R. flammula*. • Creeping spearwort is a variable species, as is often the case in amphibious plants. It has been divided into a number of varieties based on the shape and size of the leaves. • On the Isle of Skye and in many parts of the Scottish Highlands, lesser spearwort was used to raise blisters. To do this, the leaves are well bruised in a mortar and then applied in 1 or more limpet shells to the skin where the blister is being raised. This treatment was also used in the 14th century under the name of 'flame' for curing 'cankers' (ulcers). An alcoholic extract of lesser spearwort was thought to cure ulcers. • The name 'spearwort' aptly describes the shape of the leaves. Also called 'banewort' in the old days, because it was supposed to 'bane' (poison) sheep by causing ulcerated entrails.

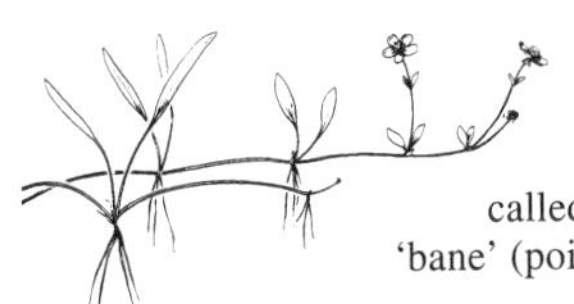

GOLDTHREAD • *Coptis trifolia*

GENERAL: Low perennial (5–15 cm tall) with slender, **thread-like, bright yellow rhizomes**.

LEAVES: All **at stem base, evergreen, shiny**, slenderly stalked, **divided into 3 short-stalked leaflets** which are egg-shaped with a wedged base, slightly 3-lobed and **sharply toothed**.

FLOWERS: Single on slender stalks; white with yellowish base, 10–15 mm across; 5–7 sepals, narrowly elliptic to lance- or spoon-shaped; 5–7 petals, **reduced to small, fleshy, club-shaped staminodia, nectar-bearing at tip**.

FRUITS: 3–9 long-stalked, spreading pods (follicles), 5–10 mm long, with **prominent, erect beak** 2–4 mm long; seeds large, shiny, black.

WHERE FOUND: Damp, mossy woods; scattered across boreal forest north to southernmost N.W.T.

NOTES: Also called *C. groenlandica*. • Although widespread, goldthread is not commonly encountered in our region. However, where it occurs there are often large collections of many individual plants. • Goldthread is said to relieve the craving for alcohol and it has been used in preparations to combat alcoholism. The roots are very astringent and were either chewed or boiled to make medicinal teas for sore throats, canker sores and other mouth irritations, which earned it the name 'canker-root.' It was also used as an eyewash and as a topical anesthetic on the gums of teething children. • Goldthread contains the alkaloid berberine, which is anti-inflammatory, antibacterial, astringent, anticonvulsant, immunostimulant and a mild sedative. Goldthread was listed in the U.S. Pharmacopoeia from 1820–82, and at the turn of the century 500 g of the root fetched about $1. Today, these attractive, shade-tolerant plants are grown as ground cover in wooded areas. • The genus name comes from the Greek *kopto*, 'to cut,' in reference to the compound leaves. The name 'goldthread' comes from the distinctive, fine, yellow rhizomes.

RED AND WHITE BANEBERRY • *Actaea rubra*

GENERAL: Perennial from fleshy rhizomes, essentially hairless; 1 to several stems, erect, **0.3–1 m tall**, branched, leafy.

LEAVES: Alternate, few, all from stem, large, **2–3 times divided in 3s; segments coarsely sharp-toothed and lobed**, 2–10 cm long.

FLOWERS: In **many-flowered, rounded clusters** on long stalks, **white**; sepals and petals 2–3.5 mm long, soon fall off.

FRUITS: Several-seeded, glossy red or white **berries**, 6–8 mm long; **poisonous**.

WHERE FOUND: Moist woods across our region, north and west to Great Bear River and western Alaska.

NOTES: All parts of red and white baneberry, especially the roots and berries, are **poisonous**. As few as 6 berries can cause severe cramps, headache, vomiting or dizziness that last for hours. European species are said to have caused deaths among children, but no deaths have been reported from North America. Although the berries are poisonous to humans, many small birds and animals eat them. • The Cree used a tea made with a small piece of the root to slow heavy menstrual flow. The root tea was also used by settlers and native peoples to relieve uterine pains, especially during and after childbirth. To the east, the Chippewa used only plants with white berries for women's complaints, while those with red berries were used for men's diseases. The Cree boiled the root with the tops of 'spruce fir' to make a medicinal tea to treat stomach problems. They also used the plant, particularly the fruit, as a strong purgative, which it most certainly is. Some native groups used a decoction of baneberry root to treat sick horses. • Baneberry is easily grown in cooler climates and makes a handsome addition to northern gardens. • The common name 'baneberry' refers to the plant's severely poisonous nature and comes from the Anglo-Saxon word *bana*, meaning 'murderous.'

TALL LARKSPUR • *Delphinium glaucum*

GENERAL: Stout perennial, from **thick, short, tough rootstock**; **usually several hollow stems**, unbranched below flowers, 1–2 m tall (sometimes only 0.3 m), usually hairless, with bluish white bloom.

LEAVES: Many, alternate, 15–20 cm wide, smaller higher on stem, palmately 5–7 lobed, hairless to sparsely hairy, with bluish white bloom; lobes sharply lobed or toothed; stalks 1.5–19 cm long.

FLOWERS: In simple to branched, long, loose, many-flowered clusters at stem tips; **deep blue to purple**, 6–12 mm long, **with prominent, straight spurs**.

FRUITS: Erect, hairless to slightly hairy, brownish, many-seeded pods (follicles), usually 12–14 mm long.

WHERE FOUND: Moist open woods, thickets, meadows, and streambanks; in our region from Saskatchewan to Alaska; common in northwest.

NOTES: Mountain monkshood (*Aconitum delphinifolium*) also has palmately lobed, deeply cut leaves, but its 1–10 deep blue to purple flowers have a distinctive hooded upper sepal. Monkshood grows in moist meadows (especially subalpine) in the boreal mountains. • Tall larkspur contains **poisonous** alkaloids that can cause upset stomach, nervousness, depression, and even suffocation and death if eaten in large quantities. Toxicity decreases as the plants age, reaching half strength at the time of flowering and 1/16 by the time the fruits have matured (although the **seeds are very toxic**). Many cattle have been poisoned, but sheep and some wildlife can tolerate high levels of these toxins, so tall larkspur can provide fair to good forage for these animals. • The seeds and flowers have been used to make preparations to kill lice, fleas and similar pests. These preparations should never be applied if the skin is broken or abraded as enough of the alkaloids could be absorbed to make a person ill. Tall larkspur was one of the many unsuccessful 'cures' for tuberculosis. • Burning, tree clearing, or overgrazing of range containing tall larkspur can lead to great increases in the density of this species.

BLUE COLUMBINE • *Aquilegia brevistyla*
SMALL-FLOWERED COLUMBINE

GENERAL: Perennial from stout, fibrous taproot; stems simple, slender, erect, 20–50 cm tall (sometimes to 80 cm), little branched, sparsely hairy and gland-bearing towards top.

LEAVES: Alternate, mainly at stem base, long-stalked, twice divided in 3s, **often bluish tinged**, paler and with bluish white bloom below; stem leaves few, nearly stalkless, lobed.

FLOWERS: Usually 2–3, spreading to drooping, **12–20 mm long, blue and cream**; with 5 bluish, **hooked spurs**, 6–8 mm long.

FRUITS: Usually 5, erect, hairy pods (follicles), 2–2.5 cm long.

WHERE FOUND: Open woods, meadows and rocky slopes; widespread across our region; west to interior Alaska; uncommon in south and east.

NOTES: The slender, backwards pointing spurs contain the fragrant nectaries of these beautiful flowers. The spurs are so long that only long-tongued insects and hummingbirds can reach the nectar, so they are responsible for pollinating the flowers of blue columbine. • The common name is derived from the Latin *columbina* meaning 'dove-like.' Look carefully at the flower. Can you see the 5 tiny doves perched around the rim of a central drinking dish? The flower's sepals form their wings and the spurs are their heads and necks. A quintet of doves arranged in a ring around a dish was a favourite device of ancient artists.

CANADA COLUMBINE • *Aquilegia canadensis*
WILD COLUMBINE

GENERAL: Perennial, 30–80 cm tall; stems stout, erect, branched towards top, hairless or nearly so.

LEAVES: On slender stalks at stem base and low on stem, twice divided in 3s; leaflets 3-lobed, wedge-shaped at base, bluntly notched, pale beneath; upper stem leaves lobed or divided, stalkless.

FLOWERS: Large, nodding, 2.5–5 cm long; 5 scarlet sepals; 5 yellow petals, **lengthened into slender, nearly straight spur**, thickened at end; **stamens and styles protrude from flower**.

FRUITS: Erect clusters of slightly spreading pods (follicles), **tipped with slender beak nearly as long as pod**.

WHERE FOUND: Thickets and open woods; fairly common in boreal forest from southern Manitoba to eastern Saskatchewan.

NOTES: Canada columbine has been used medicinally in the past, but all columbines are probably somewhat **poisonous**. In Europe, results were described as 'very unsatisfactory,' and children sometimes died from doses of columbine seeds. • In North America, the young plants and the roots were eaten by some native groups. The roots were also chewed or made into tea to increase perspiration and to treat urinary, stomach, or bowel troubles. Small amounts of the seeds were crushed in hot water and taken for headaches and fevers and as a love charm. The ripe seeds were also rubbed into the hair to control lice. • The genus name comes from the Latin *aquila*, an eagle, whose claws were likened to the tubular spurs of these flowers.

VEINY MEADOW RUE • *Thalictrum venulosum*

GENERAL: Perennial, 20–70 cm tall; stems erect, little branched, usually hairless.

LEAVES: Alternate, compound, mostly along stem, long-stalked; twice divided into 3s, leaflets stalked, with toothed lobes, 1–2 cm long, rounded or wedge-shaped at base, **often bluish green, strongly veined and pale below**; upper leaves stalkless.

FLOWERS: In a narrow, dense, many-flowered cluster at stem tip, small, with **sexes on separate plants**; 4 or 5 sepals, greenish, 2–4 mm long, soon fall off; no petals.

FRUITS: Oblong-egg-shaped, hairless achenes, 3–5 mm long, **distinctly ribbed**; **tipped with beak about 2 mm long**, in small clusters.

WHERE FOUND: Moist prairie, thickets and open woods; fairly common across our region, north to lower Mackenzie valley.

NOTES: The seeds and leaves of these plants keep their pleasant fragrance, and they may have been stored with clothing and other belongings as insect repellent. • The genus name comes from the Greek *thallo*, 'I flourish,' supposedly in reference to the plant's ability to spread rapidly once established. The species name, *venulosum*, refers to the prominent veins on the underside of the leaves.

female

male

TALL MEADOW RUE • *Thalictrum dasycarpum*

GENERAL: Robust perennial, 0.6–1.2 m tall; stems stout, simple or commonly branched towards top, **often purplish**, hairless.

LEAVES: Alternate, compound, large, mostly along stem; twice divided in 3s; leaflets stalked, egg-shaped, 2–3 cm long, rounded to wedge-shaped at base, **3-toothed or lobed at tip**, dark green above, pale, **strongly veined and generally short-hairy below**; upper leaves stalkless.

FLOWERS: In many-flowered, dense clusters at stem tips, small; **sexes on separate plants**; 4 or 5 sepals, greenish white, soon fall off.

FRUITS: Achenes, **brownish hairy**, 4–6 mm long, strongly ribbed, tipped with beak about 3 mm long; in clusters of about 10.

WHERE FOUND: Damp meadows, thickets and rich, moist woods; across southern boreal forest and northern parkland of prairie provinces; rare in Alberta, more common in east.

NOTES: This is the largest of the meadow rue species found in our region. • Blackfoot girls believed that meadow rue was a powerful love medicine. They tied bunches of the flowers or seeds in their hair, in the hope of capturing the heart of the first attractive male who saw them. Other native groups also used these plants in love potions. The seeds were hidden in the food of quarrelling couples, so that love would triumph. • The seeds were also used medicinally in poultices to relieve cramps. Root tea was taken to reduce fevers, or often the roots were chewed and the juices swallowed to reduce phlegm, improve circulation and relieve heart palpitations, diarrhea, and vomiting. Occasionally meadow rue roots were used as a cure-all. The roots of tall meadow rue were listed in the U.S. Dispensatory in 1916 as a medicine for purging the system and increasing urination. • The species name, *dasycarpum*, is from the Greek *dasys*, 'hairy,' and *karpos*, 'fruit.'

FLAT-FRUITED MEADOW RUE • *Thalictrum sparsiflorum*
FEW-FLOWERED MEADOW RUE

GENERAL: Slender perennial from rhizomes with yellow roots; stems erect, 30–100 cm tall, branched towards top.

LEAVES: Mainly on stem, roughly triangular, 6–20 cm long, 2–3 times divided in 3s; leaflets firm, round to wedge-shaped, 5–20 mm long, 3-lobed to coarsely round-toothed, stalked, glandular below.

FLOWERS: Few to many in loose clusters from leaf axils and stem tips; small, **perfect** (bisexual); no petals; 4–5 sepals, **whitish, petal-like**, soon fall off.

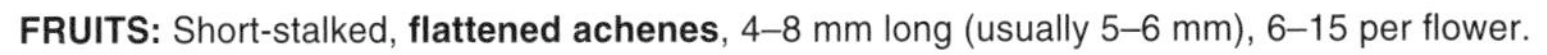

FRUITS: Short-stalked, **flattened achenes**, 4–8 mm long (usually 5–6 mm), 6–15 per flower.

WHERE FOUND: Moist meadows, thickets and woods; scattered across boreal forest from northern Manitoba to Alberta, central Yukon and western Alaska; also in northeastern Asia (amphi-Beringian).

NOTES: On the plains, the Cheyenne used meadow rue flowers or dried, ground plants as a stimulant for their horses, to make them more spirited, long-winded and enduring. The plants and flowers were also used as a perfume. • The species name *sparsiflorum* means 'few-flowered.' 'Rue' is derived from the Latin *ruta*, which means 'bitter leaved.'

Buttercup Family Silhouettes

Ranunculus reptans p. 125

Caltha palustris p. 121

Ranunculus abortivus p. 122 (see also *R. cymbalaria* p.124)

Coptis trifolia p. 125

Anemone parviflora p. 119 (see also *R. lapponicus* p. 124)

Ranunculus scleratus p. 122

Anemone riparia p. 120

Ranunculus macounii p. 123

Anemone cylindrica p. 119

Ranunculus acris p. 123

Delphinium glaucum p. 126

Aquilegia spp. p. 127

Thalictrum venulosum p. 128

Thalictrum dasycarpum p. 128

Thalictrum sparsiflorum p. 129

Rosaceae (Rose Family)

The rose family is a large family—nearly 3,400 species worldwide—and its members include trees, shrubs (often thorny) and herbs.

Members of the rose family have alternate, simple or pinnately compound leaves, usually with **stipules** at the base of the leaf stalk. The flowers are radially symmetrical, usually with 5 sepals that join at the base, 5 petals that arise from a cup- or saucer-like **hypanthium** atop the flower stalk, many stamens in several whorls, and 1 compound ovary or several to many simple ones. The fruits of the rose family are very diverse, and are dry (achenes or follicles) or fleshy (apple-like pomes, cherry-like drupes or raspberry-like drupelet clusters). Strawberries are false fleshy fruits—they have dry achenes embedded in a tasty hypanthium.

The rose family is of considerable economic importance—it gives us most of the important bush and tree fruits of temperate regions. Fruit-producers include apple (*Malus*); pear (*Pyrus*); quince (*Cydonia*); cherry, plum, prune, peach, nectarine, apricot and almond (all *Prunus* species); blackberry, raspberry and loganberry (all *Rubus* species); and strawberry (*Fragaria*). There are also notable ornamental trees and shrubs within the rose family, such as, spiraea (*Spiraea*), ninebark (*Physocarpus*), cotoneaster (*Cotoneaster*), hawthorn (*Crataegus*), flowering quince (*Chaenomeles*), mountain-ash (*Sorbus*), Japanese cherry (*Prunus*), rose (*Rosa*) and shrubby cinquefoil (*Potentilla fruticosa*).

Many boreal species in this family are woody plants and they are described in the 'Shrubs' section. The only true aquatic species from this family is marsh cinquefoil (*Potentilla palustris*), which is described in the 'Aquatics' section.

Key to the Rose Family

1a. Fruits fleshy, like little raspberries 2

- **2a.** Leaves merely lobed (not compound) .... ***Rubus chamaemorus***
- **2b.** Leaves compound (divided into leaflets) 3
  - **3a.** Flowers pink; plants without stolons .... ***Rubus acaulis***
  - **3b.** Flowers white; plants with stolons .... ***Rubus pubescens***

1b. Fruits dry (but with the 'receptacle' becoming fleshy at maturity in *Fragaria* [strawberries]) 4

- **4a.** Leaves compound, with 3 leaflets 5
  - **5a.** Plants with stolons; flowers white; fruits strawberries 6
    - **6a.** End tooth on leaf longer than adjacent teeth; leaves yellow-green on upper surface .... ***Fragaria vesca***
    - **6b.** End tooth on leaf shorter than adjacent teeth; leaves bluish green on upper surface .... ***Fragaria virginiana***
  - **5b.** Plants without stolons; fruits achenes 7
    - **7a.** Flowers yellow; annual to short-lived, perennial herb .... ***Potentilla norvegica***
    - **7b.** Flowers white; woody-based, evergreen perennial .... ***Potentilla tridentata***
- **4b.** Leaves pinnately or palmately compound, usually more than 3 leaflets 8
  - **8a.** Floral tube bur-like in fruit, with hooked bristles; flowers very small; calyx simple, 5-lobed .... ***Agrimonia striata***
  - **8b.** Floral tube not armed with hooked bristles; flowers larger, fairly showy; calyx double, with 2 rings of lobes, outer lobes smaller than inner lobes 9
    - **9a.** Style feathery or hooked, much longer than fruits 10
      - **10a.** Styles feathery, not jointed, end portion does not fall off; flowers pink to purplish, drooping .... ***Geum triflorum*** (see *G. rivale*)
      - **10b.** Styles jointed, not feathery, end portion falls off 11
        - **11a.** Flowers purple or maroon, drooping .... ***Geum rivale***
        - **11b.** Flowers yellow, erect or spreading 12

12a. End leaflet of basal leaves large, deeply 3-lobed and coarsely toothed, wedge-shaped at base; petals much longer than sepals; upper portion of style hairy .... ***Geum aleppicum***

12b. End leaflet of basal leaves merely toothed to 3-lobed, rounded to nearly heart-shaped at base; petals about as long as sepals; upper portion of style not hairy .... ***Geum macrophyllum*** *(see G. aleppicum)*

9b. Style usually falls off 13

13a. Flowers dark red to purple .... ***Potentilla palustris*** (see 'Aquatics' section)

13b. Flowers white or yellow 14

14a. Flowers white, plants with gland-tipped hairs; basal leaves pinnate, with odd number of leaflets .... ***Potentilla arguta*** (see *P. gracilis*)

14b. Flowers yellow 15

15a. Creeping plant with stolons; flowers single .... ***Potentilla anserina***

15b. Plant without stolons; flowers in clusters at stem tips 16

16a. Basal leaves palmate .... ***Potentilla gracilis***

16b. Basal leaves pinnate, with odd number of leaflets *17*

17a. 5–7 leaflets, each finely divided in narrow segments .... ***Potentilla multifida*** (see *P. gracilis*)

17b. 7–15 leaflets, each lobed or toothed but not finely divided into narrow segments .... ***Potentilla pensylvanica*** (see *P. gracilis*)

Rose Family Silhouettes

CLOUDBERRY • *Rubus chamaemorus*
BAKED-APPLE BERRY

GENERAL: Low perennial from creeping rhizomes; flowering stems erect, **unbranched**, 5–20 cm tall.

LEAVES: Long-stalked, 1–3 per stem, 2–7 cm across, round to kidney–shaped, **shallowly 5–7-lobed**, somewhat leathery; edges saw-toothed.

FLOWERS: Single at stem tips; **white**, long-stalked; **sexes on different plants**; petals about 1 cm long.

FRUITS: Large drupelets in raspberry-like clusters; hard, **soft**; **reddish at first**, **amber to yellow when mature**; edible; sepals bend back from mature fruit.

WHERE FOUND: Bogs, usually growing with sphagnum moss; widespread across boreal forest, north past treeline to Arctic coast.

NOTES: In many parts of the north, cloudberry was a very important fruit, second only to blueberries in some areas. The soft fruits are best eaten fresh, as they soon turn mushy if they are carried for any distance, but too many cloudberries in an empty stomach can cause cramps. In the past, northern native peoples preserved this fruit in oil or fat. Today, cloudberries are usually cooked with a bit of sugar, frozen, or canned. • The Chipewyan commonly used cloudberry leaves as a tobacco stretcher. Large quantities of leaves were boiled with a plug of tobacco and then dried. These leaves were said to taste just as good as tobacco, so 1 plug became many. Sometimes cloudberry leaves were mixed with kinnikinnick leaves or the inner bark of red osier dogwood for smoking. • The Woods Cree mixed cloudberry root with other plants and boiled to make a tea called 'woman's medicine.' The roots and lower stems were boiled alone to make a tea that was said to cure barrenness in women. They also made tea by boiling the whole plant and used this to help women in hard labour. The Cree mixed cloudberry plants and Labrador tea to make a medicinal tea to treat fevers, colds and other general ailments. • Cloudberry plants are either male or female, but not both—a most unusual situation for a raspberry. Female plants have flowers with dwarf, sterile, male parts; male plants have flowers with dwarf, sterile, female parts.

DEWBERRY • *Rubus pubescens*
TRAILING RASPBERRY, RUNNING RASPBERRY

GENERAL: Trailing perennial, with **slender runners**, 10–100 cm long, rooting at tips; flowering stems erect, up to 30 cm tall, with soft, long hairs.

LEAVES: 5–15 cm long, compound; **3 leaflets** (occasionally 5), oval to diamond-shaped, pointed, sharply double-toothed.

FLOWERS: 1–3 from crown of plant or from nodes on runners; **white** to pale pink, erect; petals 6–10 mm long; stalks often covered with gland-tipped hairs.

FRUITS: Dark red drupelets in clusters (**raspberry**) about 1 cm across; edible.

WHERE FOUND: Rich moist woods and openings; widespread and common across our region, west to Mackenzie delta and central B.C.

NOTES: Although these little aggregate fruits are sweet tasting and juicy, they are usually too small and too few to gather in quantity. • Dewberry was not widely used in the north, but native peoples to the south and east boiled the roots to make medicinal teas to relieve stomach problems, diarrhea, dysentery and other bowel complaints. Some native groups used parts of the plants to treat women threatened with abortion caused by overwork, while others used them to induce abortions or relieve irregular menstruation.

DWARF RASPBERRY • *Rubus acaulis*
DWARF NAGOONBERRY, STEMLESS RASPBERRY

GENERAL: Low, somewhat tufted perennial from slender creeping rhizomes; flowering stems erect, 5–15 cm tall, finely hairy, with 2–4 leaves.

LEAVES: Round- to heart-shaped in outline, compound; **3 leaflets**, rounded, coarsely toothed, more or less hairy.

FLOWERS: Usually single; showy, **pink to reddish pink**; petals 1–1.5 cm long, distinctly narrowed at base.

FRUITS: Fleshy **red** drupelets in spherical clusters of 20–30 (**raspberry**) about 1 cm across.

WHERE FOUND: Wet woods thickets, meadows, and peatlands; widespread across boreal forest and northern parkland, north slightly past treeline.

NOTES: Also called *R. arcticus* ssp. *acaulis*. • Dwarf raspberries usually disappear as quickly as they ripen, probably because small mammals and birds find them as appetizing as we do. However, they make a delicious nibble if you are lucky enough to find some. A refreshing drink can be made by soaking ripe berries in vinegar for 1 month, then straining the liquid, diluting it with water and adding sugar to taste. Linnaeus ranked it as the choicest of all European berries, both in taste and smell. • *Rubus* means 'red' and refers to the colour of the fruits of many species in this genus (including this one!).

WOODLAND STRAWBERRY • *Fragaria vesca*

GENERAL: Perennial with scaly rhizome and long, slender trailing stolons; leaf stalks and flower stems greenish or very lightly tinged reddish purple, lightly to densely hairy.

LEAVES: At stem base, compound; 3 leaflets, 10–30 mm long, stalkless, strongly toothed; silky-hairy below, **yellow-green** and somewhat silky above; **end tooth projects beyond adjacent teeth**.

FLOWERS: 3–15 on stems that are 5–25 cm long (**usually longer than leaves**); white; 5 petals, 8–11 mm long.

FRUITS: Small achenes, buried in fleshy, juicy, delicious, red berries; **fruit stems longer than leaves**.

WHERE FOUND: Dry to moist open woods, streambanks, and meadows; widespread across boreal forest, north and west to southern N.W.T. and eastern B.C.

NOTES: Native peoples used the roots of woodland strawberry to make a tea for stomach problems, jaundice, or heavy menstruation. The root was also used occasionally as a 'chewing stick' (toothbrush). • In Europe, a strong tea of the leaves or roots was taken with honey to purify the blood and to treat intestinal and urinary tract problems. It was also used as a gargle for strong gums and as a wash to relieve eczema, sunburn, and other skin conditions. A 17th century herbalist wrote, 'The berries cool the liver, blood and spleen, or a hot choleric stomach. They refresh and comfort fainting spirits and quench the thirst. They are good for inflammations, but it is best to refrain from them in a fever, lest they putrefy in the stomach and increase the fits.' Modern herbalists prescribe woodland strawberry leaf tea to stimulate appetite, and scientific investigations suggest that this claim may be valid. • These plants usually reproduce by sending out runners, so picking their delicious berries should have little effect on next year's crop. • The name 'strawberry,' from the Anglo-Saxon *streowberie*, could derive from the dried runners being scattered on the ground, or from the Old English word for straw, which also meant 'mote' or 'chaff'—here in reference to the small achenes embedded in the surface of the berry.

WILD STRAWBERRY • *Fragaria virginiana*
SMOOTH WILD STRAWBERRY

GENERAL: Perennial from short, scaly rhizome, with several slender trailing runners (stolons); leaf stalks and flowering stems sparsely hairy, usually not reddish tinged.

LEAVES: At stem base, compound; 3 leaflets, very short-stalked, strongly toothed; **bluish green, hairless above**; naked to silky-hairy below, sometimes with whitish bloom; **end tooth narrower and shorter than adjacent teeth**.

FLOWERS: White; 2–15 in open clusters on 3–15 cm long stems (**usually shorter than leaves**); 5 petals, 6–8 mm long (sometimes to 12 mm).

FRUITS: Tiny achenes embedded in surface of juicy, delicious, red berries; much richer in flavour than domestic species; **fruit stem shorter than leaves**.

WHERE FOUND: Dry to moist open woodlands and clearings, often in disturbed areas; widespread across our region.

NOTES: The Cree used the roots, leaves and runners of wild strawberries to treat diarrhea and dysentery. They also mixed the roots of strawberry and yarrow plants, boiled them in water and cooled the decoction to make a medicinal tea for curing insanity. Strawberry leaf tea was also used as a wash to treat sores, eczema, and other skin problems. When using dried leaves, make sure that they have dried completely. Partially wilted leaves can contain a **poison**, but it is rendered harmless by drying. The stems and roots were collected in the fall and boiled to make a mouthwash for sore throat and mouth sores. The fruits have been used to make a stain- and tartar-removing mouthwash. They are held in the mouth for a few minutes and then rinsed off with warm water. They are also said to whiten sunburned skin. • The delicious fruits of wild strawberry are very popular among most northern residents. They are usually eaten raw, but they can also be collected in quantity and frozen, dried, or made into jams and jellies. The leaves are often steeped in boiling water to make tea. Like the fruits, the leaves are high in vitamin C. • This species is the original parent of 90% of all the cultivated strawberries now grown. The related coastal strawberry (*F. chiloensis*) is the source parent of the remaining cultivars.

THREE-TOOTHED CINQUEFOIL • *Potentilla tridentata*

GENERAL: Tufted, low-growing, much-branched perennial with long creeping roots, 10–20 cm tall; stems slender, reddish brown to grey-black, with flattened hairs and scaly covering of leafless leaf stalks.

LEAVES: Alternate, compound, **evergreen**, most near stem base; 3 leaflets, firm, leathery, 1–2.5 cm long, oblong to narrowly spoon-shaped, taper gradually to base; **3 teeth at blunt tip**; dark green above, paler below, usually hairless on both surfaces; leaf stalks up to 3 cm long; stipules linear-lance-shaped.

FLOWERS: In open, stiff, few- to several-flowered clusters (cymes); **white** or rarely pinkish, 10–15 mm across; bractlets lance-shaped; sepals egg-shaped to triangular; petals longer than sepals.

FRUITS: Small cluster of densely hairy achenes surrounded by calyx.

WHERE FOUND: Dry sandy or gravelly places, cliffs, ledges, rock crevices, lakeshores and sandy pine woods; fairly common in appropriate habitats across boreal forest of the prairie provinces, north and west to southern N.W.T.

NOTES: Cinquefoil roots have been widely used to produce red dyes. Chrome mordants give a red-brown dye, and iron mordants give a purple-red dye. Because of their high tannin content, the roots of some cinquefoils were used for tanning leathers. • *Potentilla* is derived from the Latin *potens*, which means 'powerful,' 'strong,' or 'potent,' in reference to the effectiveness of these plants in stopping bleeding and dysentery in both man and beast.

ROUGH CINQUEFOIL • *Potentilla norvegica*

GENERAL: Annual or biennial, 20–60 cm tall; stems erect or somewhat spreading, simple or branched towards top, **often reddish**, leafy, covered with stiff spreading hairs.

LEAVES: Alternate, compound; **3 leaflets**, 2–6 cm long, obovate to elliptic, coarsely toothed, hairy.

FLOWERS: In fairly dense, leafy clusters (cymes) at stem tips; pale yellow, 6–12 mm across; **petals slightly shorter than sepals**.

FRUITS: Many ridged achenes, enclosed in much enlarged calyx.

WHERE FOUND: Moist meadows, lakeshores, streambanks, roadsides, waste places and recently disturbed soil; common and widespread across our region; circumpolar.

NOTES: Although rough cinquefoil was not widely used in the north, several native groups in more southern regions used it to make a medicinal tea to relieve stomach cramps, sore throats and other pains. The plants were burned to treat aching heads, bones and eyes. • Rough cinquefoil can become a troublesome weed in disturbed areas. It is often a pioneer species on recently disturbed moist soils and in clearings or burns.

GRACEFUL CINQUEFOIL • *Potentilla gracilis*

GENERAL: Somewhat tufted perennial from thick, scaly, woody base, highly variable; several stems, erect, **30–70 cm tall**, more or less branched.

LEAVES: Many at stem base, long-stalked, **palmately divided into 5–7 coarsely toothed leaflets**, hairy to hairless above, **usually paler and more densely hairy (often woolly) below**; few stem leaves, smaller.

FLOWERS: Yellow, showy; **several to many**, in flat-topped clusters, 15–20 mm across, at stem tips; petals notched.

FRUITS: Many achenes, 1–1.5 mm long, surrounded by calyx.

WHERE FOUND: Open woods, grasslands, and waste places; widespread across southern boreal forest and parkland, north and west to central Yukon and Alaska.

NOTES: The leaves of graceful cinquefoil are divided into 5–7 leaflets that are white-hairy below and arranged like fingers on a hand. This distinguishes it from **white cinquefoil** (*P. arguta*, inset photo), which has pinnately compound leaves (with 7–9 leaflets) that are not white-woolly below. White cinquefoil also has sticky-hairy stems and leaves and pale yellow to whitish flowers. **Prairie cinquefoil** (*P. pensylvanica*) has yellow petals and 5–9 pinnate leaflets that are pale green or grey-hairy below. Both white and prairie cinquefoils grow in grassy meadows and openings and on rocky slopes across our region. White cinquefoil is found mainly in the southern boreal forest, but prairie cinquefoil is more widespread and occurs north almost to the treeline and west to Alaska. • Another species, **branched cinquefoil** (*P. multifida*) is distinguished by its less hairy, deeply toothed or lobed pinnate leaflets, with their edges rolled under. It occurs in open meadows and on gravel bars in the northern boreal forest from Manitoba to the Mackenzie River delta and interior Alaska.

P. arguta

P. gracilis

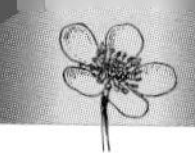

SILVERWEED • *Potentilla anserina*

GENERAL: Tufted perennial from **thick fleshy roots**; long, **strawberry-like runners**, rooting and forming new plants at nodes.

LEAVES: Tufted at base, compound, usually 10–20 cm long; **7–21 oval to oblong, coarsely toothed leaflets**, interspersed with smaller leaflets; **silky, white-woolly below**, green above, sometimes hairy on both sides.

FLOWERS: Single; bright **yellow**; on **3–28 cm long, leafless stems**; petals 7–16 mm long.

FRUITS: Corky achenes, 1.5–2 mm long, grooved on back and top; in dense cluster surrounded by enlarged calyx.

WHERE FOUND: Lakeshores, riverbanks and damp meadows; widespread across our region, north and west to Mackenzie delta and interior Alaska.

NOTES: The starchy rhizomes of silverweed are said to taste like parsnips, sweet potatoes or chestnuts. They can be eaten raw, but usually they are roasted, boiled or fried. The roots were said to taste best in the cold months, so most native groups collected them in the fall. They could also be dried and stored for the winter. • All parts of silverweed, but especially the roots, are high in tannins, making them astringent. Several herbalists recommend silverweed tea as a remedy for cramps and diarrhea and as an external application to stop bleeding and heal sores. Usually it is suggested as a gargle for problems in the mouth and throat. A distilled water of silverweed used to be popular for removing freckles, spots and pimples and for restoring sunburned skin. It has also been used since the Middle Ages in European folk medicine as a treatment for cancer. • Silverweed apparently has a stimulant action on the uterine muscle, so **pregnant women** should not use it. • Silverweed provides food for many species of wildlife, and it is also a favourite food of cattle, horses, pigs, goats and geese. • These attractive perennials transplant well to rock gardens and borders, but be forewarned—they are notorious for reaching out to cover new ground!

AGRIMONY • *Agrimonia striata*

GENERAL: Perennial from stout rhizome and fibrous roots, 30–80 cm tall; stems erect, stiff, sometimes forked towards top, **with soft, fine brownish hairs**.

LEAVES: Alternate, pinnate, 2–7 cm long; 5–9 main leaflets interspersed with smaller ones, lance-shaped to narrowly egg-shaped, coarsely toothed, smooth above, paler, veiny, soft hairy and gland-dotted below; stipules large, deeply toothed, up to 2 cm long.

FLOWERS: In slender, spike-like clusters at stem tips; small, yellow; 5 petals, about 3 mm long; 5-lobed calyx; gland-dotted hypanthium with 3–4 rows of hooked bristles nearly as long as sepals.

FRUITS: 2 seed-like achenes, surrounded by gland-dotted, bristly hypanthium.

WHERE FOUND: Edges of aspen woods, cutlines and roadsides; across southern boreal forest and parkland of prairie provinces.

NOTES: The seeds of agrimony are distributed as 'stick-tights'—their hooked bristles cling readily to the fur or fabric of passers-by. • In North America, native peoples and settlers considered agrimony a very successful herbal remedy for fevers. • Agrimony is similar to the widely used European agrimony (*A. eupatoria*). The two were used in similar ways and for many years they were not recognized as separate species. • Agrimony has no narcotic properties, but it was widely believed that if a piece of the plant was put under someone's head, that person would fall into a deep sleep that could only be broken by removing the magical plant. • The origin of the genus name is unclear. It may be derived from the Greek *argema*, 'cataract of the eye,' which agrimony or some other related plant was supposed to cure. The species name refers to the 'striae' (furrows) on the base of the flowers.

YELLOW AVENS • *Geum aleppicum*

GENERAL: Perennial from short rhizome, 0.4–1 m tall; stems coarse, erect, usually hairy throughout.

LEAVES: Alternate, pinnately compound, broadest at tip, to 15 cm long, hairy; 5–7 leaflets, toothed or notched, **end leaflet wedge-shaped at base**, often with a few, very small, interspersed leaflets; stem leaves smaller, with 3–5 leaflets; large, broad stipules.

FLOWERS: Several in leafy cluster (cyme); yellow, 10–25 mm across; sepals lance-shaped, bent downwards, 5–8 mm long; petals as long as or slightly longer than sepals; upper part of style hairy.

FRUITS: Achenes, flattened, elliptic, hairy, 3–4 mm long; with hooked, **hairy** beak about twice as long as body; in spherical to egg-shaped head 10–15 mm across.

G. macrophyllum

WHERE FOUND: Moist to wet meadows, streambanks, thickets and open woods; common across our region north and west to southern Yukon.

NOTES: Yellow avens could be confused with **large-leaved avens** (*G. macrophyllum*, includes *G. perincisum*), but that species has leaves with 5-15 leaflets, and the end leaflet of the basal leaves is large, kidney-shaped, and toothed, notched or deeply 3-cleft. Its petals are no longer than the sepals, and the upper portion of the style is not hairy. Large-leaved avens is fairly common across our region. It grows in moist meadows, thickets, and open forest north to the Arctic Circle. • The Woods Cree boiled yellow avens roots to make a medicinal tea to treat sickness associated with teething, and to relieve sore teeth or sore throats. A similar decoction, with other herbs added, was taken to make a person sweat. • The Cree called yellow avens 'jealousy plant' because the tiny seeds with their hooked beaks catch onto the clothing of anyone brushing against them, and are carried away, often in large numbers. It was explained this way: 'When that person is walking by, and someone is jealous, (it's) just as if someone is trying to hurt you with those little spurs—they stick to your clothing instead.'

G. aleppicum

PURPLE AVENS • *Geum rivale*

GENERAL: Slender, **more or less hairy** perennial from stout rhizome; stems erect, little branched, 30-60 cm tall (sometimes to 1 m).

LEAVES: Pinnately divided **at stem base**, with **3 major lobes** (uppermost lobe largest) and additional **small, scattered leaflets**; stem leaves **alternate**, small, irregularly divided; edges sharply **toothed**.

FLOWERS: Few to several in **nodding** clusters at stem tips, 15-20 cm across; petals **flesh-coloured to yellowish with purple veins**; sepals 7-10 mm long, purplish.

FRUITS: Dense head of spreading achenes with **long feathery styles**.

WHERE FOUND: Streambanks, marshes and wet ground in southern boreal forest across Canada; also across Eurasia.

G. rivale

NOTES: Purple avens could be confused with **three-flowered avens** (*G. triflorum*), but that species has only a few, opposite stem leaves, rather than several alternate stem leaves like purple avens. Three-flowered avens usually has 3 pink-purple, nodding flowers and long, feathery styles. It grows in dry grassy areas east of the Rockies and mostly along the southern fringe of our region, although it does extend north to Great Slave Lake. • The aromatic, purplish root of purple avens was used like chocolate to make beverages, and the plant was sometimes called 'chocolate root' or 'Indian chocolate.' In Alberta, the Blackfoot reportedly used the crushed seeds as a perfume. • Three-flowered avens is also known as 'prairie smoke.' When there are many plants growing together the colour and 'plumage' of the fruits suggest a haze of low-lying smoke.

G. triflorum

Fabaceae (Pea Family)

With about 17,000 species worldwide, the pea family (Fabaceae, also known as Leguminosae) is probably the third largest plant family in the world, after the Asteraceae and Orchidaceae. *Astragalus* and *Oxytropis* are the largest pea genera in our region.

The boreal members of the pea family are all herbs that usually have alternate, pinnately compound (sometimes palmate) leaves with stipules at the base of the leaf stalk. Their flowers are bilaterally symmetrical, with 5 fused sepals, 5 petals, 10 stamens, and 1 superior ovary. The arrangement of the unequal petals gives the flower a butterfly-like appearance. There is an upright **standard**, a side pair of **wings**, and 2 inner, joined petals that form a **keel**, which envelops the stamens and ovary (see illustration in glossary). The fruit is a 1-chambered pod (legume) with several to many seeds in 2 alternating rows.

The pea family is one of the most economically important families. Its species provide food, fodder, dyes, gums, resins, oils, medicines and timber. Food plants include garden pea (*Pisum*), lentil (*Lens*), peanut (*Arachis*), bean (*Phaseolus*) and soybean (*Glycine*); fodder and forage plants include clover (*Trifolium*), alfalfa (*Medicago*), lupine (*Lupinus*), vetch (*Vicia*), bird's-foot trefoil (*Lotus*) and sweet-clover (*Melilotus*). Members of over 150 genera are grown as ornamentals, including wisteria (*Wisteria*), sweet pea (*Lathyrus*), lupine, redbud (*Cercis*), acacia (*Acacia*) and mimosa (*Mimosa*). Caragana (*Caragana*) is a thorny shrub commonly planted on the Canadian prairies to create windbreaks.

Key to the Pea Family

1a. Leaves with even number of opposite leaflets, end leaflet represented by tendril 2

2a. Hairs around tip of style, like bottlebrush .... ***Vicia americana***

2b. Hairs at tip of style arranged on 1 side, like toothbrush (the peavines) 3

3a. Flowers creamy white to yellowish, in clusters of 5-10; pods not evidently veined .... ***Lathyrus ochroleucus***

3b. Flowers purple, in clusters of 10-20; pods veined .... ***Lathyrus venosus***

1b. Leaves with odd number of opposite leaflets (including end leaflet), or palmately compound 4

4a. Leaves palmately compound (leaflets radiate from common point, like fingers of hand) .... ***Lupinus arcticus***

4b. Leaves pinnately compound or with 3 leaflets (trifoliate) 5

5a. Leaves with 3 leaflets; flowers often fragrant 6

6a. Flower clusters long and narrow (the sweet-clovers) 7

7a. Flowers white .... ***Melilotus alba*** (see *M. officinalis*)

7b. Flowers yellow .... ***Melilotus officinalis***

6b. Flowers clustered in dense, usually rounded heads (the clovers) 8

8a. Flowers red to pink, in heads 2-3.5 cm across; plants hairy .... ***Trifolium pratense*** (see *T. hybridum*)

8b. Flowers white to pinkish red, in heads 1.5-2.5 cm across; plants hairless (or nearly so) 9

9a. Flowers white to pinkish white; leaflets notched at tip; plants creeping, rooting at nodes .... ***Trifolium repens*** (see *T. hybridum*)

9b. Flowers pink to pinkish red; leaflets usually not notched at tip; plants ascending to upright, not rooting at nodes .... ***Trifolium hybridum***

5b. Leaves pinnately compound 10

10a. Pods articulated, constricted between seeds, break crosswise; keels equal to or longer than wings (the sweet-vetches) 11

11a. Flowers pink to reddish purple, 1–1.5 cm long; leaflet veins conspicuous .... ***Hedysarum alpinum***

11b. Flowers deep purple, 2.5–3 cm long; leaflet veins inconspicuous .... ***Hedysarum boreale*** (see *H. alpinum*)

10b. Pods not obviously constricted between seeds, open lengthwise; keels usually shorter than wings 12

12a. Plants usually without leafy stems (but see *O. deflexa*); keel of flower narrowed to a beak-like point (the locoweeds) 13

13a. With leafy stem; pods hang down .... ***Oxytropis deflexa***

13b. No leafy stem; pods spreading to erect 14

14a. Plants sticky hairy; flowers purple .... ***Oxytropis leucantha*** (see *O. splendens*)

14b. Plants not sticky hairy 15

15a. Leaflets in whorls of 3 or 4, with dense, silky hairs; flowers dark blue to pinkish purple .... ***Oxytropis splendens***

15b. Leaflets opposite or alternate, sparsely hairy; flowers white to yellowish .. ***Oxytropis campestris* var. *gracilis*** (see *O. deflexa*)

12b. Plants usually with leafy stems; keel of corolla not beaked (the milk-vetches) 16

16a. Plants low (less than 30 cm tall), straggling 17

17a. Flowers and pods erect; flowers blue or purple 18

18a. Leaflets, flowers and pods generally more than 1 cm long 19

18b. Leaflets, fllowers and pods generally less than 1 cm long .... ***Astragalus bodinii*** (see *A. alpinus*)

19a. Plants with slender rootstalks; hairs on the leaflets attached by their bases ..... ***Astragalus agrestis***

19b. Plants with a taproot and a stout stem base; hairs on the leaflets attached by their middles .... ***Astragalus asdurgens*** var. ***robustior*** (see *A. agrestis*)

17b. Flowers and pods hanging; flowers bluish-purple to almost white .... ***Astragalus alpinus***

16b. Plants upright, generally taller (more than 30 cm tall) 20

20a. Flowers usually purple .... ***Astragalus eucosmus***

20b. Flowers whitish 21

21a. Flowering head densely flowered 22

22a. Flowers yellowish-white; plants from creeping rhizomes; 13–29 leaflets; pods erect, usually less than 1.5 cm long, woody .... ***Astragalus canadensis***

22b. Flowers greenish-white; plants from stout stem base; 7–15 leaflets; pods hanging, usually more than 1.5 cm long, not woody .... ***Astragalus americanus***

21b. Flowering head loosely flowered 23

23a. Flowers yellowish-white, tipped with purple; wing petals deeply 2-toothed; 7–15 leaflets ..... ***Astragalus aboriginum*** (see *A. eucomus*)

23b. Flowers white; wing petals not toothed; 11–21 leaflets ..... ***Astragalus tenellus*** (see *A. americanus*)

WILD VETCH • *Vicia americana*
AMERICAN VETCH

GENERAL: **Perennial**, trailing or climbing, **often in tangled masses**; stems 0.3–1 m tall.

LEAVES: Alternate, compound, **simple or forked tendrils** at tip; **8–14 leaflets**, elliptic to oblong, usually 15–35 mm long; stipules 3–11 mm long, sharply toothed.

FLOWERS: **3–9 in loose clusters** at stem tips; pea-like, **bluish to reddish purple**; petals **1.5–2 cm long**.

FRUITS: Hairless pods, flat, 2–4 cm long.

WHERE FOUND: Fields, thickets, and open woods; widespread across our region, north and west to Mackenzie delta.

NOTES: Although some species of *Vicia* are edible, many vetches contain compounds that produce hydrocyanic acid and cause **cyanide poisoning**. Never eat a vetch unless you are certain it is not poisonous. The seeds in the pods in particular may attract a child's curiosity, as they resemble tiny peas. Wild vetch is not recorded as poisonous, nor is it generally considered edible. • When an insect lands on a vetch flower in search of nectar, the tiny hairs on the style of the flower brush pollen out from between the 2 lower petals and dust it onto the belly of the visitor. The insect then carries its load of pollen to the next flower. • This common plant was not widely used by native peoples in the boreal forest. However, in the East, Iroquois women made a tea from the roots as a love medicine. In the south, the Navaho-Kayenta smoked wild vetch near a horse to increase the animal's endurance. The Navaho-Ramah considered it a 'life medicine.' • The name 'vetch' is from the Latin *vicia*, which is thought to be derived from the Latin verb *vincio* ('to bind'), in reference to the climbing habit of these plants.

PURPLE PEAVINE • *Lathyrus venosus*
VEINED PEAVINE

GENERAL: Perennial, up to 1 m tall; stems slender, **climbing, strongly 4-sided**, finely hairy.

LEAVES: Alternate, pinnately compound, **8–12 leaflets, elliptic to oblong-egg-shaped**, 2–5 cm long; hairless above, finely hairy and **very veiny below**; tendrils from leaf tips, usually branched; **stipules small, narrowly lance-shaped to almost arrow-shaped**.

FLOWERS: In 10–20-flowered clusters at stem tips; purple, 10–18 mm long.

FRUITS: Hairless pods, 3–5 cm long, **very veiny**.

WHERE FOUND: Shrub thickets and deciduous woods; across southern boreal forest and parkland, north and west to northeastern B.C.; most common in east.

NOTES: The anthers of peavines are hidden inside a pocket formed by the 2 lower petals (the keel). When insects land on the flowers, a special brush on the style sweeps out through the keel, dusting the lower side of the visitor with pollen. In peavines, these bristles are all on 1 side of the style, like a toothbrush, but in vetches (*Vicia* spp.) they surround the style, like a bottlebrush. • Purple peavine provided a valuable source of forage and hay in the early days of settlement of the northern bushlands. • *Venosus*, meaning 'with veins,' refers to the highly visible veins on the underside of the leaves and on the fruits.

CREAMY PEAVINE • *Lathyrus ochroleucus*
CREAM-COLOURED VETCHLING

GENERAL: Slender, climbing perennial from creeping rhizomes; stems somewhat flat-sided, 0.3–1 m tall, hairless.

LEAVES: Alternate, compound, tendril at tip; **6–10 leaflets**, egg-shaped to elliptic, 2–5 cm long; tendrils well-developed, usually branched; **stipules broad, oval, often half as long as leaflets**.

FLOWERS: 5–10 in clusters at stem tips; pea-like; petals **white to yellowish white**, about 15 mm long.

FRUITS: Hairless pods, about 4 cm long.

WHERE FOUND: Open woods, thickets, and clearings; widespread across our region, north and west to southern N.W.T. and northern B.C.

NOTES: Many plants of this genus are eaten by livestock and have been used successfully in various parts of the world. However, they are generally viewed with suspicion because some cause a type of **poisoning** called lathyrism, which results from eating too much vetchling seed over long periods of time. Epidemics of lathyrism date back to ancient Greece, but cases in humans usually occurred during famines when people were forced to eat vetchling almost exclusively. After 10 days to 4 weeks, this can cause progressive loss of coordination, ending in irreversible paralysis. Vetchling is not generally considered poisonous but these plants should be approached with caution. • *Lathyrus* is the ancient Greek name for 'a pea' or 'a pulse' (legume).

ALPINE HEDYSARUM • *Hedysarum alpinum*
ALPINE SWEET-VETCH

GENERAL: Perennial from woody taproot; stems 20–70 cm tall, reclining to erect, branched towards top, sparsely hairy.

LEAVES: Compound; 15–21 leaflets, oblong to lance-shaped, 1–3 cm long, pointed tips; sparse, grey hairs and **prominent veins** below.

FLOWERS: Several to many, somewhat nodding in compact to long clusters at stem tips; pea-shaped, 1–1.5 cm long; petals **pink to reddish purple, paler at base**; sepals broadly triangular, upper shorter than lower.

FRUITS: Flattened pods with constrictions between each of 2–5 seeds (**like short strings of flattened beads**), essentially **hairless**, stalked.

WHERE FOUND: Moist open woods, meadows, slopes, and disturbed areas; widespread across our region, north across tundra to Arctic coast.

H. boreale

NOTES: Do not confuse alpine hedysarum with **northern hedysarum** (*H. boreale*, also called *H. mackenzii*) (inset photo). The 2 species are easily distinguished by their leaves. Alpine hedysarum has leaves with prominent veins on the lower surface, whereas the thicker leaves of northern hedysarum mask the veins, making them more difficult to see. Northern hedysarum also has narrower (linear oblong) leaflets, inconspicuous greyish stipules, and larger (2.5-3 cm long) deep purple flowers. Northern hedysarum grows on open gravelly areas, often on riverbanks and lakeshores. It is widespread from northern Manitoba to western Alaska, but less common in the prairie provinces. • Young alpine hedysarum plants can be eaten raw, boiled or roasted, but usually the roots are eaten. These are dug from fall to spring, as long as the ground is thawed, but during the summer they become dry and woody. They can be eaten raw, or cooked by boiling, baking, or frying. The young roots have a sweet taste (like licorice) in the spring, and once they are cooked they taste like young carrots.

AMERICAN MILK-VETCH • *Astragalus americanus*
AMERICAN RATTLEPOD

GENERAL: Perennial from somewhat woody base, 50–100 cm tall; stems usually single, sparsely hairy to hairless.

LEAVES: Alternate, pinnately compound; **7–15 leaflets, elliptic-oblong**, 2–4 cm long, 7–15 mm wide; hairless above, sparsely hairy below; **stipules conspicuous, oblong**, bent backwards, **not joined at base**.

FLOWERS: In long-stalked, 12–20-flowered clusters at stem tips; white, turn yellowish or brownish with age, 10–15 mm long, **at first ascending, later bent downwards**; **flower stems shorter than leaves**.

FRUITS: Pods; **inflated, hanging, shiny**, hairless or very sparsely hairy, **almost round in cross section**, 1.5–2 cm long, stalked.

WHERE FOUND: Streambanks and moist open woods; scattered across our region north and west to Great Bear Lake and interior Alaska.

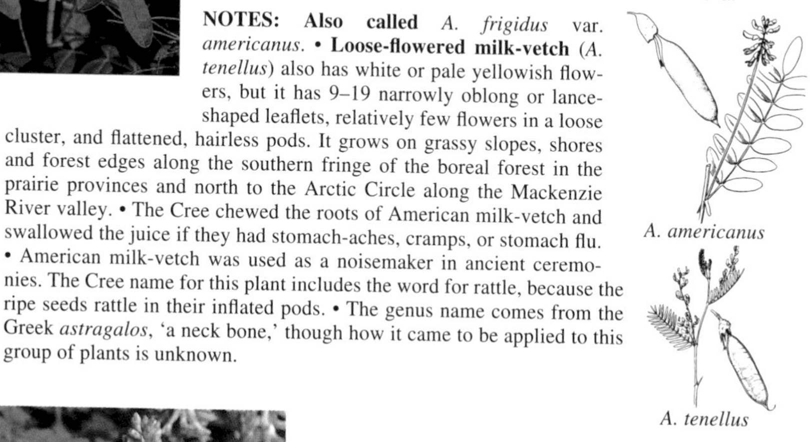
A. americanus

A. tenellus

NOTES: Also called *A. frigidus* var. *americanus*. • **Loose-flowered milk-vetch** (*A. tenellus*) also has white or pale yellowish flowers, but it has 9–19 narrowly oblong or lance-shaped leaflets, relatively few flowers in a loose cluster, and flattened, hairless pods. It grows on grassy slopes, shores and forest edges along the southern fringe of the boreal forest in the prairie provinces and north to the Arctic Circle along the Mackenzie River valley. • The Cree chewed the roots of American milk-vetch and swallowed the juice if they had stomach-aches, cramps, or stomach flu. • American milk-vetch was used as a noisemaker in ancient ceremonies. The Cree name for this plant includes the word for rattle, because the ripe seeds rattle in their inflated pods. • The genus name comes from the Greek *astragalos*, 'a neck bone,' though how it came to be applied to this group of plants is unknown.

CANADIAN MILK-VETCH • *Astragalus canadensis*

GENERAL: Perennial from creeping rhizome, 40–100 cm tall; stems simple, **stout**, branched towards top, hairless or thinly hairy.

LEAVES: Alternate, pinnately compound; **13–29 leaflets, elliptic to oblong**, 2–4 cm long, 5–18 mm wide, usually hairless above, sparingly hairy below; **stipules lance-shaped, pointed, joined at base**.

FLOWERS: In long-stalked, **dense**, 30–100-flowered cluster at stem tip; greenish yellow to white, 10–15 mm long, **spreading to drooping**.

FRUITS: Erect pods, **stalkless, somewhat woody**, hairless or nearly so, **rounded heart-shaped in cross section**, 10–15 mm long, abruptly sharp-pointed, **crowded, remain after seeds disperse**.

WHERE FOUND: Moist open woods, banks and shores; across our region, north to Liard River, N.W.T.

NOTES: Canadian milk-vetch is a robust and showy species with a distinctive, dense cluster of seed pods that remain on the plant after the seeds have dispersed. • Many milk-vetches contain the **toxic** base, locoine, which causes a type of insanity called 'locoism.' These plants may also absorb selenium and molybdenum from the soil. Selenium causes 'blind staggers' and molybdenum causes poor growth, brittle bones and anemia. • Although a few species of *Astragalus* were used as medicines in frontier days, most species are still known as 'locoweed' because of their history in poisoning livestock. The name 'locoweed' more properly refers to *Oxytropis* species (pp. 144-45)—a genus that is very similar to *Astragalus* and often confused with it.

PRETTY MILK-VETCH • *Astragalus eucosmus*

GENERAL: Stems single, or several from base, slender and stiffly erect, 30–60 cm tall, hairless or hairy.

LEAVES: Alternate, pinnately compound; 9–15 leaflets, oblanceolate to elliptic, 8–25 mm long, sparsely hairy below, essentially hairless above.

FLOWERS: In compact spikes 2–3 cm long (lengthening to about 10 cm in fruit); **purple** to nearly white, 6–8 mm long; **calyx black-hairy**.

FRUITS: Densely hairy pods, egg-shaped to elliptic, about 1 cm long, stalkless, **bent downwards**.

WHERE FOUND: Moist meadows, gravel bars, rocky slopes and sheltered shores; across boreal forest, north and west to Mackenzie delta and southern Alaska; uncommon in southeast.

NOTES: Indian milk-vetch (*A. aboriginum*) is distinguished by its yellow-white flowers tipped with purple and by its wing petals with conspicuously notched tips. It is a western species of streambanks, meadows and woodland clearings from Saskatchewan to Great Bear Lake and interior Alaska. • The name 'milk-vetch' came from the belief that the milk supply of goats was increased when they ate these plants.

A. eucosmus *A. aboriginum*

ALPINE MILK-VETCH • *Astragalus alpinus*

GENERAL: Low, mat-forming perennial with widespread rhizomes; stems slender, **leafy**, spreading or ascending, 5–30 cm tall.

LEAVES: Compound, 5–15 cm long; **11–25 leaflets**, oval to oblong-elliptic, rounded or shallowly notched at tip, 5–20 mm long.

FLOWERS: 5–17 in crowded, 1–4 cm long cluster at stem tips; sepals black-hairy; petals **light bluish or pinkish purple to almost white**, side petals often whitish, 7–12 mm long.

FRUITS: Black-hairy pods, about 1 cm long, short-stalked, bent downwards on stalk.

WHERE FOUND: Forest edges, meadows, and open sandy or gravelly places; widespread across boreal forest, north through much of arctic archipelago; circumpolar.

NOTES: Yukon milk-vetch (*A. bodinii* or *A. yukonis*), also called Bodin's milk-vetch, is a western species of riverbars, gravelly terraces, meadows, claybanks, and roadsides, west from Great Slave Lake. It is mat-forming like alpine milk-vetch but more sprawling, and it has smaller (5–12 mm long), narrower, lance-shaped leaflets, small (less than 1 cm long) bluish purple flowers, and erect pods 6–10 mm long. • Alpine milk-vetch is considered good forage for animals. • The flowers, leaves and stems have been used to produce yellow to greenish yellow dyes. • The pods of alpine milk-vetch do not open. Instead, they are blown away by the wind and eventually torn to pieces.

PURPLE MILK-VETCH • *Astragalus agrestis*

GENERAL: Perennial, **often forms large patches** from slender rhizomes, 5–20 cm tall; stems slender, **reclining to erect**, sparsely hairy to hairless.

LEAVES: Alternate, pinnately compound; **11–19 leaflets, linear-oblong to elliptic**, 1–2 cm long; hairless or nearly so above; hairy below; **stipules linear to egg-shaped, joined at base**.

FLOWERS: In many-flowered clusters, 2–3 cm long, at stem tips; purple (drying bluish), rarely white, 14–18 mm long, **erect**; **calyx often with black hairs**.

FRUITS: Erect pods, **stalkless, egg-shaped to oblong**, about 1 cm long, **densely hairy**.

WHERE FOUND: Moist prairies, meadows, slopes and open areas; across our region north and west to Yukon.

A. agrestis

NOTES: Also called *A. dasyglottis*, *A. danicus* and *A. goniatus*. • **Ascending purple milk-vetch** (*A. adsurgens* var. *robustior*, also called *A. striatus*) somewhat resembles purple milk-vetch, but is tufted, 10-30 cm tall, and has pointy-tipped leaflets. It also grows on grassy slopes, in openings and open woods, from the southern Yukon east, especially along the southern parkland fringe and in the prairie. • The pods of most of our milk-vetches split open at maturity and eject their seeds. If this happens late in the season, the round seeds are readily carried over the surface of windblown snow. Pods that do not open are usually inflated and bladder-like. They too can be carried over the snow by the wind for great distances in winter before being torn open.

A. adsurgens

SHOWY LOCOWEED • *Oxytropis splendens*

GENERAL: Tufted perennial from deep taproot with branching top; stems leafless, erect, 10-30 cm long, bearing **showy flower clusters**.

LEAVES: At stem base, 3-26 cm long; **many leaflets, mostly in whorls** of 3-4, lance-shaped, 1-2.5 cm long, **silky-hairy**.

FLOWERS: 12-35 in showy, **spike-like clusters**, 2-10 cm long; petals dark **blue to pinkish purple**, 1-1.5 cm long; sepals densely long-silky.

FRUITS: Egg-shaped pods 8-12 mm long (sometimes to 17 mm), densely **silky-hairy**.

WHERE FOUND: Open woods, clearings and riverbanks; common in prairie and parkland; occasional in boreal forest, north to Great Bear Lake and east-central Alaska.

NOTES: A western species, **viscid locoweed** (*O. leucantha* or *O. viscida*) is sticky-hairy rather than silky, and has opposite leaflets. It is found on dry banks and hillsides in the boreal forest from Great Slave Lake to northern Alaska. • Many species of locoweed (*Oxytropis*) are **poisonous** to livestock. The poison varies from 1 species to the next and, to further complicate things, some locoweeds become poisonous by taking up selenium from the soil. Some locoweeds produce classical 'loco' poisoning in horses, sheep, and cattle. To be lethal, large amounts of toxic locoweed would have to be eaten over a long time, however, these plants can be habit-forming with some animals, and many deaths have resulted. Unfortunately, there has been no comprehensive study to determine exactly which species contain the poisonous alkaloid locoine, and which species are non-poisonous. It is probably best to avoid using locoweeds altogether.

REFLEXED LOCOWEED • *Oxytropis deflexa*

GENERAL: Tufted perennial from strong taproot; stems 10–40 cm long, **reclining to ascending**, loosely hairy.

LEAVES: Alternate, pinnately compound, 5–20 cm long, **appear flattened; 15–25 leaflets, lance-shaped to elliptic-lance-shaped**, 5–15 mm long, **hairy on both sides; stipules lance-shaped, joined to leaf stalk for part of their length**.

FLOWERS: In 5–25-flowered clusters; bluish purple or white with purple tip, 6–10 mm long; **nearly erect in bud, later nod to 1 side**.

FRUITS: Pods; 10–18 mm long, **bent downwards** on short stalks, **often black-hairy**.

WHERE FOUND: Open woods, moist thickets, banks, shores and gravel bars; found across our region north and west to interior Alaska.

NOTES: Late yellow locoweed (*O. campestris* var. *gracilis*, includes *O. varians* and *O. monticola*) (inset photo) has white to yellowish flowers, does not have a leafy stem, and has erect or spreading pods. It is a highly variable circumpolar species that has been divided into many subspecies and varieties by taxonomists. • Despite the poisonous nature of many locoweeds (see showy locoweed), some have been used as medicine by native groups in western North America to treat asthma, sore throats and bronchial problems. Whole plants were made into hot compresses to treat sores, swellings, rheumatism, sore muscles and painful joints. Water in which the plants had been boiled was used to wash the head, hair, and body. • Many locoweeds, including reflexed locoweed, have nodules on their roots that contain bacteria capable of 'fixing' nitrogen (converting atmospheric nitrogen into a form that can be used by plants), even in cold climates. This can be very important in enriching the nitrogen-poor soils of the Arctic and subarctic.

ALSIKE CLOVER • *Trifolium hybridum*

GENERAL: Hairless perennial from taproot; 1–several stems, usually branched, 15–80 cm tall, ascending to erect, **not rooting at nodes**.

LEAVES: Compound; **3 leaflets**, oval to elliptic, 1–4 cm long (usually 2–3 cm), rounded or shallowly notched at tip, finely sharp-toothed.

FLOWERS: Many, in **dense**, 15–25 mm long, **spherical heads**, on 2–8 cm long stalks; petals **pinkish** to reddish or white, 5–9 mm long.

FRUITS: Small, stalked pods (often hidden by sepals); 2–3 (sometimes 4) seeds per pod, lens-shaped, greenish to black.

WHERE FOUND: Introduced weed of cultivated and waste ground; widespread across most of North America, north to Churchill, Man., Great Slave Lake and interior Alaska; more or less circumpolar.

NOTES: Alsike clover is a hybrid between **red clover** (*T. pratense*) and **white clover** (*T. repens*) Inset photo). Red clover has erect or spreading stems and large (20–35 mm long) heads of red to pink flowers immediately above 1 or 2 short-stalked leaves. White clover has creeping, freely rooting stems and long-stalked heads of white to pale pink flowers. Both species have been introduced and naturalized from Europe, and have escaped from cultivation or seed mixes for lawns and road rights-of-way. They grow in meadows, along roads and in other disturbed areas, scattered across our region, more commonly in the south. • Clover is an important crop throughout the world for fodder in pastures and hay, for honey production, and for improving nitrogen-poor soils. Clover roots have nodules that contain bacteria capable of taking nitrogen from the air and converting it into a form that can be used by the plant. This helps the clover to grow, and when the plant dies, the nitrogen is left in the soil for other plants. • Dried, ground alsike flowers were once listed in the U.S. Dispensatory for use in anti-asthma cigarettes. Research at the Mayo Clinic in Rochester, Minnesota, revealed a blood thinning agent in clover that could be useful in treating heart disease.

T. pratense

YELLOW SWEET-CLOVER • *Melilotus officinalis*

GENERAL: Sweet-smelling **annual or biennial**; stems 0.5–1.5 m tall (sometimes to 2.5 m), branched, leafy.

LEAVES: Compound; **3 leaflets**, **oblong**, 1–2.5 cm long, finely sharp-toothed.

FLOWERS: Many, **in narrow, 20–50-flowered clusters**, 3–15 cm long, from leaf axils or stem tips; pea-like, yellow, 4–7 mm long.

FRUITS: Egg-shaped to nearly spherical, **wrinkled pods**, 3–5 mm long, yellowish brown, hang when ripe; 1–2 seeds per pod, often olive green or purple spotted.

WHERE FOUND: Introduced, weedy species of cultivated and waste ground; widespread across most of North America; north to Churchill, Man., Great Slave Lake, central Yukon and interior Alaska; more or less circumpolar.

NOTES: White sweet-clover (*M. alba*) is a similar introduced weed in the boreal forest. It is easily distinguished by its smaller (4–5 mm) white flowers and unwrinkled (but net-veined) pods with yellowish seeds. It also grows in disturbed habitats across our region, most commonly in the prairie provinces. • Fresh or dried sweet-clover is harmless and sweet smelling, a favourite fodder of horses and cattle. A substance called coumarin gives it its vanilla-like taste. If the plants are allowed to mold, the coumarin becomes dicoumarol, a chemical that can cause uncontrollable bleeding. Synthetic compounds such as warfarin (which prevents blood clotting in rats) have been developed from such coumarins. • Sweet-clover was traditionally dried and used in teas to treat headaches, painful urination, nervous stomachs, colic, diarrhea with gas, painful menstruation, and aching muscles. It was also used to make poultices to relieve rheumatism, reduce inflammation, and heal ulcers and wounds, especially in tender areas such as the eyes. Some people smoked the leaves to relieve asthma. The flower clusters and leaves can be made into a pleasant, vanilla-flavoured tea that was a popular beverage and has the added advantage of relieving gas.

ARCTIC LUPINE • *Lupinus arcticus*

GENERAL: Bushy, perennial herb with few to several erect stems from branching rhizome; stems 15-50 cm tall, slender, hollow, long-hairy.

LEAVES: Mainly at stem base, long stalked, compound; **6–8 leaflets arranged like fingers on a hand**, pointed, 1.3–5 cm long.

FLOWERS: In **showy**, 4–14-cm-long **clusters**; **bluish purple, pea-like**, 1.5–2 cm long.

FRUITS: Yellowish hairy pods, 2–4 cm long, twist after opening; 5–8 seeds.

WHERE FOUND: Open areas and grassy slopes; widespread from northern B.C. and central N.W.T. to western Alaska.

NOTES: Many species of lupine provide excellent forage, but some are **poisonous** to sheep, cattle and horses. The fruits are most poisonous. The rhizomes and leaves of some species have been eaten, but only after they are cooked. Raw, they contain toxic alkaloids that cause a 'drunken sleep,' and they can be fatal in large quantities. • When a Yukon mining engineer discovered a hoard of large seeds in some ancient lemming burrows in the permafrost, he saved them and eventually sent some to the National Museum of Canada. Carbon-dating showed the seeds to be **10,000 to 15,000 years old**. Six of the seeds successfully germinated within 48 hours, and 1 year later one of the plants flowered and produced seed. • These attractive plants are **hardy and easy to cultivate**. They are adapted to growing in northern climates, and they make an excellent addition to northern gardens. Lupines can grow on and enrich soils that are low in **nitrogen**. Bacteria in their root nodules can take nitrogen from the air and change it into a form that can be used by the plant. Then, when the lupine dies, this nitrogen is returned to the soil for use by other plants.

EARLY BLUE VIOLET • *Viola adunca*

GENERAL: Tufted perennial from **slender rhizomes**; stems 4–15 cm tall, **short in the spring**, later 10–15 cm tall.

LEAVES: Alternate, egg-shaped, to 3 cm long, heart-shaped base, finely round-toothed, hairy to hairless; stipules reddish brown or with reddish brown flecks, narrowly lance-shaped, smooth-edged to slender-toothed.

FLOWERS: Single, from leaf axils; **blue to deep violet, 8–15 mm long**; **slender spur**, 4–6 mm long; lower 3 petals often white at base, side pair white-bearded.

FRUITS: 4–5 cm long capsules, open explosively by 3 valves; seeds dark brown.

WHERE FOUND: Meadows, open woods, and thickets; widespread across much of our region, north to Great Slave Lake and north-central Yukon.

NOTES: Many violets produce 2 types of flowers. In the spring and early summer, their showy blooms (the flowers we recognize as typical violets) appear. However, only a few of these are fertilized each year. If no seed is produced, the plant produces much smaller flowers in the fall. These closed blooms are seldom seen as they are hidden away among the leaves, at or below ground level. Their petals are reduced to tiny scales and their sepals are half normal size. There are only 2 stamens, instead of 5 and only 2 pollen sacs, instead of 4. They have no scent and they never open. Instead, these flowers fertilize themselves and produce an abundance of seed. The seeds are normal in appearance, but they tend to develop more quickly than seeds produced by cross-pollination. • When violet seed capsules mature, the seeds are explosively shot out to several centimetres away from the parent plant. Violet seeds have special oily bodies called elaiosomes. These attract ants which carry them to their nests, thus dispersing the seeds farther.

BOG VIOLET • *Viola nephrophylla*

GENERAL: **Stemless** perennial from shallow, fleshy, spreading rhizomes.

LEAVES: Simple, 3–5 cm wide, rounded to kidney-shaped, heart-shaped at base, prominently toothed, hairless or nearly so, tip blunt to bluntly-pointed; stalks 5–25 cm long.

FLOWERS: Single, at end of stalks exceeding leaves; violet-purple, 1–2 cm long; **lower 3 petals whitish at base and bearded**; spur 2–3 mm long.

FRUITS: Capsules, 8–10 mm long; seeds olive-brown.

V. nephrophylla

WHERE FOUND: Moist meadows, streambanks, willow thickets and moist open woods; common across our region north to Great Bear Lake, N.W.T.

NOTES: **Marsh violet** (*V. palustris*) has violet to lilac flowers (sometimes white), usually paler than those of bog violet, and arises from slender creeping rhizomes and above-ground runners. Marsh violet occurs in swamps, fens and wet forests and on streambanks, from northern British Columbia east. To the west, it is replaced by the closely related **dwarf marsh violet** (*V. epipsila* ssp. *repens*), which some consider to be part of *V. palustris*. Dwarf marsh violet is very similar to marsh violet but it produces only 2 leaves with the flowers, and these leaves are smaller. • **Great-spurred violet** (*V. selkirkii*) also has pale violet flowers but with 5–8 mm long spurs, and it has thin rhizomes but no runners, and leaves that are hairy on the upper surface. It is uncommon, scattered in moist forests across our region. • When a bee or butterfly visits the flower, lines on the petals act as honey-guides, directing the visitor to the nectaries. The insect must push past the anthers in order to probe for nectar. At the same time, hairs on the side petals and the style comb the insect's body for pollen from other plants.

V. palustris

WESTERN CANADA VIOLET • *Viola canadensis*
CANADA VIOLET

GENERAL: Perennial from **thick, ascending rhizomes**, often with slender, underground **stolons** producing plants at tips; **stems leafy**, 10–40 cm tall.

LEAVES: Heart-shaped, long-stemmed, up to 10 cm across, **sharply pointed**, smaller and simpler towards stem tip, usually hairy on 1 or both surfaces, edged with coarse, rounded teeth.

FLOWERS: Single, from upper leaf axils; **white** to purplish or pinkish; petals **8–12 mm long, yellow at base**, lower 3 have purple lines, upper 2 purple-tinged on back, **side petals bearded**.

FRUITS: Egg-shaped capsules; seeds brown, 1.5–2 mm long.

WHERE FOUND: In moist to fairly dry woods and clearings; widespread across much of our region, north and west to southwestern N.W.T. and southern Yukon.

NOTES: Also called *V. rugulosa*. • The Cree mixed fresh violet leaves in fresh lard and heated this at a low temperature for an hour. The cooled fat was then used as a salve. Violet leaves, soaked in hot water and drained were applied directly to sores or wounds as a poultice. They were replaced every 2 hours if the wound was infected. A relaxing tea was made by pouring boiling water over a handful of leaves and letting the mixture cool. This was said to give restful sleep, to moderate anger, and to act as a mild laxative. The Cree steeped violet plants, including the roots and flowers, in hot water to make a summer tonic. They also chewed the leaves and flowers for this purpose.

KIDNEY-LEAVED VIOLET • *Viola renifolia*

GENERAL: Perennial from short, scaly, ascending rhizomes; **no stolons or leafy, erect stems**.

LEAVES: At stem base, **heart- to kidney-shaped**, 2–5 cm across, rounded at tip, sharp- to **round-toothed, hairy, at least below**.

FLOWERS: Single, from leaf axils; **pure white**; stalks often shorter than leaf stems; petals 8–10 mm long, beardless, lower 3 purple-lined; spur very short.

FRUITS: Purplish, nearly spherical capsules; seeds brown.

WHERE FOUND: Cool, moist to swampy woods; widespread across our region, north and west to Great Bear Lake and southern Alaska.

NOTES: All species of violets are edible, even the garden varieties such as johnny-jump-ups and pansies. The leaves and flowers can be eaten raw in salads, used as potherbs or thickeners or made into tea. Violets are high in vitamins A and C; 125 ml (1/2 cup) of violet leaves contains as much vitamin C as 4 oranges. The flowers have been used as a garnish (fresh or candied) or as a flavouring and colouring in vinegar. They have also been made into jelly and syrup. However, the rhizomes, fruits and/or seeds of some violets are **poisonous**, causing severe stomach and intestinal upset, as well as nervousness and respiratory and circulatory depression. • Violet flowers and leaves have long been used in various herbal remedies as poultices, as laxatives for children and to relieve coughs and lung congestion. • The species name *renifolia*, from the Latin *rens*, 'a kidney,' and *folium*, 'a leaf,' refers to the kidney-shaped leaves typical of this plant.

FIREWEED • *Epilobium angustifolium*

GENERAL: Erect perennial from **rhizome-like roots**, often forms extensive colonies; stems usually unbranched, 0.3–2 m tall (sometimes to 3 m), upper part often purplish and short-hairy.

LEAVES: Alternate, narrowly lance-shaped, 5–15 cm long, slightly paler and **veiny below**.

FLOWERS: More than 15 in long clusters at stem tips; pink to purple (rarely white), 1.5–3.5 cm across; 4 petals; bract below each flower.

FRUITS: Linear seed pods (capsules), green to red, 4–10 cm long; split lengthwise to release 100s of seeds with fluffy, white tufts of hair.

WHERE FOUND: Open woods, burned over forests, waste ground and roadsides; widespread and common across our region, north past treeline to Arctic coast; circumpolar.

NOTES: Broad-leaved willowherb (*E. latifolium*) (lower inset photo), also called river beauty, is a gorgeous, circumpolar plant that is common from the Northwest Territories to Alaska, especially on gravel bars and floodplains. Broad-leaved willowherb has fewer (1–7), larger (2–6 cm across) pink-purple flowers in leafy clusters. It is also shorter (15–70 cm tall) than fireweed and its leaves often have a whitish grey bloom and indistinct veins on the undersurface. • Broad-leaved willowherb is high in vitamins A and C, and it can be eaten raw or as a potherb or in tea. It is sometimes preferable to fireweed, as it is usually less bitter with age. The plants or their roots have been used as poultices on sores and wounds, and dried plants have been powdered and used to stop bleeding. Unlike fireweed, this species does not produce rhizomes. Instead, it reproduces only by seed. Most plants produce large amounts of fluffy-parachuted seeds each year. • The Chipewyan boiled fireweed plants to make a worm-medicine. The Woods Cree chewed the inner root or raw stems and used this as a poultice to draw out infection from boils, abscesses and other wounds. The fresh leaves were used to plaster bruises. Alaskan native peoples took fireweed tea to relieve stomachaches. • Although young plants are sometimes eaten by children, most Chipewyan now consider fireweed inedible. Historically, fireweed was a popular food of native peoples throughout most of its range. The flowers make a colourful addition to salads, but usually it was the young tender plants and later the leaves that were eaten like other greens, either cooked and served with butter and seasoning or added to soups and stews. Fireweed is very high in vitamins A and C. Some people find fireweed slightly laxative, so start with small quantities. The Dena'ina mix cooked plants with the food for their dogs. • In Canada, Russia and England, fireweed has been widely used to make tea. • Fireweed is often very plentiful and it was sometimes laid out to provide a working surface for cleaning fish—old-timer's plywood. Some native groups used the stem fibres to make thread. • Because of its abundance and long flowering season, fireweed is an important source of nectar for honeybees. Mild fireweed honey is available across Canada. • Fireweed is widespread in burned areas, where it is one of the first plants to colonize. It spreads rapidly by seed and by underground networks of rhizomes, and it is often very important in controlling erosion of disturbed areas. Soon after World War II, in which bombs levelled much of London, fireweed appeared in the heart of the city for the first time in generations. • Fireweed is the floral emblem of the Yukon Territory.

PURPLE-LEAVED WILLOWHERB • *Epilobium glandulosum*

GENERAL: Erect **perennial** from small bulb-like buds (offsets), leafy rosettes or remains of previous year's plants; stems 5–120 cm tall, simple or branched, glandular-hairy towards top and in flower cluster.

LEAVES: **Opposite**, 10–14 pairs, lance-shaped to oval or elliptic, 3–12 cm long, pointed, toothed; 4–8 conspicuous side veins; stalk short or stalkless.

FLOWERS: Few to many in erect clusters at stem tips; **white, pink or rose-purple**; petals **1.5–12 mm long**.

FRUITS: Hairy capsules, 3–10 cm long; seeds 0.8–1.6 mm long, ridged, with tuft of soft white hairs 2–8 mm long.

WHERE FOUND: Moist habitats, including streambanks, lakeshores, moist woods, meadows, and disturbed sites; widespread across our region.

NOTES: Northern willowherb (*E. watsonii*, also called *E. adenocaulon*, *E. glandulosum* var. *adenocaulon* or *E. ciliatum*) (inset photo) is usually branched towards the top, has relatively narrow and uncrowded stem leaves, and has smaller flowers. The typically unbranched purple-leaved willowherb has broader stem leaves that are often crowded around the larger flowers, and gland-tipped hairs in the flower cluster. It also has bulb-like offsets from its rhizome-like roots. • The genus name comes from the Greek, *epi*, 'upon,' and *lob*, 'a pod,' in reference to the flower growing upon the pod-like ovary.

E. watsonii

MARSH WILLOWHERB • *Epilobium palustre*

GENERAL: Perennial, 10–60 cm tall; **thread-like runners end in tiny winter buds**; stems simple or branched towards top; densely hairy towards top; **lines of hairs extend down stem from lower leaf stalks**.

LEAVES: Simple, opposite, **flat, linear to lance-shaped**, 2–6 cm long, 4–8 mm wide, stalkless, **blunt**, toothless or finely toothed, hairless except for **short, stiff, curly hairs on edges and veins**.

FLOWERS: Few to many in cluster at stem tip; pink or whitish, **often nodding in bud**; petals 2–9 mm long, **notched**.

FRUITS: Hairy capsules, 3–6 cm long; seeds minutely roughened, with tuft of white to tawny hairs.

WHERE FOUND: Marshes, fens, ditches and pond edges; widespread but seldom common across our region, north to Arctic coast.

NOTES: Narrow-leaved willowherb (*E. leptophyllum*) is as tall as marsh willowherb, but has narrower (linear) leaves with inrolled edges and dense, short hairs on their upper surface. It is found in wet meadows, fens and marshes and has a more restricted distribution, extending across the boreal forest of the prairie provinces to northern British Columbia and southwestern Yukon. • These plants reproduce by small, bulb-like offsets, called 'turions,' at the base of their stems.

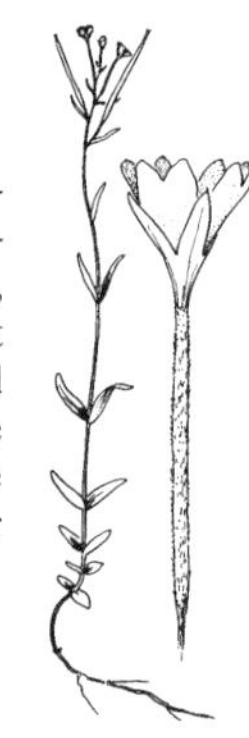

SMALL ENCHANTER'S-NIGHTSHADE • *Circaea alpina*

GENERAL: Delicate, **fleshy perennial**, from **slender rhizomes with tuberous thickenings**, 5–25 cm tall; stems hairless near base, somewhat hairy above.

LEAVES: Opposite, **egg- to heart-shaped**, 2–5 cm long, coarsely toothed, wavy, short-hairy, especially on lower surface; **stalks long, with narrow wings**.

FLOWERS: 8–12 in cluster at stem tip, on slender stalk; **white**; **2 petals**, deeply 2-lobed, 1–3 mm long; 2 sepals, bent backwards.

FRUITS: Club- to pear-shaped capsules (about 2 mm long), covered with hooked, soft hairs; 1 (sometimes 2) seeds per capsule.

WHERE FOUND: Moist, cool woods and thickets; widespread across boreal forest, scattered in parkland, north and west to southwestern N.W.T. and west-central Alaska; circumpolar.

NOTES: The name 'enchanter's nightshade' results, in part, from some confusion. The genus *Circaea* was named after the Greek goddess Circe—the enchantress—who supposedly used this plant 'as a tempting powder in some amorous concerns.' The application of 'nightshade' to this plant is, however, a mystery. It was originally applied to deadly nightshade (*Atropa belladonna*), and somehow it was later transferred to *Circaea*. The origin of the word 'nightshade' was, in itself, a mistake. An old herbal text describes *A. belladonna* as a *solatrum* or 'soothing anodyne.' In translation this was mistaken for the words *solem atrum*, which mean 'black sun' or 'eclipse'—hence the name 'nightshade.'

YELLOW EVENING-PRIMROSE • *Oenothera biennis*

GENERAL: Biennial from stout taproot; forms leafy rosette in first year; second year **stem erect, stiff**, 0.5–1.5 m tall, often branching, greyish or sometimes with reddish hairs.

LEAVES: Alternate, ascending or spreading, **narrowly lance-shaped to oblong**, 2–15 cm long, sparsely to densely hairy, essentially stalkless.

FLOWERS: In spikes at stem tips; **large, bright yellow, open in evening**; buds erect; sepals united at first, ultimately separate and bend back; petals 12–25 mm long.

FRUITS: Erect, hairy capsules, 2–3 cm long.

WHERE FOUND: On lighter soils in dry open areas; often quite weedy; most common in prairie and parkland, extends into southern boreal forest of prairie provinces along some major river valleys.

NOTES: The showy yellow flowers open at night to attract nocturnal insects such as the sphinx moth. Each flower lasts only 1 night before fading, to be replaced by another. • Research suggests that the oil extracted from the seeds may be useful for treating a variety of ailments, including eczema, asthma, migraines, inflammations, premenstrual syndrome, breast problems, metabolic disorders, diabetes, arthritis, and alcoholism. Some extracts act as antispasmodics, and the oil relieved inflammation in animal studies. These plants also appear to have an anti-clotting factor that could be useful in preventing heart attacks. • Native peoples made a tea from the roots to treat obesity and bowel pains. They also used crushed roots as poultices to heal piles and bruises, or rubbed them on muscles to give athletes strength. • Herbalists use the plant mucilage internally to help inhibit coughing and externally to heal sores, pimples, and wounds. • The young shoots can be used in salads, and the roots are said to be wholesome and nutritious when boiled. • Yellow evening-primrose was introduced into Europe as early as 1614, where it was widely cultivated. Later, it was recommended to American gardeners under the name 'German rampion.'

Apiaceae (Carrot Family)

The carrot family, also known as Umbelliferae, is a medium-large family (nearly 3,000 species) mostly of the northern hemisphere.

Carrot family plants are mostly biennial or perennial, aromatic herbs. Their stems usually have a large pith that shrivels at maturity, which leaves the stem hollow between nodes. Their leaves are alternate, often arise from the stem base, and are usually pinnately or palmately compound. The base of the leaf stalk usually sheathes the stem.

The flowers are usually in many-flowered, compact, simple or compound **umbels**. There is often a set of bracts (**involucre**) below the flower cluster. The flowers are individually small and radially symmetrical and have: a single, inferior, 2-chambered ovary; 5 sepals, which are fused to the ovary; 5 distinct petals; and 5 stamens, which alternate with the petals and arise from a disc atop the ovary.

The fruit (**schizocarp**) is dry and splits in 2 when it is mature. Each half (**mericarp**) is 1-seeded, flattened or rounded, often ribbed and sometimes winged or spiny. It is suspended by a slender wiry stalk after the fruit splits.

Key to the Carrot Family

1a. Leaves palmately compound, 5–7 large leaflets; fruit has short, hooked spines ...... ***Sanicula marilandica***

1b. Leaves pinnately compound, 3-parted (trifoliate), or simple; fruits not spiny 2

2a. Leaves at stem base simple, toothed ...... ***Zizia aptera***

2b. Leaves at stem base compound or deeply cut 3

3a. 3 leaflets, large (at least 10 cm across) ...... ***Heracleum lanatum***

3b. More than 3 leaflets, usually less than 10 cm across 4

4a. Plants from taproots or stout, vertical rhizomes ...... ***Osmorhiza depauperata***

4b. Plants from fibrous or fleshy, thickened, clustered roots 5

5a. Leaves once-pinnate; veins do not run to notches along leaflet edges ...... ***Sium suave***

5b. Leaves twice-pinnate; veins run to notches between teeth along edges of leaflets 6

6a. Small bulbs present in axils of upper leaves; leaflets narrow, linear ...... ***Cicuta bulbifera***

6b. No bulbs in leaf axils; leaflets lance-shaped 7

7a. Fruit broader than long; leaflet 10 times or more longer than wide; tuberous roots hardly developed ...... ***Cicuta mackenzieana*** (see *C. maculata*)

7b. Fruit longer than broad; leaflet about 5 times longer than wide; stem base thickened, hollow, with cross-partitions; some roots thickened like tubers ...... ***Cicuta maculata***

HEART-LEAVED ALEXANDERS • *Zizia aptera*

GENERAL: Perennial, 20–60 cm tall; stems clustered.

LEAVES: Simple at stem base, long-stalked, broad, heart-shaped, toothed, 2.5–10 cm long, 1.5–8 cm wide; **stem leaves mostly 3-parted**, with shorter stalks; upper leaves more coarsely, irregularly and sharply toothed than lower ones.

FLOWERS: In compact, densely-flowered, umbrella-shaped clusters; small, bright yellow; without bracts.

FRUITS: Oblong to broadly elliptic, 2–4 mm long, hairless, ribs prominent but not raised.

WHERE FOUND: Moist meadows, streambanks and low ground; across our region in prairie provinces, north to southern Yukon.

NOTES: Heart-leaved Alexanders is also known as meadow-parsnip, suggesting the root looks like or was used like parsnip. However, there are no records of these plants being used as food, and related species are considered **potentially toxic**. Therefore, heart-leaved Alexanders should not be eaten. Foster & Duke (1990) warn, 'Amateurs playing with plants in the parsley family are playing herbal roulette.' • This genus was named for German botanist Johann Baptist Ziz (1779–1829). The species name is from the Greek *a*, 'without,' and *pteryx*, 'a wing,' in reference to the wingless fruits of this plant.

SPREADING SWEET-CICELY • *Osmorhiza depauperata*
BLUNT-FRUITED SWEET-CICELY

GENERAL: Perennial from thick, sweet-scented roots; stems 20–60 cm tall, branched towards top.

LEAVES: Twice divided in 3s; leaflets notched or lobed, coarsely toothed, hairy.

FLOWERS: In loose umbrella-shaped clusters; **rays of clusters wide-spreading or bent downwards**; greenish white, inconspicuous; involucral bracts absent.

FRUITS: Black, 8–13 mm long, baseball-bat-shaped, with bristly hairs, **tip round-tapered**.

WHERE FOUND: Moist deciduous woods; found in southern boreal forest and parkland across prairie provinces into northern B.C. and southern N.W.T.

NOTES: Also called *O. chilensis* var. *cupressimontana* and *O. obtusa*. • A closely related species, **smooth sweet-cicely** (*O. longistylis*), is also found in the southern boreal forest of the prairie provinces. It is distinguished from spreading sweet-cicely by its larger fruits (12–22 mm long) and by the presence of involucral bracts. Although spreading sweet-cicely was not widely used in the north, the Blackfoot ground the roots to make flour. • Spreading sweet-cicely was fed to mares during winter to improve their condition for foaling. The Blackfoot used closely related species to make medicinal teas to treat tickling throats, colds and pneumonia. Historical reports of sweet-scented roots that were made into necklaces or boiled to make a wash for newborn babies probably refer to sweet-cicely.

COW-PARSNIP • *Heracleum lanatum*

GENERAL: Tall, coarse, hairy perennial from **stout taproot or fleshy fibrous roots**; smells pungent when mature; stems single, leafy, hollow, 1–2.5 m tall.

LEAVES: Compound, stalked; **3 leaflets, 10–30 cm wide, variously lobed, coarsely toothed, often maple-like**; woolly-hairy at least when young; smaller and simpler upwards; base of leaf stalk noticeably **inflated and winged**.

FLOWERS: Many in flat-topped, umbrella-like cluster, 10–30 cm across, at stem tip, with 1–4 smaller clusters from side shoots; **white**, small; bracts at base of clusters, soon fall off.

FRUITS: Flattened, egg- to heart-shaped, 7–12 mm long, 1-seeded, **fragrant** sunflower-like seed; side ribs **broadly winged**.

WHERE FOUND: Streambanks, moist woods, clearings and ditches; widespread across our region, north and west to approximately 64°N in western N.W.T. and Alaska.

NOTES: Also called *H. maximum* or *H. sphondylium*. • The Cree gathered young cow-parsnip stalks (about 50 cm high) and roasted them on hot coals. When the stalks went limp the pith was scraped out and eaten. In Alaska, the stem marrow, which is said to have a sweet licorice taste, was eaten raw or cooked. The young roots were also cooked as a vegetable, similar to a rutabaga or parsnip. The stalks of fresh leaves were peeled and eaten fresh, like celery, or they were be boiled, with the water changed 2–3 times. The ashes of the leaves and stem bases have been used as a salt substitute. • The Cree mashed fresh cow-parsnip roots, or made a paste of dried, grated roots and warm water. They put it on swollen legs twice a day for about half an hour (it caused blistering if left on too long). A mixture of ground cow-parsnip root, sweet flag and yellow pond-lily was applied to aching limbs, or to the head in the case of severe headaches, to relieve the pain. It was also used to cure *mancos* (worms of the flesh). A wash made from the same plants was also used with this poultice. • Be careful not to confuse cow-parsnip with the extremely **poisonous water-hemlocks**. The roots and outer skin of cow-parsnip contain furanocoumarin, which **irritates** sensitive skin and can blister the lips. • Dene hunters chewed cow-parsnip roots and used them to bait their bear snares. • Deer, elk, bears and marmots feed on cow-parsnip stems and leaves, especially after the first frosts. In the spring, bears sometimes eat the flowering heads, and in the fall, many birds feed on the large heads of seeds.

SNAKEROOT • *Sanicula marilandica*
BLACK SANICLE

GENERAL: Perennial, 0.4–1.2 m tall; with fibrous roots; stems single, erect, usually only branched towards top.

LEAVES: Long-stalked at stem base and low on stem, **palmately 5–7-parted**, sharply toothed; 2 side segments often deeply 2-parted; usually several stem leaves, smaller and stalkless upwards.

FLOWERS: In rounded, 15–25-flowered, umbrella-shaped clusters; **greenish white**; a few minute, greenish bractlets below flowers.

FRUITS: Egg-shaped, 4–6 mm long, 3–5 mm wide, **covered with many hooked prickles with thickened bases**.

WHERE FOUND: Moist rich meadows, thickets and open woods; across southern boreal forest and parkland, north and west to northern B.C.

NOTES: Native peoples used snakeroot to treat menstrual irregularities, pain, kidney problems, rheumatism, fevers, and a variety of skin conditions. The effects of snakeroot were said to be similar to those of valerian, soothing the nerves and relieving pain. Some herbalists considered it a 'cure all,' capable of cleansing and healing the system, inside and out. Not only would it stop bleeding, diminish tumours, and heal wounds, it would seek out the problem area and focus its effects there. Wherever the trouble lay, this plant would find it. The roots were used as poultices on snakebites; hence the common name snakeroot. Strong snakeroot tea with honey was used as a gargle for sore throats, and the fresh juice was taken to cure dysentery. The root tea was used as a wash for skin diseases and chapped or bleeding skin. Snakeroot contains tannin, which is astringent and this could account for its use in gargles for sore throats. • The name *Sanicula* comes from the Latin *sanare*, meaning 'to heal.'

WATER-PARSNIP • *Sium suave*

GENERAL: Erect, hairless, perennial from short, solid (**not chambered**), stem base with spindle-shaped, **fibrous roots**, sometimes with tuberous thickenings; stems stout, leafy, hollow, strongly ridged, usually branched towards top, 0.5–1.2 m tall.

LEAVES: Alternate, **once divided**; **5–15 lance-shaped to linear leaflets**, 5–10 cm long, saw-toothed; main side veins not directed towards base of teeth; stalk bases sheath stem; underwater leaves much more frequently and finely divided.

FLOWERS: In dense, umbrella-like heads, 5–18 cm across, at stem tips; white, small; **6–10 bracts** at base of flower clusters, **narrow, bent downwards**.

FRUITS: Oval to elliptic, somewhat flattened, 2–3 mm long, **prominently ribbed**.

WHERE FOUND: Marshes, swamps, sloughs and lakeshores; widespread across our region, north and west to interior Alaska.

NOTES: The Woods Cree gathered the roots of water-parsnip and ate them raw, roasted or fried. Richardson (1852) reported that the Chipewyan gave water-parsnip roots to the Franklin expedition. Although water-parsnip stems and roots are edible, raw or cooked, the flower tops are believed to be **poisonous**. • It was widely believed that water-parsnip roots collected along the Churchill River were edible, while those from other areas were poisonous. Today the Chipewyan consider water-parsnip roots to be poisonous. This belief probably comes from the close resemblance of water-parsnip to the very **poisonous water-hemlocks**. • **Warning:** If there is any question at all about whether a plant is water-hemlock or water-parsnip, it should be considered **poisonous**. Water-hemlock is recognized by its yellowish, strong-smelling roots (rather than whitish and sweeter smelling), its smooth (rather than ribbed) stems and its twice-divided leaves. Water-parsnips were collected in the spring and fall, before the plants had started to grow, and at that time of the year they had to be identified by the characteristics of the roots and the previous year's stems.

WATER-HEMLOCK • *Cicuta maculata* var. *angustifolia*

GENERAL: Stout perennial from taproot or cluster of tuberous roots, 0.5–1.8 m tall; stems single or a few together from **tuberous, thickened, chambered base**, leafy, hairless.

LEAVES: At stem base and along stem, **divided 1–3 times**, into many **lance-shaped to oblong leaflets**, sharply pointed and toothed; **side veins end at base of teeth**.

FLOWERS: In compound umbels; 18–28 main branches, each bearing small, compact cluster (umbellet) of 12–25 small white to greenish flowers; **few or no bracts**.

FRUITS: Egg-shaped to circular, usually longer than broad, 2–4 mm long, hairless, with **corky, thickened, unequal ribs**.

WHERE FOUND: Marshes, edges of streams, ponds and lakes, wet ditches and clearings; common and widespread across boreal forest, parkland and prairie north to upper Mackenzie valley with an isolated station in interior Alaska.

NOTES: Also called *C. douglasii* (in part). • **Narrow-leaved** or **Mackenzie's water-hemlock** (*C. mackenzieana* or *C. virosa*) (inset photo) is similar, but has narrower, longer leaflets, 9-21 rays in the primary umbel and 50 or more in the umbellets, and smaller fruits that are about as long as broad or broader. It has a more northern distribution across the boreal forest of the prairie provinces north to western Alaska and grows on lakeshores and in wetlands and shallow water. • **Warning:** All parts of water-hemlock (especially the rhizomes) are extremely **poisonous** and can cause sickness and **death** if they are eaten. Symptoms of water-hemlock poisoning include: stomach pains, nausea and vomiting, weak and rapid pulse, and violent convulsions. Vomiting should be induced at once, followed by the use of a laxative. One rhizome is enough to kill a cow, and a piece the size of a marble will kill a human. Even children making pea-shooters from the hollow stems have been poisoned. • The chambered rhizome and the stem both exude a strong-smelling, oily, yellow liquid, characteristic of this plant. Everyone should be familiar with this plant, even if they do not plan to eat wild greens. An important feature that aids in identification of water-hemlock and serves to distinguish it from other similar plants (e.g., *Angelica, Heracleum* and *Sium*) is the arrangement of the leaf veins. The side veins of its leaflets end at the base of the leaf-edge teeth rather than at the points. Also, the thickened stem base, when cut lengthwise, clearly reveals the chambers and an evil-looking, orange-yellow resin. • In the past water-hemlock was sometimes used for suicides. • During John Palliser's exploration of western Canada, he once heard a noise coming from the swamp near his camp. The Metis explained that the sound came from the poisonous water-hemlock, because of its miraculous nature.

BULB-BEARING WATER-HEMLOCK • *Cicuta bulbifera*

GENERAL: Perennial (sometimes without thickened roots), 0.3–1 m tall; stem single, relatively slender, **not thickened at base**.

LEAVES: Alternate, all borne on stem; middle and lower leaves usually divided into linear segments, toothless or with a few small teeth, 0.5–4 cm long, 0.5–4 mm wide; upper leaves smaller, simpler, with fewer segments, **many bearing 1 or more bulbils in axil**.

FLOWERS: In small umbrella-shaped clusters; small, white; **often absent or infertile**.

FRUITS: Circular, 1-seeded; ribs are broader than narrow intervals.

WHERE FOUND: Marshes, fens, wet meadows, and shallow standing water; widespread but rarely abundant across boreal forest north and west to Great Bear River and interior Alaska.

NOTES: The genus *Cicuta* has been described as the **most poisonous** in North America. The roots are especially toxic and the entire plant should be treated with caution. See water-hemlock (above) for more information. • The fruits of these plants are probably dispersed in the mud by sticking to the feet of waterfowl. Bulb-bearing water-hemlock is unique in this genus for replacing most of its flowers with vegetative bulbils.

ONE-FLOWERED WINTERGREEN • *Moneses uniflora*
SINGLE DELIGHT

GENERAL: Delicate, **evergreen** perennial from slender, creeping rhizomes; flowering stems simple, 3–10 cm tall (sometimes to 15 cm).

LEAVES: Mainly at stem base, opposite or in whorls of 3, **egg-shaped to round**, 1–2.5 cm long, thin, veiny, toothless to **finely toothed**.

FLOWERS: Single, nodding atop long, leafless stalk (with 1 or 2 small bracts); **white, waxy**, fragrant, 1–2.5 cm across; 5 spreading petals; 10 stamens; large, prominent, 5-lobed stigma.

FRUITS: Spherical, erect capsules, 6–8 mm across; split open lengthwise from tip into 5 parts; many tiny seeds.

WHERE FOUND: Cool, moist woodlands, often with deep moss; widespread across boreal forest and northern parkland; circumpolar.

NOTES: Also called *Pyrola uniflora*. • When it does not have flowers, one-flowered wintergreen can be confused with one-sided wintergreen, whose leaves are usually larger with fewer veins. • One-flowered wintergreen plants are edible and contain vitamin C. They have been used occasionally to brew tea. Also, the seeds and their capsules are reported to be edible raw, roasted, parched, ground, etc. However, one-flowered wintergreen is very small, and it is rarely abundant enough to warrant gathering. • The tea has been taken to relieve colds, sore throats, upset stomachs and lung troubles such as tuberculosis. The flowers are reported to be good for treating rashes, bunions, and corns. Simply use a cloth bandage to tie the flowers in place on the affected area. • *Moneses* is a monotypic genus—one-flowered wintergreen is its only species. • It is worth getting down on your hands and knees to take a closer look at the beautiful waxy-looking blossoms of one-flowered wintergreen, and to smell their lovely fragrance.

ONE-SIDED WINTERGREEN • *Pyrola secunda*

GENERAL: Small, **evergreen** perennial from long, slender, much-branched rhizomes; stems single, often woody towards base, with several bracts above, 5–20 cm tall.

LEAVES: Several to many, **mostly at stem base**; blades longer than stalks, 1.5–4 cm long, elliptic-oval to nearly round, **small-toothed** to nearly toothless, green above, paler below.

FLOWERS: 4–20 in long, narrow cluster, **all directed to 1 side**; pale greenish to white, bell-shaped, nodding, 5–6 mm across; **style straight**, projects from flower.

FRUITS: Spherical, 5-chambered capsules.

WHERE FOUND: Woods and thickets; widespread across our region, north to Arctic coast.

NOTES: Also called *Orthilia secunda*. • The Chipewyan mashed the leaves of one-sided wintergreen and mixed them with lard to make a salve that was applied to cuts for 3 days, to stop bleeding and promote healing. The leaves were also chewed to relieve the pain of toothaches. The Woods Cree did not use one-sided wintergreen for medicine. Their name for it means 'grouse leaf,' perhaps because grouse feed on these plants. • The common name 'wintergreen' refers to the evergreen leaves of these plants. The species name *secunda* refers to the one-sided (secund) arrangement of flowers and fruits.

LESSER WINTERGREEN • *Pyrola minor*

GENERAL: Perennial **evergreen** from slender rhizomes; **flowering stem single, 5–20 cm tall**.

LEAVES: In rosette at stem base, 1–4 cm long, **broadly oval to round**, dull, dark, green, thin, with **small, rounded teeth**; base rounded to heart-shaped; stalks usually about as long as blades.

FLOWERS: 5–20 in long cluster at stem tip; **white to flesh-coloured or pink**, nodding, about 6 mm across; **style short, straight**, does not protrude from flower.

FRUITS: Spherical capsules, much longer than short, straight styles.

WHERE FOUND: Moist woods and thickets; widespread across northern boreal forest; circumpolar.

NOTES: Lesser wintergreen is similar to green wintergreen (below), but that species has greenish white flowers with styles that bend downwards and stick out beyond the petals. • Lesser wintergreen flowers have a soft, almond scent. The pollen is released through tiny openings at the end of the anthers. • The genus name *Pyrola* comes from the Latin word *pyrus*, meaning 'pear,' because *Pyrola* leaves were thought to resemble those of a pear tree.

GREEN WINTERGREEN • *Pyrola virens*
GREENISH-FLOWERED WINTERGREEN

GENERAL: Perennial **evergreen** with long, slender rhizomes and leafy, sterile shoots; flowering stems 8–25 cm tall.

LEAVES: 1–few in rosette at stem base; **blades round to broadly oval**, 1–3 cm long, **leathery**, dull green above, darker below, **slightly round-toothed** to toothless; stalks longer than blades.

FLOWERS: 3–10 in long cluster at stem tip; pale **yellowish to greenish white**, nodding, 8–12 mm across; **style curved**, protrudes from flower.

FRUITS: Spherical capsules with 4–7 mm long, curved style.

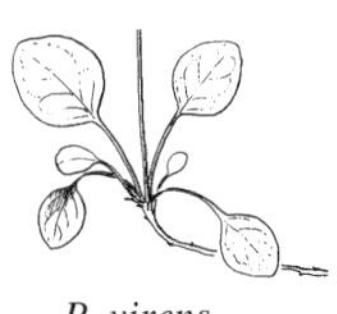

P. virens

WHERE FOUND: Moist woods and thickets; widespread across our region, north and west to southern Alaska; circumpolar.

NOTES: Also called *P. chlorantha*. • **White wintergreen** (*P. elliptica*) has white, often green- or pink-veined flowers not unlike those of green wintergreen, but its leaf blades are broadly elliptic to oblong, 3-7 cm long, and mostly longer than their stalks. • Green wintergreen could also be confused with lesser wintergreen (above), but lesser wintergreen has a nearly straight style and white to pinkish flowers. • The leaves are high in methyl salicylate, a natural painkiller, and they can be used in emergency first aid by chewing the leaves and applying them to wounds as a poultice. • Wintergreens grow best in their natural habitat. They produce very few feeding roots along their spreading underground stems, and these grow in association with special fungi. Transplanting destroys the delicate roots, and the garden environment is unlikely to have the same fungi that are found in the forest.

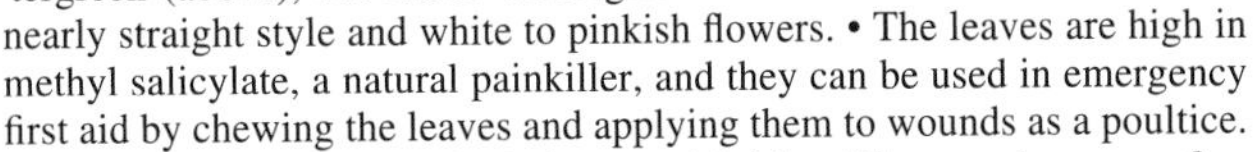

P. elliptica

COMMON PINK WINTERGREEN • *Pyrola asarifolia*

GENERAL: Evergreen perennial from long branched rhizomes; flowering stems **10–30 cm tall**, with a few papery bracts.

LEAVES: Numerous in **basal rosette; blades broadly elliptic to kidney-shaped, 3–6 cm long, leathery, shiny, finely toothed**, dark green above, purplish beneath; stalks usually longer than blades.

FLOWERS: 7–15 in long loose terminal cluster; usually nodding, 8–12 mm across, **pale pink to purplish red**, bell- or cup-shaped; style long, **curved, bends downwards**.

FRUITS: Spherical capsules, 5-chambered, with arching, 5–10 mm long style.

WHERE FOUND: Moist woods and thickets; widespread across our region.

NOTES: If not in flower, **arctic wintergreen** (*P. grandiflora*) (inset photo) could be mistaken for common pink wintergreen. Arctic wintergreen's flowers are creamy white. It is very common in open forests and on sheltered, sunny tundra slopes from the Northwest Territories to Alaska. • The Cree used the leaves of common pink wintergreen to stimulate urine production and clear blockages of the urinary tract. The Chipewyan mashed the leaves with lard and used this paste on cuts to stop bleeding and promote healing. They also chewed the leaves to relieve the pain of a toothache. The Woods Cree steeped the leaves in boiling water to make a wash for sore eyes. They also boiled the whole plant to make a medicinal tea to stop coughing up of blood. The tea was also used as a gargle for sore throats and a wash for skin diseases, cancerous sores and painful tumours. • The Cree called common pink wintergreen 'beaver ears,' because its round leaves were thought to resemble the ears of a beaver.

PRINCE'S-PINE • *Chimaphila umbellata*
PIPSISSEWA

GENERAL: Stout, slightly woody, **dwarf evergreen**, 10-30 cm tall, from long, creeping rhizomes; stems greenish, hairless.

LEAVES: Usually in **whorls of 3-8**, **evergreen**, dark green and **shiny** above, 2-7 cm long, lance-shaped, broadest above middle, **sharply toothed**.

FLOWERS: In small, loose cluster of 3-7; **nodding**, **saucer-shaped**, faintly perfumed; petals light **pink to rose**, **waxy**, 4-6 mm long.

FRUITS: Erect, rounded capsules, 5-7 mm across, split open from tip.

WHERE FOUND: Dry coniferous forests or clearings; scattered across southern part of our region, rarely north of 60°N; circumpolar with large gaps.

NOTES: The Woods Cree used prince's-pine alone or with other herbs to make teas to relieve pain or fever resulting from heart conditions, backache, and coughing up of blood. The leaves were eaten raw or boiled with the roots to make a tonic high in vitamin C. Some native peoples and European herbalists traditionally used prince's-pine as a remedy for fluid retention and kidney or bladder problems. It was often preferred to other herbs because it did not upset the stomach. Some tribes also used it to treat coughs and sore eyes, and to purify the blood. Studies have confirmed that prince's-pine has value as a mild urinary antiseptic. • Prince's-pine is used to flavour candy, soft drinks (especially root beer) and traditional beers. • Prince's-pine leaves were sometimes smoked as a tobacco substitute. • The Cree call it *pipsisikweu*, meaning 'it breaks into small pieces,' because they believed the leathery evergreen leaves dissolved kidney stones.

SALINE SHOOTING STAR • *Dodecatheon pauciflorum*
FEW-FLOWERED SHOOTING STAR

GENERAL: Hairless perennial from pale roots; flowering stems leafless, 5–50 cm tall.

LEAVES: Oblong lance- to spoon-shaped, 3–20 cm long, **gradually narrow** to winged stalks; smooth-edged to weakly toothed.

FLOWERS: Magenta to lavender; 1–25 (usually 3–12) in umbel at stem tip; 5 petals, 10–20 mm long, **swept backwards**, joined at base in yellowish collar with wavy purplish ring; 5 stamens, united into **yellowish to orange** tube, 0.5–3.5 mm long; tipped with 5 anthers, 3–8 mm long, yellowish to reddish purple.

FRUITS: Capsules, cylindrical to egg-shaped, about 12 mm long, 1-chambered, **split up to tip**; seeds numerous.

WHERE FOUND: Salty flats, wet meadows and streambanks; widespread across our region from southern Manitoba to Mackenzie River delta and eastern Alaska.

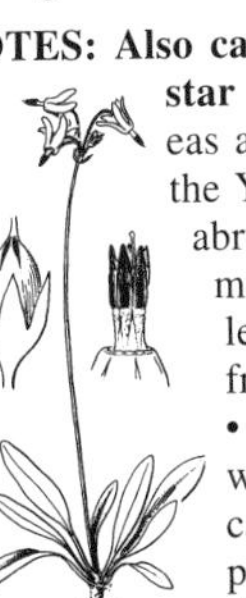

NOTES: Also called *D. pulchellum* and *D. radicatum*. • **Northern shooting star** (*D. frigidum*) commonly grows in wet meadows, seepage areas and on streambanks, from northern British Columbia through the Yukon and Alaska. It has oval to egg-shaped leaves that taper abruptly to the stalk and it has a purplish red tube of stamen filaments. • Saline shooting star is a highly variable species. Its leaves can be densely hairy or hairless and the tube of anthers that projects from the centre of the flower can be yellow, orange or even reddish purple. • The anthers point downwards and have open pores at their tips. The pollen, which is very fine and powdery, is released from the anthers by the vibrations caused by bees with very fast wing movements (up to 300 cycles a second)—a process known as 'buzzing.' • Saline shooting star grows easily from seed in rich, moist, partly shaded areas. It will readily seed itself in future years.

MEALY PRIMROSE • *Primula incana*

GENERAL: Perennial; leafless stems, 10–40 cm tall.

LEAVES: In rosette at stem base, narrowly spoon-shaped, 2–6 cm long, stalkless or with short, winged stalks, shallowly toothed; **conspicuous powdery coating below**.

FLOWERS: Up to 12 in flat-topped cluster at end of leafless stalk; pale lilac with yellow centre, 6–10 mm across; **stalks usually covered with powdery coating**.

FRUITS: 5-chambered elliptic capsules, slightly longer than sepals.

WHERE FOUND: Moist ground, saline meadows and shores; scattered across our region north and west to interior Alaska.

NOTES: Several other primroses grow in this region; all are generally smaller (less than 30 cm tall) and less robust than mealy primrose, and they have little or no powder on their leaves. **Greenland primrose** (*P. egaliksensis*) has oblong to spoon-shaped leaves with smooth edges. Its flowers are violet to lilac with yellow centres and 5-9 mm across, and its sepals are edged with gland-bearing hairs. **Dwarf Canadian primrose** (*P. mistassinica*), also called **bird's-eye primrose**, rarely reaches 15 cm in height and has lance- to egg-shaped leaves with coarse teeth. Its flowers are 1-2 cm across, pink, bluish purple or white, with yellow centres and their sepals have no fringe of hairs. **Erect primrose** (*P. stricta*) has lance- to egg-shaped leaves with blunt-toothed or smooth edges, and lilac flowers that are 5-8 mm across. Its sepals lack glandular hairs and are sac-like at the base. All 3 species grow in wet meadows and on wet, usually calcium-rich lakeshores and riverbanks, scattered across the region. • Never take less than a handful of (cultivated) primroses when visiting a friend who raises ducks or chickens. A report from the 1800s tells of a serious quarrel between 2 elderly women: one apparently sent her neighbour's child home with a single primrose, believing that would cause only 1 egg of every clutch to hatch. The recipient of the flower maintained that the charm had worked, and sought retribution for the chicks she had lost.

SEA MILKWORT • *Glaux maritima*

GENERAL: Hairless, **fleshy perennial** from slender, spreading rhizomes; pale, often with whitish bloom; stems erect or spreading, leafy, 3–12 cm tall, much branched.

LEAVES: Many, **opposite (sometimes alternate towards top), oval to oblong, blunt-tipped**, stalkless, jointed at base, usually 6–12 mm long.

FLOWERS: Single, stalkless in leaf axils near middle of stem; **white or pinkish** to purplish, 3–5 mm long; no petals; **sepals fused into cup** with 5 petal-like lobes.

FRUITS: Spherical capsules, 2–3 mm long; split into 5 parts at top; seeds few, flattened, about 1.5 mm long.

WHERE FOUND: Salt flats and saline wetlands; scattered across southern boreal forest, parkland and prairie, from Manitoba and Saskatchewan north and west to southwestern N.W.T. and southern Yukon; circumpolar.

NOTES: Some native peoples ate the boiled roots of sea milkwort to induce sleep. The roots were also used to make a medicinal tea for nursing mothers to increase their milk supply. • The succulent, salty leaves of sea milkwort were used in Europe as a pickle. • The genus name *Glaux* comes from the Greek *glaucos*, 'bluish green,' in reference to the colour of these plants. The species name, *maritima*, 'belonging to the sea,' and the common name, 'sea milkwort,' both refer to the usual seashore habitat of the these plants. Inland, in the boreal forest, sea milkwort is restricted to salty lakeshores and sloughs. This plant is called 'milkwort' (along with the real milkwort—*Polygala vulgaris*) because it was thought to increase milk production in both livestock and humans.

FRINGED LOOSESTRIFE • *Lysimachia ciliata*

GENERAL: Perennial from slender rhizome, 30–100 cm tall; stems erect, simple or sparingly branched towards top, hairless.

LEAVES: Simple, **opposite**, egg-lance-shaped to egg-shaped, 5–10 cm long, 3–7 cm wide; smooth-edged; pointed tip, usually rounded at base; 5–20 mm long **leaf stalks with hairy fringe on 1 side**.

FLOWERS: Bright yellow, 1.5–2.5 cm across, bell-shaped, deeply 5-lobed, **reddish and glandular at base**; **1–3 on slender stalks in axils of upper leaves**.

FRUITS: Many-seeded egg-shaped capsules, longer than calyx.

WHERE FOUND: Moist thickets, woods and shores; widespread, but not abundant, in southern boreal forest and parkland across prairie provinces and into central B.C.

NOTES: Also called *Steironema ciliatum*. • The common name 'loosestrife' derives from a mistranslation of the genus name *Lysimachia*, as if it were from the Greek *lusi*, 'losing' (i.e. 'ending'), and *makhos*, 'strife' or 'battle.' The genus was, in fact, named for King Lysimachus, a Macedonian general who ruled Thrace from 323–281 B.C. There was an ancient belief that loosestrife would calm savage beasts. It was said to be especially effective with oxen, and a piece hung from their yokes would be enough to appease any strife or unruliness. Loosestrife appears to repel gnats and flies, so when it was placed under the yokes of oxen it could keep these bothersome pests away from their faces. It was also burned in houses to keep flies out. Loosestrife was also believed to be effective against snakes and serpents, which would disappear the moment they encountered the fumes of burning loosestrife. • The species name *ciliata* and the common name 'fringed' loosestrife refer to the fine hairs (cilia) on the leaf stalks.

TUFTED LOOSESTRIFE • *Lysimachia thyrsiflora*

GENERAL: Erect perennial from creeping rhizomes; stems leafy, erect, unbranched, 20–60 cm tall.

LEAVES: Opposite, scale-like on lower stem, larger upwards, 3–10 cm long, **lance-shaped to narrowly elliptic**, stalkless, **dotted with red to purplish black glands**.

FLOWERS: In 4–6 **dense bottlebrush-like clusters on long stalks from axils** of leaves near middle of stem, **yellow**, about 6 mm across, often spotted or streaked with purple; 5 linear-lance-shaped, petal-like lobes; 5 stamens, stick out beyond petals.

FRUITS: Rounded capsules, 2.5 mm across, dotted with glands; few seeds, pitted.

WHERE FOUND: Marshes, lakeshores, ponds and wet meadows; widespread across much of our region, scattered north and west to interior Alaska.

NOTES: Tufted loosestrife has been used to stop bleeding and diarrhea, and it was made into a gargle for sore throats and an eyewash for treating sore eyes or snow blindness. Salves and lotions made with loosestrife were used to heal sores and ulcers and to remove spots, marks and scabs. • The young leaves are said to be edible. • *Thyrsiflora* comes from the Greek *thyrsi*, 'densely branched,' and *flora*, 'flowers,' in reference to the dense, much-branched flower clusters of tufted loosestrife.

NORTHERN STARFLOWER • *Trientalis borealis*

GENERAL: Perennial from slender creeping rhizomes, stems simple, erect, 5–20 cm tall, hairless or with tiny glands.

LEAVES: Simple, **in whorls of 5–9 at stem tip**, with a few small, scale-like leaves below, lance-shaped, 3–10 cm long, stalkless or short-stalked, thin, toothless or finely toothed.

FLOWERS: Single, or sometimes 2 or 3, on slender stalks from centre of leaf cluster, white, 8–14 mm across; 5–9 petals, lance-shaped to egg-shaped, long-pointed.

FRUITS: Spherical, 5-chambered, few-seeded capsule.

WHERE FOUND: Moist woods; across boreal forest of prairie provinces.

T. europaea

NOTES: Also known as *T. latifolia*. **Arctic starflower** (*T. europaea* ssp. *arctica*, also called *T. arctica*) replaces northern starflower from the Northwest Territories to Alaska. Arctic starflower has broader, less-pointed leaves and petals and larger stem leaves. Its white flowers are on stalks that are longer than the leaves in the end cluster (2-5 cm long), and it has smaller leaves spread out along the stem below. It grows in moist, mossy forests from central Alberta to western Alaska. • In the eastern boreal forest, the Montagnais steeped northern starflower plants in boiling water to make a medicinal tea to treat general sickness and later for consumption (tuberculosis). • In Latin, *triens* means 'the third part.' One explanation for the derivation of the genus name is that the plant is about 4" high, or 1/3 of a foot!

Lamiaceae (Mint Family)

The mint family is a fairly large (about 3,000 species) cosmopolitan family mostly of annual or perennial herbs.

Mints usually have opposite leaves, aromatic oils, and stems that are square in cross-section. The flower cluster usually appears to be whorled and is often spike- or head-like at the top of the stem, although *Scutellaria* has single flowers in the leaf axils. The flowers usually have strong bilateral symmetry (except in *Mentha* and *Lycopus*) with 5 fused sepals and 5 petals that are usually fused into a 5-lobed, 2-lipped tube—hence the alternative family name Labiatae, meaning 'lips.' The flowers have 2 or 4 stamens, the ovary is superior and 4-lobed, and the single style arises from the navel at the **base** of the ovary. The fruit typically consists of 4 nutlets.

The mint family shares its distinctive 4-lobed ovary and basal style with the borage family (e.g., tall lungwort, *Mertensia paniculata*), but borages generally have round stems, alternate, rough-hairy leaves and radially symmetrical flowers.

The mint family is an important source of volatile, aromatic oils and garden ornamentals. Aromatic oil producers include sage (*Salvia*), lavender (*Lavendula*), rosemary (*Rosmarinus*), and mint (*Mentha*). Other important culinary herbs prized for flavour or aroma include pot marjoram (*Origanum*), hyssop (*Hyssopus*), basil (*Ocimum*), thyme (*Thymus*), and savory (*Satureja*). Horehound (*Marrubium*) is used in medicine and confectionery. Principal garden ornamentals in this family include: salvia (*Salvia*), bugloss (*Ajuga*), dragonhead (*Dracocephalum*), bee-balm (*Monarda*), and skullcap (*Scutellaria*), as well as species of *Coleus*, *Nepeta*, *Stachys*, *Teucrium*, *Thymus* and *Lavendula*.

WILD MINT • *Mentha arvensis*

GENERAL: Aromatic perennial from creeping rhizomes; stems leafy, ascending to upright, 4-sided, hairy, 15–50 cm tall.

LEAVES: Opposite, lance- to egg-shaped, 1–8 cm long, short-stalked, usually gland-dotted, **saw-toothed**, slightly smaller upwards.

FLOWERS: Many, in dense, **whorled clusters in middle and upper leaf axils, pink to pale purple or white**; petals **4–6 mm long**, fused into 4–5-lobed tube; 5 sepals, 2–3 mm long, fused for about half their length, with gland-tipped hairs; 4 stamens, stick out of flowers.

FRUITS: 4 small, egg-shaped nutlets.

WHERE FOUND: Streambanks, lakeshores, wet meadows and clearings; widespread across our region south of 66°N; circumpolar.

NOTES: The Cree took strong mint tea each morning to alleviate bad breath. A slightly milder tea was taken as a spring tonic. They chewed mint leaves or drank mint tea to cure hiccups, and they also used the tea to reduce giddiness in children. The Chipewyan boiled wild mint plants to make a medicinal tea for coughs. The Woods Cree put the ground flowers of wild mint and yarrow in a cloth, and this was moistened and used to wash the pus from infected gums. They used mint tea to soothe upset stomachs, to relieve headaches and fevers, and to cure and prevent colds. Ground leaves or leafy stems were also placed on the gums to relieve toothaches. If someone had a bad nosebleed, mint was tightly wadded and inserted into the nostril. • The Chipewyan mixed mint leaves with beaver castor to make red fox bait, and they also used mint in lynx bait. • Most northern native groups have used wild mint in teas and as a seasoning, and it is still widely used. The Woods Cree added wild mint leaves to regular tea to improve the flavour. They also mixed wild mint with sturgeon oil to sweeten its smell. • 'Mint' comes from the Latin *mentha* (the source of the genus name), which was from the Greek *minthe*. Minthe was once a nymph, but in a fit of jealousy Persephone turned her into a mint plant.

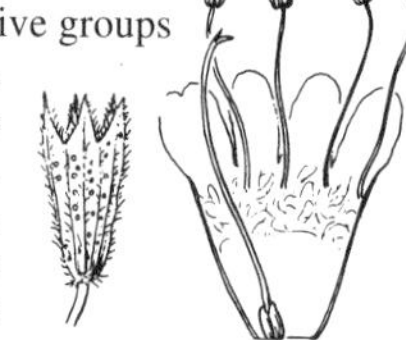

NORTHERN WATER-HOREHOUND • *Lycopus uniflorus*
BUGLEWEED

GENERAL: Perennial from long slender rhizomes and stolons, **often with tubers**; stems erect, single, 10–60 cm tall, **4-sided**, leafy, usually unbranched, finely hairy.

LEAVES: Opposite, lance-shaped to elliptic, 2–7 cm long, short-stalked, only slightly smaller upwards, coarsely and irregularly **toothed**.

FLOWERS: Several to many clustered in upper leaf axils, **white or pinkish, about 3 mm long**, stalkless; nearly regular, 4-lobed, hairy within; 2 stamens; **calyx lobes blunt.**

L. uniflorus

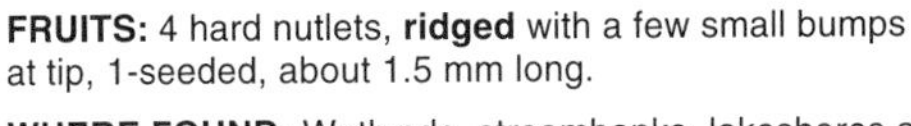

FRUITS: 4 hard nutlets, **ridged** with a few small bumps at tip, 1-seeded, about 1.5 mm long.

WHERE FOUND: Wetlands, streambanks, lakeshores and wet thickets; scattered across boreal forest; north and west to southern N.W.T. with a disjunct population in interior Alaska.

NOTES: Also called *L. virginicus* var. *pauciflorus*. • **Western water-horehound** (*L. asper*) resembles northern water-horehound, but has sharp-pointed calyx lobes. **American water-horehound** (*L. americanus*) is distinguished by its deeply lobed leaves. Both western and American water-horehounds grow in marshes and on streambanks and lakeshores, mostly east of the Rockies and along the southern edge of our region. • Like true horehound (*Marrubium vulgare*), the water-horehounds were used as a folk remedy for coughs. The Blackfoot mixed water-horehounds with other herbs to make a cold remedy for children.

L. americanus

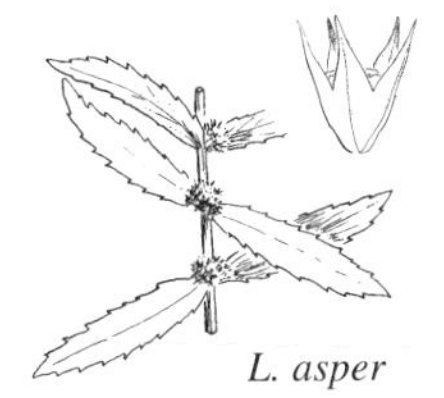

L. asper

MARSH SKULLCAP • *Scutellaria galericulata*

GENERAL: Perennial with slender rhizomes; 1-few stems, erect, **4-sided**, leafy, simple to branched, 10–60 cm tall.

LEAVES: Opposite, lance-shaped to oblong egg-shaped, 2–5 cm long, very short-stalked, hairless above, short hairy below, blunt-toothed.

FLOWERS: In pairs (or single) in upper leaf axils, on stalks about 2 mm long; petals **blue or pink-purple**, streaked with white, **12–20 mm long**, fused into 2-lipped tube; upper lip hooded; **sepals fused into 2 small lips**, raised **bump** on upper lip; 4 stamens.

FRUITS: 4 nutlets, yellowish, warty, raised on small stalk.

WHERE FOUND: Wet meadows, thickets, streambanks, lakeshores, and roadside ditches; widespread across our region, north and west to interior Alaska; circumpolar.

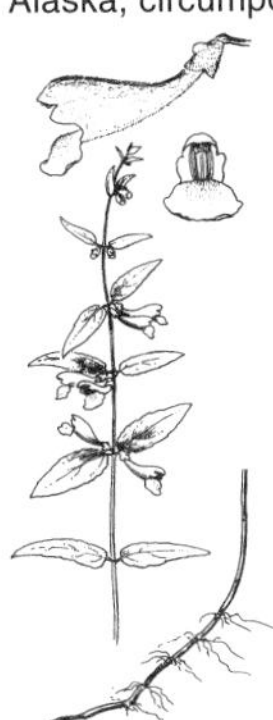

NOTES: Also called *S. epilobiifolia*. • Marsh skullcap contains a flavonoid called scutellarin that has sedative and anti-spasmodic properties. Warm tea made from the dried leaves of these plants has been used by herbalists for over 250 years to treat nervous disorders including insomnia, headaches, hysteria, convulsions, hydrophobia, St. Vitus's dance, epilepsy, and even hiccups. More recently, it has been used to help wean addicts from barbiturates, Valium and meprobamate abuse. In these cases it is thought to relieve some of the latter stage convulsions and frenzies. Similarly, skullcap is mixed with American ginseng to treat people suffering from the delirium tremens of alcoholism. • Although it is not considered poisonous, skullcap should be used in **moderation**. Too much can have an effect similar to caffeine, causing **excitability** and wakefulness.

HEMP-NETTLE • *Galeopsis tetrahit*

GENERAL: Coarse, weedy **annual from taproot**; stems 20–80 cm tall, leafy, erect, **square**, **stiff-hairy**, often swollen below joints.

LEAVES: Opposite, egg- to lance-shaped, 3–12 cm long, stalked, **coarsely toothed**, more or less coarse-hairy.

FLOWERS: Many in 2–6 **dense clusters in leaf axils**, **pale purple, pink or whitish**, **stalkless**; sepals fused into ribbed tube with 5 spine-tipped lobes, become 5–11 mm long; petals commonly have 2 yellow spots, hairy, 15–23 mm long, fused into 2-lipped tube, lower lip 3-lobed and with **2 bumps near base**; 4 stamens, hairy.

FRUITS: 4 nutlets, egg-shaped, 3–4 mm long, grey-brown with white warts.

WHERE FOUND: Common weed of fields, waste places, clearings, roadsides, and gardens; naturalized from Europe; widespread across North America, north to Fort Simpson, N.W.T.; circumpolar.

NOTES: Also called *G. bifida* (in part). • **Handle with care.** The bristly hairs on the stems are strong enough to penetrate the skin, as are the spiny flower clusters. • In the past, hemp-nettle was used mainly to control bleeding. An extract of the whole plant was taken by patients spitting blood or suffering from dysentery. This was also used as a wash to stop the bleeding of wounds and sores. The flowers were boiled to make a medicinal tea for bruises, burns, wounds, rashes, eczema, and other skin problems. Herbalists reported that hemp-nettle was an exhilarating herb, capable of making the heart merry, driving away melancholy and quickening the spirits. • Hemp-nettle was introduced from Eurasia. It appears to have originated as a hybrid between 2 European species of *Galeopsis*. • Hemp-nettle flowers have been baked in sugar and used as a condiment and decoration, and the young leaves were used as a potherb.

AMERICAN DRAGONHEAD • *Dracocephalum parviflorum*

GENERAL: Annual or biennial, 20–60 cm tall; **stems 4-sided**, simple or branched, single or clustered, slightly hairy to nearly hairless.

LEAVES: Simple, **opposite**, **lance- to egg-shaped**, 2–6 cm long, stalked, **sharply toothed**, hairy below, pointed at tip.

FLOWERS: Crowded in dense spikes at branch tips, light blue to violet or pinkish, **irregular**, small, **only slightly longer than calyx**; **calyx stiff and spiny, giving heads a prickly appearance**.

FRUITS: 4 small nutlets.

WHERE FOUND: Open woods and dry disturbed areas; often one of the first species to appear on recently disturbed sites; common and widespread across our region north and west to interior Alaska.

NOTES: Also called *Moldavica parviflora*. • American dragonhead was not widely used by the native peoples of the boreal forest. However, in the American Southwest, the seeds were used for food and the plants were taken as a 'life medicine' and to treat infants with diarrhea. They were also used to make an eyewash. • This hardy annual is an excellent addition to a northern garden. If the seeds are sown in early April, the plants will start blooming in July. The flowers and leaves give off a pleasant lemony scent, and they remain fragrant and fresh for several weeks. • The genus name comes from the Greek *draco*, 'a dragon,' and *cephale*, 'a head,' because of the shape of the flowers. The species name, *parviflorum*, 'small-flowered,' is appropriate since this species has very small flowers.

MARSH HEDGE-NETTLE • *Stachys palustris*
SWAMP HEDGE-NETTLE

GENERAL: Perennial from creeping rhizomes, 30–80 cm tall; **stems 4-sided**, erect, simple or sparingly branched, **hairy throughout, often with glands**.

LEAVES: Simple, **opposite, lance-shaped, oblong or oblong-egg-shaped**, 4–8 cm long, 1–3 cm wide, stalkless or short-stalked, broadly rounded to somewhat heart-shaped at base, toothed, hairy on both sides.

FLOWERS: Leafy-bracted clusters of 6–10 **form interrupted spike at stem tip**, pinkish purple, **mottled with lighter and darker spots**, 10–15 mm long, **strongly 2-lipped**; upper lip hairy.

FRUITS: 4 dark brown nutlets, about 2 mm long.

WHERE FOUND: Moist meadows, streambanks and lakeshores; common and widespread across prairie provinces north and west to interior Alaska.

NOTES: The plump, crisp, tuberous roots of marsh hedge-nettle, collected in the fall, are edible and have a rather agreeable flavour. They can be eaten raw, boiled, baked or pickled. They were also dried, ground and used to make bread in times of necessity. The young shoots can be cooked as a vegetable (like asparagus), but they have a disagreeable smell. • For many years, marsh hedge-nettle had a great reputation as a poultice and as an ingredient in ointments and syrups used to stop bleeding and promote healing of wounds, both inside and out. Apparently, its popularity sprang from a treatise written by a physician to one of the Roman emperors, and this, plus the 'medicinal' smell of the plant perpetuated the belief in its powers. However, no scientific evidence supports these claims. Marsh hedge-nettle was also used, more recently, to relieve gout, cramps, painful joints and vertigo.

GIANT HYSSOP • *Agastache foeniculum*

GENERAL: Perennial with creeping rootstocks, 30–80 cm tall; stems erect, **4-sided**, little branched, hairless (hairy in flower cluster).

LEAVES: Simple, **opposite**, **egg-shaped**, 2–7 cm long, short-stalked, pointed, coarsely toothed, green above, **whitish below with fine, dense hairs**.

FLOWERS: In dense, sometimes interrupted, 2–10 cm long **spikes at stem tips**, blue, 2-lipped **(irregular)**, 6–12 mm long.

FRUITS: 4 small nutlets, held in calyx, each with 1 erect seed.

WHERE FOUND: Moist meadows, thickets and open deciduous woods; in southern boreal forest and parkland across prairie provinces north to southernmost N.W.T.

NOTES: Giant hyssop has a pleasant anise-like scent and it was usually used to make tea or to flavour cooked dishes. The Woods Cree added the leaves of giant hyssop to store-bought tea to improve the flavour. • Herbalists used the leaf tea to treat fevers, colds, and coughs, to cause sweating, and to strengthen a weak heart. The Woods Cree mixed the stems and leaves with other plants and steeped or boiled them to make a medicinal tea for people who were coughing up blood. • The genus name comes from the Greek *agan*, 'very much,' and *stachys*, 'a spike,' in reference to the many-flowered spikes.

Scrophulariaceae (Figwort Family)

The figwort family in our region is represented by herbs with alternate or opposite leaves that are simple and have smooth or lobed edges. Some of the plants in this family (from the *Melampyrum*, *Pedicularis* and *Castilleja* genera) are partially parasitic, even though they contain chlorophyll.

The flowers typically are strongly bilaterally symmetric, with a tube of fused petals. They are often conspicuous (*Pedicularis*), 2 lipped, and 4-5 lobed. *Veronica* is a weakly bilaterally symmetric, short-tubed exception to the rule. Brightly coloured bracts occur in *Castilleja*. Stamens commonly number 4; *Veronica* has only 2 stamens. The ovary is single and superior. The fruit is typically a 2-chambered capsule; seeds are numerous, smooth or variously roughened, angled, or winged.

The figwort family includes the drug plant *Digitalis*, but otherwise has little economic importance except as highly-valued garden flowers. Notable ornamentals are the snapdragons (*Antirrhinum*), speedwells (*Veronica*), beardtongues (*Penstemon*), monkey-flowers (*Mimulus*), foxgloves (*Digitalis*), and slipperflowers (*Calceolaria*).

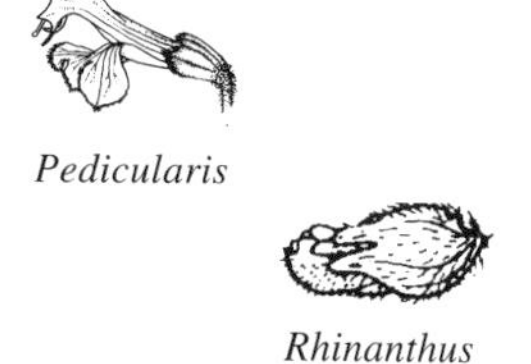

Pedicularis

Rhinanthus

Castilleja

Veronica

Melampyrum

LABRADOR LOUSEWORT • *Pedicularis labradorica*

GENERAL: Biennial or short-lived perennial from spindly taproot, 10–30 cm tall; stems **single** or several, **usually branched**, leafy, **hairy**.

LEAVES: Stem leaves alternate, **deeply notched, coarsely toothed**; smaller, simpler upwards.

FLOWERS: 5–10 in small, loose clusters at stem tips or single in axils of leaves, about 12 mm long, **yellowish**; upper lip hooded with **2 slender teeth near tip, often tinged with crimson**; lower lip 3-lobed, fringed with tiny hairs.

FRUITS: Hairless, flattened, curved capsules, 10–13 mm long.

WHERE FOUND: Dry to moist woodlands, thickets, bogs, heath and tundra; across boreal forest from northern Manitoba to Alaska; north to the western Arctic coast; rare in Saskatchewan.

NOTES: All louseworts are partial parasites. Although they have green leaves that are capable of photosynthesizing food for the plant, they also have sucker-like attachments to join their roots to the roots of neighbouring plants. Dwarf birch is a frequent host to lousewort. Once attached, the lousewort can draw water and nutrients from the host plant. Because of this interconnection, louseworts can be very difficult to transplant.

SWAMP LOUSEWORT • *Pedicularis parviflora*
SMALL-FLOWERED LOUSEWORT

GENERAL: Annual or biennial from slender **taproot**; stems single, leafy, **branched towards top**, hairless, 10–60 cm tall.

LEAVES: Mostly on stem, usually alternate, stalkless, **deeply lobed**, blunt-toothed.

FLOWERS: In short clusters at stem tips, with leafy bracts, or from leaf axils, **purple**; 10–15 mm long; upper lip rounded, **slightly hooded**, with or without small teeth near tip.

FRUITS: Hairless capsules, 8–17 mm long, partially enclosed by dry, expanded and partly fused sepals.

WHERE FOUND: Fens and wet meadows; scattered across boreal forest from central Manitoba to British Columbia; north and west to Great Slave Lake and Alaska.

NOTES: Also called *P. macrodonta.* • Louseworts have typical bee-pollinated flowers. Each flower has a landing platform for the bee and produces large amounts of nectar. Most flowers are brightly coloured. It is interesting to note the ranges of louseworts and bumblebees coincide, and some botanists believe that louseworts are specifically adapted for fertilization by bumblebees. • Each plant produces very few seeds, which are viable for only a short time and only while they remain moist. • The species name, *parviflora*, means 'small-flowered.'

ELEPHANT'S-HEAD • *Pedicularis groenlandica*

GENERAL: Erect, hairless perennial from rhizomes; stems unbranched, often clustered, 20–50 cm tall, often reddish purple.

LEAVES: At stem base, lance-shaped, 5–20 cm long, deeply lobed (**fern-like**) and sharply toothed; several stem leaves, smaller and simpler upwards.

FLOWERS: Many in dense, 5–15 cm long **spike** at stem tip, **pink-purple to reddish**; upper lip about 1 cm long, **strongly hooded and beaked, resembles head and trunk of elephant**; lower lip broadly 3-lobed, 2 side lobes resemble elephant's ears.

FRUITS: Hairless, flattened, curved capsules, 8–14 mm long.

WHERE FOUND: Fens, streambanks and wet meadows; scattered across boreal forest; south of 60°N.

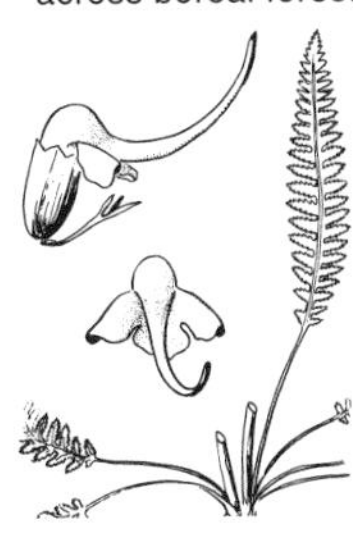

NOTES: All louseworts (*Pedicularis* species) are said to be edible, either raw or cooked as a potherb. The roots and young flowering stems were usually used as food. However, all louseworts contain enough **poisonous** glycosides to cause severe illness if they are eaten in quantity. Most animals will not eat lousewort, and few incidents of poisoning have been reported. The potency of the poison varies with the species. Symptoms of lousewort poisoning include a befuddled lethargy and interference with motor control, particularly in the legs. • The common name 'elephant's-head' describes the appearance of the flowers.

YELLOW RATTLE • *Rhinanthus borealis*
RATTLEBOX

GENERAL: Annual; stems leafy, erect, simple or few-branched towards top, usually 30–60 cm tall.

LEAVES: Opposite, stalkless, lance-shaped to linear, 2–5 cm long, finely toothed, with rough hairs.

FLOWERS: In 1-sided, leafy bracted spike-like clusters at stem tips, **yellow**, showy, 1–2 cm long, 2-lipped; upper lip hooded; lower lip 3-lobed; sepals hairy, membranous, 4-toothed, 1–2 cm long, **fused, expand like flattened balloon** in fruit; 4 stamens.

FRUITS: Capsules, flattened, narrowly winged, round, 5–15 mm long, enclosed in sepal balloon; many seeds, 4–6 mm long.

WHERE FOUND: Meadows, open woods, roadsides, and moist clearings; widely scattered across northern boreal forest from Hudson Bay to Great Bear River and southwest Alaska.

NOTES: Also called *R. minor* and *R. crista-galli* in part (according to some). • Yellow rattle is semi-parasitic on nearby grasses. Although it has normal leaves and can produce its own food through photosynthesis, it also joins its roots to the roots of its neighbours to steal some of their food and water. Yellow rattle roots are equipped with relatively large outgrowths (haustoria) that clasp the host root. The growth of the grasses is often stunted by this parasitism, and in times of drought, it is not uncommon to see a healthy yellow rattle surrounded by drooping host plants from which it is drawing extra water. • When yellow rattle flowers mature, the calyx inflates and dries out to make a hollow, balloon-like container for the many, small seeds. The loose seeds rattle around in this container—hence the common name 'yellow rattle.' Eventually, the seeds are flung from their container as the wind blows the seed heads back and forth. The seeds must find a suitable site quickly, as they soon die if they dry out. However, if they remain moist, they have enough food stored to allow the seedling to grow extensively in search of a suitable host plant.

COMMON RED PAINTBRUSH • *Castilleja miniata*

GENERAL: Perennial from woody base; stems erect or ascending, **20–60 cm tall**, hairless to short-hairy or somewhat sticky-hairy.

LEAVES: Linear to lance-shaped, usually smooth-edged, hairless to fine-hairy; upper leaves sometimes 3-lobed.

FLOWERS: In dense spikes at stem tips, largely hidden by showy bracts; petals 2–3.5 cm long, greenish with red edges, fused into tube with projecting hooked upper lip and tiny lower lip; bracts showy, **bright red to scarlet, usually toothed** and hairy.

FRUITS: Capsules; many net-veined seeds.

WHERE FOUND: Open woods and meadows, thickets, gravel bars, roadsides; widespread across southern boreal forest and parkland from Manitoba to north-central British Columbia.

NOTES: Purple paintbrush (*C. raupii*) has narrow, indistinctly 3-ribbed leaves and a purplish spike with flowers in which the lower lip has distinct lobes. It grows along forest edges, streambanks and lakeshores, from northern Manitoba to the Mackenzie River delta and southern Alaska. It is common in the Mackenzie River valley. • In Alberta, some native groups drank a mixture of red paintbrush and warm water to stimulate urination. Women suffering from abnormal bleeding drank red paintbrush tea in large quantities and also rubbed it on their abdomens. Similarly, it was given to people who were spitting up blood, and was rubbed on their chests. In the southern U.S., red paintbrush flowers were put into love charms or made into tea to be used as a love medicine. • Like yellow rattle, paintbrushes join roots with their neighbours to steal nutrients. Many paintbrush species have almost no root hairs, reflecting their dependence on neighbouring plants.

COW-WHEAT • *Melampyrum lineare*

GENERAL: Slender annual; stems few-branched or simple, 10–40 cm tall, fine-hairy to hairless, glandular (especially above).

LEAVES: Opposite, short stalked, 2–6 cm long, linear or lance-shaped, mostly smooth-edged but sometimes with bristle-tipped teeth near base.

FLOWERS: Single in upper leaf axils or in leafy spikes at stem tips, **white or pinkish**; petals tubular, 8–12 mm long, 2-lipped with yellow patch in throat, surrounded by **green, leafy bracts**.

FRUITS: Curved, asymmetrical, flattened capsules; 1–4 seeds 2–4 mm long, blackish with light coloured tips.

WHERE FOUND: Dry, sandy sites, often in pine woodlands; widespread across boreal forest from Manitoba to B.C.

NOTES: The name 'cow-wheat' was originally applied to the European species *M. arvense*, which was 'freely cropped by passing cattle, to which it was fed in times of scarcity.' Linnaeus claimed that the yellowest butter was produced from the milk of cows that had eaten cow-wheat. Another source says cow-wheat was given this name because its seed resembled wheat, although it was useless to man. Our cow-wheat is much too small and scattered to be a forage plant. • The genus name *Melampyrum* comes from the Greek *melas*, 'black,' and *pyros*, 'wheat,' because when cow-wheat seeds were mixed with grains of wheat and ground into flour, the resulting bread tended to be black.

AMERICAN BROOKLIME • *Veronica americana*

GENERAL: Hairless perennial from shallow **creeping rhizomes or rooting stems**; stems weak, somewhat succulent, simple, trailing to ascending, 10–60 cm long.

LEAVES: Opposite, 2–7 cm long, egg-shaped to oblong-lance-shaped, short-stalked, sharply pointed, toothed.

FLOWERS: In long, loose clusters from upper leaf axils, **blue to violet, lilac** or white, saucer-shaped, 3–6 mm across; 2 large, spreading stamens.

V. americana

FRUITS: Round capsules, sometimes slightly notched; many seeds.

WHERE FOUND: Wetlands, ditches, ponds and streambanks; scattered across boreal forest from southern Manitoba north and west to central Yukon and interior Alaska.

V. scutellata

NOTES: Also called *V. beccabunga* ssp. *americana*. • **Marsh speedwell** (*V. scutellata*) grows across our region, in swamps, wet thickets and ditches, and around springs. It has narrower (linear), stalkless leaves that are smooth-edged or weakly toothed. • All *Veronica* species are edible. Usually these small plants are eaten raw, like watercress, in salads. However, older plants become bitter, and are best cooked. American brooklime is high in vitamin C. Care should be taken to **avoid** plants growing in **polluted** water. • Some flowers were said to bear markings resembling those on the handkerchief of St. Veronica after she used it to wipe the face of Jesus as he carried the cross (the *vera iconica*, meaning 'true likeness')—hence the name *Veronica*. *Veronica* grows in wet mud where birds may become trapped or 'limed'— hence the name 'brooklime'.

Asteraceae (Aster Family)

The composites—plants of the aster family (Asteraceae, also known as Compositae)—make up the largest plant family in the world (or perhaps the second largest, after the orchids), with over 21,000 species worldwide.

The family is cosmopolitan in range but especially well adapted to fairly dry, temperate and cooler climates. It has considerable economic importance through food plants such as lettuce (*Lactuca*), globe artichoke (*Cynara*), endive and chicory (*Cichorium*), salsify (*Tragopogon*), and sunflower (*Helianthus*). The contact insecticide pyrethrum comes from *Chrysanthemum coccineum*. Many species are noxious weeds (e.g., Canada thistle, *Cirsium arvense*), while others are used in medicinal preparations or herbal teas or as ornamentals.

A key to species is provided. Many more species occur in our region than are covered in this guide. The best additional references are Hulten (1968), Porsild and Cody (1980) and Looman and Best (1979).

Composites are easily recognized by their flower cluster, which is often mistaken for a single large flower, but is actually made up of many individual flowers on the broadened top of the stem. These tiny flowers can be of 2 forms: tubular (**disc flowers**); or strap-shaped (**ray flowers**). The flower clusters can be composed entirely of disc flowers (e.g., pineappleweed, p. 176) or of ray flowers (e.g., common dandelion, p. 174), but in the typical 'daisy,' the central 'button' is made up of disc flowers and the outer 'petals' are ray flowers. When you play 'loves me, loves me not,' you are plucking ray flowers. Each individual flower has 5 fused petals, 5 stamens, and an inferior, 1-chambered ovary. What originally were the sepals are interpreted to have been modified into hairs (collectively called the **pappus**) that crown the single-seeded fruits (achenes) and assist in wind-dispersal (as in the downy parachute of a dandelion). Less commonly the sepals are modified into scales, awns, or hooks. Attached to the rim of the head is an **involucre**, consisting of scale-like or somewhat leaf-like **involucral bracts**.

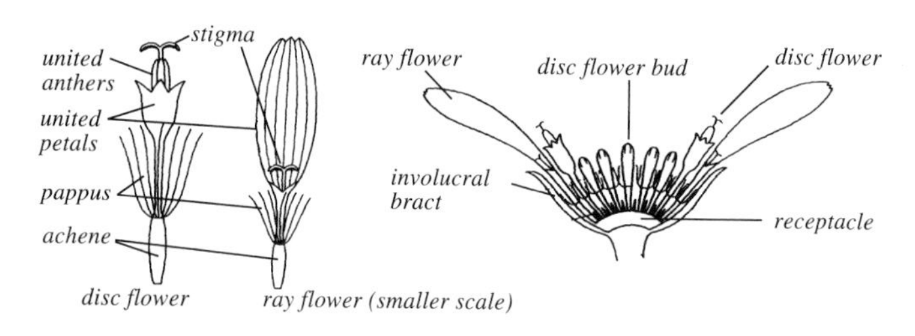

Key to the Aster Family

1a. Flower heads only have strap-like ray flowers, no tubular disc flowers; plant juice milky .... **Group I**

1b. Flower heads have tubular disc flowers, ray flowers may be absent; plant juice usually watery 2

2a. Flower heads composed of disc flowers only 3

3a. Pappus at tip of achene made of hair-like or feathery bristles .... **Group II**

3b. Pappus at tip of achene made of scales or awns, or a mere crown .... **Group III**

2b. Flower heads have both ray and disc flowers 4

4a. Ray flowers white, pink, purple or blue .... **Group IV**

4b. Ray flowers yellow or orange 5

5a. Pappus at tip of achene chaffy (when present), made of firm bracts .... **Group V**

5b. Pappus at tip of achene made of mostly hair-like, sometimes feathery, bristles .... **Group VI**

Group I: Flower heads only have strap-like ray flowers, no tubular disc flowers; plant juice milky

1a. All leaves at stem base; flower heads solitary (the dandelions) 2

2a. Leaves deeply toothed or lobed along entire length, without larger end segment; achenes reddish or purplish .... ***Taraxacum laevigatum*** (see *T. officinale*)

2b. Leaves less deeply toothed or lobed, often with larger end segment; achenes yellowish to brown .... ***Taraxacum officinale***

1b. Stems leafy; several flower heads 3

3a. Stem leaves narrow, usually less than 1 cm wide, smaller higher on stem .... ***Hieracium umbellatum***

3b. Stem leaves well-developed, broad, usually more than 1 cm wide (if less, flowers are blue) 4

4a. Stem leaves do not clasp stem; flowers blue; achenes beaked .... ***Lactuca tatarica***

4b. Stem leaves have ear-like, clasping lobes (auricles) at theirbases; flowers yellow; achenes beakless (the sow-thistles) 5

5a. Annuals or biennials from short taproots; flower heads mostly less than 3 cm across 6

6a. Leaves stiff, spiny, mostly unlobed; rounded flanges at leaf-base clasp stem .... ***Sonchus asper*** (see *S. arvensis*)

6b. Leaves soft, not spiny, pinnately lobed; sharp-pointed flanges at leaf-base clasp stem .... ***Sonchus oleraceus*** (see *S. arvensis*)

5b. Perennials from deep roots; flowers heads mostly more than 3 cm across 7

7a. Involucral bracts and flower stalks have coarse, gland-tipped hairs .... ***Sonchus arvensis***

7b. Involucral bracts and flower stalks smooth or slightly hairy .... ***Sonchus uliginosus*** (see *S. arvensis*)

Group II: Flower heads have disc flowers only; pappus made of hair-like or feathery bristles

1a. All flowers have both male and female parts (perfect) 2

2a. Involucral bracts in 1 row; flowers yellow 3

3a. Plants annual, small (to 30 cm tall), usually unbranched; leaves not much smaller higher on stem; some involucral bracts black-tipped .... ***Senecio vulgaris***

3b. Plants perennial, tall (to 80 cm tall), sometimes branched towards top; leaves smaller and simpler higher on stem; some involucral bracts purple-tipped .... ***Senecio indecorus*** (see *S. pauperculus*)

2b. Involucral bracts in 2 or more rows; flowers pink to purple 4

4a. Tip of flowering stem (where flowers attach) smooth; leaves whorled, toothed; flower heads small, about 6 mm across .... ***Eupatorium purpureum* var. *maculatum***

4b. Tip of flowering stem (where flowers attach) densely bristly; leaves alternate, spiny; flower heads large, 3.5–5 cm across .... ***Cirsium drummondii*** (see *C. arvense*)

1b. Male and female flowers usually on separate plants 5

5a. Leaves spiny; tip of flowering stem (where flowers attach) densely bristly; flower heads 12–25 mm across .... ***Cirsium arvense***

5b. Leaves smooth-edged, mostly at stem base, white-woolly, at least underneath; tip of flowering stem (where flowers attach) smooth; flower heads less than 10 mm across (the pussytoes) 6

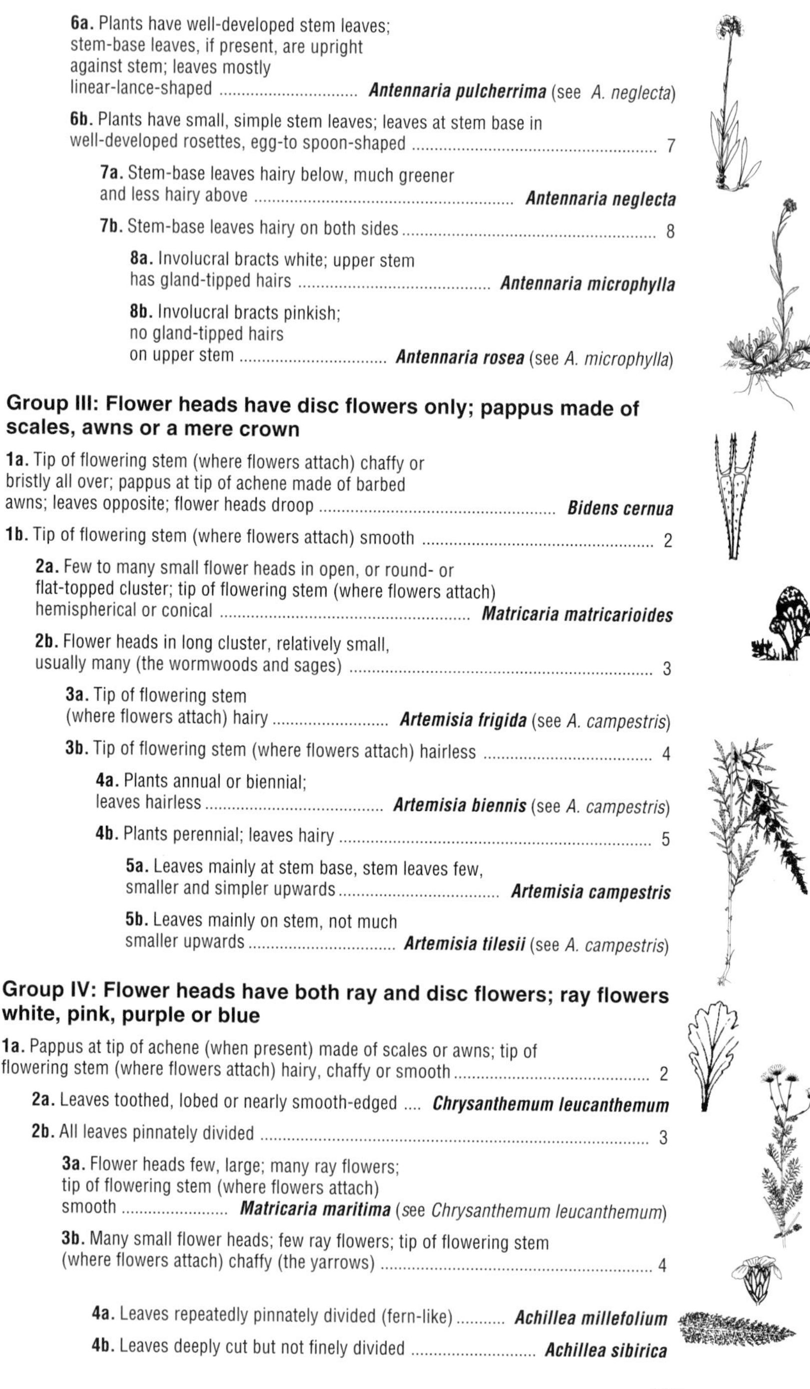

6a. Plants have well-developed stem leaves; stem-base leaves, if present, are upright against stem; leaves mostly linear-lance-shaped .... ***Antennaria pulcherrima*** (see *A. neglecta*)

6b. Plants have small, simple stem leaves; leaves at stem base in well-developed rosettes, egg-to spoon-shaped 7

7a. Stem-base leaves hairy below, much greener and less hairy above .... ***Antennaria neglecta***

7b. Stem-base leaves hairy on both sides 8

8a. Involucral bracts white; upper stem has gland-tipped hairs .... ***Antennaria microphylla***

8b. Involucral bracts pinkish; no gland-tipped hairs on upper stem .... ***Antennaria rosea*** (see *A. microphylla*)

Group III: Flower heads have disc flowers only; pappus made of scales, awns or a mere crown

1a. Tip of flowering stem (where flowers attach) chaffy or bristly all over; pappus at tip of achene made of barbed awns; leaves opposite; flower heads droop .... ***Bidens cernua***

1b. Tip of flowering stem (where flowers attach) smooth 2

2a. Few to many small flower heads in open, or round- or flat-topped cluster; tip of flowering stem (where flowers attach) hemispherical or conical .... ***Matricaria matricarioides***

2b. Flower heads in long cluster, relatively small, usually many (the wormwoods and sages) 3

3a. Tip of flowering stem (where flowers attach) hairy .... ***Artemisia frigida*** (see *A. campestris*)

3b. Tip of flowering stem (where flowers attach) hairless 4

4a. Plants annual or biennial; leaves hairless .... ***Artemisia biennis*** (see *A. campestris*)

4b. Plants perennial; leaves hairy 5

5a. Leaves mainly at stem base, stem leaves few, smaller and simpler upwards .... ***Artemisia campestris***

5b. Leaves mainly on stem, not much smaller upwards .... ***Artemisia tilesii*** (see *A. campestris*)

Group IV: Flower heads have both ray and disc flowers; ray flowers white, pink, purple or blue

1a. Pappus at tip of achene (when present) made of scales or awns; tip of flowering stem (where flowers attach) hairy, chaffy or smooth 2

2a. Leaves toothed, lobed or nearly smooth-edged .... ***Chrysanthemum leucanthemum***

2b. All leaves pinnately divided 3

3a. Flower heads few, large; many ray flowers; tip of flowering stem (where flowers attach) smooth .... ***Matricaria maritima*** (see *Chrysanthemum leucanthemum*)

3b. Many small flower heads; few ray flowers; tip of flowering stem (where flowers attach) chaffy (the yarrows) 4

4a. Leaves repeatedly pinnately divided (fern-like) .... ***Achillea millefolium***

4b. Leaves deeply cut but not finely divided .... ***Achillea sibirica***

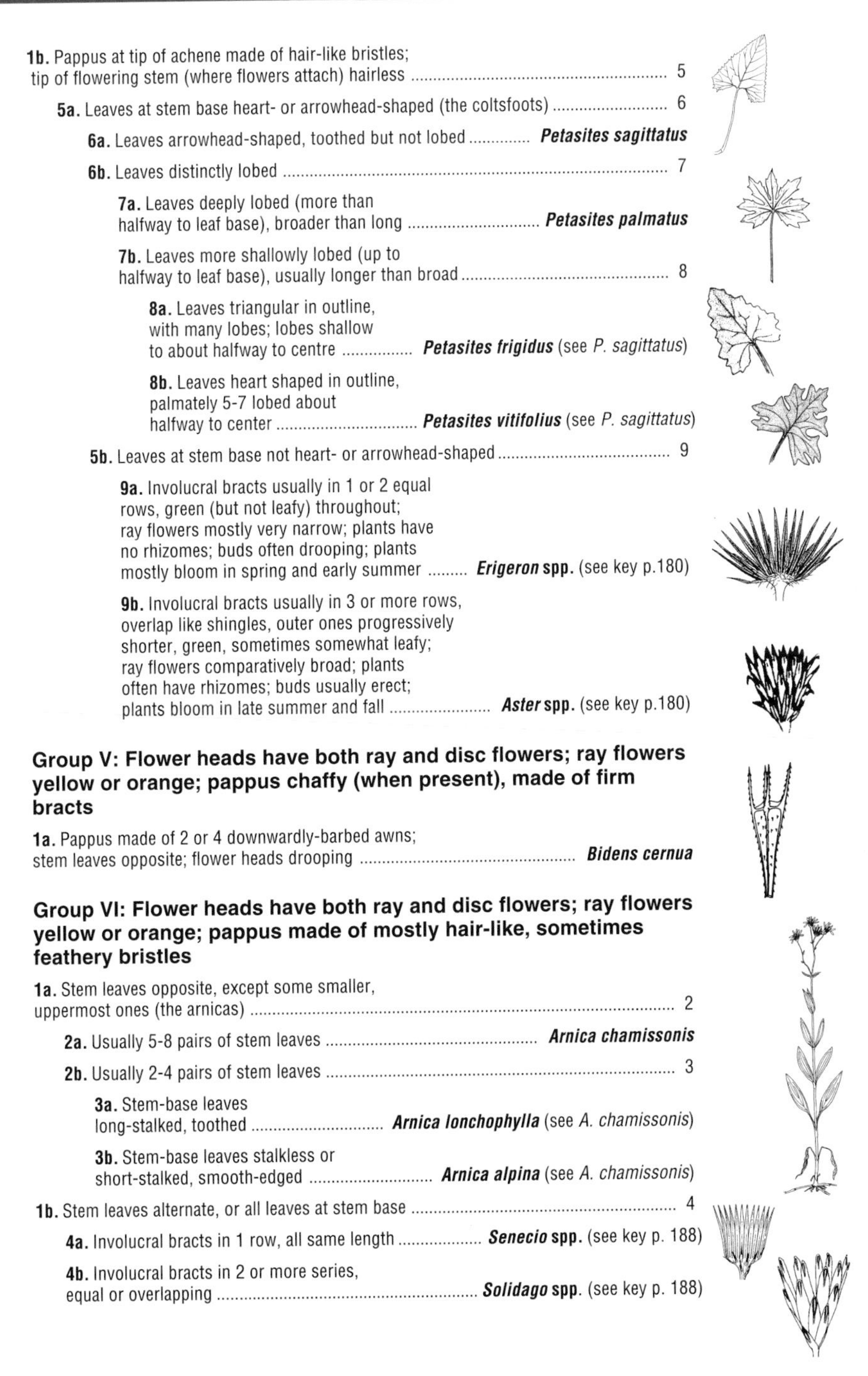

1b. Pappus at tip of achene made of hair-like bristles; tip of flowering stem (where flowers attach) hairless 5

5a. Leaves at stem base heart- or arrowhead-shaped (the coltsfoots) 6

6a. Leaves arrowhead-shaped, toothed but not lobed .... ***Petasites sagittatus***

6b. Leaves distinctly lobed 7

7a. Leaves deeply lobed (more than halfway to leaf base), broader than long .... ***Petasites palmatus***

7b. Leaves more shallowly lobed (up to halfway to leaf base), usually longer than broad 8

8a. Leaves triangular in outline, with many lobes; lobes shallow to about halfway to centre .... ***Petasites frigidus*** (see *P. sagittatus*)

8b. Leaves heart shaped in outline, palmately 5-7 lobed about halfway to center .... ***Petasites vitifolius*** (see *P. sagittatus*)

5b. Leaves at stem base not heart- or arrowhead-shaped 9

9a. Involucral bracts usually in 1 or 2 equal rows, green (but not leafy) throughout; ray flowers mostly very narrow; plants have no rhizomes; buds often drooping; plants mostly bloom in spring and early summer .... ***Erigeron* spp.** (see key p.180)

9b. Involucral bracts usually in 3 or more rows, overlap like shingles, outer ones progressively shorter, green, sometimes somewhat leafy; ray flowers comparatively broad; plants often have rhizomes; buds usually erect; plants bloom in late summer and fall .... ***Aster* spp.** (see key p.180)

Group V: Flower heads have both ray and disc flowers; ray flowers yellow or orange; pappus chaffy (when present), made of firm bracts

1a. Pappus made of 2 or 4 downwardly-barbed awns; stem leaves opposite; flower heads drooping .... ***Bidens cernua***

Group VI: Flower heads have both ray and disc flowers; ray flowers yellow or orange; pappus made of mostly hair-like, sometimes feathery bristles

1a. Stem leaves opposite, except some smaller, uppermost ones (the arnicas) 2

2a. Usually 5-8 pairs of stem leaves .... ***Arnica chamissonis***

2b. Usually 2-4 pairs of stem leaves 3

3a. Stem-base leaves long-stalked, toothed .... ***Arnica lonchophylla*** (see *A. chamissonis*)

3b. Stem-base leaves stalkless or short-stalked, smooth-edged .... ***Arnica alpina*** (see *A. chamissonis*)

1b. Stem leaves alternate, or all leaves at stem base 4

4a. Involucral bracts in 1 row, all same length .... ***Senecio* spp.** (see key p. 188)

4b. Involucral bracts in 2 or more series, equal or overlapping .... ***Solidago* spp.** (see key p. 188)

COMMON DANDELION • *Taraxacum officinale*

GENERAL: Robust perennial with **milky juice**, from thick **taproots**; 1-several leafless flowering stems, 5–40 cm tall (sometimes to 60 cm).

LEAVES: At stem base, lance- to spoon-shaped, 5–40 cm long, 1–10 cm wide, **pinnately lobed or divided** with large end lobe and backwards-pointing, triangular segments, taper to more or less winged stalks.

FLOWERS: Single yellow heads, 35–50 mm across; ray flowers only; involucre 15–25 mm high, hairless **outer bracts bent back**.

FRUITS: Achenes, yellowish or pale grey to olive brown, 3–4 mm long, beaked, ribbed, spiny above; pappus white, stalked.

WHERE FOUND: Weedy introduced species of cultivated and waste ground; widespread and common across our region, northwest to Great Slave Lake and interior Alaska; circumpolar.

NOTES: Red-seeded dandelion (*T. laevigatum*, also called *T. erythrospermum*) is similar, but its leaves have a small terminal lobe, its involucral bracts have a horned appendage at the tip, and its achenes are red to reddish purple and spiny all over. Red-seeded dandelion is an introduced weed of disturbed sites in the southern part of our region. • After its introduction to North America, the Cree gathered large dandelion roots in the autumn, dried and grated them, and mixed them with lard or oil to make a paste. This was then used as a salve for eczema. The Cree ate the leaves or boiled them into a medicinal tea to purify the blood and to treat anemia, jaundice, nervousness and eczema. Northern native groups boiled the root for a long time to make a strong antiseptic. • Many people eat young dandelions raw or cooked as wild greens. These common weeds are more nutritious than most of the vegetables growing in the gardens they invade. They are especially high in vitamins A and C. The roots can be dried, roasted and ground as a passable coffee substitute. The flowers can be added to pancakes and fritters but perhaps the most famous dandelion-flower product is dandelion wine. Even the seed-like fruits can be eaten in an emergency.

PERENNIAL SOW THISTLE • *Sonchus arvensis*

GENERAL: Perennial herb with **milky juice**, deep roots and extensive, creeping rhizomes; stems 0.4–2 m tall, hollow, with coarse, stalked, yellow glands near top.

LEAVES: Alternate, pinnately lobed, **prickly along edges**, clasp stem, 5–40 cm long, 1–15 cm wide, smaller and simpler upwards.

FLOWERS: Several **dark yellow** heads 3–5 cm wide in loose, flat- or round-topped clusters; **ray flowers only**; involucres 14–24 mm high, cup-shaped, with coarse, gland-tipped hairs.

FRUITS: Flattened, ribbed, cross-wrinkled achenes, 2.5–3.5 mm long.

WHERE FOUND: Introduced weed of cultivated and waste ground; widespread across North America, rarely north of 60°N; circumpolar.

S. arvensis

NOTES: Smooth perennial sow thistle (*S. uliginosus*) is hairless, but otherwise much like perennial sow-thistle and with a similar range. • Two annual species are found in our region. **Prickly annual** or **spiny-leaved sow thistle** (*S. asper*), has stiff, spiny, mostly unlobed leaves with rounded flanges at the base that clasp the stem. **Annual** or **common sow thistle** (*S. oleraceus*), has soft, non-spiny, pinnately lobed leaves with sharp-pointed clasping flanges. • Sow-thistles can be separated from the true thistles (*Cirsium* species) by breaking their stems. Sow-thistles have a milky latex; true thistles do not. The young leaves of sow-thistle can be eaten raw in salads or cooked as a vegetable. • Pigs apparently like to eat these plants, hence the common name 'sow-thistle.'

S. uliginosus

NARROW-LEAVED HAWKWEED • *Hieracium umbellatum*

GENERAL: Erect perennial, with **milky juice**, from **short, woody rhizomes**; 1–few stems, 0.2–1 m tall, leafy, essentially hairless towards base, often **short, star-like hairs near top**.

LEAVES: Alternate, egg- to **lance-shaped**, 3–8 cm long, stalkless, smooth-edged or somewhat toothed, with short, stout, **almost cone-shaped hairs** or nearly hairless, **few near stem base**, soon fall off; much simpler and smaller, somewhat clasping upwards.

FLOWERS: Few to many 2–2.5 cm **yellow** heads, on arching stalks in **flat-topped cluster** at stem tip; ray flowers only; involucres hairless or nearly so, often with a few glands, smoky green to blackish.

FRUITS: Ribbed achenes, 2–3 mm long.

WHERE FOUND: Open forests, meadows, clearings and disturbed ground; widespread across our region, north and west to Great Bear Lake and west-central Yukon.

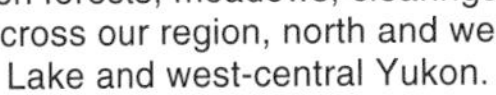

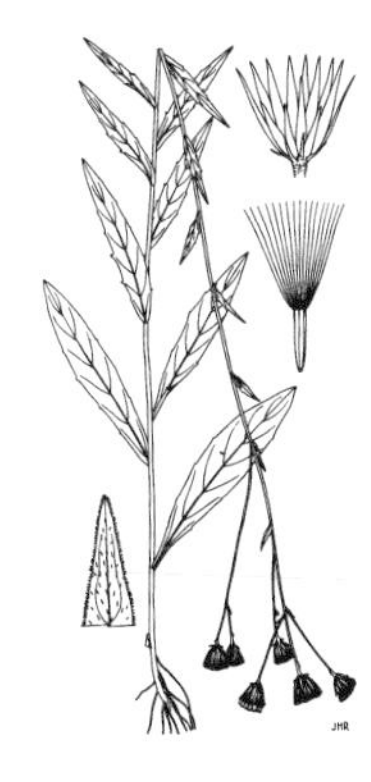

NOTES: Also called *H. scabriusculum* and *H. canadense*. • Fresh narrow-leaved hawkweed plants, with their coagulated juice, were used by several native groups as a substitute for chewing gum. • Although hawkweeds produce showy flowers that attract insects, fertilization is very rare. Instead, most plants develop seeds without fertilization. Each seed produced in this way is genetically identical to the parent plant. Hawkweeds can produce large numbers of plants with the characteristics of a single parent, which can be a real problem for botanists trying to classify species. If the offspring of 1 plant multiply and become very widespread, they may be recognized as a microspecies. In Britain there are said to be 400 species of hawkweed, and in Scandinavia 2,000 minor species have been described. • According to Pliny, a Roman writer and naturalist (23–79 A.D.), hawks ate these plants to sharpen their eyesight—hence the common name 'hawkweed.' Later, European physicians extended this power to encompass man, and used these plants to make lotions and drops to treat the eyes.

COMMON BLUE LETTUCE • *Lactuca tatarica*

GENERAL: Deep-rooted perennial, **pale bluish green**, hairless, spreads by white, running rhizomes, forms patches; stems erect; 20–60 cm tall (sometimes to 1 m), branching towards top.

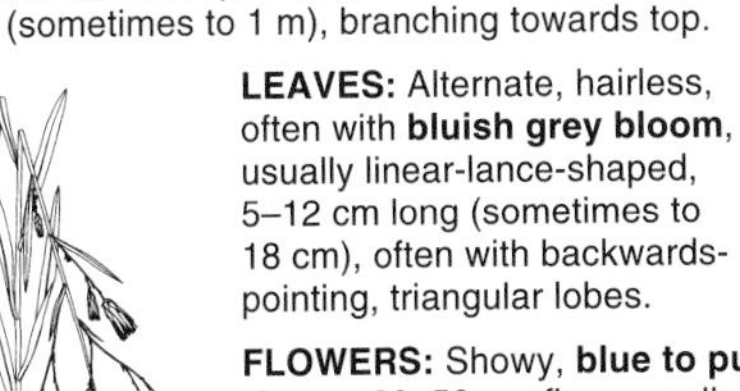

LEAVES: Alternate, hairless, often with **bluish grey bloom**, usually linear-lance-shaped, 5–12 cm long (sometimes to 18 cm), often with backwards-pointing, triangular lobes.

FLOWERS: Showy, **blue to purplish heads**, nearly 25 mm across, few in open cluster; 20–50 ray flowers, all ribbon-like.

FRUITS: Thin achenes, 4–7 mm long, several veins, **short, stout beak**; pappus of soft, slender, white hairs, attached to disc at top of achene.

WHERE FOUND: Weedy species of moist woods and clearings, usually found on cultivated and waste land; widespread across parkland and prairie; uncommon in boreal forest; reaches Churchill, Man., Great Bear Lake and interior Alaska to north.

NOTES: Also called *L. pulchella*. • Although blue lettuce is not classified as toxic, many species of this genus are known to cause skin irritation or internal poisoning. • *Lactuca* is Latin for 'lettuce,' and *pulchella* means 'pretty.' Common blue lettuce can grow to be 1 m tall, but it is small in comparison to other lettuce species.

OX-EYE DAISY • *Chrysanthemum leucanthemum*

GENERAL: Perennial from well-developed, somewhat woody rootstock, usually forms patches; stems simple or once-branched, 20-60 cm tall.

LEAVES: Toothed or lobed, in rosette at stem base, spoon-shaped, 4-15 cm long; stem leaves oblong, stalkless.

FLOWERS: Single white daisies, 2-5 cm across; 1-2 cm **yellow button of disc flowers at centre**; ray flowers white, 10-15 mm long; involucral bracts green with narrow, dark purplish or brownish edges.

FRUITS: Black, round achenes with about 10 ribs; no pappus.

WHERE FOUND: Waste places, meadows and roadsides; introduced garden plant, escaped as a **weed**, widespread in boreal forest and parkland; cosmopolitan.

C.leucanthemum

NOTES: At first glance, ox-eye daisy could be confused with **scentless chamomile** (*Matricaria maritima*, also called *M. perforata* or *M. inodora*), a Eurasian weed that is locally abundant across our region in fields, waste places and along roadsides. However, scentless chamomile is easily identified by its deeply and finely divided leaves and by the many, smaller, white flowering heads (usually about 2 cm across) produced by each plant. • Following the Doctrine of Signatures, the daisy (day's eye) was used to treat sore eyes in the 16th and 17th centuries. Later, it was used to stop bleeding of the bladder, haemorrhoids and stomach ulcers. • Ox-eye daisies are permeated with an acrid juice that is said to repel insects. However, the fragrance of these flowers, the smell of stale perspiration, is common to many species that are pollinated by small flies. • 'Daisy' is from the Old English *dæges eage*, 'day's eye,' since the disc of the flower is revealed in the morning.

Matricaria martima

PINEAPPLEWEED • *Matricaria matricarioides*

GENERAL: Leafy, hairless, **pineapple-scented annual**; stems branching, 5–40 cm tall.

LEAVES: Basal leaves withered by flowering time; stem leaves usually compact and abundant, 1–5 cm long (sometimes to 8 cm), **1–3 times divided into short, narrow segments**.

FLOWERS: Few to many yellowish, short-stalked, **cone-shaped heads**, 5–10 mm across in loose clusters at stem tips; **disc flowers only**; involucral bracts oval, greenish yellow with broad, translucent edges.

FRUITS: Slightly angular achenes, weakly 4-veined, hairless; pappus reduced to **small crown of short bristles**.

WHERE FOUND: Introduced weed of disturbed (often compacted) soils and roadsides; widespread across our region; more or less circumpolar.

NOTES: Also called *M. discoidea*. • People susceptible to hay fever should avoid pineappleweed. • Because of its citrus-like smell, some Chipewyan thought that this introduced weed first sprang from orange peels that had been discarded around the Hudson's Bay Company store. • Native peoples used pineappleweed to treat constipation. They also boiled it, either alone or mixed with Labrador tea, to make a medicinal tea to soothe sore throats or to help people relax. Pineappleweed tea has a calming effect on babies and is also a good drink at bedtime. The flowers make a pleasant tea similar to chamomile tea, but the stems and leaves add a bitter taste. To make an eye wash, the Cree boiled 2 flower heads of about the same size (to resemble eyes). • The dried plants and flowers have been used in potpourris, for lining drawers and linen cupboards, and as an aromatic addition to bath water. Pineappleweed is also recommended as a deodorant. After cleaning fish, try rubbing your hands with fresh pineappleweed to get rid of the fish smell.

CANADA THISTLE • *Cirsium arvense*

GENERAL: Spiny perennial from deep running rootstocks, usually in large patches, either male or female; **stems** 30-100 cm tall, **not winged**.

LEAVES: Alternate, usually stalkless, lance-shaped, 5-15 cm long, **deeply lobed** and **spiny toothed, wavy-edged**, green above, sometimes densely white-hairy beneath.

FLOWERS: In open clusters of **many pink-purple male or female heads** 12-25 mm across; florets all tubular; involucre 1-2 cm high, the **bracts spine-tipped**.

FRUITS: Plentiful, oblong, flattened, ribbed achenes, 3-4 mm long; pappus brownish to white, feathery.

WHERE FOUND: Common introduced **weed** of waste places, fields and roadsides.

NOTES: Drummond's thistle (*C. drummondii*) occasionally grows in meadows and open woods in the southern boreal forest and parkland. It does not have creeping rhizomes and its flower heads are 3.5-5 cm wide, with both male and female flowers. • Some native groups ate the roots of Canada thistle, boiled or roasted. Although they have little flavour, they are nutritious and make a good emergency food because they are usually abundant and easy to identify. Peel off the thistles with a sharp knife to extract the juicy, often sweet, inner stalk. The flower buds are said to be excellent steamed with lemon butter—like artichokes.

PLAINS WORMWOOD • *Artemisia campestris*

GENERAL: Biennial, or more often perennial, with taproot, **slightly aromatic**, 20–60 cm tall; usually several stems from crown.

LEAVES: Most in rosette at stem base, pinnate, 2–10 cm long, 1–4 cm wide, 2–3 times divided into linear segments, silvery hairy on both sides or nearly hairless above; stem leaves smaller and less divided.

FLOWERS: Many round, 3–4 mm heads in branched, narrow cluster; florets all tubular, outer ones female and fertile, inner ones perfect (male and female) or sterile; involucres 2–4.5 mm high.

FRUITS: Hairless achenes.

WHERE FOUND: Sandy pine woods, shallow soil over rocks, sandy banks and disturbed areas; widespread across our region in dry habitats.

NOTES: Also called *A. canadensis*, *A. caudata* and *A. borealis*. • **Pasture sagewort** (*A. frigida*)(inset photo) is also found across our region in dry habitats, but it is very aromatic and has finely divided, silvery-grey, hairy leaves. **Mountain sagewort** (*A. tilesii*) has large, pinnately lobed, coarsely toothed leaves that are green above and grey-hairy below. It typically grows on sandy or gravelly riverbars, terraces and beaches in the northwest, from Great Slave Lake to Alaska. **Biennial sagewort** (*A. biennis*) is a weedy, taprooted annual or biennial that is scattered across our region on disturbed ground near human settlements. Its hairless leaves are divided into narrow, toothed, segments. • Pasture sagewort can be used, dried or fresh, to make an aromatic tea, to season rice, and to flavour stuffing for poultry, fish, or game. However, many *Artemisia* species contain volatile oils that are **poisonous** if they are ingested in large amounts. • The Cree used sagewort as a ceremonial plant and burned it for incense. Some tribes hung it in their homes for protection and it was stuffed into the noses of people working with the dead to protect them from spirits. Because of its strong smell, sagewort has been used as an insect repellent, an alternative for mothballs and a fragrant addition to sachets. The Cree often used it as an ingredient in their trap lures. The distinctive scent comes from a strong-tasting, toxic, volatile oil. This seems to inhibit grazing, and heavily used pastures often take on a bluish tinge, as sagewort is left behind, while other grasses and herbs are selectively grazed.

A. biennis

COMMON YARROW • *Achillea millefolium*
MILFOIL

GENERAL: Aromatic perennial from spreading rhizomes; stems 30–70 cm tall.

LEAVES: Alternate, **fern-like**, 2–3 times **pinnately divided** into segments 1–2 mm wide, 3–10 cm long, stalked towards base, stalkless upwards.

FLOWERS: In many heads, in short, **flat or round-topped cluster**; about 5 ray flowers, **white (rarely pink or reddish), 2–4 mm long; 10–30 disc flowers**, yellowish.

FRUITS: Hairless, flattened achenes.

WHERE FOUND: Meadows, woods, clearings and disturbed ground; common and widespread across our region.

NOTES: Also called *A. borealis* and *A. lanulosa*. • The medicinal value of this aromatic herb was recognized long ago. In fact, fossils of yarrow found in Neanderthal burial caves suggest humans have used this plant for more than 60,000 years. • The Cree used yarrow tea to cool burns and soothe earaches. They boiled the plant and used it as a poultice on infections and to relieve aches and pains. To relieve a headache, the Chipewyan mixed yarrow flowers with tobacco and smoked it, or they burned the dried plant on a hot rock and inhaled the smoke. Yarrow smoke is still used to fumigate a room where someone has been sick. Yarrow leaves have a sage-like taste, and some say the tea is not only nourishing, but it will also purify the blood and relieve colds, diarrhea, fever and even diabetes. However, others consider it **toxic** and best avoided. The leaf tea was used as a hair rinse and as a wash for sore eyes, pimples and mosquito bites. • The Cree added the fragrant leaves and dried flowers to their lynx bait. • Yarrow has insecticidal properties and European starlings, which reuse the same nests year after year, often line their nests with yarrow to discourage parasitic insects that would otherwise infest their young. A study of yarrow oil found that it killed 98% of the mosquito larvae exposed to it. A water-based solution of yarrow has antibacterial activity against *Staphylococcus aureus*.

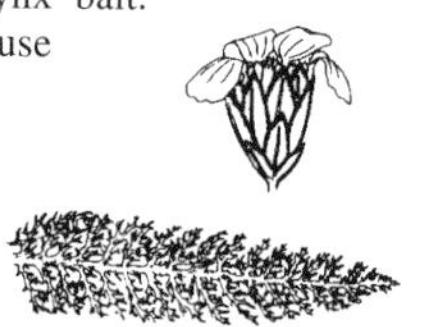

MANY-FLOWERED YARROW • *Achillea sibirica*
SIBERIAN YARROW

GENERAL: Perennial from rhizomes, 30–80 cm tall; stems simple or freely branched towards top.

LEAVES: Alternate, stalkless or nearly so, 5–10 cm long, linear or linear-lance-shaped, pointed, **pinnately divided or deeply notched**; divisions sharply toothed.

FLOWERS: Several to many heads in flat or round-topped clusters; 6–12 ray flowers, white, **1–2 mm long; 25–30 disc flowers**.

FRUITS: Hairless, flattened achenes.

WHERE FOUND: Moist thickets, streambanks, lakeshores and ditches. Widespread across our region.

NOTES: Many-flowered yarrow has sharply toothed leaves, whereas those of common yarrow are divided 2–3 times into smaller segments. • Like common yarrow, many-flowered yarrow was also used medicinally by the Cree. When young children were suffering from teething, flowers from either yarrow and from the wild mint were crushed and wrapped in a cloth. This was then dipped in water and used to wash pus from the gums. Chewed pieces of root were sometimes applied to sores on the gums or the root was mixed with other herbs and boiled to make a tea that was taken to treat sickness related to teething.

SMALL-LEAVED PUSSYTOES • *Antennaria microphylla*
SMALL-LEAVED EVERLASTING

GENERAL: **Mat-forming** perennial, from leafy stolons; flowering stems 10–25 cm tall, **upper stems with glandular hairs**.

LEAVES: Spoon-shaped or broadly lance-shaped at stem base, 1–2 cm long, densely grey- or white-woolly above and below; 7–12 stem leaves, linear, about 10 mm long.

FLOWERS: 3–20 small heads in dense cluster at stem tip; heads **white**, with 3–20 disc flowers; involucral bracts woolly below, **white above**.

FRUITS: Achenes, usually hairless.

WHERE FOUND: Dry, open areas; widespread across our region north and west to Great Bear Lake and east-interior Alaska.

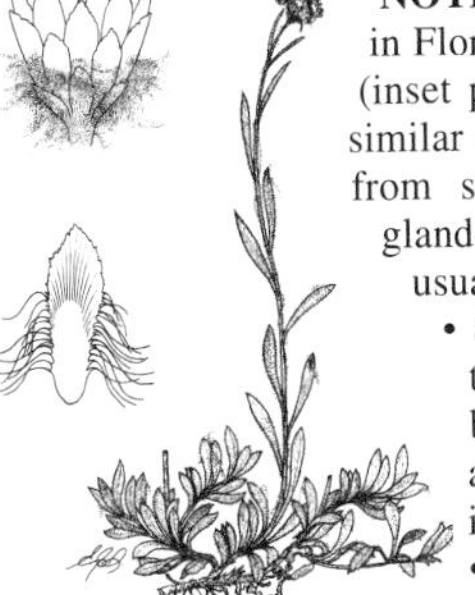

NOTES: Also called *A. nitida*; *A. parvifolia* in Flora of Alberta • **Rosy pussytoes** (*A. rosea*) (inset photo) is also a common species with a similar distribution. It is most easily separated from small-leaved pussytoes by the lack of gland-tipped hairs on its upper stem and the usual pinkish colour of its flower heads. • Some *Antennarias* are known as 'Indian tobacco' because they were added to tobacco mixtures. The leaves and stalks are also chewed, as their juice makes a pleasing gum which some claim is nourishing. • The tiny flowers grow in clusters of several tight, downy heads. These clusters resemble the soft paws of a tiny kitten, with the tight heads as the toes—hence the fanciful name, 'pussytoes.'

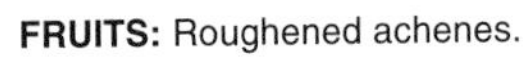

BROAD-LEAVED PUSSYTOES • *Antennaria neglecta*
FIELD PUSSYTOES

GENERAL: Mat-forming perennial, with well-developed stolons; flowering stems 5–40 cm tall.

LEAVES: Basal leaves spoon-shaped to broadly lance-shaped or elliptic, 2–5 cm long, 0.7–2 cm wide, densely **white-woolly below**, thinly woolly to hairless, **green above**; stem leaves linear to lance-shaped, smaller upwards.

FLOWERS: 5–9 heads in **compact** to somewhat open cluster at stem tip; heads greenish white, with disc flowers only; involucral bracts with **whitish tips**; involucres of female heads 6–10 mm high.

FRUITS: Roughened achenes.

WHERE FOUND: Open woods and grasslands; widespread across boreal forest and parkland; reaches southern N.W.T. and northeastern British Columbia to north and west.

NOTES: Also called *A. campestris*, *A. canadensis*, *A. howellii*, *A. neodioica* and *A. obovata*. • **Showy pussytoes** (*A. pulcherimma*) has long (4–15 cm), erect, prominently 3-veined leaves at the stem base, and similar lower stem leaves. It grows on moist river flats and meadows, especially on lime-rich soils, and is widespread in our region. • Pussytoes reproduces by seed, but often the seeds are produced without fertilization (apomictic), meaning all of the offspring are genetically identical to the parent. Male plants are often rare, and in a few cases only female plants have been found. Perhaps it's not surprising that the taxonomy of this genus can be confusing.

Key to Fleabanes (*Erigeron*) and Asters (*Aster*)

The fleabanes and asters are two of the more difficult genera when it comes to distinguishing the species. Fleabanes usually flower earlier than asters, but some of the weedy species flower well into autumn. Most fleabanes have involucral bracts that are nearly equal in length and are arranged in a single row. Their bracts are also relatively long and narrow, and they seldom have green tips. The ray flowers of fleabanes are usually more numerous and narrower than those of asters. Asters are generally taller, have leafier stems, and have more flower heads per stem than fleabanes.

1a. Involucral bracts usually in 1 or 2 equal rows, green (but not leafy) throughout; ray flowers mostly very narrow; plants without rhizomes; buds often drooping; plants mostly bloom in spring and early summer (the fleabanes) 2

2a. Ray flowers simple, small (up to 3 mm long), white or pink 3

3a. Flower cluster flat-topped; involucre often gland-bearing; leafy rosette usually present at stem base .... ***Erigeron acris***

3b. Flower cluster spike-like; involucre not gland-bearing; no leafy rosette at stem base .... ***E. lonchophyllus*** (see *E. acris*)

2b. Ray flowers well-developed 4

4a. Upper stem leaves broad, clasp stem; many (more than 150), narrow (to 0.6 mm wide) ray flowers, pink to white .... ***E. philadelphicus***

4b. Stem leaves narrow, do not clasp stem; fewer ray flowers, wider (to 1 mm across), pink, blue or purple .... ***E. glabellus***

1b. Involucral bracts usually in 3 or more rows, overlap like shingles, outer ones progressively shorter, green, sometimes somewhat leafy; ray flowers comparatively broad; plants often have rhizomes; buds usually erect; plants bloom in late summer and fall (the asters) 6

5a. Involucre and flower stalks gland-bearing .... ***Aster conspicuus***

5b. Involucre and flower stalks without glands 7

6a. Involucral bracts usually have purple tips and edges; leaves rough, at least below .... ***A. sibiricus*** (see *A. conspicuus*)

6b. Involucral bracts lack purple tips nor edges; leaves not rough below 8

7a. Stem-base and lower stem leaves heart-shaped, distinctly stalked .... ***A. ciliolatus***

7b. Stem-base and lower stems leaves not heart-shaped, usually stalkless (or nearly so) 9

8a. Plants hairless throughout; lower leaves wing-stalked, upper leaves clasp stem .... ***A. laevis*** (see *A. ciliolatus*)

8b. Plants hairy on stems and branches 10

9a. Pappus double, inner series of bristles longer than outer series .... ***A. umbellatus***

9b. Pappus single 11

10a. Leaves clasp stem with ear-like flanges .... ***A. puniceus***

10b. Leaves do not clasp stem 12

11a. Leaves narrow (usually less than 7 mm wide), linear or linear-lance-shaped 13

12a. Plants slender; leaves essentially hairless; few flower heads (generally less than 10) .... ***A. borealis***

12b. Plants more robust; leaves usually hairy; many flower heads .... ***A. ericoides*** (see *A. umbellatus*)

11b. Leaves wider (usually more than 10 mm wide), not linear 14

13a. Leaves not toothed (or barely so); ray flowers blue or purple .... ***A. hesperius***

13b. Leaves toothed; ray flowers white .... ***A. simplex*** (see *A. hesperius*)

NORTHERN DAISY FLEABANE • *Erigeron acris*
BITTER FLEABANE

GENERAL: Biennial or short-lived perennial from simple or branched stem base and taproot, usually gland-bearing towards top; stems single to several, **5–80 cm tall**.

LEAVES: Leaves broadly to narrowly spoon-shaped, 5–10 cm long, stalked, smaller and stalkless upwards.

FLOWERS: In **few to many** heads in flat-topped cluster; 3 types of florets: outer female florets with inconspicuous (to 4 mm long) **pink to purplish or white** rays; inner female florets tubular; central disc flowers yellow, bi-sexual; involucres 5–12 mm high, bracts narrowly lance-shaped, **with glands and/or hairs**.

FRUITS: Sparsely hairy, 2-veined achenes; **pappus hairs straw-coloured** or pinkish to white.

WHERE FOUND: Moist woods and openings, lakeshores, river terraces, streambanks, and roadsides; widespread across boreal forest.

NOTES: Hirsute or **spear-leaved fleabane** (*E. lonchophyllus*) is similar to northern daisy fleabane but its flower heads are on short, erect branches, its involucres have no glands, all of its female flowers have rays and its stem leaves are linear. Hirsute fleabane grows in wet meadows and on moist shores throughout our region. • Fleabanes produce a turpentine-like oil that repels fleas, and these plants have been dried, powdered and sprinkled in dog kennels to drive away fleas. Scientific studies have proven the usefulness of fleabane as an insecticide. • *Erigeron* comes from the Greek *eri*, meaning 'spring,' and *geron*, meaning 'an old man,' in reference to the hairy seed heads or possibly to some hairy spring-flowering species.

E. acris

SMOOTH FLEABANE • *Erigeron glabellus*

GENERAL: Biennial or perennial, 20–50 cm tall; stems single or few from crown, erect, with flattened or spreading hairs.

LEAVES: Basal leaves lance-shaped, broadest above middle, 10–15 cm long, 1–2 cm wide, winged stalks, toothless or irregularly toothed, usually hairy; stem leaves lance-shaped or linear, smaller upwards.

FLOWERS: Commonly in 1–3 heads, 2–3 cm across; 125–175 ray flowers, pink or blue, occasionally white, 8–15 mm long, 1 mm wide; disc flowers yellow; involucres 5–9 mm high, hairy, bracts linear, **often with obvious, brown midvein**.

FRUITS: Hairy achenes; double pappus dries reddish brown, outer bristles very short.

WHERE FOUND: Moist prairies, meadows and open woods; widespread from southern Manitoba to interior Alaska.

NOTES: Fleabanes are gathered when flowering, dried and sold as an herb for treating fluid retention and bleeding. In British Columbia, the Thompson Indians toasted fleabane plants, crushed them and mixed the powder with grease to make a salve for painful or swollen areas. They also sprinkled powdered fleabane on sores and wounds to aid healing, and they chewed the fresh plant to relieve sore throats. • According to Norse mythology, fleabane was a flower of the god Odin. Later it was associated with St. Christopher.

PHILADELPHIA FLEABANE • *Erigeron philadelphicus*

GENERAL: Biennial or short-lived perennial from **taproot or fibrous roots**, usually with long spreading hairs, sometimes hairless; stems 20–70 cm tall.

LEAVES: Narrowly spoon-shaped at stem base, 2–8 cm long, coarsely round-toothed or lobed, **hairy**, short-stalked; stem leaves smaller, narrower, stalkless, **clasping upwards**.

FLOWERS: In **few to many** heads 12–25 mm across, in open cluster at stem tip; usually **over 150** ray flowers, pink to pinkish purple or white, 5–10 mm long, **very narrow**; disc flowers yellow; involucres 4–6 mm high, bracts oblong-lance-shaped, with broad, clear edges and more or less hairy midvein.

FRUITS: Sparsely hairy, 2-veined achenes; pappus whitish.

WHERE FOUND: Moist, open woods, thickets, streambanks, roadsides, and clearings; widespread from southern Manitoba to southern N.W.T. and northern British Columbia.

NOTES: A volatile oil collected from the leaves and flower clusters of Philadelphia fleabane was used by doctors in the early 1900s to stimulate contractions of the uterus.

FRINGED ASTER • *Aster ciliolatus* LINDLEY'S ASTER

GENERAL: Stout perennial from long creeping rhizomes; stems erect, usually unbranched, 20–100 cm tall.

LEAVES: Narrowly **heart-shaped**, 5–10 cm long, sharply toothed, **long, hairy-edged winged stalks**; smaller, simpler and stalkless upwards; lower leaves usually fall off early.

FLOWERS: Usually in few heads 15–30 mm across, in open cluster at stem tip; 12–25 ray flowers **pale blue**, 8–15 mm long; disc flowers yellow; involucres 5–8 mm high, bracts linear to oblong-lance-shaped, **usually hairless, green at tip, whitish edges.**

FRUITS: Several-veined achenes.

WHERE FOUND: Open woodlands, thickets and clearings; widespread and common across much of our region, north and west to Great Bear Lake and northeastern British Columbia.

NOTES: Also called *A. lindleyanus.* • **Smooth aster** (*A. laevis*) (inset photo) has hairless, lance-shaped lower leaves and narrow upper leaves with a clasping base. Smooth aster ranges from northern British Columbia eastwards, mostly along the southern edge of our region, in dry grassy areas. • The Cree called fringed aster 'small love medicine.' They steeped 1 root in 1/2 cup of boiling water and used the cooled liquid for eyedrops. • The roots of fringed aster are strongly scented. In the past, different asters were burned and the smoke was inhaled to revive patients who were being treated in a sweat bath. Sometimes a paper cone was used to force the smoke from a smouldering aster into the nose of a patient who had lost consciousness.

A. laevis

SHOWY ASTER • *Aster conspicuus*

GENERAL: Perennial from strong, creeping rhizomes; stems usually single, unbranched, **30–90 cm tall**, with glands and stiff hairs towards top.

LEAVES: Egg-shaped to **elliptic, firm, sharp-toothed, usually stalkless**, clasping, stiff-hairy, **'sandpapery' above** when mature; lowest leaves smaller, soon withering.

FLOWERS: In round to flat-topped cluster at stem tip, with few to many heads about 4 cm across on gland-bearing stalks; 12–35 ray flowers, **blue to violet**, 10–15 mm long; disc flowers yellow; involucres 9–12 mm high, bracts lance-shaped, **densely glandular**, green at tip, whitish at base.

FRUITS: Hairless, several-veined achenes; whitish pappus hairs.

WHERE FOUND: Open woods and clearings; widespread across southern boreal forest and northern parkland; from British Columbia to Saskatchewan.

NOTES: Arctic aster (*A. sibiricus*) is a low plant with lance-shaped leaves and 1 to few flower heads. Although it is an arctic-alpine species, arctic aster also grows at low elevations on gravelly flats, riverbars, rocky meadows, or even in open forest. It is common from Alberta and the District of Mackenzie westwards. • Rub your fingers on a leaf in the spring and sniff them to smell a wonderful fragrance. • Children have been known to use large asters like these prophetically, as they would a daisy, repeating 'loves-me, loves-me-not' as they pluck the rays. • Much like fireweed, showy aster can maintain and spread itself in sterile conditions under a closed forest canopy by means of spreading rhizomes. After a fire or other disturbance removes the canopy, these species respond with vigorous growth and profuse flowering.

PURPLE-STEMMED ASTER • *Aster puniceus*

GENERAL: Perennial from rhizomes, 0.5–1.5 m tall; **stem stout**, simple or much-branched towards top, **reddish purple, with spreading hairs**.

LEAVES: Alternate, simple, lance-shaped to oblong, 6–16 cm long, 1–2 cm wide, stalkless, **clasp stem**, rough hairy to hairless above, **hairy below on midvein**.

FLOWERS: Many heads in leafy cluster, relatively large and showy; 30–60 ray flowers, blue or purplish, 8–16 mm long; disc flowers yellow; involucres 6–12 mm high, bracts narrow, loose, scarcely overlapping, some outer ones often leafy.

FRUITS: Hairy achenes, pappus hairs white.

WHERE FOUND: Marshy ground; often found in wet, grassy, roadside ditches; fairly common in boreal forest across prairie provinces north to Lake Athabasca.

NOTES: Chipewyan elders still dry purple-stemmed aster roots and mix them with tobacco for smoking, or grind them into powder to be inhaled to treat headaches. Their name for this plant means 'headache medicine,' and the same word is used to refer to Aspirin. The stems, leaves and flowers can also be used to treat headaches if they are collected when the plant is in flower. Purple-stemmed aster roots are also used as heart medicine. The Woods Cree boil the roots to make a medicinal tea to make feverish patients sweat to reduce the fever. They also chew the roots to relieve the pain of a toothache. • The genus name comes from the Greek *aster*, 'a star,' after the shape of the flowers. The species name, *puniceus*, means 'reddish purple,' in reference to the colour of the stems.

FLAT-TOPPED WHITE ASTER • *Aster umbellatus*

GENERAL: Perennial from rhizomes, 0.5–1.5 m tall; stems erect, leafy towards top, hairy.

LEAVES: Alternate, simple, lance- to egg-shaped, 6–15 cm long, 2–3 cm wide, short-stalked or nearly stalkless, toothless, hairy below, at least on veins.

FLOWERS: Many heads in large, flat-topped cluster at stem tip; 10–15 ray flowers, white, wide-spreading; disc flowers yellow; involucres 3–5 mm high, relatively few bracts, lance-shaped, pointed or rounded, hairy on back.

FRUITS: Slightly hairy, veined achenes; pappus nearly white, **double**; inner bristles firm, some thickened towards tip; outer bristles less than 1 mm long.

WHERE FOUND: Moist thickets, woods and swampy ground; an eastern species, found in our region in boreal forest and parkland of Manitoba, north to about 52°N; rare in Saskatchewan and Alberta.

NOTES: Tufted white prairie aster (*A. ericoides*, also called *A. pansus*) (inset photo) also has white flower heads but its flowers have a single series of pappus bristles, and its heads are in a long, one-sided cluster. It has clustered stems and many small, narrow, grey-green, rough-hairy leaves. Tufted white prairie aster grows in dry open areas from southern Manitoba north and west along the Mackenzie River valley to the Arctic Circle; it is also reported from the southern Yukon and interior Alaska. • Tufted white prairie aster is known to the Cree as 'Big Love Medicine.' The highly scented roots of these plants are dried and then pounded into a powder that is dusted on cuts to stop bleeding. The Cree also boil the root in water for 5 minutes until the water turns green, and use the cooled liquid as an eyewash. • The name *umbellatus* refers to the distinctive flat-topped flower cluster, unusual for an aster.

MARSH ASTER • *Aster borealis*
RUSH ASTER

GENERAL: Perennial from **long slender rhizomes**; stems slender, single, usually unbranched, 10–80 cm tall, **hairy in lines** downwards from upper leaf bases.

LEAVES: Linear to linear-lance-shaped, 2–8 cm long, 2–7 mm wide, stalkless, usually slightly clasping, hairless, smooth-edged; lower leaves short-stalked, smaller and simpler, soon fall off.

FLOWERS: In clusters of **1 to few heads** at stem tips, **15–20 mm across**, on hairy stalks; 20–50 ray flowers, **white to pale blue** or rose, 7–15 mm long; disc flowers yellow; involucres 5–7 mm high, bracts oblong, overlapping in several rows, **greenish, often with purple tips and edges, hairless**.

FRUITS: Hairy achenes with several veins.

WHERE FOUND: Marshy ground and fens; widespread across our region, north to the Arctic Circle.

NOTES: Also called *A. junciformis*.

WESTERN WILLOW ASTER • *Aster hesperius*

GENERAL: Perennial from creeping rhizome; stems 0.5–1.2 m tall, usually branching towards top, **hairs in lines down from upper leaf bases**.

LEAVES: Simple, alternate, **linear to broadly lance-shaped**, 5–15 cm long, usually **smooth-edged**, hairless or somewhat rough hairy, usually stalkless; lower leaves stalked, fall off by flowering time.

FLOWERS: Many heads in leafy cluster; **ray flowers 6–14 mm long, usually bluish or purple**, sometimes white; disc flowers yellow; involucres 5–8 mm high, bracts linear-lance-shaped, overlapping or of nearly equal lengths.

FRUITS: Achenes; pappus hairs white.

WHERE FOUND: Streambanks, ditches, edges of marshes and moist ground; across our region from Manitoba to British Columbia.

NOTES: Panicled aster (*A. simplex*) is a similar eastern species. It is often taller than western willow aster, with narrower, saw-toothed leaves, shorter, white ray flowers, and 3-6 mm high involucres. Panicled aster is found in moist meadows and thickets, and on lakeshores and streambanks in the southern boreal forest, parkland and prairie of Manitoba and eastern Saskatchewan. Where the ranges of these 2 species overlap, it is very difficult to separate them with certainty. • *Hesperius* refers to 'the west' or 'evening,' and is quite appropriate for western willow aster, which is primarily a western species in North America.

SPOTTED JOE-PYE WEED • *Eupatorium purpureum* var. *maculatum*

GENERAL: Perennial, 0.6–1.5 m tall; stems stout, **purple or purple-spotted**, soft hairy near top.

LEAVES: In whorls of 3–6, lance- to egg-shaped, 6–20 cm long, 2–7 cm wide, narrow at base, sharp-pointed at tip, coarsely toothed; somewhat hairy below.

FLOWERS: Many heads, in conspicuous, **rounded to flat-topped clusters** 15–20 cm across, pink to purple, **disc flowers only,** 9–22; involucres 6–9 mm high, bracts pinkish, overlapping.

FRUITS: Gland-dotted achenes; pappus hairs white.

WHERE FOUND: Marshy ground and moist open woods; boreal forest and parkland of Manitoba and Saskatchewan north to about Lake Athabasca; rare in Alberta.

NOTES: Also called *E. maculatum.* • Native peoples used Joe-pye weed tea to stimulate urination in treatment of fluid retention and bladder and kidney infections. The root tea was used to relieve fevers, colds, chills, diarrhea, liver and kidney ailments and even a sore womb after childbirth. It was also used as a wash for rheumatism. Although Joe-pye weed was once famous as a medicine for breaking fevers, no experimental evidence has been found to support this use. • 'Joe-pye' was named either for a 19th century Caucasian 'Indian theme promoter,' or for a native medicine man famous throughout New England for using this plant to induce sweating and thereby break fevers during a typhus outbreak. The genus is named in honour of Mithridates Eupator, a 1st century B.C. king of Pontus, which was a country just south of the Black Sea. He was supposedly the first person to use a plant of this genus in medicine. *Maculatum* means 'spotted,' in reference to the purple spots on the stem.

PALMATE-LEAVED COLTSFOOT • *Petasites palmatus*

GENERAL: Perennial from slender, creeping rhizomes; flowering stems stout, 10–50 cm tall, **appear before leaves**.

LEAVES: Round to **heart- or kidney-shaped** at stem base, 5–20 cm wide, deeply divided (more than halfway to centre) into **5–7 toothed lobes, green, essentially hairless above, thinly white-woolly below**; stem leaves reduced to alternate bracts.

FLOWERS: In clusters of several to many white, 8–12 mm wide heads on glandular, often white-woolly stalks, mostly female or mostly male; ray flowers **creamy white**; disc flowers **whitish to pinkish**; involucres 7–16 mm high, bracts lance-shaped, hairy at base.

FRUITS: Hairless, linear achenes, about 2 mm long, 5–10 ribs; pappus soft, white.

WHERE FOUND: Moist woods and openings, and wetlands; widespread across boreal forest and northern parkland to Great Bear Lake and eastern Yukon.

P. palmatus

P. frigidus

NOTES: Also called *P. frigidus* var. *palmatus*. • The leaves of **arctic coltsfoot** (*P. frigidus*) are triangular and coarsely toothed to shallowly (not deeply) lobed. It grows on streambanks and lakeshores and in seepage areas, wet meadows and thickets, from the Mackenzie River valley westwards. • The Cree used palmate-leaved coltsfoot leaves as poultices on infected sores. The leaves were also crushed or boiled to make a salve for relieving insect bites, inflammation, swellings, burns, sores and some skin diseases. The leaves could be dried and stored and later soaked in lukewarm water to soften them for use. Some native groups chewed the roots or made them into a tea to treat chest ailments (tuberculosis and asthma), rheumatism, sore throats, and stomach ulcers. Coltsfoot leaves and flowers were steeped in hot water to make a tea for people suffering from diarrhea. • Strong doses have been reported to cause **abortion.** • Coltsfoot has been widely used as a medicine over the years. It was once the official sign of the French apothecaries.

ARROW-LEAVED COLTSFOOT • *Petasites sagittatus*

GENERAL: Perennial from **slender** creeping rhizomes; flowering stems more or less white-woolly, 10–50 cm tall, appear before leaves.

LEAVES: Triangular to broadly arrowhead-shaped at stem base, 10–30 cm long, 10–20 cm wide, **usually with 20 or more teeth per side** (sometimes nearly smooth-edged), long-stalked, green, thinly hairy above, densely to thinly white-woolly below; stem leaves reduced to alternate bracts.

FLOWERS: In dense clusters of several to many 20–35 mm high heads, on gland-bearing and/or white-woolly stalks, either mainly male or mainly female; ray flowers small, **whitish**, female; disc flowers whitish, bisexual; involucres 7–10 mm high, bracts lance-shaped, gland-tipped hairs at base.

FRUITS: Hairless achenes, 3–3.5 mm long.

WHERE FOUND: Moist sites and wetlands; widespread across our region, north to the Arctic coast.

NOTES: Vine-leaved coltsfoot (*P. vitifolius*) resembles arrow-leaved coltsfoot, but it has heart-shaped leaves that are palmately 5-7-lobed halfway to the centre—it is intermediate in leaf form (perhaps a hybrid) between palmate-leaved coltsfoot and arrow-leaved coltsfoot and some consider it the same species as arctic coltsfoot (*P. frigidus*). Vine-leaved coltsfoot grows in wet meadows, openings and shaded, wet forest, north and west to northern British Columbia and the southern Yukon. • The ash of coltsfoot leaves and stems was widely used as a salt substitute. The leaves were either rolled in tight balls and dried, to increase the consistency of the ash, or they were fired while encased in balls of clay to contain the ash. The roots, leaves, and young flower shoots can be eaten as a vegetable, either roasted or boiled. However, **pregnant women** should not eat large quantities. • The Cree used coltsfoot leaves medicinally as a poultice on itching skin and to kill worms eating the flesh. They believed that the flower would cause a rash if it was touched.

NODDING BEGGARTICKS • *Bidens cernua*

GENERAL: Annual from fibrous roots; stems leafy, erect (sometimes with prostrate, rooting base), simple to freely branched, hairless to sparsely hairy, 10–80 cm tall.

LEAVES: Opposite, stalkless, joined to the stem, hairless, linear-lance-shaped, 5–15 cm long, saw-toothed to almost smooth-edged.

FLOWERS: In **hemispheric**, single heads at stem tips, 20–35 mm across, nodding with age; 6–8 ray flowers, yellow, sometimes absent; disc flowers yellow; involucre in 2 rows; inner bracts egg-shaped, erect, dark streaked; 5–8 outer bracts, narrowly lance-shaped, green, leafy, unequal, **spreading or bent backwards**.

FRUITS: Achenes, narrowly wedge-shaped, 5–7 mm long, 4-sided, flattened, ribbed, **convex and leathery** at tip, tan; pappus of **2–4 barbed bristles**.

WHERE FOUND: On marshy ground and in shallow water; widespread across southern boreal forest and parkland from southern Manitoba north and west to Great Slave Lake and interior Alaska.

NOTES: The fruits (sometimes called seeds) or 'beggarticks' provide a classic example of adherent or 'stick-tight' dispersal, as you will discover if you walk through a patch in the autumn. 'It is as if you had unconsciously made your way through the ranks of some countless but invisible Lilliputian army, which in their anger had discharged all their anger at you, though none of them reached higher than your legs' (H.D. Thoreau, *The Dispersion of Seeds*). • *Bidens* means '2 teeth'—the achenes of some species have 2 sharp, barbed teeth or spikes at their tips (nodding beggarticks often has 4).

LEAFY ARNICA • *Arnica chamissonis*
MEADOW ARNICA

GENERAL: Perennial from slender rhizome, 30–60 cm tall; stems single, leafy, hairy, **with glands towards top**.

LEAVES: Lower leaves taper to winged stalk, clasp stem at base; stem leaves opposite, mostly 5–8 pairs, stalkless, **not much smaller upwards**, lance-shaped, often broadest above middle, 5–15 cm long, 1–4 cm wide, slightly toothed or smooth-edged.

FLOWERS: 3–7 heads 2–5 cm across; 10–15 ray flowers, pale to lemon yellow, 15–20 mm long; disc flowers deep yellow; involucres 8–12 mm high, bracts blunt to sharp-pointed, with tuft of long hairs at tip.

FRUITS: Achenes, with short hairs and glands to almost hairless; pappus hairs tawny or straw-coloured.

WHERE FOUND: Moist meadows and thickets; boreal forest and northern parkland, north and west to southeastern Alaska.

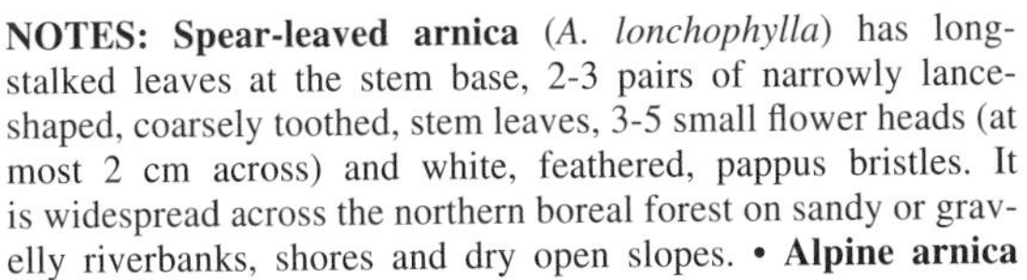

NOTES: Spear-leaved arnica (*A. lonchophylla*) has long-stalked leaves at the stem base, 2-3 pairs of narrowly lance-shaped, coarsely toothed, stem leaves, 3-5 small flower heads (at most 2 cm across) and white, feathered, pappus bristles. It is widespread across the northern boreal forest on sandy or gravelly riverbanks, shores and dry open slopes. • **Alpine arnica** (*A. alpina* ssp. *attenuata*)(inset photo) grows in similar habitats across the northern part of our region. It has a white pappus, narrow, stalkless or short-stalked leaves at the stem base, 3-5 pairs of linear, smooth-edged stem leaves, and 3-5 large flower heads. • Arnicas are **poisonous** and should never be eaten. • The rootstocks and flowers were used to make tinctures and teas for external use (**never on broken skin**) to help heal and relieve the pain of sprains, bruises and swollen feet.

Key to the Ragworts (*Senecio*) and Goldenrods (*Solidago*)

1a. Flower heads have disc flowers only 2

2a. Plants annual, low (to 30 cm tall), usually unbranched; leaves not much smaller higher on stem; some involucral bracts black-tipped .... ***Senecio vulgaris***

2b. Plants perennial, taller (to 80 cm tall), sometimes branched towards top; leaves smaller and simpler higher on stem; some involucral bracts purple-tipped .... ***Senecio indecorus*** (see *S. pauperculus*)

1b. Flower heads have both ray and disc flowers 3

3a. Involucral bracts in 1 equal series 4

4a. Plants annual or biennial, with short, fibrous roots and hollow, somewhat fleshy stem; often woolly hairy .... ***Senecio congestus***

4b. Plants perennial, with stout rootstocks 5

5a. Stem-base leaves pinnately lobed, upper stem leaves similar (not much smaller); plants robust, to 1 m tall .... ***Senecio eremophilus***

5b. Stem-base leaves not pinnately lobed; stem leaves smaller upwards; plants to 60 cm tall 6

6a. Stem-base leaves thin, elliptic to oblanceolate, bluntly toothed, tip rounded .... ***Senecio pauperculus***

6b. Stem-base leaves thick, firm, elliptic to nearly round, bluntly toothed to shallowly lobed; leaf base wedge-shaped; tip straight, often with 2 or 3 coarse teeth .... ***Senecio streptanthifolius*** (see *S. pauperculus*)

3b. Involucral bracts in 2 or more series, equal lengths or overlapping 7

7a. Stem-base leaf stalks have conspicuous, hairy fringe ... ***Solidago multiradiata*** (see *S. spathulata*)

7b. Stem-base leaf stalks lack conspicuous, hairy fringe 8

8a. Leaves spoon-shaped, blunt-tipped; flower heads in spike at stem tip .... ***Solidago spathulata***

8b. Leaves pointy-tipped; flower heads in pyramidal or flat-topped cluster at stem tip 9

9a. Leaves linear, up to 10 cm long, 0.7–1 cm wide; flower heads in flat-topped cluster .... ***Solidago graminifolia***

9b. Leaves long but wider, lance-shaped; flower heads in pyramidal cluster .... ***Solidago canadensis***

COMMON GROUNDSEL • *Senecio vulgaris*

GENERAL: Robust **annual or biennial** from fibrous roots; stems stout, hollow, leafy, 10–55 cm tall, much-branched.

LEAVES: Oblong to lance-shaped in outline, 2–10 cm long, wavy to deeply cut with irregularly toothed lobes, smaller, simpler and soon withering at stem base, stalkless and somewhat clasping upwards.

FLOWERS: In loose clusters of several to many yellow heads 6 mm wide; **no ray flowers**; involucres 5–8 mm high, with 2 rows of **black-tipped**, linear-lance-shaped **bracts**, outer row much shorter.

FRUITS: Rounded achenes, 5–10-veined, hairy.

WHERE FOUND: Introduced weed of moist disturbed ground; widespread across the southern part of our region; less common north of 60°N; circumpolar.

NOTES: Of the 50 or so species of ragwort growing in North America, 7 are considered **poisonous**, and common groundsel is one of these. No ragwort is palatable to livestock, so poisoning is rare. However, young plants, and especially young leaves, are most poisonous, and the toxic alkaloids are not destroyed by drying for hay. Poisonous ragworts contain alkaloids that affect the liver, and they have caused acute illness and death in livestock in many parts of the world.

MARSH RAGWORT • *Senecio congestus*

GENERAL: Annual or biennial, 0.2–1 m tall; stems stout, soft, **hollow**, unbranched except in flower cluster, **densely hairy when young**.

LEAVES: Simple, alternate; lower leaves lance- to spoon-shaped, 5–15 cm long, **edges wavy,** smooth, toothed or shallowly lobed, **stalks winged**; upper leaves smaller, narrower, stalkless, somewhat lobed or toothed, **clasp stem**.

FLOWERS: Several to many heads 1–2 cm across, **in dense to open cluster** at stem tip; ray flowers pale yellow, short; disc flowers yellow; involucres about 8 mm high, hairy.

FRUITS: Hairless achenes; pappus white, longer than achene.

WHERE FOUND: Marshes, ponds and ditches; common across our region, north to arctic islands; circumpolar.

NOTES: Marsh ragwort often forms a solid belt around small sloughs and ponds, producing a spectacular display of colour when the plants are in flower. Clouds of fluffy seeds follow in the late summer. Many people with allergies are very sensitive to the pollen of these plants. • Most marsh ragwort plants live for only 1 year and flower once during that time. However, in the North, where short summers can end suddenly, plants in which flowers were killed by early frost may live through the winter to flower and fruit again the following year. • The seeds of marsh ragwort have been carried on the wind for 200 km or more during storms. • The young leaves and flowering stems of marsh ragwort can be eaten raw in salads or cooked as a potherb or vegetable. However, care should be taken to **avoid** plants from areas with **polluted** water. • Marsh ragwort is sometimes called 'mammoth plant,' an appropriate name for such a large, hairy northern plant.

CUT-LEAVED RAGWORT • *Senecio eremophilus*

GENERAL: Perennial, 0.3–1 m tall; stems stout, hairless, simple to freely branched, **leafy up to flower cluster**.

LEAVES: Simple, alternate, 5–15 cm long, 1–4 cm wide, deeply lobed, usually toothed, hairless to sparsely hairy; lower leaves stalked, upper leaves smaller, stalkless.

FLOWERS: Many heads 10–25 mm across, **in large, flat-topped cluster**; 7–9 ray flowers, pale yellow, 6–10 mm long; 40–60 disc flowers, yellow; involucres 6–8 mm high, bracts linear-lance-shaped, **often black-tipped**.

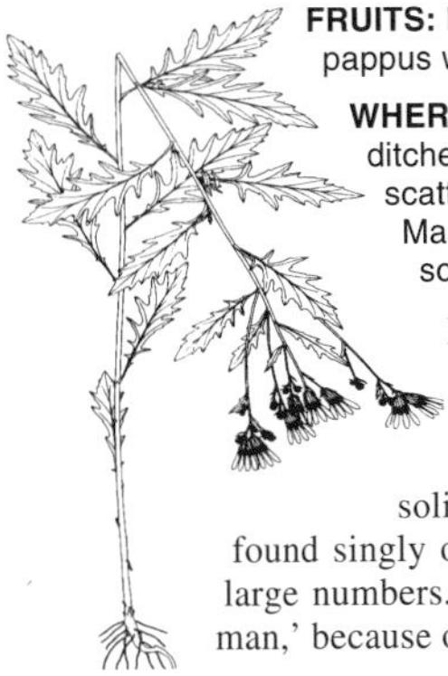

FRUITS: Hairless or slightly hairy achenes; pappus white.

WHERE FOUND: Moist open woods, ditches, roadsides and rich disturbed soil; scattered across our region from southern Manitoba to eastern British Columbia and southern N.W.T.

NOTES: Ragwort was once used widely as a charm against recurring chills and fever. • The species name, *eremophilus*, comes from the Greek meaning 'fond of solitude,' in reference to the fact that plants of this species are often found singly or only a few together. Only in disturbed areas do they occur in large numbers. The genus name *Senecio* comes from the Latin *senex*, 'an old man,' because of the fluffy, white seed heads.

BALSAM GROUNDSEL • *Senecio pauperculus* CANADIAN BUTTERWEED

GENERAL: Perennial from fibrous roots and short, woolly stem bases; stems slender, single, 10–60 cm tall (usually 20–40 cm), essentially **hairless** but with woolly tufts.

LEAVES: Basal leaves thin, **broadly to narrowly spoon-shaped or oblong**, 2–10 cm long, usually **toothed**, slender-stalked, hairless; stem leaves lance-shaped, pinnately toothed or lobed, much smaller and simpler upwards, stalkless.

FLOWERS: 2–10 or more long-stalked heads, 10–15 mm across, in loose, umbel-like cluster; ray flowers yellow, usually showy, **5–10 mm long**; disc flowers yellow; involucres 3–9 mm high, bracts green or purple-tinged at tip.

FRUITS: Hairless or finely hairy achenes.

WHERE FOUND: Moist open woods, meadows, streambanks, lakeshores, and roadsides; widespread across boreal forest and parkland; reaches Great Bear Lake and interior Alaska to north and west.

NOTES: Rayless ragwort or rayless mountain butterweed (*S. indecorus*) has finely toothed leaves at the stem base, and pale yellow flower heads with no ray flowers. It grows in moist meadows and forests and on streambanks across our region. • **Northern ragwort** or Rocky Mountain butterweed (*S. streptanthifolius*) has somewhat fleshy leaves, which are toothed above the middle, mostly at the stem base, and it has bright yellow flower heads with rays. It grows on shores, in moist to dry meadows or open coniferous forest in our region, from northern Saskatchewan to the Yukon.

CANADA GOLDENROD • *Solidago canadensis*

GENERAL: Robust perennial from **long, creeping rhizomes**; stems slender, leafy, 30–90 cm tall.

LEAVES: Many, crowded, **only gradually smaller and simpler upwards**, narrowly lance-shaped, 5–10 cm long, 3-veined, stalkless, saw-toothed (rarely smooth-edged), rough-hairy, green; **lower leaves soon fall off**.

FLOWERS: Many small heads in **dense pyramidal terminal clusters**, often turned to 1 side of ascending branches; 10–17 ray flowers, yellow, 1–3 mm long; disc flowers yellow; involucres 3–5 mm high, bracts narrow, pointed, **overlapping, sometimes with sticky glands**.

FRUITS: Short-hairy achenes.

WHERE FOUND: Moist open woods, meadows and roadsides; widespread across our region, less common in the northwest, but scattered north to the Arctic Circle.

NOTES: Goldenrod is much maligned as a cause of hay fever. Some people are sensitive to it in close contact, but the pollen is too heavy to be carried by the wind. Instead, it is transported by bees and other flying insects. Unfortunately goldenrod flowers at the same time as the real hay fever culprit, ragweed. The inconspicuous flowers of ragweed are easily overlooked beside the bright yellow goldenrod flowers. • A strong yellow dye can be made from goldenrod flower clusters. When the flowers dry, they are an attractive golden colour, an excellent addition to dried herb and flower arrangements. • Goldenrod was reputedly carried into battle during the Crusades and was commonly used as a substitute for the highly taxed English tea during the American Revolution.

MOUNTAIN GOLDENROD • *Solidago spathulata*
SPIKE-LIKE GOLDENROD

GENERAL: Resinous, aromatic perennial from short stout rhizomes or prostrate stem bases; stems 10–50 cm tall, single or tufted, **often reddish-tinged**.

LEAVES: Broadly lance- to spoon-shaped, 2–10 cm long, usually round-toothed towards tip, 1 vein, hairless, fewer, **smaller and simpler upwards**.

FLOWERS: Many heads about 8 mm high, in **long, narrow**, dense, often interrupted clusters at stem tips; usually 8 ray flowers, **yellow**; about 13 disc flowers, yellow; involucres 4–6 mm high, **5–7 mm broad**, bracts narrow, broad-tipped, **overlapping, often sticky**.

FRUITS: Densely hairy achenes, about 2 mm long.

WHERE FOUND: Open woods, meadows, riverbanks and terraces; widespread across our region to interior Alaska.

NOTES: Also called *S. decumbens*. • The leaf stalks of **alpine** or **northern goldenrod** (*S. multiradiata*)(inset photo) are fringed with hairs, and the flower cluster is more compact and rounded than that of mountain goldenrod. Alpine goldenrod is widespread on gravel bars and open grassy slopes from central Manitoba to Alaska. • Goldenrods may be confused with yellow-flowered species of *Hieracium* or *Senecio*, but can be identified by their many small heads, each usually less than 1 cm across. • Goldenrod plants were boiled to make a medicinal tea that was used to clean wounds, sores, and ulcers on humans, and saddle sores on horses. Milder teas were taken internally to relieve sore throats, throat constrictions, and nasal congestion and to promote perspiration. Ground, dried goldenrod leaves were sprinkled on infections as an antiseptic powder. Sometimes the leaves were mixed with soap to make a plaster for binding on sore throats. All goldenrods contain small quantities of natural rubber in their sap. • Goldenrod tea has been used by herbalists for many years to relieve intestinal gas and cramps, and pharmacologists have found support for this use. • *Solidago* comes from the Latin *solidus*, 'whole,' and *ago*, 'to make'—i.e., 'to make whole' (or cure)—in reference to its medicinal uses.

FLAT-TOPPED GOLDENROD • *Solidago graminifolia*

GENERAL: Perennial from creeping rhizomes, 30–70 cm tall; stems leafy, **marked with fine lines along length**, hairless or nearly so.

LEAVES: Lower leaves soon fall off; **many** stem leaves, **somewhat crowded, linear-lance-shaped**, 2–10 cm long, **10–15 times as long as wide**, stalkless, smooth-edged, hairless.

FLOWERS: Small yellow heads in compact, **flat-topped clusters**; 12–20 yellow ray flowers; 8–12 yellow disc flowers; involucres 3–5 mm high, **slightly sticky**, bracts blunt or rounded, yellowish.

FRUITS: Hairy achenes.

WHERE FOUND: Riverbanks, lakeshores and wet meadows; fairly common in our region north to upper Mackenzie valley.

NOTES: Also called *Euthamia graminifolia*. Livestock usually avoid eating goldenrod, but occasionally they develop an appetite for these plants, and some cases of poisoning have been reported. However, this was the result of a rust (*Coleosporium*) infestation rather than toxins in the plant itself. Diterpenes in some species may be poisonous to sheep. • There are very few records of these common plants being used as food. However, goldenrod leaves and flowers have been used fresh or dried to make tea. The plants have also been cooked as a vegetable or potherb, and the seeds have been used to make mush and thicken soups. In the U.S., at the time of the Revolution, wild goldenrod flowers were collected, dried and shipped to China, where they brought a high price as a tea. • The species name *graminifolia*, from the Latin *gramen*, 'grass,' and *folium*, 'a leaf,' refers to the long, narrow, grass-like leaves characteristic of this goldenrod.

BASTARD TOADFLAX • *Comandra umbellata*
PALE COMANDRA

GENERAL: Erect, hairless perennial from creeping, white rhizomes; **roots parasitic on neighbouring plants**; stems usually clustered, branched, leafy, usually 10–25 cm tall.

LEAVES: Many, **alternate**, nearly stalkless to short-stalked, linear to lance-shaped or narrowly oblong, usually 1–3 cm long.

FLOWERS: Several to many **3–5-flowered clusters**, form egg-shaped cluster at stem tip; no petals; 5 petal-like sepals, greenish white to purplish, 2–5 mm long.

FRUITS: Dry to somewhat fleshy, egg-shaped to spherical, with remains of calyx at tip, **green or drab purplish brown**, 3–8 mm long, 1-seeded.

WHERE FOUND: Dry, open woods, gravelly slopes and grassland; widely scattered across southern boreal forest and parkland from southern Manitoba to southern Yukon.

NOTES: Also called *C. pallida* and *Geocaulon umbellatum*. • When the urn-shaped fruits of bastard toadflax are fully grown, but still slightly green, they are sweet and oily and make a delicious nibble. Although they are seldom available in quantity, they were eaten as a snack by western native groups. Too many will cause nausea. Although they are best green, the fruits are still edible when fully mature and brown. • Bastard toadflax is a semi-parasite. Although it has green leaves capable of producing food and roots capable of taking up nutrients, it also joins to the roots of neighbouring plants, robbing them of water and nutrients. • Because of its attractive white flowers, bastard toadflax is sometimes transplanted to gardens, but it needs to be placed near other plants that it can parasitize.

NORTHERN BASTARD TOADFLAX • *Geocaulon lividum*
NORTHERN COMANDRA

GENERAL: Perennial, from creeping, thread-like, reddish rhizomes; stems usually unbranched, 10–25 cm tall.

LEAVES: Alternate, thin, **oval**, 1–3 cm long, **blunt-tipped**, bright green, frequently yellow streaked.

FLOWERS: In slender-stalked, 2–4-flowered clusters from leaf axils; usually centre flower female and outer flowers male; petal-like sepals 1–1.5 mm long, greenish purple, inconspicuous.

FRUITS: Scarlet to fluorescent orange, berry-like drupes, juicy, 5–8 mm across; edibility questionable, not recommended.

WHERE FOUND: Moist woods; widespread across boreal forest, north and west to interior Alaska.

NOTES: Also called *Comandra livida*. • Although the fruit of northern bastard toadflax is edible, it is not very palatable and is not usually used as food. • Northern bastard toadflax is parasitic, taking nutrients from the roots of a variety of other plants, including bearberry (p. 73) and asters (pp. 182–85). Its thin, spreading underground stems reach out to nearby plants, and tiny, sucker-like organs on the rootlets attach themselves to the roots of other plants. In this way, northern bastard toadflax can supplement the food it produces with additional nutrients from its neighbours. • The yellow streaking often seen on the leaves is caused by the lodgepole pine's *Comandra* blister rust, of which northern bastard toadflax is the alternate host. • The Chipewyan considered the fruits inedible, but a few were swallowed once a year to relieve chronic chest problems (perhaps tuberculosis). The Cree boiled the leaves and bark to make a medicinal tea to induce vomiting and purge the system. They also used the chewed stems and leaves as a poultice on wounds.

LAMB'S-QUARTERS • *Chenopodium album*

GENERAL: Stout **annual**, **greenish to greyish**, with small flaky scales; stems erect, 20–100 cm tall, often with many short branches, become reddish to purplish with age.

LEAVES: Lance- to diamond-shaped, 2–12 cm long, alternate, somewhat fleshy, **greyish green and densely mealy** below; edges of larger leaves wavy to irregularly toothed or lobed.

FLOWERS: In **dense clusters in spikes from leaf axils or stem tips**; tiny, greenish (often tinged blue); no petals; 5 sepals, ridged, usually white-mealy.

FRUITS: Black, shiny, flattened, circular seeds, 1.1–1.6 mm long, in thin, white, papery envelope.

WHERE FOUND: Aggressive, introduced weed of disturbed or cultivated land; widespread near settlements across our region; circumpolar.

NOTES: Includes *C. berlandieri*. • Many people eat lamb's-quarters as a wild green but, like other members of this genus, it should always be eaten in **moderation**. • Some native groups ground the seeds into flour for making bread or mush. The seeds can also be added to soups and stews, or they can be eaten raw as a nibble or served with milk and sugar, like cold cereal. Young plants are a rich source of vitamins A and C and calcium. They can be eaten raw or cooked as a potherb and are very good in soups and stews. • Lamb's-quarters is commonly thought to be naturalized from Europe, but it is probably native. Caches of seeds dating from about 1,000 B.C. have been found in archaeological sites in western North America. • Lamb's-quarters is a prolific weed. A single plant can produce over 70,000 seeds, each capable of surviving for many years. In fact, archeologists and geologists have reported the discovery of living seeds buried in layers up to 10,000 years old.

STRAWBERRY BLITE • *Chenopodium capitatum*

GENERAL: Succulent, yellowish green annual, hairless; stems erect to ascending, 10–50 cm tall, usually branched.

LEAVES: Broadly triangular, 2–12 cm long, toothless or wavy-toothed, often turn reddish; lower leaves often have 2 diverging lobes.

FLOWERS: Tiny, in dense, **bright red, pulpy, spherical**, 1–1.5 cm wide clusters from leaf axils, or in interrupted spikes at stem tips; no petals; 3–5 sepals, fleshy, red when mature.

FRUITS: Erect, black, lens-shaped seeds, 0.7–1.2 mm long.

WHERE FOUND: Disturbed soil in clearings, burns, river bars, cultivated soil and waste places; widespread across our region.

NOTES: Neither **maple-leaved goosefoot** (*C. hybridum* var. *gigantospermum*, also called *C. gigantospermum*) nor **red goosefoot** (*C. rubrum*) have the red, juicy fruits of strawberry blite. Maple-leaved goosefoot is a weedy, fresh-green annual with loose, branched flower clusters, and with jaggedly-lobed leaves shaped somewhat like those of a maple. Red goosefoot is a native, reddish-tinged annual with spikes of flowers in leaf axils, and with egg-shaped to triangular leaves with wavy-toothed edges. Both of these species are most common in the southern prairie provinces; maple-leaved goosefoot in thickets, open forests, on shores and in disturbed areas near settlements; red goosefoot in moist saline or alkaline flats, or in disturbed habitats. • Strawberry blite is enjoyed as a wild green, but it should not be eaten in large quantities. It is high in calcium, protein, and vitamins A, B_1, B_2, B_6 and C. The young leaves can be eaten raw or cooked as a potherb or added to soups and stews. The red flower clusters can also be eaten. • Some native groups used the red flower clusters to make ink or to dye porcupine quills, clothes, hides, basket materials, implements and even their own skin. The colour is bright red at first, but eventually it darkens to purple or maroon.

STINGING NETTLE • *Urtica dioica*
COMMON NETTLE

GENERAL: Erect **perennial** from strong, spreading **rhizomes**, armed with stinging hairs; stems **0.5–2 m tall**, slender, sharp-angled.

LEAVES: Opposite, narrowly lance- to heart-shaped, 4–15 cm long, saw-toothed; stalks slender, 1–6 cm long.

FLOWERS: In **drooping clusters** from leaf axils, **sexes on separate plants or in separate spikes on same plant**; greenish, inconspicuous; no petals; 4 sepals, 1–2 mm long.

FRUITS: Flattened, lens-shaped achenes, 1–2 mm long.

WHERE FOUND: Moist woodlands, thickets, open areas, streambanks and disturbed sites; widespread across our region, north and west to Great Slave Lake and interior Alaska.

NOTES: Also called *U. gracilis.* • Stinging nettle is covered with tiny, hollow, pointed hairs. The swollen base of each hair contains a tiny droplet of formic acid, and when the hair tip pierces you, the acid is injected into your skin. This can cause itching and burning for a few minutes to a couple of days. Rubbing nettle stings with the plant's own roots is said to help to relieve the burning. • Tender young nettle plants can be boiled and eaten like spinach or in soups and stews. The acid is destroyed by cooking or drying, but eating large quantities of cooked nettles can still cause a burning sensation. Young plants can also be used to make nettle tea, wine or beer. Older plants become fibrous and gritty from an abundance of small crystals. • The Cree used nettle leaf tea as a blood purifier and as a remedy for kidney stones, phlegm in the lungs, diarrhea and worms. • Nettle fibres were used for many years to make rope, paper and cloth. The fibres were considered superior to cotton for velvet or plush, and were said to be more durable than linen. • Galen, a physician in ancient Greece, said nettle seeds, soaked in wine and cooked, would excite a person to games of love. He also prescribed a tablespoon of powdered seeds mixed with jam or honey as a cure for impotence.

BICKNELL'S GERANIUM • *Geranium bicknellii*

GENERAL: Annual or biennial, from slender taproot; stems much branched, 15–60 cm tall, with spreading hairs, sometimes gland-bearing towards top.

LEAVES: Heart-shaped to round, 2–7 cm wide, **deeply 5-lobed**; lobes divided into pointed, narrow segments; hairs and glands on veins and stalks.

FLOWERS: Usually 2 per stalk, from upper leaf axils; **pale pink-purple**; petals 3.5–7 mm long, slightly longer than hairy, **slender-pointed sepals**; stalks have gland-tipped hairs.

FRUITS: Linear, 5-parted capsules, 14–20 mm long, with slender, **3–6 mm long beak**; 5 seeds, thick, cylindrical.

WHERE FOUND: Open woods, clearings, streambanks and disturbed areas; widespread across our region, north and west to Great Slave Lake and interior Alaska.

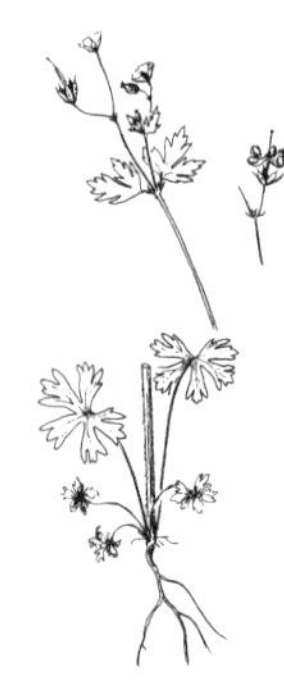

NOTES: The roots of Bicknell's geranium are rich in tannin, and they make an astringent medicinal tea that was gargled to help heal mouth sores and sore throats, and swallowed to relieve urinary problems, excessive menstruation, diarrhea and intestinal diseases such as dysentery and cholera. Geranium tea was also given to cattle to cure 'bloody water.' • Bicknell's geranium, together with pink corydalis and golden corydalis (opposite), is often very abundant after disturbances such as wildfire or clear-cutting followed by slash-burning. Its seeds apparently remain on the forest floor for decades, and germination is triggered by disturbance, exposure, and warming of the soil. • The capsules open explosively, splitting lengthwise from the bottom, and flinging seeds away from the parent plant.

GOLDEN CORYDALIS • *Corydalis aurea*

GENERAL: Annual or biennial; **stems low, often prostrate**, 10–30 cm long, soft, with watery juice, **much branched**, hairless.

LEAVES: Alternate, much divided.

FLOWERS: In many-flowered, loose end clusters; **golden yellow, irregular**; petals 12–15 mm long; spur 4–5 mm long.

FRUITS: Pod-like capsules, 15–25 mm long, spreading or somewhat drooping, constricted between seeds, usually curved; seeds black, shiny.

WHERE FOUND: Open woods, clearings, banks, shores and roadsides; often weedy; common across our region north and west to Great Bear Lake and Alaska.

NOTES: Pink corydalis (*C. sempervirens*) (inset photo) also grows in open areas across the boreal forest. It is a much taller plant (30–60 cm) with erect, slender, less-branched stems, and loose clusters of delicate pink to purplish flowers with yellow tips. • *Corydalis* species are classified as **poisonous** because they contain isoquinoline and other alkaloids. Some poisoning of cattle and sheep has been reported. • If a visiting insect wants to reach the nectary, it must force its way between the 2 lips formed by the petals. By taking this route, the insect picks up pollen from the anthers and fertilizes the flower. Only long-tongued insects can reach the nectar by this route, but some short-tongued bees have learned how to rob corydalis flowers by chewing holes in the petals and stealing the nectar without ever entering the flower.

SPOTTED TOUCH-ME-NOT • *Impatiens capensis*

GENERAL: Annual, 0.3–1.5 m tall; stems branched, **succulent**, often tinged orange-red.

LEAVES: Simple, alternate, stalked, lance- to egg-shaped, 2–10 cm long, coarsely toothed, hairless.

FLOWERS: Showy, bright orange to pale yellow, **with reddish brown or purplish spots**, 20–25 mm long; lower sepals sac-like, **abruptly constrict into spur bent back parallel to sac**.

FRUITS: Pod-like capsules, burst open when touched.

WHERE FOUND: Wet woods, streambanks, lakeshores and beaver dams; across our region, north and west to southernmost N.W.T.

NOTES: Also called *I. biflora.* • **Western jewelweed** (*I. noli-tangere*) is the western counterpart of spotted touch-me-not. It grows in moist clearings and thickets and around springs, from Manitoba to Alaska, but it is uncommon at boreal latitudes. Western jewelweed has light green or straw-coloured stems and pale yellow flowers that are unspotted or have many small dots and that taper more gradually to the spur. • Most of the seed capsules arise not from the large, showy flowers but from very small flowers that do not expand and are fertilized without opening. Stand back when touching ripe seed capsules. They will fly open explosively, throwing their seeds in all directions. • A poultice of crushed touch-me-not leaves, a tea made from the leaves (fresh or frozen in cubes), and the sticky juice from the stem (harvested before flowering) have all been used to treat recent poison-ivy rash. A 1957 study of 115 patients found touch-me-not to be effective in relieving poison ivy rash within 2–3 days in 108 cases. The leaf poultice was also used to treat bruises, burns, cuts, eczema, insect bites, sores, sprains, warts, and even ringworm. • Young succulent stems of spotted touch-me-not are said to be edible if collected in spring and early summer.

WILD SARSAPARILLA • *Aralia nudicaulis*

GENERAL: Perennial from **widely spreading rhizomes** (often forms large clonal colonies); **leaf and shorter flowering stalk** arise from stout, woody base.

LEAVES: Single, long-stalked, 30–60 cm long, **compound**; **3 main divisions, each with 3–5 leaflets**; leaflets egg- to lance-shaped, 3–15 cm long, **finely toothed**.

FLOWERS: In **2–7** (usually 3) **umbrella- or ball-shaped clusters**, 2–5 cm across, on 10–40 cm tall, **naked, flowering stems**, **hidden under leaf**; 5 petals, greenish white, 2–3 mm long; flowers perfect or with sexes on separate plants.

FRUITS: Plump berries, greenish, turn dark purple, about 6 mm long, edible but not very tasty.

WHERE FOUND: Woodlands; widespread and common across our region, north and west to southern N.W.T. and northeastern B.C.

NOTES: Bristly sarsaparilla (*A. hispida*) (inset photo) is found on dry sites in the boreal forest of Manitoba and Saskatchewan. It is easily recognized by its leafy, bristly stems. • Some native peoples on long war or hunting journeys carried bundles of sarsaparilla roots that they chewed for energy. The Chipewyan considered the berries inedible, but they collected the roots in the autumn and chewed them or made a tea to treat heart pain, stomach upset and liver problems. The Woods Cree used boiled teas to treat infected gums of teething children, venereal disease, pneumonia and low milk production in nursing mothers. Poultices made from bruised bark or chewed, woody roots were applied to wounds to promote healing and draw out infection.

SPREADING DOGBANE • *Apocynum androsaemifolium*

GENERAL: Erect perennial with **milky sap**, from rhizomes; stems **branched** towards top, **20–100 cm tall**.

LEAVES: Opposite, simple, egg-shaped to oblong, 2.5–8 cm long, abruptly sharp-pointed, **spreading and drooping**, short-stalked.

FLOWERS: In showy clusters from stem tips and leaf axils; **pink, bell-shaped**, sweet-scented; petals 6–9 mm long, bend back.

FRUITS: 8–12 cm long, narrow (5 mm thick), paired cylindrical pods; **many seeds**, 2.5–3 mm long, **each with long tuft of cottony hairs**.

WHERE FOUND: On well-drained, open sites in woods and on roadsides, open hillsides and ridges; widespread across our region, north and west to Great Slave Lake and interior Alaska.

NOTES: Indian-hemp (*A. sibiricum*, also called *A. cannabinum* var. *hypericifolium*) has ascending rather than drooping leaves, and 2–4 mm long, greenish to whitish petals. It grows on dry, rocky slopes and gravelly meadows and thickets, scattered across our region, north and west to northeastern B.C. and the southern N.W.T. • The innocent-looking flowers of spreading dogbane can be a death-trap for unsuspecting insects. A scale covered with horny teeth is located above each of the nectaries in the throat of the flower. When these scales are touched, they spring inwards, imprisoning the intruder. Visiting butterflies, bees and bumblebees are usually strong enough to pull away, but small insects are trapped and die. • The tough, fibrous bark was used by native peoples to make twine, fishing nets, and thread. This thread was said to be finer and stronger than cotton thread, and 3 strands plaited together were used as bowstrings. • Dogbane contains bitter, **toxic** glycosides that discourage browsing. The milky juice can raise **blisters** on sensitive skin, but most people can handle it without effect. • The Cree boiled dogbane to make a medicinal tea to increase lactation. They apparently also used it to treat sore eyes. Ground root, boiled in water and applied to the hair as a final rinse, is said to irritate the follicles and dilate small blood vessels, thereby stimulating hair growth.

FRINGED MILKWORT • *Polygala paucifolia*

GENERAL: Perennial, 5–15 cm tall; stems branched towards top, hairless.

LEAVES: Simple, alternate, most clustered near top of stem, elliptic to egg-shaped, 1–5 cm long, 1–2 cm wide, short-stalked or stalkless, pointed.

FLOWERS: Usually 3 or 4 in axils of upper leaves; pink to rose-purple, irregular; 5 sepals, 3 are 3–6 mm long, 2 are 10–17 mm long; 3 petals, 10–18 mm long, **middle petal boat-shaped with fringed crest** about 6 mm long.

FRUITS: Small 2-seeded capsules, split lengthwise.

WHERE FOUND: Moist coniferous woods; occasional across southern boreal forest from southeastern Manitoba to northeastern Alberta; rare in Alberta.

NOTES: Fringed milkwort also produces flowers that are fertilized without opening on short underground, side branches. • At first glance the flowers of fringed milkwort might look as though they belong to the pea family (Fabaceae, pp. 138–146), but they are really very different. Some people call them 'Snoopy flowers,' because the keel and the 2 wing petals resemble the rounded nose and flying ears of a dancing Snoopy (the cartoon dog). • The genus name *Polygala* is derived from the Greek *poly*, 'much' or 'many,' and *gala*, 'milk,' in reference to the profuse secretions of some plants of this genus, and the belief that a diet high in milkwort would increase the milk production of both domestic cows and nursing mothers. *Paucifolia* comes from the Latin *paucus*, 'few' and *folium*, 'leaf,' in reference to the small number of leaves on these little plants.

BUNCHBERRY • *Cornus canadensis*

GENERAL: Low perennial from spreading rhizomes, often forms colonies; stems 5–15 cm tall; somewhat **woody base**.

LEAVES: Evergreen, 4–7 in whorl at stem tips, oval-elliptic, 2–8 cm long, green above, whitish below, veins parallel.

FLOWERS: Dense **cluster of small, greenish white to purplish flowers** surrounded by **4 white to purplish tinged, petal-like bracts**.

FRUITS: Bright **red, fleshy, berry-like drupes**, 6–8 mm across, pulpy, sweet, in dense cluster at stem tip.

WHERE FOUND: Moist woods and clearings; widespread and common across our region.

NOTES: The scarlet fruits can be eaten fresh as a snack. Some people consider them mealy and tasteless, but others say they are juicy and full of crunchy, little, poppy-like seeds. The fruits can also be cooked and used in puddings, sauces, or preserves. • The Cree call bunchberry 'itchy chin berry,' perhaps because the leaves feel bristly against the skin, and the Chipewyan say that if you rub your face with the berries it feels as though you have 'needles in your skin.' Children do this on purpose. • It is easy to overlook the true flowers of bunchberry. The striking white 'petals' are really large bracts surrounding a cluster of tiny flowers at their centre. The showy white bracts make the flowers more noticeable to passing insects and lure them to the plant. In autumn, the flowers mature into a cluster of scarlet berry-like drupes—hence the name bunchberry. These fruits are a favourite food of many small mammals and birds. • Bunchberry is reported to have an explosive pollination mechanism, whereby the petals of the mature, but unopened, flower buds suddenly bend back and the anthers spring out simultaneously, catapulting their pollen loads into the air. The trigger for this explosion appears to be a tiny 'antenna,' just over 1 mm long, that projects from near the tip of one of the 4 petals in a bud.

NORTHERN GENTIAN • *Gentianella amarella*
FELWORT

GENERAL: Annual or biennial from taproot; stems erect, 5–50 cm tall, with strongly ascending branches.

LEAVES: Elliptic to spoon-shaped at stem base; **5–8 pairs on stem, linear to lance-shaped**.

FLOWERS: Few to many in dense clusters at stem tips or in 1–3s from upper leaf axils; petals **violet to pale blue or rarely white**, 1–1.5 cm long, **form slender tube with hairy fringe at throat**.

FRUITS: Cylindrical capsules; seeds egg-shaped.

WHERE FOUND: Moist woods, clearings, meadows and roadsides; widespread across our region, south of Arctic Circle; circumpolar.

NOTES: Also called *Gentiana amarella* and *Gentiana acuta*. • **Four-parted gentian** (*Gentianella propinqua*) (inset photo) is also a small annual or biennial, but its deep-blue, 4-parted flowers have bristle-tipped lobes and do not have a frilly throat. It grows in open gravelly clearings, meadows and forest edges from Hudson Bay across the Northwest Territories to western Alaska. • Because they contain extremely bitter alkaloids, gentians have been used as tonic medicines since the 1st century A.D. Some medical authorities suggest that gentians simply irritate the mucous membrane of the digestive tract, but others say they can stimulate appetite in people with chronic conditions such as anorexia, old age, prolonged illness or indigestion. Although it was considered one of the most valuable vegetable tonics, it was also recommended that it be used with a purgative, so that the patient could avoid its debilitating effects. An overdose will cause nausea and vomiting, but considering the extreme bitterness of these plants, only the most grimly resolute masochists could manage eat a large amount. People with **high blood pressure** and **pregnant women** should not take gentian root.

FRINGED GENTIAN • *Gentianella crinita* ssp. *macounii*

GENERAL: Annual, 15–50 cm tall; stems stiffly erect, **somewhat 4-sided**, simple or with few upright branches.

LEAVES: Spoon-shaped at stem base; **stem leaves** in 2–4 **pairs**, 2–5 cm long, linear-lance-shaped, **pointed**.

FLOWERS: Usually single, on long, erect stalks, from leaf axils and stem tips; blue, rarely whitish, 2–4 cm long; **4 spreading lobes have conspicuous hairy fringe; keels of calyx have nipple-shaped projections** (papillae).

FRUITS: Spindle-shaped capsules.

WHERE FOUND: Moist meadows, shores and calcium-rich fens; occasional in southern boreal forest and parkland of prairie provinces; widely separate population near Great Slave Lake.

G. crinita

NOTES: Also called *Gentiana macounii*. • **Raup's fringed gentian** (*G. detonsa* ssp. *raupii* or *Gentiana raupii*) resembles fringed gentian but has fewer stem leaves and has blunt-tipped sepals that are not ridged on the back and lack nipple-shaped projections at their bases. It grows in floodplain meadows, shores and marshes in the Mackenzie River basin from northern Alberta to the Mackenzie delta. • Gentians have long been used as bitters in many forms, including powder, tea, syrup, alcohol extract and wine. In the 18th century, gentian wine was served as an aperitif. Today, herbalists usually use gentian as a bitter stomach tonic for aiding digestion. Commercially produced bitters often contain gentian, but some suggest that it is just the alcohol base that stimulates the appetite. • This genus was named after Gentius, a king of Illyria (an ancient country near present-day Albania), who was supposedly the first to use it medicinally. *Crinita* means 'provided with long hair,' an apt reference to the hairy fringe around the petals.

G. detonsa

SPURRED GENTIAN • *Halenia deflexa*

GENERAL: Annual or biennial, 10–50 cm tall; stem erect, simple or branched above.

LEAVES: Lower leaves spoon-shaped, stalked; stem leaves opposite, lance- to egg-shaped, 2–4 cm long, pointed, stalkless or nearly so, smaller upwards.

FLOWERS: In clusters at stem tips and from leaf axils; **purplish green or bronze**, up to 1.5 cm long; 4 petals, **each lengthened at base into slender spur**.

FRUITS: Narrowly-oblong capsules.

WHERE FOUND: Moist woods, thickets and clearings; fairly common in southern boreal forest and parkland of prairie provinces.

NOTES: Spurred gentian can be distinguished from true gentians by the 4 hollow spurs that project downwards from the base of the flowers. • The genus was named for Jonas Halenius, a Swedish botanist (1727–1810). The species name, *deflexa*, is from the Latin *de*, 'downwards,' and *flexus*, 'bending,' in reference to the downwards-pointing spurs at the base of each flower.

NARROW-LEAVED COLLOMIA • *Collomia linearis*

GENERAL: Erect, finely hairy annual; stems 5–40 cm tall, leafy, usually branching.

LEAVES: Mostly alternate, **linear-lance-shaped**, 1–6 cm long, **smooth-edged**, stalkless, sometimes clasp stem.

FLOWERS: In head-like, leafy clusters at stem tips; petals **pink or purplish to whitish, fused into slender trumpet**, 8–15 mm long.

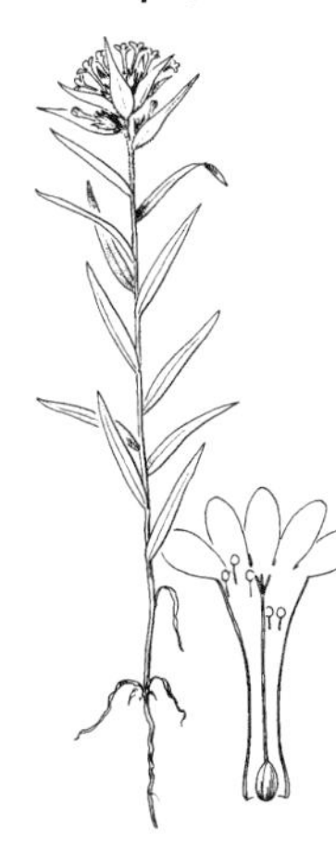

FRUITS: Capsules, split open along 3 lines; many seeds, become sticky and produce spiralling threads when wet.

WHERE FOUND: Weedy on moist, disturbed ground; native to most of Canada and U.S. but introduced in Yukon, N.W.T. and Alaska; scattered across our region, north and west to Great Slave Lake and interior Alaska.

NOTES: The small, dull pink flowers of collomia are self-fertilizing. When their seed capsules mature, they split into 3 sections. These segments remain enclosed within the calyx, pressing out against the fused sepals. Eventually, the calyx dries and shrinks, tightening even more around the capsule. Finally the tension becomes too great, and the enclosed capsule splits open the calyx, shooting out its seeds by force. When the seeds get wet, they exude a sticky, adhesive material from their husks that cements them to the soil. • *Collomia* is from the Greek *kolla*, 'glue,' in reference to the sticky seeds.

FRANKLIN'S SCORPIONWEED • *Phacelia franklinii*

GENERAL: Biennial or winter annual, 20–50 cm tall; stems erect, usually single, **soft-hairy and often with glands**.

LEAVES: Alternate, **hairy**; lower leaves stalked, **pinnately divided into linear-oblong toothed segments**; upper leaves nearly stalkless, smaller, less divided.

FLOWERS: In **dense, coiled cluster** at stem tip; blue, occasionally almost white, 6–9 mm long, 8–12 mm wide, **broadly bell-shaped**; **stamens and styles protrude slightly**.

FRUITS: Stalkless capsules, egg-shaped; many brown seeds.

WHERE FOUND: Rock outcrops, slopes and dry open areas; often pioneering on sandy roadsides and in burned or cutover pine woods; common in boreal forest across prairie provinces north and west to Great Bear Lake and southern Yukon.

NOTES: The hairy stems, leaves and flower clusters of most scorpionweeds are very irritating to the skin and can cause **allergic reactions** in some people. • The genus name, *Phacelia*, derived from the Greek *phakelos*, 'a bundle' or 'a fascicle,' refers to the congested flower clusters with their tight groups of flowers. The species name, *franklinii*, honours Sir John Franklin (1786–1847), an English arctic explorer.

NORTHERN VALERIAN • *Valeriana dioica*
MARSH VALERIAN

GENERAL: Perennial with fibrous roots and stout, **musky-scented** rhizomes; stems slender, 30–70 cm tall, hairless or nearly so.

LEAVES: Rosette at stem base, **simple**, oblong to lance-shaped, 2–5 cm long; **2–4 pairs** of stem leaves, 1.5–5 cm long, divided into **3 or more pairs** of oval to lance-shaped **lobes**; smaller, stalkless upwards.

FLOWERS: Many, in tight, hemispherical to flat-topped head that lengthens with age; corolla funnel-shaped to circular-flattened, about 3 mm across, white, **2–3.5 mm long**; protruding styles and stamens give head fluffy appearance.

FRUITS: Achenes, 3–5 mm long, each with **feathery plume** for wind dispersal.

WHERE FOUND: Moist meadows, grassy openings and wetland edges; widespread across boreal forest, north and west to Great Bear Lake and central Yukon.

NOTES: Also called *V. septentrionalis*. • **Capitate valerian** (*V. capitata*) has unlobed or merely 3-lobed upper stem leaves, and a head-like cluster of pink to white flowers. It grows on turfy shores and in peatlands and open forest, from the western Northwest Territories to western Alaska. • The Woods Cree chewed the root of northern valerian, wrapped it in a cloth and placed it in their ears to relieve earaches. It was also rubbed on the head and temples of people suffering from headaches and applied externally to patients suffering from seizures. The root was powdered to treat menstrual problems and it was added to smoking mixtures to relieve cold symptoms. Valerian root tea was considered an all purpose medicine, either alone or mixed with other herbs. • In Europe, common valerian (*V. officinalis*) has been used to calm nervousness and hysteria for at least 1000 years. It has a tranquilizing effect, with fewer side effects than some synthetic sedatives. Large doses can cause **vomiting**, stupor, and dizziness, and continued use may lead to depression.

TALL LUNGWORT • *Mertensia paniculata*
TALL BLUEBELLS

GENERAL: Erect **perennial** from woody base; **1–several** stems, 20–80 cm tall (sometimes to 1 m), hairy.

LEAVES: With **prominent veins**, coarsely rough-hairy above or on both sides; basal leaves **long-stalked, egg-** to heart-shaped; stem leaves narrower, 3–15 cm long, short-stalked; smaller, stalkless upwards.

FLOWERS: Few to many in branched clusters at stem tips or from upper leaf axils, congested at first, more open later; petals **blue, sometimes pink or white**, bell-shaped, 8–15 mm long, drooping, with protruding styles.

FRUITS: Four nutlets, 2.5–5 mm long, **wrinkled**.

WHERE FOUND: Moist woods, thickets, meadows and streambanks; widespread across boreal forest and northern parkland.

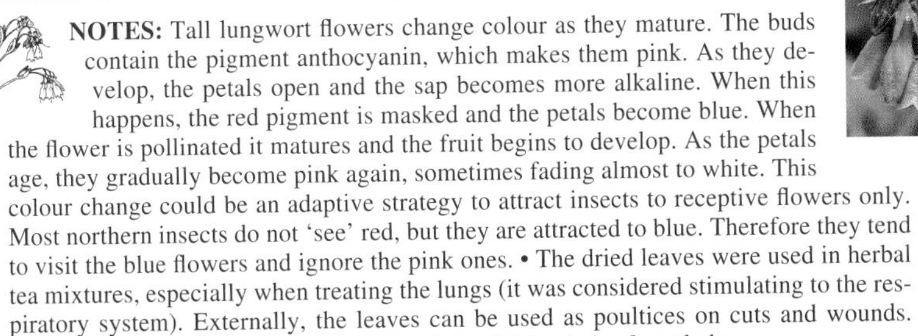

NOTES: Tall lungwort flowers change colour as they mature. The buds contain the pigment anthocyanin, which makes them pink. As they develop, the petals open and the sap becomes more alkaline. When this happens, the red pigment is masked and the petals become blue. When the flower is pollinated it matures and the fruit begins to develop. As the petals age, they gradually become pink again, sometimes fading almost to white. This colour change could be an adaptive strategy to attract insects to receptive flowers only. Most northern insects do not 'see' red, but they are attracted to blue. Therefore they tend to visit the blue flowers and ignore the pink ones. • The dried leaves were used in herbal tea mixtures, especially when treating the lungs (it was considered stimulating to the respiratory system). Externally, the leaves can be used as poultices on cuts and wounds. • Lungwort can be used as a potherb, but it is a bit too hairy for salads.

COMMON PLANTAIN • *Plantago major*
WHITEMAN'S FOOT

GENERAL: Stemless perennial from short, thick rhizomes.

LEAVES: All at stem base, prostrate, egg-shaped, 3–15 cm long, strongly 5–7-ribbed, smooth-edged to slightly toothed, **taper abruptly** to winged stalk.

FLOWERS: In dense, 3–30 cm long spikes on 10–50 cm tall stalks; many, **tiny**; petals yellowish white, translucent, 4 spreading lobes, fused at base; sepals 2 mm long; bracts small, egg-shaped, brownish.

FRUITS: Egg-shaped, membranous capsules, 2–4 mm long; **top splits off like lid**; several to many seeds, flat-sided, about 1 mm long.

WHERE FOUND: Introduced weed of cultivated and waste ground; widespread across our region and most of North America; circumpolar.

P. eriopoda

NOTES: Saline plantain (*P. eriopoda*) is a native plantain of saline or alkaline flats and slough edges from Manitoba to eastern Alaska. It is easily distinguished from common plantain by its narrower (lance-shaped), ascending leaves, which gradually taper to the stalk, and by its brownish- to reddish-woolly root crown. • The Cree chewed plantain leaves to relieve toothaches, and ate them to stop internal bleeding. Fresh juice from the leaves was taken to relieve stomach problems and expel worms. A medicinal tea was used to treat kidney, bladder, gastrointestinal, and respiratory problems. Plantain was highly recommended for healing the bites of poisonous reptiles, insects and spiders. Sometimes plantain root was carried as a charm against snake bites. • Plantain is edible, and its leaves are rich in vitamins A, C and K. The seeds can be ground into meal or flour for use in bread or pancakes. • 'Plantain' is from the Latin *planta*, 'sole of the foot.' Common plantain was sometimes called 'whiteman's foot' because the leaves resemble a footprint and they seemed to follow the white man wherever he settled.

NORTHERN BEDSTRAW • *Galium boreale*

GENERAL: Erect perennial from slender, brown rhizomes; stems slender, mostly 20–60 cm tall, clustered, simple or few-branched, **square**, usually smooth.

LEAVES: In whorls of 4, narrowly lance-shaped, 2–6 cm long, **blunt-tipped, strongly 3-veined**.

FLOWERS: Many, in showy, repeatedly 3-forked clusters **at stem tips**; 4 petals, white to slightly creamy, about 2 mm long.

FRUITS: Dry nutlets, 1.5–2 mm long, in pairs, covered with short, **whitish hairs**.

WHERE FOUND: Open woods, clearings, meadows, and roadsides; widespread across our region; circumpolar.

NOTES: Northern bedstraw is a member of the coffee family, and the seeds can be dried, roasted, and ground as a coffee substitute. Young northern bedstraw plants were cooked and used as a potherb. The leaves and roots were used to make teas. Northern bedstraw tea should not be used continually as it will start to **irritate** the mouth and tongue. It has also been suggested that people with **poor circulation** or with a tendency towards **diabetes** should avoid this plant. • Bedstraw plants have been used to make sweet-smelling hot compresses to stop bleeding and to soothe sore muscles. Bedstraw juice or tea has been used as a wash for skin problems including rashes, cuts, bites, sunburn, freckles, stretch marks, pimples, eczema and even ringworm. It was also mixed with butter to make a salve. Bedstraw tea stimulates urination and it was given to patients suffering from bladder infections or kidney stones. It was said to be mildly astringent and has been taken to relieve diarrhea and internal bleeding. Some have even used bedstraw tea as a weight loss treatment. It is said to speed up the metabolism of stored fat and to reduce weight in about 6 weeks.

SWEET-SCENTED BEDSTRAW • *Galium triflorum*

GENERAL: Perennial from slender rhizomes; stems square, 20–80 cm long, **reclining** or sometimes ascending, simple or somewhat branched.

LEAVES: In **whorls of 5–6** on main branches, **narrowly elliptic**, 1–6 cm long, abruptly bristle-tipped, **1-veined**, sweet, **vanilla-scented**; hooked bristles along edges and below on midvein.

FLOWERS: 3 per stalk in loose, open clusters at stem tips and from upper leaf axils, become wide-spreading; 2–3 mm across; 4 petals, **greenish white**.

FRUITS: Nutlets, 1.5–3 mm long, in pairs, covered with hooked bristles.

WHERE FOUND: Moist places and damp woods; widespread across our region, north and west to southern N.W.T. and southwestern Alaska; circumpolar.

NOTES: Sweet-scented bedstraw contains coumarin compounds that smell strongly of vanilla. These compounds are indirect anticoagulants by blocking vitamin K, and have a delayed effect on the blood. • Because these plants are sweet-smelling when dried, they were often used to stuff mattresses—hence the common name 'bedstraw.'

SMALL BEDSTRAW • *Galium trifidum*

GENERAL: Slender perennial from very slender, creeping rhizomes; many stems, square, 5–50 cm tall, **much branched**, slender, **weak**, tend to scramble on other vegetation, **often matted**, usually rough with sharp, stiff hairs along edges.

LEAVES: In **whorls of 4**, stalkless, **linear to narrowly elliptic**, 6–14 mm long, blunt-tipped, **1-veined**; sharp, stiff hairs along edges.

FLOWERS: 1–3 on long, slender stalks **at stem tips or from leaf axils**; **3 petals**, **greenish white**, about 1.5 mm across.

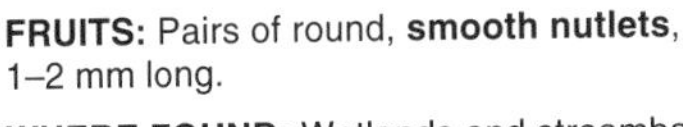

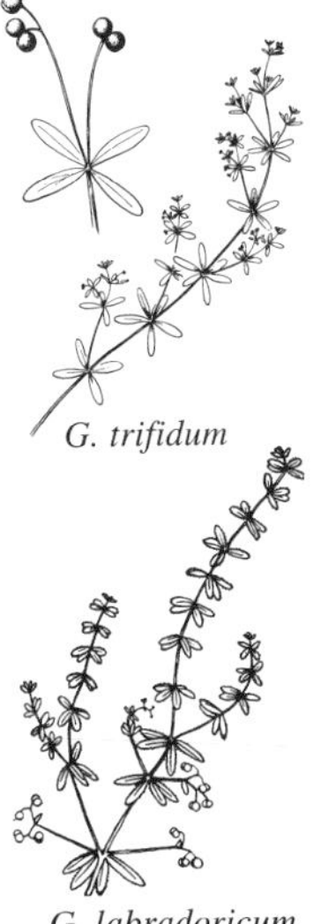

G. trifidum

G. labradoricum

FRUITS: Pairs of round, **smooth nutlets**, 1–2 mm long.

WHERE FOUND: Wetlands and streambanks; widespread across our region; circumpolar.

NOTES: Labrador bedstraw (*G. labradoricum*) also has leaves in whorls of 4, but its stems and leaf edges are smooth and its flowers have 4 petals. It grows in moist mossy forests, thickets and fens, north and west to the southern N.W.T. and northeastern British Columbia. • Native peoples used bedstraw to dye porcupine quills and other decorations red and yellow. The Cree mixed bedstraw with cranberry and strawberry juice to get a scarlet dye. Bedstraw leaves with an alum mordant give a yellow dye. However, the roots are even more popular for dying because they give reddish colours - no mordant - brownish pink; alum - light red; chrome - purplish red. For deeper colours, vinegar can be added to the roots and yarn and simmered gently for 1 hour. • The Ojibwa name for this plant means 'male genitalia,' and a close look at the fruits will explain this. The genus name, *Galium*, comes from the Greek *gala*, 'milk,' because bedstraws were widely used to curdle milk. They also gave the cheese its rich yellow colour.

MOSCHATEL • *Adoxa moschatellina*

GENERAL: Delicate perennial with **musky odour**, 5–20 cm tall, from short, scaly rhizome.

LEAVES: Long-stalked at stem base, divided into 3 long-stalked, somewhat 3-cleft leaflets; 1 pair of opposite leaves borne just above middle of flowering shoot, smaller, less divided.

FLOWERS: 3–6 in head-like cluster at flowering shoot tip; bell-shaped, 4–6-lobed, 5–8 mm wide.

FRUITS: Small, green, dry drupes, with 3–5 nutlets.

WHERE FOUND: Cool, mossy, rich deciduous woods; mainly along southern edge of boreal forest north to Yukon and Alaska.

NOTES: This little plant is rare or uncommon through much of its range, but it is too widespread to be considered threatened. The small, pale green flowers are often arranged in a cube, Moschatel emits a musk-like odour in the evening, but this scent disappears when the plant is bruised. • If not in flower, moschatel could easily be mistaken for an anemone or a buttercup. There really are no clues except for moschatel's musky odour and perhaps its white, scaly rhizome. But moschatel is a peculiar, sporadic and rare plant, so do not dig it up to look. • The common name 'moschatel' comes from the Italian *moscato*, 'musk,' in reference to the musk-like odour of this plant. The genus name comes from the Greek *adoxos*, meaning 'obscure' or 'of humble origin.'

TWINFLOWER • *Linnaea borealis*

GENERAL: Trailing, somewhat hairy evergreen perennial; stems slender, semi-woody runners, 15–75 cm long, in loose mats with erect, leafy branches 3–10 cm tall.

LEAVES: Opposite, firm, **evergreen**, broadly egg-shaped, 1–2 cm long, with **a few shallow teeth along upper half**, short-stalked.

FLOWERS: In pairs on slender, 3–10 cm tall, Y-shaped stalks; **nodding**, fragrant; petals whitish to **pink, trumpet-like**, 6–15 mm long, hairy within.

FRUITS: Dry nutlets, egg-shaped, 1.5–3 mm long; **hooked bristles** readily catch onto fur or feathers.

WHERE FOUND: Open woods, thickets, meadows; widespread and often common across our region, north slightly beyond treeline.

NOTES: The fruits of twinflower are sticky little burs that catch onto the fur or clothing of passers-by and are carried to new locations. Twinflowers also spread and form colonies by sending out long, spreading stems that send up shoots along their length. • Native groups in eastern North America used twinflower to treat women who were pregnant or had difficult menstruation, people with inflamed limbs or colds, and children suffering from cramps, fever, or crying. • The small flowers produce a very sweet perfume that is strongest near evening. • The Dena'ina called this plant *k'ela tl'lia*, 'mouse's rope,' because of its slender, trailing, rope-like stems. • This genus was named in honour of Carolus Linnaeus (1707–1778), a Swedish botanist who is considered the originator of modern taxonomy (the classification of plants and animals). Gronovius, one of his teachers, was aware of Linnaeus' fondness for these delicate little flowers and named the plant in his honour in 1737. *Borealis* means 'northern.'

COMMON HAREBELL • *Campanula rotundifolia*
BLUEBELL

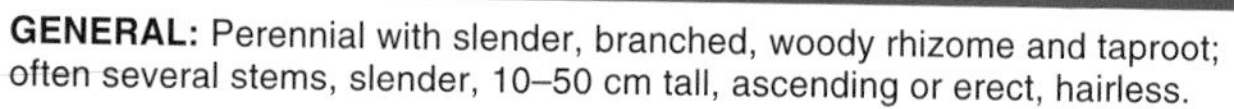

GENERAL: Perennial with slender, branched, woody rhizome and taproot; often several stems, slender, 10–50 cm tall, ascending or erect, hairless.

LEAVES: Many on stem, thin, **linear or narrowly lance-shaped**, 1–5 cm long; few at stem base, long-stalked, **egg- to heart-shaped**, 1–2.5 cm long, **coarsely toothed**, usually **wither before flowers appear**.

FLOWERS: 1–8 nodding on slender, wiry stalk in loose cluster at stem tip; corolla purplish blue, rarely white, 1.5–2.5 cm long, bell-shaped.

FRUITS: Nodding, oblong, cone-shaped capsules; open by pores near base.

WHERE FOUND: Dry, open sites in woods, meadows, hillsides and streambanks; widespread across our region; circumpolar.

NOTES: The Chipewyan and the Cree chewed harebell roots to relieve heart palpitations. The Chipewyan also used the roots to make a medicinal tea for flu, fever, and lung or heart problems. Dried, chopped roots were used in a compress to stop bleeding, relieve swelling and aid the healing of cuts or other wounds. • When the seeds mature, the fruit develops small slits or holes near the base of the nodding capsules. These openings close quickly in damp weather, protecting the seeds from excessive moisture. On dry, windy days the capsules swing to and fro on their long, elastic stalks, broadcasting their seeds over quite a distance by the 'censer mechanism.' • Harebell is the famous 'bluebells-of-Scotland,' so beloved by poets. But in some parts of Scotland it was called 'the aul' man's bell' and was avoided as a plant of ill omen, commonly left unpulled. Do not let the fragile-looking blossoms fool you—harebell is a very hardy plant. • The species name, *rotundifolia*, refers to the round leaves at the stem base, but do not be surprised if they are missing.

MARSH BELLFLOWER • *Campanula aparinoides*

GENERAL: Perennial, 20–50 cm tall, **stems 3-sided, thin, slender, weak**, recline on other vegetation, **rough with short, stiff, downwards-pointing hairs**.

LEAVES: Simple, alternate, lance- to linear-lance-shaped, 2–5 cm long, 1–3 mm wide, stalkless, smooth-edged or slightly toothed, **rough with stiff hairs on edges and midvein**.

FLOWERS: At stem tips; white or bluish-tinged, 10–15 mm long, bell-shaped.

FRUITS: Nearly spherical capsules, open by pores near base.

WHERE FOUND: Wet meadows and marshes; in boreal forest and parkland of southern Manitoba and Saskatchewan.

NOTES: Includes *C. uliginosa*. When these bell-shaped flowers bend downwards, their petals shield the pollen from getting wet. Many small flies spend their nights in the relatively warm shelter of these blossoms, picking up and depositing pollen in the process. • The genus name is derived from the Latin word *campana*, 'a church bell,' in reference to the shape of the flowers. The species name comes from the Greek *apairo*, 'take,' or 'lay hold of,' in reference to the plant's habit of climbing over other plants by means of its stiff, downwards-pointing hairs.

KALM'S LOBELIA • *Lobelia kalmii*

GENERAL: Biennial or short-lived perennial, 10–30 cm tall; stems slender, simple or with a few erect branches, hairless.

LEAVES: Spoon- or egg-shaped at stem base, 10–25 mm long, stalked, usually hairy, **often fall off early**; 4–15 stem leaves, alternate, **linear to lance-shaped**, 1–5 cm long, 1–5 mm wide.

FLOWERS: In loose, few-flowered, **often 1-sided clusters** at stem tips; blue with white centre or rarely completely white, **irregular**, 8–15 mm long.

FRUITS: Capsules, 4–8 mm long.

WHERE FOUND: Calcium-rich fens, shores, ditches and wet meadows; in boreal forest of prairie provinces north to southern N.W.T.; rare in parkland.

NOTES: Many lobelias are considered potentially **toxic** and therefore these plants should never be taken internally. The Blackfoot ate lobelia to induce vomiting and some native peoples were reported to have used it as a cure for syphilis, perhaps when less severe 'remedies' had failed. • The genus *Lobelia* is named in honour of Belgian botanist Matthias de L'Obel (1538–1616). The species name honours Pehr Kalm (1716–1779), a Finnish pupil of Linnaeus who travelled and collected in Canada.

SLENDER ARROW-GRASS • *Triglochin palustris*

GENERAL: Perennial, 15–50 cm tall, with **short, bulb-like rhizomes** and **slender runners**.

LEAVES: All at stem base, 10–30 cm long, **narrowly linear, somewhat fleshy**, sheathing at base.

FLOWERS: In few-flowered, **long-stalked spikes much longer than leaves**; small, **greenish**, 6-parted.

FRUITS: Capsules, 5–7 mm long, **narrowly club-shaped**, split open from base upwards.

WHERE FOUND: Brackish marshes, ditches and mud flats; widespread across our region.

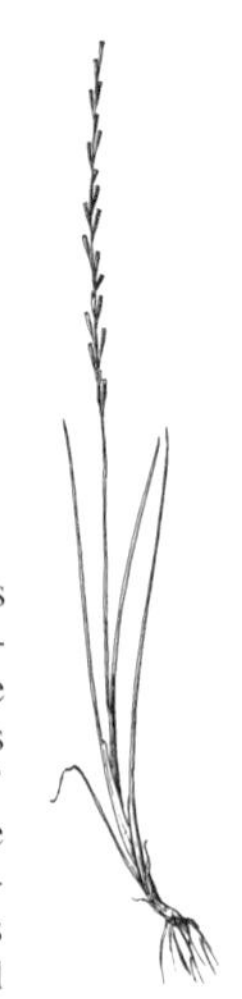

NOTES: Although slender arrow-grass is smaller and less common than seaside arrow-grass (below), it is equally **poisonous** to cattle and sheep. • The genus name *Triglochin* comes from the Latin *tri,* 'three,' and *glochis*, 'a point,' in reference to the 3-pointed fruits of some species, which are also the source of the common name 'arrow-grass.' *Palustris* means 'of marshes,' in reference to the preferred habitat of these plants.

SEASIDE ARROW-GRASS • *Triglochin maritima*

GENERAL: Hairless perennial from **stout rhizomes**, sometimes in large clumps; flowering stems leafless, 20–120 cm tall, bases often covered with whitish, fibrous, old leaf-sheaths.

LEAVES: All at stem base, rather fleshy, upright to spreading, **linear**, up to 30 cm long, 3 mm wide, **half-round to somewhat flattened**, hairless; sheathing bases.

FLOWERS: Many in 10–40 cm long, **narrow, spike-like clusters**; short-stalked, small, **inconspicuous, greenish** with feathery reddish stigmas, 6-parted.

FRUITS: Dry, egg-shaped capsules, 5–6 mm long, 2–3 mm across; split lengthwise into 3–6 1-seeded segments.

WHERE FOUND: Brackish marshes, fens and wet meadows; widespread across our region, north past treeline to Arctic coast.

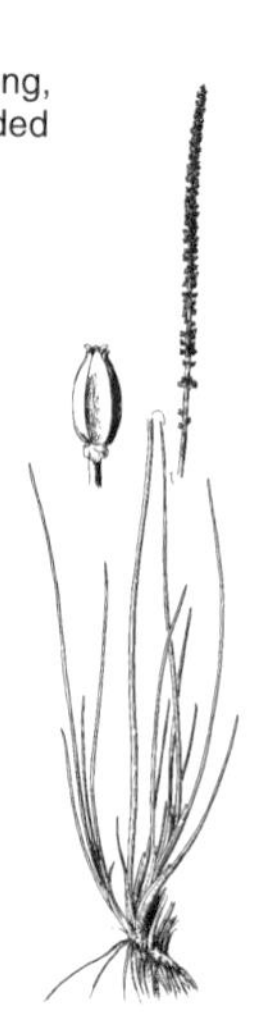

NOTES: Arrow-grasses produce hydrocyanic acid, which can cause **cyanide poisoning** in cattle and sheep. Mature plants should never be eaten. Plants growing in water are 5–10 times less toxic than plants from dry habitats. • In late spring coastal native groups gathered the fleshy, whitish leaf bases (not the flower-stalk bases which caused a headache if eaten) for a mild, sweet vegetable, similar to cucumber. Sometimes they were canned for winter use. The seeds were gathered later in the year, parched and eaten or ground as a coffee substitute.

Eating For A Living

Most plants are green with chlorophyll and use this pigment to produce their own food through photosynthesis. However, some plants do not look particularly green. These species often have little or no chlorophyll and must find other ways of getting food. Also, some green plants that grow in very nutrient-poor environments are able to survive only by supplementing their diet with nutrients from other plants and animals. These unusual plants have been grouped in this special section.

Parasitic plants take their food from other living plants. Although this can greatly weaken the plant under attack, our species do not kill their hosts. True parasites, such as dwarf mistletoe, have no chlorophyll and rely entirely on their host for nutrients. There are also many semi-parasites in the boreal forests. These plants appear normal, with green leaves capable of photosynthesis, but underground they join with the roots of neighbouring plants and steal nutrients. These semi-parasites are described with the other members of their families elsewhere in the book. Several examples can be found in the figwort family, including paintbrushes (*Castilleja*) and yellow rattle (*Rhinanthus borealis*), Toad-flaxes (*Comandra umbellata* and *Geocaulon lividum*), in the sandwood family, are also partial parasites .

Indian-pipe (*Monotropa uniflora*), is connected to the roots of coniferous trees by a specialized combination of fungal filaments and plant roots called a **mycorrhiza**, which means 'fungus-root.' Nutritionally it is a parasite, but a very special one, because it is not connected directly to its host.

Saprophytic plants take their nutrients from dead and decaying plants. They usually grow on soils that are rich in decomposing organic material. The saprophyte treated here, **pinesap** (*Hypopitys monotropa*), is an unusual member of the wintergreen family (Pyrolaceae). A close examination of such pale 'apparitions' will reveal the characteristic bell-shaped flower of a wintergreen.

Some orchids are saprophytic—represented by the coralroots (*Corallorhiza*) in our region—but these species have been described with the other members of the orchid family in the 'Wildflowers' section.

Carnivorous plants in the boreal forest trap and digest their prey using specialized leaf structures. Usually insects are taken, but aquatic species, such as the **bladderworts** (*Utricularia* species) can trap a wide variety of invertebrates, ranging from large protozoans to nematodes and small crustaceans. Many of these plants live in cold, acidic environments (usually bogs) where nitrogen, an essential nutrient, is not readily available. By trapping and digesting insects, these plants find alternative sources of nitrogen and are able to grow and reproduce where other species cannot.

Arceuthobium

Monotropa

Drosera

Sarracenia

Pinguicula

Utricularia

DWARF MISTLETOE • *Arceuthobium americanum*

GENERAL: Small, hairless, greenish yellow, **leafless, fleshy**; **parasitic on trunks and branches of pines**; **stems emerge from host**, slender, 1–2 mm thick, 2–10 cm long, **segmented**, usually tufted, often much-branched, with 2-several branches in whorls.

LEAVES: Opposite, reduced to tiny, fused scales.

FLOWERS: 1-several in axils of scales; greenish yellow, inconspicuous, no petals; sexes on separate plants; male flowers have 3 sepals, 3 anthers; female flowers have 2 sepals.

FRUITS: Greenish or bluish berries, sticky, egg-shaped, 2–3 mm long, on short stalk, explosively eject 1 sticky seed.

WHERE FOUND: Parasitic almost exclusively on pines; widespread across southern boreal forest, generally south of 55°N, but north to Lake Athabasca.

NOTES: Dwarf mistletoe causes disorganized growths on pines, called 'witch's broom.' Mistletoe infections can cause significant reductions in the growth of both jack and lodgepole pines. • The common name appears to be derived from the Old English word *mistletan*. *Tan* means 'twig' and some sources say *mistl* means 'different'—hence 'different-twig.' Others say *mistl* comes from Old German *mist*, meaning 'dung,' since birds eat the berries and then deposite the seeds on trees in their droppings. • The name *Arceuthobium* is from the Greek *arkeuthos*, 'juniper,' and *bios*, 'life,' because some mistletoes parasitize junipers.

INDIAN-PIPE • *Monotropa uniflora*

GENERAL: Perennial, **fleshy, waxy-white or rarely pinkish (blackens with age)**, from fleshy ball of small roots; flowering stems single or in **clusters**, unbranched, 5–30 cm tall.

LEAVES: Alternate, linear to oval, **scale-like**, up to 1 cm long.

FLOWERS: Single; **nodding at first, white, narrowly bell-shaped**; usually 5 petals, erect, 15–20 mm long, inner surface hairy.

M. uniflora

FRUITS: Erect, oval to circular capsules, 5–7 mm wide, brown, split open when mature; many seeds.

WHERE FOUND: Rich woodland; rare and scattered across southern boreal forest, north to Lake Athabasca.

Hypopitys monotropa

NOTES: Pinesap (*Hypopitys monotropa*, also called *M. hypopitys*) also lacks chlorophyll, but it is yellowish to bronze-pink, and has several flowers on each stem. It grows in mossy, coniferous forests in Alberta and Saskatchewan in our region. • The Woods Cree chewed Indian-pipe flowers as a remedy for toothaches. Other native groups used the juice of the plant to treat inflamed eyes, bunions and warts, or used it to make a medicinal tea to relieve aches and pains from colds. The roots were also used to make a sedative tea for treating convulsions, fits and epilepsy. • Water extracts have been found to kill bacteria. However, the safety of Indian-pipe has yet to be determined. It contains several glycosides and therefore could prove **toxic**. • The roots of Indian-pipe are connected by fungi to the roots of nearby trees. In this way, Indian-pipe, which lacks chlorophyll and cannot produce its own food, obtains nutrition from the efforts of another plant. • The common name 'Indian-pipe' refers to the pipe-like flowers. It is also called 'ghost-flower' and 'corpse plant'—names inspired by the unusual colour and texture of the plant. Indian-pipe has been called 'ice plant' because it resembles frozen jelly, and 'melts' when handled. • *Monotropa* comes from the Greek *monos*, 'one,' and *tropos*, 'direction,' in reference to the flowers, which turn to one side.

ROUND-LEAVED SUNDEW • *Drosera rotundifolia*

GENERAL: Small, **insect-eating** perennial, 5–18 cm tall.

LEAVES: Covered in **sticky glands**, spreading in **rosette at stem base**, usually 3–6 cm long and 1 cm wide, **round to broadly egg-shaped** or wedge-shaped, fringed with **long, reddish, gland-tipped hairs (tentacles)** that exude sticky drops; stalks hairless.

FLOWERS: 2–8 in long 1-sided cluster that gradually uncoils; petals **white**, 4–6 mm long, fully open only in strong sunlight.

FRUITS: Partitioned capsules, 5–7 mm long; many seeds, with loose seed coat and lengthwise markings.

WHERE FOUND: Bogs, swamps and fens, frequently in sphagnum moss; widespread across boreal forest, north to or slightly beyond treeline; circumpolar.

NOTES: Round-leaved sundew has been used to treat many ailments of the respiratory system, including whooping-cough, tuberculosis, bronchitis and asthma. The juice of the leaves was used to cure warts, bunions and corns and was mixed with milk to treat freckles, pimples and sunburn. • In British Columbia, the Kwakiutl are said to have used sundews as a love charm—'medicine to make women love-crazy.' In a similar vein, sundews caused farmers no end of trouble in the early 1800s, because 'sheep and other cattell, if they do but only taste it, are provoked to lust.' • Sundews are commonly pollinated by mosquitoes, midges, and gnats—the same insects they use for food! • The sap is acrid and has a reputation for curdling milk. Fresh leaves were used in Europe in the preparation of cheeses and junkets.

OBLONG-LEAVED SUNDEW • *Drosera anglica*
LONG-LEAVED SUNDEW

GENERAL: Small, stemless, **insect-eating** perennial, 5–17 cm tall; flower stalk erect, 4–10 cm high.

LEAVES: In rosette at stem base, often reddish, **ascending to erect, spoon-shaped to oblong or linear**, 1–3 cm long, 3–4 mm wide, covered with sticky hairs and **fringed with 'tentacles.'**

FLOWERS: Usually 2–7 in long cluster on long, leafless stalk that uncoils as it matures; 4–8 petals, **white** or pale pink, 6 mm long.

FRUITS: Capsules, 5–7 mm long; many seeds, black, spindle-shaped.

WHERE FOUND: Fens and pond edges; widely scattered across boreal forest, north to treeline; circumpolar.

NOTES: The leaves of **slender-leaved sundew** (*D. linearis*) (inset photo) are linear, even slimmer than those of oblong-leaved sundew. Slender-leaved sundew grows in similar habitats (often with oblong-leaved sundew), and it is scattered across our region north to the southern District of Mackenzie. • Sundew leaves are covered with special hairs, each tipped with a shiny droplet. These hairs do 3 things: they secrete sticky mucilage to attract, catch, and overcome prey and to aid in digestion; they secrete enzymes to dissolve all but the exoskeletons of their prey; and they absorb most of the nutrient rich fluid produced by this digestion. When an insect touches the hairs, the leaf bends inwards, bringing the insect into contact with the denser inner hairs and exuding more enzyme-containing liquid. It can take 24–48 hours to completely enfold a victim. When the prey has been digested, all that is left is the dry exoskeleton, so the leaf uncurls and the remains blow away. Darwin discovered that sundew leaves responded to being touched by a tiny piece of meat or a human hair, but hardly responded to something inorganic, such as a pebble.

PITCHER-PLANT • *Sarracenia purpurea*

GENERAL: Insect-eating perennial; flowering stems leafless.

LEAVES: Curved, ascending, hollow leaves (**pitchers**) in rosette at stem base, 10–30 cm long, green with purple veins, broadly winged on upper side, hood-like top has downwards-pointing, bristly hairs.

FLOWERS: Single at end of 20–40 cm long stalk; **deep purple** or sometimes yellow, nodding, nearly round, 5–7 cm across; 5 spreading sepals; 5 petals, obovate, curve inwards over style; many stamens; **style** slender at base, greatly enlarged and **umbrella-like above**.

FRUITS: Many-seeded capsules.

WHERE FOUND: Bogs and fens; across boreal forest of prairie provinces north to northeastern B.C. and southern N.W.T.; rare in B.C. and N.W.T.

NOTES: The hollow pitcher-plant leaves are designed to catch and digest insects. The hood-like, purplish lip and the upper part of the inner surface of the 'pitcher' are covered with tiny, stiff, downwards-pointing hairs. Insects are drawn to the colour and pattern of the pitcher lip. Once they land on the surface, they can easily move inwards. Gradually the surface curves down and by the time the visitor tries to retreat it is too late. The stiff, smooth hairs make it impossible to climb back out and the insect falls into the pool of liquid at the bottom of the cup. Unable to climb the smooth lower walls it drowns and decomposes, and its nutrients are absorbed by the plant. Like many carnivorous plants, pitcher-plant grows in cool, acidic habitats where the supply of nitrogen and other plant nutrients is low. • The Cree regarded pitcher-plant as a very special plant, and they credited it with saving many lives. They added it to other herbs in medicinal teas for the very sick.

COMMON BUTTERWORT • *Pinguicula vulgaris*

GENERAL: Insect-eating perennial from fibrous roots; flowering stems 4–12 cm tall.

LEAVES: In rosette at stem base, greenish yellow, succulent, egg-shaped to elliptic, 1.2–5 cm long, 1–2 cm wide, **greasy-slimy** above; **edges rolled inwards**.

FLOWERS: Single on long, leafless stalk; resemble violets; 5 petals, **lavender purple** (rarely white), usually 1.5–2 cm long; form funnel-like tube with white hairs in throat and slender, 6–8 mm long, nectar-producing **spur** behind.

FRUITS: Erect, round capsules, 4–6 mm long, split into 2 pieces.

WHERE FOUND: Moist sites, often on bare, calcium-rich soil and/or near water; scattered but widespread across boreal forest, north and west to Arctic coast; circumpolar.

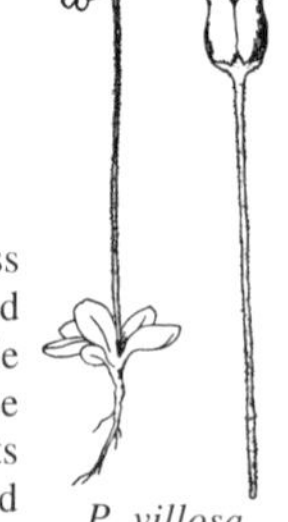

P. villosa

NOTES: Hairy butterwort (*P. villosa*)(inset photo) is a less common, smaller species with 2–5 cm tall flowering stems densely covered in gland-tipped hairs. Its flowers are smaller (up to 8 mm long) and pale blue-violet. It typically grows in bogs on sphagnum hummocks and at the edges of pools and rivulets, across the northern boreal forest. • Butterworts supplement their diet with passing insects (mostly ants, mosquitos and midges) and other small invertebrates. The 'greasy' upper surface of the leaves is covered with small glands. When an insect becomes stuck, its struggles stimulate the leaf to secrete large amounts of relatively acidic mucilage and enzymes. The acid and enzymes reduce the soft parts of the insect's body to a nutritive liquid that is absorbed by the stalkless glands. The nitrogen gained by digesting small insects could be very important to these plants, as they often grow on leached, acidic soils, which are low in nitrogen. Significant amounts of nitrogen could be obtained from their prey. Close to 500 small flies have been seen stuck to a single large butterwort plant. The same insects that these plants digest also visit their flowers, carrying the pollen necessary for cross-fertilization and seed production.

FLAT-LEAVED BLADDERWORT • *Utricularia intermedia*

GENERAL: Aquatic perennial; stems very slender, often creeping on mud in shallow water.

LEAVES: Alternate, many, 5–20 mm long, finely divided into thread-like, flattened, toothed segments; few **bladders** about 5 mm long **on separate branches**.

FLOWERS: 2–4 at end of leafless, upright stalk above water; yellow, 1–1.5 cm long, 2-lipped; lower lip large, 8–12 mm long; upper lip 4–6 mm long; spur nearly as long as lower lip.

FRUITS: Many-seeded capsules, split irregularly.

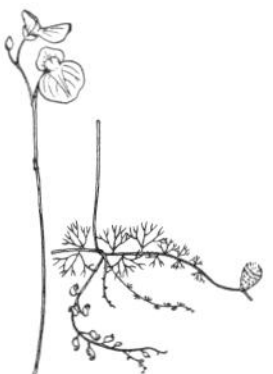

U. intermedia

U. cornuta

WHERE FOUND: Ditches, ponds, lakes and fens; widespread across our region; circumpolar.

NOTES: Horned bladderwort (*U. cornuta*) grows in the eastern part of our region, north and west to northeastern Alberta. It tends to be semi-aquatic, often creeping on peaty or muddy shores. It has thread-like, undivided leaves with tiny, difficult-to-see bladders, but its 1-3 yellow flowers are large (1.5-2.5 cm long). • The tiny, elastic-walled bladders of the bladderworts are the most sophisticated trapping devices among the carnivorous plants. These specialized leaves have been modified to trap small invertebrates in peat or mud (horned bladderwort) or in water (aquatic bladderworts). Usually they catch microscopic animals. Larger prey (water fleas, mosquito larvae) are caught by their heads and ingested in stages. When passing prey brush against the trigger hairs on the mouth of the bladder, a tiny flap of tissue, springs open and water rushes in, carrying the prey with it. Then the door swings closed, the plant secretes digestive enzymes and acid, and the animal is digested. Gradually, the water and nutrients are removed through absorbent cells in the bladder wall and the trap is reset. The whole cycle usually takes $^1/_2$–2 hours to complete. The nutrients from these small animals provide an important dietary supplement for bladderworts.

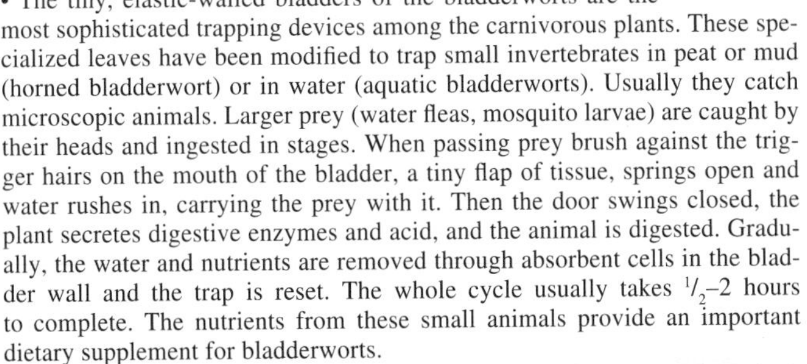

COMMON BLADDERWORT • *Utricularia vulgaris*

GENERAL: Aquatic perennial, free-floating, **non-rooting**; stems leafy, to 1 m long or more, **about 1 mm thick**.

LEAVES: Alternate, many, 1–5 cm long, submerged or floating; **divided into many, tubular, thread-like segments** with many, 3–5 mm long, buoyant, valve-lidded **bladders** that trap small animals.

FLOWERS: 5–20 in loose cluster atop stout stalk, 8–25 cm long, well above water; petals **deep yellow** with brown lines, **2-lipped (much like snapdragon)**, 12–20 mm long, lower lip has short, sac-like spur.

FRUITS: Capsules, on down-curved stalks.

WHERE FOUND: Shallow to deep water in ponds, lakes, and slow-moving streams; widespread across boreal forest, parkland and prairie; extends north to western Arctic coast and west through Alaska; circumpolar.

U. minor

NOTES: Small bladderwort (*U. minor*) is a more delicate plant with fewer bladders, on small, finely divided leaves. Its flowers are pale yellow and 4–8 mm long. Small bladderwort floats in shallow ponds or creeps on the mud of peatlands and shores, scattered across our region to Great Bear Lake and Alaska. • Like most aquatic plants, the bladderworts reproduce mainly by vegetative means. In the autumn, buds at the tips of their floating stems fall off and sink down to the mud. The following spring, they sprout and produce new plants that grow underwater until they are ready to bloom. Then the bladders fill with gases, like tiny balloons, and the plant floats to the surface. The bladders can also serve as water reservoirs in times of drought, holding enough water to keep the plant alive for some time.

Aquatics

The plants treated in this section all grow in habitats that are submerged for at least part of the growing season. Some grow entirely underwater, some float partially or entirely on the surface of the water, some have their feet in the water and their heads in the air. Although some of our horsetails (e.g., *Equisetum fluviatile*) and mosses (e.g., *Calliergon* species) and many algae are aquatic plants, we have restricted this section to flowering plants.

Despite the fact that the aquatic flowering plants belong to many often unrelated families, they share several adaptations to life in an aquatic environment. Water supports them, so aquatic plants do not need to develop the strong, reinforced stems seen in most terrestrial plants. Our aquatics are all herbs; most of the submerged and floating species collapse when removed from the water. Rigid strength is not very important but flexibility is, because aquatic plants must be able to move with water currents or waves. Leaves are often long and narrow, with a central cluster of non-thickened conducting tubes. These plants, especially those with roots submerged in muddy bottoms, sometimes have to deal with a shortage of oxygen and a surplus of carbon dioxide, so many of them have lots of spongy tissue through which gases can diffuse, and many are able to actively 'pump' oxygen to their roots.

Note that several aquatic species, including bladderworts, sedges and grasses, are described elsewhere in this guide.

Aquatic plants (including algae) form the base of the watery food pyramid and provide the food upon which all other aquatic organisms feed, directly or indirectly. Aquatic plants also provide habitat for a wide variety of organisms living in or near streams, lakes, ponds and ditches, and humans have found many uses for aquatic plants.

Key to the Aquatic Plants

1a. At least some part of plant extends out of water 2

2a. Leaves grass-like; individual flowers inconspicuous 3

3a. Flowers bisexual (with both male and female parts) 4

4a. Plants usually less than 40 cm tall; 3–12 greenish white flowers, in short, end cluster (raceme); fruits are round pods (follicles); leaves at stem base and on stem .... ***Scheuchzeria palustris***

4b. Plants usually more than 40 cm tall (often much taller); all leaves at stem base 5

5a. Flowers and fruits in spikes surrounded by green protective leaf (spathe); fruit berry-like .... ***Acorus calamus***

5b. Flowers and fruits in open, branched panicle to 15 cm long, without protective leaf; fruits are achenes 6

6a. Stems spongy, easily crushed between fingers .... ***Scirpus lacustris* ssp. *validus***

6b. Stems firm, not easily crushed between fingers .... ***Scirpus lacustris* ssp. *glaucus*** (see *S. lacustris* ssp. *validus*)

3b. Sexes in separate flowers on same plant 7

7a. Flowers in dense, unbranched, sausage-shaped spike, 20–30 cm long; fruits are hairy achenes .... ***Typha latifolia***

7b. Flowers in interrupted spike of 3–20 dense, round heads, male heads above female; fruits (achenes) in bur-like heads, not hairy .... ***Sparganium eurycarpum***

2b. Leaves various shapes, but not grass-like; individual flowers usually conspicuous 8

8a. Leaves alternate, compound (divided into leaflets) 9

9a. Plants creeping to ascending, to 1 m long; leaves pinnate, with 5–7 leaflets; flowers reddish purple, bowl-shaped .... ***Potentilla palustris***

9b. Plants upright, to 30 cm tall; leaves have 3 leaflets (trifoliate); flowers white, funnel-shaped .... ***Menyanthes trifoliata***

8b. Leaves simple (not divided into leaflets), all at stem base 10

10a. Flowers and fruits in spikes surrounded by white, petal-like, protective leaf (spathe); leaves egg- to heart-shaped .... ***Calla palustris***

10b. Flowers and fruits not in spikes surrounded by protective leaf 11

11a. Leaves egg-shaped to oblong-lance-shaped .... ***Alisma plantago-aquatica***

11b. Leaves arrowhead-shaped; flowers and fruits in loose clusters; fruits are achenes 12

12a. Achene beak upright, less than 0.5 mm long .... ***Sagittaria cuneata***

12b. Achene beak horizontal, 0.5–1.5 mm long .... ***Sagittaria latifolia*** (see *S. cuneata*)

1b. Plants grow submerged or floating (sometimes emerge from waterin Hippuris and *Polygonum*) 13

13a. Plants floating on or in water, not attached to bottom 14

14a. Plants have several roots .... ***Spirodela polyrhiza*** (see *Lemna minor*)

14b. Plants have single root 15

15a. Plants 2–5 mm across; lobes round,
with short stalks or stalkless; float on water surface .... ***Lemna minor***

15b. Plants 5–10 mm across; 3 lance-shaped lobes,
with long stalks; float beneath water surface .... ***Lemna trisulca*** (see *L. minor*)

13b. Plants submerged or floating, attached to bottom 16

16a. Leaves whorled 17

17a. Stems upright, to 30 cm tall; leaves in whorls of 6–12, not divided .... ***Hippuris vulgaris***

17b. Stems limp, leaves divided 18

18a. Leaves 2- or 3-forked, often spiny-tipped; fruit is achene .... ***Ceratophyllum demersum***

18b. Leaves finely divided, appear feather-like; fruit is 4-sectioned nutlet 19

19a. Bracts and flowers mostly
opposite or whorled .... ***Myriophyllum spicatum* var. *exalbescens***

19b. Bracts and flowers mainly alternate .... ***Myriophyllum alterniflorum***
(see *M. spicatum* var. *exalbescens*)

16b. Leaves alternate or opposite 20

20a. Leaves opposite 21

21a. Both submerged and floating leaves; submerged leaves linear; floating
leaves lance- to spoon-shaped; leaf-bases joined by small, wing-like ridge ***Callitriche verna***

21b. All leaves submerged, linear; leaf-bases
not joined by ridge or wing .... ***Callitriche hermaphroditica***
(see *C. verna*)

20b. Leaves alternate 22

22a. Plants have small (1–3 mm wide) bladders that trap
tiny animals; flowers yellow, 1–2 cm long, like snapdragon flowers .... ***Utricularia* spp.**
(see 'Eating for a Living' section)

22b. Plants do not have bladders 23

23a. Both submerged and floating leaves, 2 leaf types distinctly different 24

24a. Submerged leaves grass-like,
not divided; floating leaves elliptic to oblong,
not lobed; flowers inconspicuous, in stalked spikes .... ***Potamogeton gramineus***

24b. Submerged leaves finely divided; floating
leaves 3–5-lobed; flowers conspicuous 25

25a. Flowers yellow, at stem tips .... ***Ranunculus gmelinii***

25b. Flowers white, in leaf axils 26

26a. Leaves stiff, stalkless; 30–80 achenes
per fruiting head .... ***Ranunculus circinatus***
(see *R. aquatilis* var. *capillaceus*)

26b. Leaves limp, stalked;
15–25 achenes per fruiting head ..... ***Ranunculus aquatilis* var. *capillaceus***

23b. Leaves either submerged or floating, not of 2 distinctly different types 27

27a. Leaves submerged, occasionally emergent, but not floating 28

28a. Leaves egg-lance-shaped,
less than 15 times as long as wide ***Potamogeton perfoliatus* ssp. *richardsonii***

28b. Leaves linear, more than 20 times as long as wide 29

29a. Stipules attached to leaf-base 30

30a. Flower spike has 5–12 whorls of flowers **_Potamogeton vaginatus_** (see *P. friesii*)

30b. Flower spike has 2–6 whorls of flowers 31

31a. Leaves have pointed tips; fruits 2.5–4 mm long, with small beak .... **_Potamogeton pectinatus_** (see *P. friesii*)

31b. Leaves have blunt tips; fruits about 2 mm long, with almost no beak .... **_Potamogeton filiformis_**

29b. Stipules free from leaf-base 32

32a. Stipules fibrous, become whitish with age; flower spike interrupted, cylindrical, with 3–4 whorls of flowers ..... **_Potamogeton friesii_**

32b. Stipules delicate; flower spike more compact, with 1–3 whorls of flowers .... **_Potamogeton pusillus_** (see *P. filiformis*)

27b. All leaves floating 33

33a. Leaves grass-like; flowers tiny, single-sexed, greenish, in heads with male heads above female; fruits (achenes) in bur-like heads 34

34a. Achene beaks less than 1.5 mm long; 1 male flower head 35

35a. Leaves 1–5 mm wide, opaque, often yellow .... **_Sparganium hyperboreum_** (see *S. angustifolium*)

35b. Leaves 2–8 mm wide, translucent, usually dark green .... **_Sparganium minimum_** (see *S. angustifolium*)

34b. Achene beaks more than 1.5 mm long; at least 2 male flower heads 36

36a. Achene beaks strongly curved .... **_Sparganium fluctuans_** (see *S. angustifolium*)

36b. Achene beaks straight or slightly curved 37

37a. Usually 4–7 distinct female heads; beaks 2–5 mm long .... **_Sparganium chlorocarpum_** (see *S. angustifolium*)

37b. Usually 2–3 female heads, crowded and often not distinct; beaks 1.5–4 mm long 38

38a. Leaves usually 2–5 mm wide, rounded on back; fruiting heads 7–15 mm across . **_Sparganium angustifolium_**

38b. Leaves 5–10 mm wide, flat on back; fruiting heads 20–25 mm across .... **_Sparganium multipedunculatum_** (see *S. angustifolium*)

33b. Leaves not grass-like; flowers conspicuous, bisexual, white, pink or yellow . 39

39a. Flowers white, 2.5–4 cm across; leaves elliptic-oval, 7–10 cm long .... **_Nymphaea tetragona_** (see *Nuphar variegatum*)

39b. Flowers yellow or pink 40

40a. Flowers yellow, 4–7 cm across; leaves egg-heart-shaped, 10–25 cm long (the pond-lilies) 41

41a. Usually 6 sepals, 2.5–3.5 cm long; stamens yellow .... **_Nuphar variegatum_**

41b. Usually 9 sepals, 3.5–6 cm long; stamens reddish or purplish .... **_Nuphar polysepalum_** (see *N. variegatum*)

40b. Flowers pink, in spikes to 3 cm wide; leaves elliptic to lance-shaped, 5–15 cm long; plants occasionally terrestrial (the smartweeds) 42

42a. Flower spikes to 12 cm long, tapering, stalks hairy .... **_Polygonum coccineum_** (see *P. amphibium*)

42b. Flower spikes usually less than 4 cm long, egg-shaped to oblong, stalks hairless .... **_Polygonum amphibium_**

SCHEUCHZERIA • *Scheuchzeria palustris*

GENERAL: Yellowish green, **rush-like** perennial from branching yellowish grey rhizomes; **stems leafy**, zig-zag, 10–40 cm tall, with dead leaves at base.

LEAVES: Erect, alternate, **grass-like, tightly rolled and somewhat tubular**, with pore at tip, 10–40 cm long, much smaller upwards, with broad sheaths (up to 10 cm long) and prominent ligules (up to 12 mm long) at base.

FLOWERS: In 3–12-flowered clusters at stem tips; small, **greenish white**; 6 oblong, membranous sepals/petals, about 3 mm long; 6 stamens; usually 3 separate ovaries; stalks have sheathing bracts.

FRUITS: 3 spreading, **egg-shaped capsules**, 4–8 mm long, greenish brown; 1–2 seeds, black to dark brown, 4–5 mm long.

WHERE FOUND: Cold poor fens and lake edges; uncommon but widely scattered across boreal forest, north to 64°–65°N.

NOTES: Scheuchzeria is a close relative of the arrow-grasses (p. 206), but it is usually found in fresh water environments, while arrow-grasses prefer calcium-rich, brackish or salty habitats. • This genus was named for Swiss botanist John Jakob Scheuchzer (1672–1733). The species name, *palustris*, means 'of marshes or swamps.'

COMMON GREAT BULRUSH • *Scirpus lacustris* ssp. *validus*
TULE

GENERAL: Perennial with stout, scaly, reddish brown, extensively creeping rhizomes, 0.5–3 m tall; **stems round**, up to 2 cm thick at base, **usually soft and easily compressed**.

LEAVES: Few, mainly near stem base, commonly have well-developed sheath and poorly-developed blade, or bladeless.

FLOWERS: In spikelets in branched, spreading to **drooping** clusters, 5–10 cm long; branches 1–7 cm long, longer ones repeatedly branching; spikelets egg-shaped, 5–10 mm long, **reddish brown**; 1 greenish bract, erect, 1–7 cm long; scales more or less hairy.

FRUITS: Brownish or olive, lens-shaped achenes, 1.5–2.5 mm long, short-pointed; 4–6 slightly longer, barbed bristles from base.

WHERE FOUND: Sloughs, ponds, lakes and marshes; tolerant of alkali conditions; very common across our region, north and west to interior Alaska.

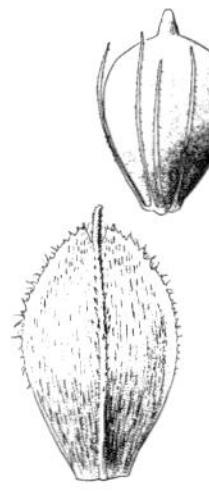

S. lacustris spp. *validus*

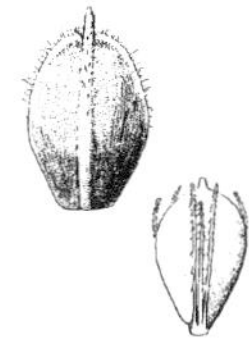

S. lacustris spp. *glaucus*

NOTES: Also called *S. validus*. • **Great bulrush** (*S. lacustris* ssp. *glaucus*, also called *S. acutus*), also called hard-stemmed bulrush, is similar, but it has harder stems, a compact, erect flower cluster, longer, paler scales with reddish brown lines and larger achenes. It forms dense colonies in water up to 1 m deep along sheltered lakeshores and sluggish rivers across our region south of 60°N. • In spring, some Chipewyan elders still enjoy eating the inside of the first 10 cm of bulrush rhizome just below the stem base. It is said to be white, tender and rich like fat. If the rhizome is red instead of white, it is too old and tough to eat. Bears also enjoy the succulent rhizomes in spring. Some native groups dried and ground the rhizomes and used the flour to make bread. The Woods Cree used a boat to collect the tender stem bases and leaf bases of great bulrush, which they ate fresh. They also ate the roots. • The Woods Cree used the stem pith of great bulrush as a compress to stop bleeding. • The long stems of these tall plants can be woven into mats or twisted to make the seats of rush-bottomed chairs. They have also been used to caulk the seams of wooden casks and to thatch the roofs of houses.

NARROW-LEAVED BUR-REED • *Sparganium angustifolium*

GENERAL: Aquatic perennial from rhizomes; stems usually submerged or floating, 0.2–1 m long.

LEAVES: Alternate, **floating, long and narrow**, 0.2–1 m long, 2–15 mm wide, usually have enlarged sheathing bases and translucent edges.

FLOWERS: Tiny, greenish, many in **spherical heads**, sexes in separate heads on same plant; **2–5 male heads**, stalkless, at stem tips; 2–5 female heads, **7–25 mm wide at maturity**; sepals/petals reduced to 3–6 linear scales; 3–5 stamens.

FRUITS: Hardened, nutlet-like achenes, spindle-shaped, 3–6 mm long (including narrow, 1.5–4 mm long beak), stalked.

WHERE FOUND: Ponds, lakeshores and sluggish streams; widespread across boreal forest.

NOTES: Many-stalked bur-reed (*S. multipedunculatum*) resembles narrow-leaved bur-reed but is larger, with ribbon-like leaves that are 5–10 mm wide. It is scattered across our region. **Slender bur-reed** (*S. minimum*, also called *S. natans*) has 1 mm long achene beaks, a single, separate male head and translucent, usually dark green leaves. It is scattered across our region in shallow pools and ponds and on lakeshores. **Water bur-reed** (*S. fluctuans*) has floating stems, flat leaves that are all floating, several male and female heads, and flattened, curved, achene beaks. It is found east of the Rockies along the southern edge of our region. **Stemless bur-reed** (*S. chlorocarpum*) has several male heads, but its female heads are widely separated. Its achenes have beaks 2–5 mm long (about as long as the body) and its stems and leaves stick up out of the water and overtop the flower cluster. It grows scattered across our region in streams and ponds and on lakeshores. • These plants are called 'bur-reeds' because of their bur-like fruits and narrow reed-like leaves.

GIANT BUR-REED • *Sparganium eurycarpum*

GENERAL: Perennial, 0.5–1.5 m tall; stems stout, erect.

LEAVES: Narrow, stiff, flat or slightly ridged, 30–80 cm long, 6–12 mm wide.

FLOWERS: In branched, 15–40 cm long cluster of **3–20 single-sexed heads**; lower 1–4 heads female, 2–3 cm wide at maturity; **female flowers mature before male flowers**.

FRUITS: Stalkless achenes, 6–9 mm long, 4–8 mm wide, narrow abruptly to stout beak, about 3 mm long.

WHERE FOUND: Wet meadows, marshes and shallow water at edges of ponds, lakes and slow-moving streams; widespread and relatively common across boreal forest and northern parkland of prairie provinces north to 60°N; separate population near Fort Norman, N.W.T.

NOTES: Giant bur-reed is the largest and most conspicuous of our bur-reed species. • The bulbous stem bases are edible, and the lower, tender sections, immediately above the root, can be eaten fresh from the water. The creeping rhizomes of some plants produce tubers late in the autumn, and these have also been used as food. However, they are small and widely spaced, so it is rarely possible to collect enough to provide a meal. • Waterfowl and marsh birds eat the seed heads, and muskrats and deer eat the plants (usually preferring the base sections). Perhaps most importantly bur-reeds provide cover for marsh birds and waterfowl. • The Greek physician, Dioscorides (40–70 A.D.), claimed that the roots and seeds of bur-reed, drunk in wine, would help neutralize snake bites. • The genus name is derived from the Greek *sparganion*, 'a band,' supposedly in reference to the long, narrow leaves.

COMMON CATTAIL • *Typha latifolia*

GENERAL: Perennial from coarse rhizomes; stems **pithy**, cylindrical, 1–2 m high.

LEAVES: Alternate, upright, grass-like, somewhat spongy, **1–2 cm wide**, greyish green; bases sheathe stem.

FLOWERS: Tiny, many, in dense **cylindrical spike** at stem tip (looks like fast-food corn-dog); lower, dark brown female portion, 15–20 cm long, 1–3 cm thick, remains all year; upper male portion smaller, yellowish, cone-shaped, disintegrates early; no petals or sepals; female flowers have 1-chambered ovary on hairy stalk; male flowers have 1–7 (usually 3) stamens and several hairs.

FRUITS: Elliptic **achenes**, about 1 mm long, with many, **long, slender hairs** at base, designed to float.

WHERE FOUND: Marshes, ponds and wet ditches in slow-flowing or standing water; widespread across our region, north and west to Great Bear Lake and interior Alaska.

NOTES: The tender, whitish leaf bases were collected in spring and early summer as a fresh vegetable. They were also boiled or roasted as they became more fibrous later in spring. The Chipewyan name for cattail is *tlh'oghk'a*, 'grass fat.' They gathered the rhizomes in spring when they were 'just like white fat inside.' The rhizomes were peeled and eaten raw or roasted, or they were dried and ground to meal for making porridge. The Cree collected the rhizomes just after the plants had bloomed, and ate them raw or dipped in boiling water. The young flowering spikes are also edible. The male flowers (at the top of the spike) produce large quantities of pollen, which can be collected and mixed with equal parts of flour to make muffins, biscuits, pancakes, and cookies. The green female flowers (at the bottom of the spike) can be boiled or roasted on the spikes and nibbled off like corn on the cob. They are said to taste like olives, globe artichokes or sweet corn. • In more southern areas, care must be taken not to confuse cattail leaves with those of the **poisonous** irises.

SWEET FLAG • *Acorus calamus*
RATROOT

GENERAL: Slender, **aromatic** plant with acrid, watery juice and thick, creeping, rhizomes; **erect stems resemble leaves** but have flower spikes and extend above spike in leaf-like bract.

LEAVES: Sword-shaped, 40-80 cm long, 1-2 cm wide, crowded at base of plant.

FLOWERS: Many, in dense, **cylindrical, yellow-brown spike**, 3-10 cm long, about 1 cm thick; no petals; 6 sepals, tiny, brownish.

FRUITS: Berry-like, hard, dry outside, gelatinous inside.

WHERE FOUND: Swamps, marshes and quiet water by streams; scattered across southern boreal forest; circumpolar; rare; northern populations may have been introduced.

NOTES: Also called *A. americanus*. • Although sweet flag is more common to the south, it is one of the most widely and frequently used herbal medicines among the Chipewyan. Experiments suggest that these roots may alleviate stomach cramps. However, an Asian variety of this species has caused cancer in rats, and in 1968 the U.S. Food and Drug Administration labelled sweet flag **'unsafe.'** • The similar rhizomes of sweet flag and the **poisonous** water arum (p. 218) often grow mixed together. Water arum rhizomes are green and more slender than the pale brown rhizomes of sweet flag. • Some people smoked the rhizome when they had a cough, but it was usually chewed and sucked and the juice swallowed to treat a cold, cough, sore throat, upset stomach, toothache or other pain, to avoid getting a cold after sweating in winter and to restore energy. The rhizome, softened in water, was used to stop bleeding, to aid the healing of cuts or to relieve the pain of a toothache or earache. A poultice of chewed rhizome was applied to relieve sore throat, sore muscles and painful rheumatic joints and muscles. The rhizomes were so widely used as medicine by native peoples, that they became a medium of exchange for some groups.

WATER ARUM • *Calla palustris*

GENERAL: Perennial, 10–30 cm tall, from long creeping rhizomes, rooting at nodes; stems leafless.

LEAVES: Simple, all at stem base, **egg-shaped to rounded heart-shaped**, 5–10 cm long; stalks 7–20 cm long.

FLOWERS: Greenish yellow, in short, cylindrical **spike** (spadix), **1.5–2.5 cm long**, on **thick, fleshy stalk** 10–20 cm long; surrounded by large, white, egg-shaped, pointed **bract** (spathe), 2.5–7 cm long.

FRUITS: Berry-like, red, few-seeded, in dense heads; inedible.

WHERE FOUND: Marshes, swamps, ditches and shallow water; widespread across boreal forest.

NOTES: The roots of water arum, properly boiled and dried to remove the acrid and **poisonous** properties, were formerly mixed with flour to make bread, but this is not recommended. • The genus name comes from the Greek *kallos*, meaning 'beautiful.'

BROAD-LEAVED WATER-PLANTAIN • *Alisma plantago-aquatica*

GENERAL: Semiaquatic perennial from **fleshy, corm-like base** with fibrous roots; flowering stalk leafless, 30–100 cm tall.

LEAVES: At stem base, ascending; blades **egg-shaped to lance-oblong**, 5–18 cm long, pointed at tip, heart-shaped to rounded at base; stalks long, sheathe stem.

FLOWERS: Many in whorls in open, branched clusters at stem tips, well above leaves; 3 petals, **white** (occasionally pinkish), **about 5 mm long**, fall off early; 10–20 1-chamber ovaries.

FRUITS: Whorls of flattened, oblong egg-shaped, grooved achenes.

WHERE FOUND: Marshes, ponds and wet ditches; usually above water, sometimes largely submerged; widespread across southern boreal forest, parkland and prairie in prairie provinces and disjunct in interior Alaska; circumpolar.

NOTES: The Woods Cree dried the stem bases of broad-leaved water-plantain for medicinal use. This was eaten directly or grated and mixed in water for treating heart troubles (including heartburn), stomachaches, cramps, stomach flu and constipation. It was also given to women during childbirth to prevent them from fainting. • Studies in China support the traditional use of these roots for stimulating urination in the treatment of urinary tract disorders and fluid retention. In experiments with animals, broad-leaved water-plantain has lowered blood pressure, reduced glucose levels in the blood, and inhibited the storage of fat in the liver. • The bruised leaves have been applied locally to bruises and swellings, and even had a reputation for healing rattlesnake bites. They redden and sometimes **blister** the skin, and they can be harmful to cattle. • The genus name *Alisma* comes from the Celtic word *alis*, 'water,' in reference to the aquatic habitat of these plants. *Plantago* refers to the similarity of the leaves to those of a true plantain (*Plantago* spp., p. 201).

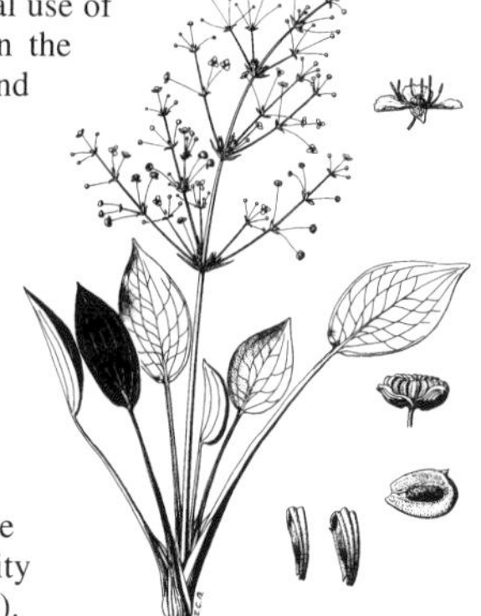

ARUM-LEAVED ARROWHEAD • *Sagittaria cuneata*

GENERAL: Perennial from rhizomes and **fleshy thickened roots (tubers)**, 20–50 cm tall; **milky juice**.

LEAVES: Simple, basal, long-stalked; blade **broadly arrowhead-shaped**, 6–12 cm long, lobes at base shorter than end part; submerged leaves linear, up to 40 cm long, 2–10 mm wide.

FLOWERS: In clusters of 3 on stalk up to 50 cm long; white, **single-sexed**; **lower female flowers develop before upper male flowers**; 3 green sepals; 3 white petals, 7–12 mm long, longer than sepals.

FRUITS: Flattened, winged achenes, 2–2.5 mm long, with short, straight beak; in nearly spherical heads.

WHERE FOUND: Marshes, ditches, muddy shores and shallow water; widespread across our region.

NOTES: The milky juice of arrowhead roots tastes unpleasant raw, but after roasting or boiling it becomes sweetish and tasty. The entire root is edible, but usually only the starchy tubers are worth collecting. The larger, more southern species, **wapato** (*S. latifolia*), can produce tubers up to 2–5 cm in diameter by autumn, but those of arum-leaved arrowhead are much smaller and have rarely been used. • Arum-leaved arrowhead sends out runners (rhizomes) from the fleshy base of the stem. The rhizomes develop tubers several feet away from the parent plant. These tubers produce new plants the following year, and send out runners of their own. The tubers are eagerly eaten by ducks, for which reason arrowheads have sometimes been planted by duck hunters. • The genus name comes from the Latin *sagitta*, 'an arrow,' in reference to the shape of the leaves.

BUCK-BEAN • *Menyanthes trifoliata*

GENERAL: Aquatic to semi-aquatic perennial from thick, scaly, submerged rhizomes; flowering stems erect, hairless, 10–30 cm tall, leafless.

LEAVES: Leaves alternate, crowded near stem base, **divided into 3 egg-shaped to elliptic leaflets**, 3–10 cm long, thick, smooth-edged; stalks 5–20 cm long, sheathing at base.

FLOWERS: Few to several in clusters at tip of stout, tall stem; unpleasant-smelling; 5 petals, **white** or tinged purplish pink, 10–15 mm long, covered with **long white hairs** on inner surface, lower half **fused into short funnel**.

FRUITS: Round to oval **capsules**, 6–10 mm long, split open irregularly; few seeds, smooth, shiny, 2–2.5 mm long, buoyant.

WHERE FOUND: Fens, ditches and lake and pond edges; widespread across boreal forest, north slightly beyond treeline; circumpolar.

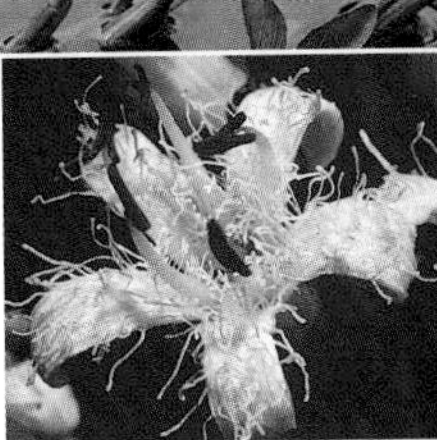

NOTES: Buck-bean was not widely used by northern native groups. However, in Europe, the powdered roots were mixed with flour to make a bitter-tasting bread. More commonly, buck-bean roots were used as a substitute for hops or other bitters in making beer, or they were boiled with honey to make a type of mead. Buck-bean roots were sometimes fed to cattle, and they had a reputation for preventing rot in sheep. However, it is doubtful that livestock ate much of these extremely bitter plants. Some sources say buck-bean can be used in summer salads, but large doses can cause **vomiting and diarrhea**. • The foliage is similar to that of broad beans, which may have given rise to the 'bean' part of the common name. 'Buck' may come from the French *bouc*, for 'goat,' since an older name for buck-bean may have been 'goat's bean.' Alternatively, since buck-bean was used to treat scurvy, 'buck' may be from *sharbock*, the German word for scurvy. • The genus name comes from the Greek *men*, 'month,' and *anthos*, 'flower,' either because the flower was said to last 1 month or because it was once used medicinally to bring on menstruation.

MARSH CINQUEFOIL • *Potentilla palustris*

GENERAL: Perennial with **long-creeping**, **often floating**, woody **rhizomes**; stems often reddish, ascending to erect, 20–60 cm tall.

LEAVES: Mainly on flowering stems, compound; **5–7 leaflets**, oblong, 2–7 cm long, coarsely toothed, **pale green** above, with whitish bloom or fine hairs below; smaller and simpler upwards.

FLOWERS: Few, in loose clusters at stem tips; **reddish purple**, about 2 cm wide; 5 sepals, purplish, longer than alternating bractlets; 5 petals, reddish, shorter than sepals.

FRUITS: Many oval to **egg-shaped achenes**, plump, smooth, brown, buoyant.

WHERE FOUND: Fens, swamps, marshes, wet meadows and lake edges; widespread across boreal forest, north to Arctic coast; circumpolar.

NOTES: Marsh cinquefoil is an unusual cinquefoil. It is the only *Potentilla* with **purple petals** and sometimes it has 6 rather than the usual 5 petals and 5 sepals. Its reddish purple flowers emit a **fetid odour** that attracts carrion-feeding insects as pollinators. • The leaves have been dried and used to make a popular tea throughout much of the Arctic. The roots, which are bitter and astringent, have been boiled to make a medicinal tea for treating stomach cramps and dysentery. It was also used as a wash for cuts and burns. • The name 'cinquefoil' literally means 'five leaves' and refers to the fact that many species of *Potentilla* have leaves divided into five leaflets.

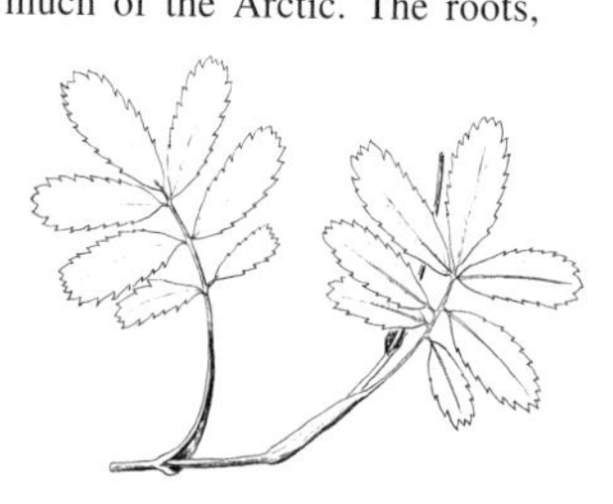

WATER SMARTWEED • *Polygonum amphibium*

GENERAL: Aquatic to terrestrial perennial with running, branched rhizomes; stems leafy, **prostrate or ascending, floating when aquatic**, 0.5–1 m long.

LEAVES: Alternate, many, usually floating, variable, **narrowly elliptic to oblong lance-shaped**, 5–15 cm long, long-stalked, somewhat leathery; **stipules sheathing**, cylindrical, 1–2 cm long, often with spreading collar.

FLOWERS: Many, in 1 or 2 upright, **oblong or egg-shaped**, 1–3 cm long spikes at tip of hairless stem; 5 petal-like segments, **bright pink** to scarlet, 4–5 mm long.

FRUITS: Achenes, lens-shaped, 2.5–3 mm long, dark brown, smooth.

WHERE FOUND: Shallow water or muddy banks and shores; widespread across our region; circumpolar.

NOTES: **Long-spiked water smartweed** (*P. coccineum*) resembles water smartweed, but is distinguished by its long (up to 15 cm) spikes, which are on hairy stalks, and its upright (not floating), lance-shaped, pointed leaves. It grows in wet meadows and marshes, on streambanks and in shallow quiet water, scattered north and west to the southwestern Northwest Territories. • All *Polygonum*s are edible. • The Woods Cree applied the fresh root of water smartweed directly to blisters in the mouth. They also dried and powdered the roots as an ingredient in a many-herb remedy that was taken for a variety of ailments. • Smartweeds were formerly called 'arsmart' because of the irritating effect of the leaves 'when used in these parts.' It's not clear why smartweeds were used medicinally on the human hindquarters, but they were, for everything from poultices for external bleeding to treating piles and itchy skin diseases.

P. coccineum

CLASPING-LEAF PONDWEED • *Potamogeton perfoliatus* RICHARDSON'S PONDWEED var. *richardsonii*

GENERAL: Aquatic perennial from slender rhizomes; stems submerged, leafy, round, green to brownish, branched, 1–2 mm thick, 0.3–1 m long.

LEAVES: Alternate, **all submerged**, many, **lance-egg-shaped**, 3–12 cm long, 13–25-veined (with 5–7 strong veins), dark green, stalkless; heart-shaped base clasps stem; **edges wavy**; stipules separate from rest of leaf, white to brownish, soon **disintegrate into tufts of white threads**.

FLOWERS: In dense, stalked 1.5–3 cm long spike of 4–12 closely crowded whorls; tiny, inconspicuous, stalkless, 4-parted; no petals or sepals.

FRUITS: Achenes, 2.5–3.5 mm long, semi-fleshy at first, then dry and hard, short-beaked.

WHERE FOUND: Ponds, lakes and sluggish streams; widespread across our region.

NOTES: Also called *P. richardsonii*. • Muskrats, beavers, moose and deer eat large quantities of pondweeds and their tubers. These plants are also a favourite food of many shorebirds, marsh birds and waterfowl. Birds usually eat the seeds, but they will also feed on the stems and leaves of more delicate species. Pondweeds also provide food, shelter and shade for fish and small aquatic invertebrates. Many insects live among their leaves, and these in turn, provide food for fish. • In the Mackenzie River delta, some of the large lakes produce very large whitefish (*Coregonus* spp.). The fish cannot migrate to and from these lakes, so they must spend the whole year there. Although there is a thick covering of ice and snow in the winter, the dense growth of clasping-leaf pondweed is thought to produce enough oxygen for these fish to survive. Natives who fished these lakes in the winter knew they would find good fishing only in those lakes 'with much grass on the bottom.'

VARIOUS-LEAVED PONDWEED • *Potamogeton gramineus* GRASS-LEAVED PONDWEED

GENERAL: Aquatic perennial from slender rhizomes that produce long tubers in autumn; stems submerged, leafy, rounded, greenish, freely branched, 0.3–1 m long.

LEAVES: Alternate, usually **of 2 types**; many **submerged leaves, linear to lance-shaped**, stalkless, 1–12 cm long, 3–10 mm wide, 3–9 veins; **floating leaves elliptic to oblong**, 2–6 cm long, leathery, shiny, 13–19 veins, long, slender stalks; stipules brownish, separate from leaf, 2–3 cm long.

FLOWERS: In stalked, 1–2.5 cm long spike of 5–10 crowded whorls; tiny, inconspicuous, stalkless, 4-parted.

FRUITS: Achenes, 2.5 mm long, semi-fleshy at first, then dry and hard, ridged, with short beaks.

WHERE FOUND: Ponds, lakes, ditches, and slow-flowing streams; widespread across boreal forest; occasional in parkland; circumpolar.

NOTES: Various-leaved pondweed is highly adaptable to local conditions, and consequently it is an extremely variable species. It can grow in water from just a few centimetres to 2 m deep, or can even be found on muddy shores, stranded by recedingwater. • Many pondweeds reproduce more freely by tubers or over-wintering buds than by seeds. Dormant buds produced in autumn sink to the mud at the bottom of the pond where they remain protected through the long, cold winter months. Each bud can grow to produce a new plant, and these small, vegetative propagules can also be picked up by the feathers or muddy feet of waterfowl and carried to new sites. Pondweed seeds sprout only after being in cold water for several months and they do not survive drying-out. • The species name *gramineus* means 'grass-like,' in reference to the submerged leaves.

FRIES' PONDWEED • *Potamogeton friesii*

GENERAL: Aquatic perennial; **stems compressed, freely branched**, 0.5–1 m long.

LEAVES: Simple, **all submerged**, thin, bright green, translucent, **linear**, 3–8 cm long, 1.5–3.5 mm wide, **mostly 5-veined, usually with 2 yellowish glands at base**, tip blunt or abruptly short-pointed; **stipules** 7–15 mm long, **free from blade, strongly fibrous, whitish, torn with age**.

FLOWERS: In 1–2 cm long spike of **3–4 slightly separated whorls** on 1.5–4 cm long stalk; tiny, inconspicuous, 4-parted.

FRUITS: Obovoid achenes, 2–3 mm long, **with short, slightly curved beak**.

WHERE FOUND: In shallow fresh to brackish water; widespread across boreal forest and northern parkland, north and west to interior Alaska.

P. pectinatus

P. friesii

NOTES: Large-sheath pondweed (*P. vaginatus*) (inset photo) has more whorls (5–9) in its flower spikes than Fries' pondweed. It has long (to 50 cm), 1–3-veined leaves that are 1-2 mm wide and have blunt tips. The stipules are united with the leaf-base to form a prominent, broadened, 2–5-cm-long sheath. Large-sheath pondweed grows scattered throughout our region in water to 2 m deep. • **Sago pondweed** (*P. pectinatus*) is distinguished from large-sheath pondweed by its inconspicuous (less than 2 cm long) stipule sheath, and very narrow (1 mm wide or less), clustered leaves that taper to a long sharply pointed tip. Sago pondweed grows in ponds, lakes and sluggish streams across our region. It is the most common of the fine-leaved pondweeds in the prairie provinces. • The roots of most pondweeds are too small to warrant collecting, but the thickened rhizomes and swollen roots of some species are eaten. The younger roots are starchiest, and these are steamed or boiled in stews or soups. The whole plant of sago pondweed is eaten, and its abundant tubers are said to be an important food for ducks.

P. vaginatus

THREAD-LEAVED PONDWEED • *Potamogeton filiformis*

GENERAL: Aquatic perennial; **stems nearly round**, 10–40 cm long, stout and thick, **branching** near base, slender, **largely unbranched** towards top.

LEAVES: Simple, many, **all submerged, thread-like to narrowly linear**, 1–10 cm long, 0.2–1.5 mm wide, blunt or short-pointed at tip, **1-veined, no glands at nodes**; **stipules joined to blade for 5–10 mm**, fused to form **sheath for most of their length**.

FLOWERS: In 1–5 cm long spike of **2–5 whorls**, separate to crowded, on slender, 5–15 cm long stalk; tiny, inconspicuous, **held above water**.

FRUITS: Beakless, egg-shaped achenes, about 2 mm long.

WHERE FOUND: Shallow water of ponds and lakes; common and widespread across our region.

P. filiformis

NOTES: Small-leaf pondweed (*P. pusillus*, also called *P. berchtoldii*) is a similar species that is widespread but less abundant in the boreal forest and parkland. It differs from thread-leaved pondweed in that its stipules are not fused to the leaf-base and soon fall off; it has 2–5 cm long, 3–5-veined leaves with a prominent midvein and 2 translucent glands at the leaf-base; and it has compact, continuous flower spikes on curved stalks. • The hard achenes are adapted to dispersal by birds. They are very difficult to digest and can remain in the gut for a long time, but their ability to germinate increases after they have passed through the gut of a bird. The seeds of at least 1 species can germinate **only** after they have passed through a bird. Pondweed seeds are often carried for well over 1,500 km by migrating waterfowl. The loss of seeds in such a trip is counterbalanced by the wide dispersal of the species and the deposition of the seeds in favourable sites such as lakes and ponds.

P. pusillus

VERNAL WATER-STARWORT • *Callitriche verna*

GENERAL: Annual or perennial, submerged, partly floating or stranded on mud; stems slender, 5–20 cm long.

LEAVES: Simple, **opposite, pale green**; **leaf-bases joined by small winged ridges**; submerged leaves linear, stalkless, 1-veined, 10–15 mm long, tip shallowly notched; **floating leaves** broadly spoon-shaped, stalked, 3-veined, up to 4 mm wide, **often crowded in rosette at stem tip**.

FLOWERS: 1–3 in leaf axils; tiny, inconspicuous, greenish, **no sepals or petals**, **pair of bracts at flower base**; **sexes in separate flowers on same plant**; male flowers have 1 stamen, female flowers have 1 pistil.

FRUITS: Small, nut-like, compressed, 4-lobed, 1–1.5 mm long; split into 4, 1-seeded portions.

WHERE FOUND: Ponds, ditches and slow-moving streams; often forms mats in shallow water or on wet mud; common and widespread across our region.

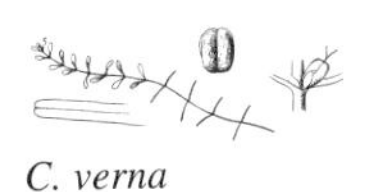

C. verna

C. hermaphroditica

NOTES: Also called *C. palustris*. • **Northern water-starwort** (*C. hermaphroditica*, also called *C. autumnalis)* is similar, but it grows fully submerged, and has only linear, dark green, underwater leaves that are not connected by winged ridges. Its flowers do not have basal bracts. It is found in similar habitats across our region but is less common. • It was once thought that the flowers of water-starworts were fertilized underwater. Pollen is sometimes disseminated underwater, but in many cases the filaments elongate as much as necessary to reach the surface of the water and project the anthers up into the air. The light 'seeds' are covered with a waxy coating that makes them ideally suited for dispersal by water currents.

COMMON MARE'S-TAIL • *Hippuris vulgaris*

GENERAL: Aquatic or amphibious, hairless perennial from creeping rhizomes; stems leafy, **upright**, usually unbranched, **5–30 cm tall**.

LEAVES: In whorls of 6–12, stalkless, **linear**, 10–35 mm long, 1–2 mm wide, **pointed**; stiff above water, limp and up to 6 cm long underwater.

FLOWERS: Single, stalkless in upper leaf axils; tiny, inconspicuous; **no petals**, 1 stamen; 1 single-celled ovary.

FRUITS: Nutlets, 1-seeded, about 2 mm long.

WHERE FOUND: Ponds, lakeshores, streambanks and mudflats, usually in shallow water; widespread across our region, north to arctic islands; circumpolar.

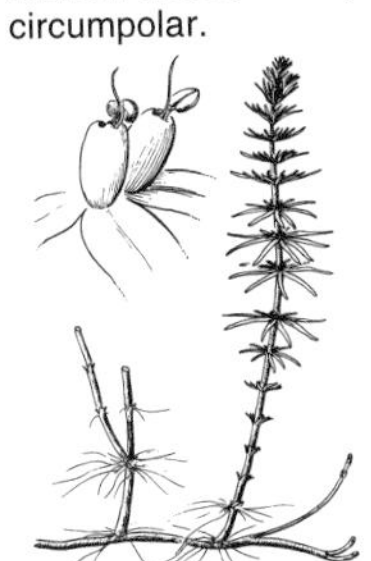

NOTES: Common mare's-tail looks much like a horsetail (*Equisetum* spp.), but is a flowering plant and doesn't have jointed, hollow stems. • The length of the leaves can vary greatly with light levels. The amount of light available for photosynthesis decreases rapidly with increasing depth in the water. Leaves above the pond surface are usually about 7–9 times longer than wide, but submerged leaves can reach lengths over 30 times longer than their width. • Common mare's-tail is one of the few aquatic food plants, though some say that it adds only roughage and salt to the diet. Alaskan native peoples gathered the shoots to use as a potherb or as an addition to soup. It is tender and edible at any time of year, and has even been collected where it protruded above the ice in winter. • Mare's-tail usually forms dense colonies in the water, and these shelter many small aquatic animals. Wildlife feed on the seeds and occasionally on the leaves. • The name 'mare's-tail' is a misinterpretation of 'female horse-tail,' the name it was given in old herbals since it was thought to be the female part of swamp horsetail.

HORNWORT • *Ceratophyllum demersum*
COONTAIL

GENERAL: Submerged, non-rooted aquatic with slender, freely-branched stems, 30–150 cm long.

LEAVES: Rigid, 0.5–2.5 cm long, **2–3 times forked into slender, often spiny-toothed segments**, in whorls of 5–12, often coated with lime.

FLOWERS: Single in leaf axils; tiny, inconspicuous, stalkless; **sexes in separate flowers on same plant**; **no sepals or petals**; **surrounded by 8–12 bracts**.

FRUITS: Flattened, elliptic achenes, 4–6 mm long; **style at tip**, 6–12 mm long; **2 divergent spines at base**, 1–6 mm long.

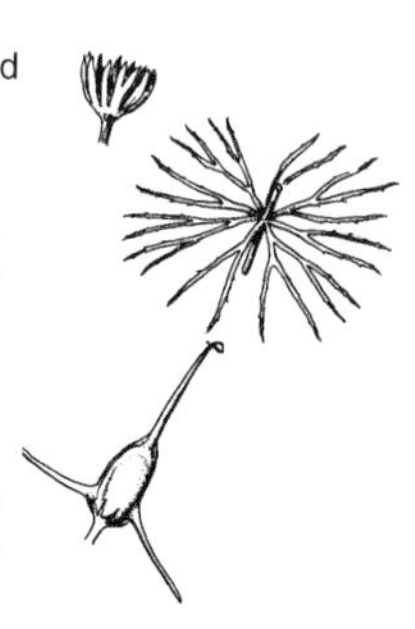

WHERE FOUND: Ponds, lakes, ditches and quiet streams; widespread and fairly common in southern boreal forest and northern parkland; less common north to Great Slave Lake and interior Alaska; circumpolar.

NOTES: Hornwort superficially resembles sterile spiked water-milfoil (below), but it is readily distinguished by its forked rather than pinnate leaves. • In nutrient-rich waters hornwort can become a pest that clogs water intakes in reservoirs, fouls boat propellers and makes swimming unappealing and possibly dangerous. • Although only recently reported north of 60°N, where it is apparently rare, well preserved *Ceratophyllum* fruits, radiocarbon dated as 5,500 years old (+/- 250 years), are known from frozen peat deposits in the upper Thelon River valley in the Northwest Territories. • Hornwort is also called 'coontail' because its leafy stem can resemble a raccoon's bushy tail.

SPIKED WATER-MILFOIL • *Myriophyllum spicatum* var. *exalbescens*

GENERAL: Aquatic perennial from rhizomes; stems leafy, limp, 2–3 mm thick, 30–100 cm long, purple, dry whitish.

LEAVES: In whorls of 3–4, 1–3 cm long, **feather-like, pinnately divided** into 13–21 **thread-like segments**.

FLOWERS: Whorled, in 2–8 cm long **spike-like clusters at stem tips** that usually stick out of water; 4 petals, 2.5 mm long, soon fall off; sexes separate on same plant; male flowers usually uppermost, 8 stamens; female flowers usually lower, 4 feathery styles.

FRUITS: Nut-like, split into 4, 1-seeded, rounded achenes, 2–3 mm long.

WHERE FOUND: Lakes, ponds, sloughs and slow-moving streams; widespread across our region, north to southern arctic islands.

NOTES: Also called *M. exalbescens* or mistakenly *M. verticillatum*. • **Alternate-flowered water-milfoil** (*M. alterniflorum*) is similar to spiked water-milfoil, but its upper male flowers are alternate rather than whorled. It is found in our region primarily south and east from Great Bear Lake. • Spiked water-milfoil has been eaten as a potherb in some areas in the boreal forest. • Most waterfowl, especially ducks, eat the fruits, stems and leaves of water-milfoil. Muskrats also feed on these plants. The many whorls of finely-divided, spreading leaves give water-milfoils the appearance of bottle-brushes. Small fish and aquatic invertebrates breed and find shelter in this frilly foliage, although they risk being eaten by passing waterfowl. Bass are said to nest among the lower stems and roots of water-milfoil. • The female flowers of a plant usually mature a few days before the male flowers. Water-milfoil frequently reproduces vegetatively by producing special, hardy, dormant buds that sink to the bottom of the pond where they overwinter in the mud. As the buds develop in spring they rise once again to the surface. These buds are also picked up by migrating waterfowl and carried to other waterbodies, sometimes over great distances.

M. alterniflorum

LARGE-LEAVED WHITE WATER-CROWFOOT • *Ranunculus aquatilis* var. *capillaceus*
WHITE WATER-BUTTERCUP

GENERAL: Aquatic, usually perennial; stems submerged, weak, sparingly branched, leafy, root at lower nodes, 1–2 mm thick, 20–100 cm long.

LEAVES: Alternate, **of 2 types; submerged leaves repeatedly divided** in 3s into linear segments, about 1–2.5 cm long, weak, collapse out of water; **floating leaves broad, flat, palmate, 3–5-lobed**.

FLOWERS: Single, 1–1.5 cm across, long-stalked; **5 petals, white**, 4–8 mm long.

FRUITS: Achenes, 15–25 in rounded cluster, asymmetrically egg-shaped, 1–2.5 mm long, very short beaked to beakless, cross-ridged.

WHERE FOUND: Shallow water in ponds, sluggish streams, lakes and ditches; widespread across our region, north past treeline; circumpolar.

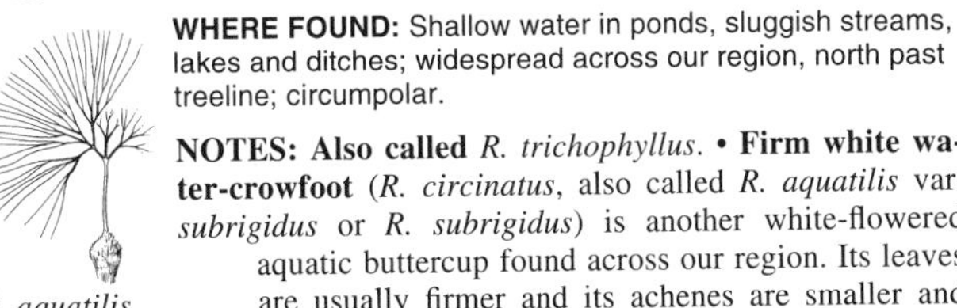

R. aquatilis submerged leaf

NOTES: Also called *R. trichophyllus*. • **Firm white water-crowfoot** (*R. circinatus*, also called *R. aquatilis* var. *subrigidus* or *R. subrigidus*) is another white-flowered aquatic buttercup found across our region. Its leaves are usually firmer and its achenes are smaller and more numerous (30–80) than those of large-leaved white water-crowfoot. • Large-leaved white water-crowfoot often forms large clumps or floating mats. • Studies have shown that the determination of leaf form in large-leaved white water-crowfoot depends partly on light level and partly on water level.

R. circinatus

YELLOW WATER-CROWFOOT • *Ranunculus gmelinii*
SMALL YELLOW WATER-BUTTERCUP

GENERAL: Perennial, **creeps on mud or floats in water**; stems 10–50 cm long, **root freely at nodes**.

LEAVES: Alternate, mainly on stem, kidney-shaped to circular in outline, **1–3 cm long, deeply divided into 3–5 lobes, each lobe divided again into linear lobes**; submerged leaves delicate, finer than floating leaves, often with almost thread-like lobes.

FLOWERS: 1–4 on long stalks; **6–15 mm across**; petals yellow, 4–7 mm long.

FRUITS: Achenes, 4–7 mm long, hairless, 50–70 in spherical heads; **short, thin beaks**, 1–1.5 mm long.

WHERE FOUND: Mudflats and shallow water in ponds and ditches, and along lakeshores; widespread across boreal forest and parkland; extends north past treeline to arctic islands and west through Alaska to Asia.

NOTES: Buttercup flowers follow a generalist strategy. Their bright yellow petals attract a wide range of insects, and an abundant supply of nectar and pollen is readily available to most visitors. • The shiny yellow of the petals is created by a thin layer of yellow pigment at the surface and an underlying layer of thick starch grains. When light passes through the upper yellow layer, it is reflected back from the white starch below, much like light behaves in a watercolour painting. • The name 'crowfoot' comes from the supposed resemblance of the leaf to the foot of a crow. The species name *gmelinii* commemorates one of the Gmelin family, a group of German botanists that lived in the 1700s and 1800s.

SMALL YELLOW POND-LILY • *Nuphar variegatum*

GENERAL: Aquatic perennial with stout, creeping rhizomes.

LEAVES: Simple, **floating**, rounded heart-shaped, 10–30 cm long, 8–15 cm wide, on long, slender, **flattened** stalks.

FLOWERS: Single on long stalks held above water; yellow, 4–7 cm across; **usually 6 sepals**, leathery; many small, inconspicuous petals; many **yellow anthers**.

FRUITS: Many-seeded, leathery, berry-like capsules, 4 cm long; split irregularly.

WHERE FOUND: Ponds, lakes and slow-moving streams; common across boreal forest, north and west to Great Bear Lake and central Yukon; occasional in northern parkland.

NOTES: Showy yellow pond-lily (*N. polysepalum*) (inset photo), also called Rocky Mountain cow-lily, resembles small yellow pond-lily but is generally larger and has 7–9 yellow sepals, reddish anthers, and round leaf-stalks. It is common in the Yukon and Alaska. • Although widespread, **white water-lily** (*Nymphaea tetragona*) is much smaller and less common than the yellow pond-lilies. It has smaller leaf blades (6-7 cm long) with pointed lobes. • The Chipewyan chewed thin slices or used a tea made from the rhizomes to relieve painful joints. The rhizomes were soaked in hot water and placed on boils or infected wounds to draw out the infection. Fresh or dried and powdered slices of these rhizomes were applied to cuts to stop bleeding and promote healing. The Woods Cree mixed the grated roots of small yellow pond-lily, sweet flag, and occasionally cow-parsnip, to make a salve to relieve headaches, sore joints, swellings and painful limbs. These same 3 ground roots were also mixed together and used as a poultice for treating *mancos* (worms in the flesh). • The Woods Cree used the dried rhizomes for food. They ate them dry, or rehydrated and cooked them. • The leaves and rhizomes are eaten by many animals, particularly beavers, muskrats and moose.

COMMON DUCKWEED • *Lemna minor*

GENERAL: Tiny, colonial, aquatic perennial, **not differentiated into leaf and stem**; small disc-like, leafy, floating fronds, 2–5 mm long; usually **1 short rootlet** hanging from lower surface.

FLOWERS: Very small, usually 3, single-sexed, without sepals or petals; 2 male flowers with 1 stamen, 1 female flower with flask-shaped ovary.

FRUITS: Small, thin-walled, bladder-like (utricle); 1–7 seeds.

WHERE FOUND: Still, fresh water; widespread and abundant across our region, north to southern N.W.T. and interior Alaska; circumpolar.

NOTES: Ivy-leaved duckweed (*L. trisulca*) has bigger (5–10 mm long), stalked, 3-lobed fronds. It forms large, submerged, tangled colonies in quiet streams and ponds across our region, north to the Mackenzie River delta. • **Larger duckweed** (*Spirodela polyrhiza*) is another floater and has several rootlets hanging beneath its oval fronds. It grows in ponds and sluggish streams in the southern boreal forest and parkland of the prairie provinces. • Duckweeds are easily picked up by the feet and feathers of waterfowl and carried to new ponds and streams. In this way, they have been dispersed through most of the world. Duckweeds usually multiply by producing tiny bulblets at the base of a parent plant. These separate and grow into new plants, or they may stay attached for some time and form small colonies. In winter, the bulblets sink to the bottom of the water and then rise to the surface in spring to produce new plants. • Duckweeds reproduce so quickly and efficiently by vegetative division, that they seldom flower. On the rare occasions when they bloom, they produce the tiniest of our wild flowers. The flowers appear in a fold on the upper surface of the plant. A male flower is simply a single stamen and a female flower is a tiny, flask-shaped pistil the size of a pin-point. The pollen grains are covered with spines, suggesting that these tiny plants are pollinated by insects.

L. minor

L. trisulca

S. polyrhiza

The Sedge Family (Cyperaceae)

Sedges (family Cyperaceae), rushes (family Juncaceae), and grasses (family Poaceae) are somewhat similar and can be grouped as graminoids—grass-like plants. They all have long, narrow, parallel-veined leaves and inconspicuous flowers with several scale-like bracts. You can most easily distinguish them by examining their stems: those of sedges are generally triangular in cross-section and solid (not hollow), with the leaves in 3 rows; those of rushes are round and solid ('pithy'); those of grasses are round, jointed, often hollow, and with the leaves in 2 rows.

The sedge family has several genera, but *Carex* is by far the largest (with over 100 species in the North American boreal region). *Carex* is distinctive in that its ovary and seed-like fruit (**achene**) are enclosed in a membranous sac (**perigynium**) in the axil of a single, scale-like bract. Sedges (*Carex*) and cotton-grasses (*Eriophorum*) are common and conspicuous plants of boreal wetlands.

More information about sedges in our region is provided in Hultén (1968), Hudson (1977), Looman and Best (1979), Porsild and Cody (1980) and Taylor (1983).

Key to the Genera of the Sedge Family

1a. Flowers single-sexed; achenes enclosed or wrapped in sac (perigynium) .... ***Carex*** (the sedges)

1b. Flowers bisexual; achenes not enclosed in perigynium 2

  2a. Styles thickened towards base, form conspicuous bump (tubercle) on achenes 3

    3a. 1 spikelet; several to many fertile flowers (or achenes) per spikelet; 0–6 perianth bristles; stems leafless .... ***Eleocharis*** (the spike-rushes)

    3b. Several to many spikelets; 1 (or 2) fertile flowers per spikelet; 8 or more perianth bristles; stems have bristle-like leaves .... ***Rhynchospora*** (the beak-rushes)

  2b. Styles not thickened to form tubercle 4

    4a. 10 or more perianth bristles, lengthen in fruit .... ***Eriophorum*** (the cotton-grasses)

    4b. 8 or fewer perianth bristles, do not usually lengthen in fruit .... ***Scirpus*** (the bulrushes)

Key to the Sedges (*Carex*)

1a. 1 spike .... **Group I**

1b. More than 1 spike, sometimes congested in head-like cluster 2

  2a. Achenes lens-shaped; 2 stigmas 3

    3a. Spikes all alike, stalkless, usually with both male and female flowers (or with sexes on separate plants) 4

      4a. Most spikes have female flowers above male (except in *Carex disperma*) .... **Group II**

      4b. Some or all spikes have male flowers above female .... **Group III**

    3b. Spikes of 2 kinds: end spike usually male; side spikes female, usually stalked .... **Group IV**

  2b. Achenes triangular; 3 stigmas 5

    5a. Perigynia hairy .... **Group V**

    5b. Perigynia hairless .... **Group VI**

Group I

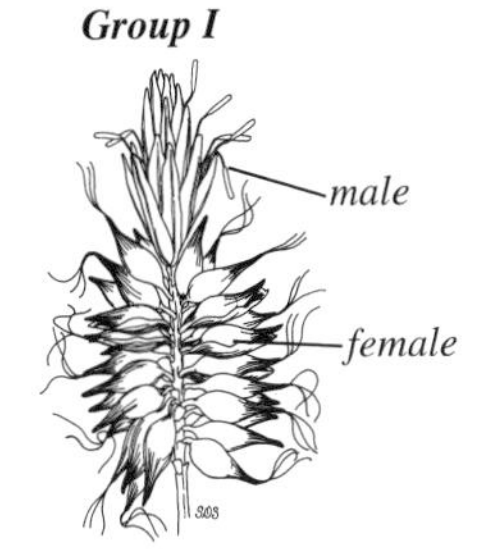

Group II *Group VI*

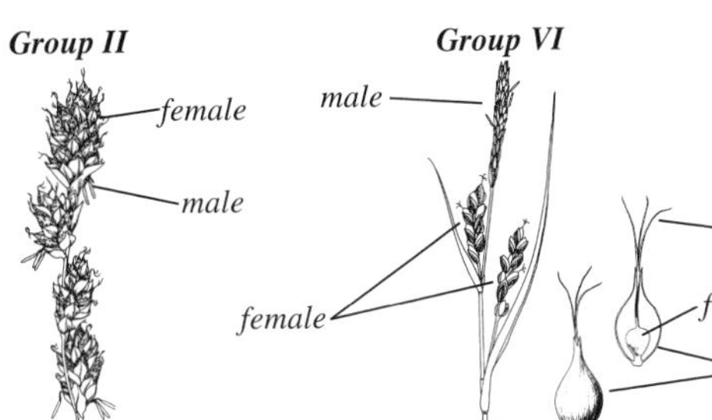

Group I: Sedges with a single spike

1a. Achenes lens-shaped; 2 stigmas 2

2a. Stems single or few together, from long creeping roots; spikes cylindrical, usually single-sexed (or with male flowers above female); perigynia spreads or bends downwards with age ... ***Carex gynocrates***

2b. Stems densely tufted; spikes nearly round, with male flowers above female; perigynia spreading but not bent downwards .... ***C. capitata*** (see. *C. gynocrates*)

1b. Achenes triangular; 3 stigmas 3

3a. Spikes single-sexed; perigynia densely hairy .... ***C. scirpoidea*** (see *C. gynocrates*)

3b. Spikes have male flowers above female; perigynia hairless 4

4a. Female scales do not fall off; edges of male scales fused nearly to middle; perigynia rounded, beakless; plants small and delicate .... ***C. leptalea***

4b. Female scales soon fall off 5

5a. Perigynia linear-lance-shaped, stalked, to 6 mm long, lower ones bend downwards at maturity; few dried up leaves from previous year .... ***C. microglochin*** (see *C. gynocrates*)

5b. Perigynia broader, stalkless, at most 4 mm long, spreading-ascending; conspicuous dried up leaves from previous year .... ***C. obtusata*** (see *C. gynocrates*)

Group II: Sedges with several spikes, side spikes stalkless, spikes usually have both male and female flowers, female flowers above male, achenes lens-shaped, 2 stigmas

1a. Perigynia wing-edged, with distinct, 2-toothed beak 2

2a. Flower cluster to about 8 cm long, often bent; lowest spike does not reach base of one above it 3

3a. Perigynia egg-shaped, 4–5 mm long, beak toothed to tip .... ***Carex aenea***

3b. Perigynia lance-shaped, 6–7 mm long, tip of beak not toothed .... ***C. praticola*** (see *C. aenea*)

2b. Flower cluster compact, usually less than 4 cm long; all spikes more or less overlap 4

4a. Perigynia lance-shaped, 3–4 times as long as wide .... ***C. crawfordii*** (see *C. bebbii*)

4b. Perigynia egg-lance-shaped or broader, at most twice as long as wide .... ***C. bebbii***

1b. Perigynia not wing-edged 5

5a. Perigynia thin-edged, spread or bend downwards at maturity, spongy-thickened at base, distinctly beaked .... ***C. interior***

5b. Perigynia with rounded edges, pressed to stem or spreading at maturity 6

6a. Perigynia at least 4 mm long, with long, slender, 2-toothed beak, not covered with small whitish dots; plants tufted .... ***C. deweyana***

6b. Perigynia less than 4 mm long, covered with small whitish dots, short-beaked; plants have creeping roots 7

7a. Spikes have male flowers at tip; usually 3 perigynia, rounded, with fine veins; beak minute, less than 3 mm long .... ***C. disperma***

7b. Spikes have female flowers at tip; perigynia half rounded 8

8a. Flower cluster compact, usually even lowest spikes overlap 9

9a. Heads silvery, about 1 cm long; perigynia elliptic-oval, nearly beakless .... ***C. tenuiflora***

9b. Heads green, become brown, to 3 cm long; perigynia egg-shaped, with flattened, toothed beak .... ***C. arcta*** (see *C. interior*)

8b. Flower cluster interrupted, only upper spikes overlap, lower ones well spaced 10

10a. Lowest bract bristle-like, many times longer than lowest spike; 2–3 spikes, all well spaced; perigynia beaked, to 3 mm long .... ***C. trisperma*** (see *C. disperma*)

10b. Lowest bract short or absent; upper spikes overlap 11

11a. 3–8 perigynia per spike, beakless, to 3 mm long; plants have creeping rhizomes .... ***C. loliacea*** (see *C. tenuiflora*)

11b. 5-30 perigynia per spike, beaked; plants tufted 12

12a. Spikes greenish to brown, 5–10-flowered; perigynia loosely spreading at maturity, conspicuous beak minutely toothed .... ***C. brunnescens*** (see *C. canescens*)

12b. Spikes whitish to silvery-brown, 10–30-flowered; perigynia ascending at maturity, beak inconspicuous .... ***C. canescens***

Group III: Sedges with several spikes, side spikes stalkless, spikes usually have both male and female flowers, male flowers above female, achenes lens-shaped, 2 stigmas

1a. Stems branched, become prostrate, send up new stems from axils of old leaves .... ***Carex chordorrhiza***

1b. Stems unbranched, erect or ascending 2

2a. Plants usually single-sexed (spikes either completely male or completely female) .... ***C. praegracilis*** (see *C. siccata*)

2b. Plants bisexual (spikes have both male and female flowers), male flowers above female 3

3a. Stems single or a few together from long creeping rhizomes; spikes often dissimilar, sometimes single-sexed 4

4a. Perigynia nearly wingless, beaks obscurely 2-toothed, about half as long as body; spikes readily distinguishable .... ***C. praegracilis*** (see *C. siccata*)

4b. Perigynia narrowly wing-edged, beaks sharply 2-toothed 5

5a. Less than 12 spikes, lowest one small and often wholly female, middle ones largely male, uppermost female near top and male below; plants slender .... ***C. siccata***

5b. Up to 20 spikes, upper ones crowded, often single-sexed, lower ones often female; plants robust .... ***C. sartwellii***

3b. Stems densely to loosely tufted, rhizomes not long and creeping 6

6a. Leaf sheaths copper-tinged at mouth; flower heads interrupted; perigynia dull, nearly concealed by scales .... ***C. prairea*** (see *C. diandra*)

6b. Leaf sheaths not copper-tinged at mouth; flower heads compact; perigynia shining, not concealed by scales .... ***C. diandra***

Group IV: Sedges with several spikes, end spike male, side spikes female and stalked, achenes lens-shaped, 2 stigmas

1a. Lowest bract sheathes stem; spikes few-flowered 2

2a. Mature perigynia golden to orange; scales widely spreading .... ***Carex aurea***

2b. Mature perigynia greenish white; scales not wide-spreading .... ***C. garberi*** (see *C. aurea*)

1b. Lowest bract does not sheathe stem; spikes many-flowered 3

3a. Perigynia short-stalked, ribbed; female scales tinged with red-brown .. ***C. lenticularis*** (see *C. aquatilis*)

3b. Perigynia stalkless, not ribbed; female scales purplish black .... ***C. aquatilis***

Group V: Sedges with several spikes, end spike male and side spikes female, achenes triangular, 3 stigmas, perigynia hairy

1a. Perigynia have short beaks cut at an angle 2

2a. Male spike to 6 mm long, usually stalkless; plants to 20 cm tall .... ***C. concinna***

2b. Male spike to 25 mm long, stalked; plants to 35 cm tall .... ***C. richardsonii*** (see *C. concinna*)

1b. Perigynia have 2-toothed beaks 3

3a. Female flower spikes more than 1 cm long, with more than 25 perigynia; plants usually more than 40 cm tall 4

4a. Leaf edges rolled inwards, leaves not more than 2 mm wide, smooth; stems somewhat triangular .... ***Carex lasiocarpa***

4b. Leaves flat, 2–5 mm wide, edges roughened; stems sharply triangular .... ***C. lanuginosa*** (see *C. lasiocarpa*)

3b. Female flower spikes less than 1 cm long, with fewer than 25 perigynia; plants less than 40 cm tall 5

5a. Stems all long; no spikes hidden among leaf bases, nor entirely female .... ***C. nigromarginata*** (see *C. deflexa*)

5b. Stems of various lengths; shorter spikes crowded among leaf bases, often entirely female 6

6a. Bract of lowest non-basal female spike scale-like, shorter than flower head; male spike usually long-stalked, to over 12 mm long .... ***C. umbellata*** (see *C. deflexa*)

6b. Bract of lowest non-basal female spike leaf-like, usually overtops flower head; female scales shorter than perigynia 7

7a. Rhizomes slender; stems usually loosely tufted; perigynia 2–3 mm long, short-beaked; male spikes up to 5 mm long, inconspicuous .... ***C. deflexa***

7b. Rhizomes stout; stems densely tufted; perigynia 3–4.5 mm long, beaks usually more than 0.5 mm long; male spikes up to 12 mm long, conspicuous ..... ***C. rossii*** (see *C. deflexa*)

Group VI: Sedges with several spikes, end spike male, side spikes female, achenes triangular, 3 stigmas, perigynia hairless

1a. Bract at base of flower cluster has long, tubular sheath 2

2a. Perigynia spreading or bent downwards, with distinct, 2-toothed beak; female spikes nearly round to short-cylindrical, upper ones usually stalkless .... ***C. viridula***

2b. Perigynia ascending, beakless or beaks not toothed; female spikes cylindrical, usually stalked 3

3a. Plants tufted; perigynia 2–4 mm long; spikes at most 1.5 cm long, at least lowest one drooping .... ***C. capillaris***

3b. Plants have rhizomes; perigynia 2.5–5 mm long; spikes up to 2 cm long, not drooping 4

4a. Female spikes short-stalked or stalkless, erect, densely-flowered; perigynia beakless .... ***C. livida*** (see *C. limosa*)

4b. Female spikes on stalks as long as spikes or longer, spreading, loosely 5–12-flowered; perigynia beaked .... ***C. vaginata***

1b. Bract at base of flower cluster sheathless or short-sheathing 5

5a. Roots covered with yellowish woolly hairs 6

6a. Male spikes 1.5–3 cm long; female scales rounded or bluntly pointed, as broad or broader than perigynia and barely longer than them, do not fall off; plants have long, slender rhizomes .... ***C. limosa***

6b. Male spikes 4–12 mm long; female scales long-pointed, much longer and narrower than perigynia, fall off early; plants densely tufted .... ***C. paupercula*** (see *C. limosa*)

5b. Roots not covered with yellowish woolly hairs 7

7a. End spike usually has female flowers above male, sometimes entirely female; perigynia roughened; beaks short 8

8a. Side spikes long stalked, often drooping .... ***C. atratiformis* ssp. *raymondii*** (see *C. capillaris*)

8b. Side spikes stalkless or short-stalked, erect or nearly so 9

9a. Stems single or a few together; female scales long-awned, as long as or longer than perigynia; beak to 0.2 mm long .... ***C. buxbaumii*** (see *C. media*)

9b. Stems loosely to densely tufted; female scales blunt, shorter than to as long as perigynia; beak to 0.5 mm long .... ***C. media***

7b. End spike usually entirely male; beaks prominent 10

10a. Perigynia round or triangular in cross-section, not ribbed .... ***C. supina* ssp. *spaniocarpa*** (see *C. deflexa*)

10b. Perigynia flattened, strongly ribbed 11

11a. Leaves and sheaths hairy; perigynia more or less leathery, with 12–25 ribs; beak prominently toothed, teeth 1.2–3 mm long .... ***C. atherodes***

11b. Leaves and sheaths hairless; perigynia not leathery, with 7–10 ribs 12

12a. Lower perigynia bend downwards at maturity; lower bracts at least 3 times as long as flower cluster .... ***C. retrorsa*** (see *C. rostrata*)

12b. Perigynia spread at maturity; lower bracts not more than twice as long as flower cluster .... ***C. utriculata***

NORTHERN BOG SEDGE • *Carex gynocrates*
YELLOW BOG SEDGE

GENERAL: Stems **single or a few together** from long, slender rhizome, 5–20 cm tall, stiff, smooth, sometimes curved, clothed with old leaves at base.

LEAVES: Clustered near base, about 0.5 mm wide, channelled, **stiff**, shorter than stems.

FLOWER CLUSTER: Single spike, bisexual, with male flowers at top, or sexes on separate plants.

PERIGYNIA: Egg-shaped, **plump**, brown, **glossy with fine lines**, short beaked, widely spreading at maturity, 2-edged; **2 stigmas**.

WHERE FOUND: Open mossy forests, bogs and fens; across our region.

NOTES: Also called *C. dioica* ssp. *gynocrates.* • **Rush-like sedge** (*C. scirpoidea*), also called single-spiked sedge, has single flower spikes that are either all male or all female, stout, scaly rhizomes, leaves to 3 mm wide, hairy perigynia and 3 stigmas. It grows in moist to dry meadows and turfy heaths, and on rocky slopes and shores throughout our region. **Capitate sedge** (*C. capitata*) is tufted (without long, creeping rhizomes) and has flower spikes that are male above and female below, never unisexual. It is a circumpolar species that favours fen habitats, and is found across the boreal forest. **Short-awned sedge** (*C. microglochin*), also called few-seeded sedge, has a single, bisexual, few-flowered spike with narrow, 4–6 mm long perigynia with 3 stigmas that turn down as they mature. It grows in loose tufts in fens, swamps and on peaty, calcium-rich shores across our region, but it is rare in the prairie provinces. **Blunt sedge** (*C. obtusata*) also has a single bisexual spike, but it has hairless, shiny brown, leathery, short-beaked, upright perigynia, purplish leaf sheaths and dark, cord-like rhizomes. It grows on dry bluffs, grassy slopes and gravelly, calcium-rich areas across our region. • The name *gynocrates* comes from the Greek *gyno*, 'female,' and *krates*, 'rule' or 'power,' perhaps because the female spikes are much more noticeable and therefore seem to be more prevalent than the inconspicuous male spikes.

BRISTLE-STALKED SEDGE • *Carex leptalea*

GENERAL: Tufted perennial from **thread-like creeping rhizomes**; stems very slender, 10–60 cm tall.

LEAVES: Shorter than flowering stems, **0.5–1.3 mm wide**.

FLOWER CLUSTER: Single, 1–10-flowered spike, 4–15 mm long, with male flowers at tip.

PERIGYNIA: 2–5 mm long, **greenish** with **distinctive, rounded tip**, hairless, **many veins**; scales much shorter than perigynia, greenish to brownish, translucent edges, green midvein; 3 short stigmas.

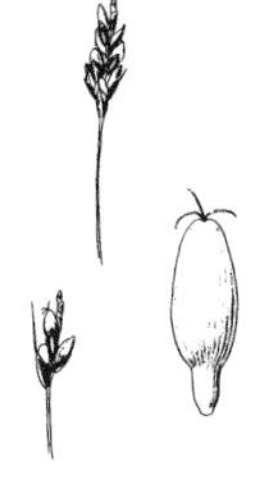

WHERE FOUND: Treed fens, moist woodlands and thickets; widespread across boreal forest.

NOTES: The species name *leptalea* comes from the Greek *lepto*, 'thin' or 'slender,' in reference to the very slender, weak stems and leaves.

SARTWELL'S SEDGE • *Carex sartwellii*

GENERAL: Stems single or a few together from well-developed, **dark brown or blackish creeping rhizomes**, 40–80 cm tall, brown at base, stiff, rough, **sharply triangular towards top**.

LEAVES: Flat, 2–4 mm wide, pointed at tip; **scattered along stem, not clustered at stem base**.

FLOWER CLUSTER: 6–20 densely clumped, stalkless spikes in narrow head 3–6 cm long; **male flowers at top of each spike**; bract at base of lowest spike shorter than flower cluster.

PERIGYNIA: Brown, elliptic-lance-shaped, 2.5–3.5 mm long, **finely veined**, thin-edged, toothed above middle, abruptly taper to short, 2-toothed beak; **scales** egg-shaped, pale brown, green midvein, whitish edges, **about as long as perigynia**; 2 stigmas.

WHERE FOUND: Shores, wet meadows, marshes, willow swamps and other wet places; locally common across boreal forest of prairie provinces from southern Manitoba north to upper Mackenzie valley.

NOTES: The seeds of many sedges are eaten and dispersed by waterfowl. Often they are the main food of these birds, along with grass and eel-grass (*Zostera*) seeds. After a heavy meal, a duck can contain as many as 8,000 seeds. The protective membrane that covers each seed slows digestion and allows some seeds to survive their stay in the gut. In this way sedges can be carried for great distances by migrating waterfowl.

HAY SEDGE • *Carex siccata*

GENERAL: Perennial, **1 to few stems** at intervals along long, brown-scaly, tough rhizomes; fertile stems slender, 15–80 cm tall, rough-edged towards top; dried leaves often clothe base; **sterile stems prominent with long, soft leaves**.

LEAVES: 1–3 mm wide, flat (occasionally somewhat channelled).

FLOWER CLUSTER: 2–4 cm long **head** of 2–12 closely clustered, stalkless spikes; lower spikes slightly spaced; end spike bisexual, middle spikes male, lower spikes female.

PERIGYNIA: 3–6 mm long, oblong-lance-shaped, **long, prominent beak**, 2-sided, veined, **wing-edged**; beak 2–3 mm long, 2-toothed; scales brownish with translucent edges and 3-veined, green centre, shorter than perigynia; 2 stigmas.

WHERE FOUND: Dry, usually sandy, open areas; widespread across southern boreal forest and parkland; north to about 62°N in Yukon and N.W.T.

NOTES: Also called *C. foenea*. • **Graceful sedge** (*C. praegracilis*), also called field sedge, is similar to hay sedge in size, but it grows in wetter habitats (grassy swales, moist meadows and fens), its leaf sheaths have a transparent membrane, its spikes tend to be more widely spaced, and its perigynia are not winged. Graceful sedge occurs along the southern fringe of the boreal forest in the prairie provinces. • Hay sedge is seldom abundant enough to make hay, but where it grows in sufficient quantity, it provides fair forage for cattle and horses. • Opinions vary about the palatability of graceful sedge, but most agree that it provides considerable winter grazing for livestock; in the North it is rated as excellent. In Montana it is known as 'nutgrass,' and cattle are said to grow fatter on it than on any grass.

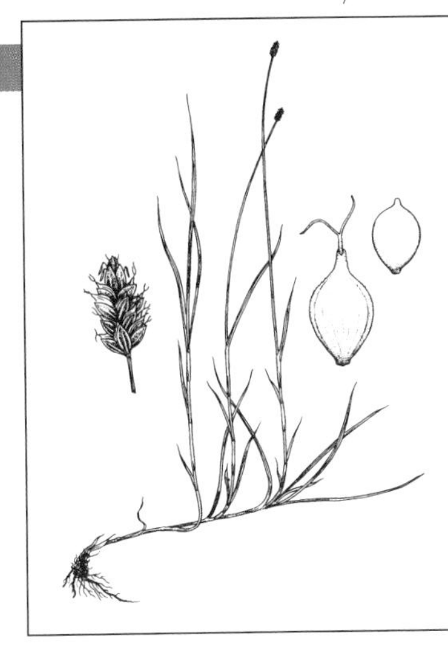

PROSTRATE SEDGE • *Carex chordorrhiza*

GENERAL: Stems scattered along creeping rhizomes up to 1 m long; fertile stems 10–30 cm tall, **leaning or reclining**, arise from end nodes, sterile stems from lower nodes; roots little developed.

LEAVES: 1–3 on fertile stems, 1–5 cm long, 1–2 mm wide, ridged, slightly shorter than stem; several much longer leaves on sterile stems.

FLOWER CLUSTER: 3–8 stalkless spikes in egg-shaped head, 5–15 mm long; **male flowers at top of each spike**.

PERIGYNIA: Egg-shaped, leathery, 2–3.5 mm long, **strongly many-veined**, thin-edged, short; smooth beak; **scales** straw-coloured to brownish, **broader than and as long as perigynia**; 2 stigmas.

WHERE FOUND: Fens, pond edges, lakeshores and shallow water; widespread across boreal forest; circumpolar.

NOTES: Although many botanical works list prostrate sedge as a sphagnum bog species, in our region it is more characteristic of intermediate to rich fens. When it occurs with sphagnum moss it is usually in wet poor fens. • Sedge rhizomes and stems have been used by many native groups as a material for weaving baskets. The stems of large species can reach over 1 m in height. Their spreading rhizomes are even longer, and these are soaked, peeled, and split into strands for weaving. • The species name *chordorrhiza* comes from the Latin *chorda*, 'a string' or 'a rope,' and the Greek *rhiza*, 'a root.' This describes the long, reclining stems of prostrate sedge, which often lie buried in peat or moss. *Chordorrhiza* has also been interpreted to mean 'scattering or dispersing roots,' an appropriate description for the long trailing growth habit.

TWO-STAMENED SEDGE • *Carex diandra*
LESSER PANICLED SEDGE

GENERAL: Tufted perennial, **in large loose clumps** from short rhizomes; stems erect, **30–80 cm tall**, stiff, with harsh, rough edges.

LEAVES: Erect to ascending, shorter than stem, greyish green, flat or channelled, narrow, **1–2.5 mm wide** (sometimes to 4 mm); **sheaths often red-dotted at top**, rough.

FLOWER CLUSTER: 6–10 spikes in **dense, dark brown head** 2–5 cm long; **male flowers at tip of each spike**.

PERIGYNIA: 2–2.7 mm long, dark brown, glossy, triangular-**egg-shaped, taper to long, flattened**, saw-toothed beak, soon become wide-spreading; scales light brown with translucent edges, wider than perigynia, often absent; 2 stigmas.

WHERE FOUND: Fens, ponds, lakeshores and wet meadows; widespread across boreal forest, north and west to Mackenzie delta and interior Alaska; circumpolar.

NOTES: Prairie sedge (*C. prairea*) resembles 2-stamened sedge, but its perigynia are concealed by the scales, and its upper leaf sheaths are copper-tinged at the mouth without red spots. It grows in wet habitats similar to those of 2-stamened sedge, but is more southern in distribution.

TWO-SEEDED SEDGE • *Carex disperma*
SOFT-LEAVED SEDGE

GENERAL: Loosely tufted perennial from long slender rhizomes; stems **very slender, weak**, 10–60 cm long, **usually arching**, clothed with old leaves at base.

LEAVES: Flat, 1–2 mm wide, soft, mostly shorter than stem, light green.

FLOWER CLUSTER: Narrow, 3–5 cm long group of **2–5 small, widely spaced**, stalkless spikes; spikes 3–6 mm long, greenish; 1–2 **male flowers at top**; 1–6 female flowers at base.

C. disperma

PERIGYNIA: Plump, greenish to brownish, shiny at maturity, 2–3 mm long, with minute beak, faint lines, 2-sided; scales usually shorter than perigynia, whitish translucent with green midvein; 2 stigmas.

WHERE FOUND: Moist woods, wetlands, streambanks and lake edges; widespread across our region; circumpolar.

NOTES: Three-seeded sedge (*C. trisperma*) has 2–3 spikes with the female flowers above the male. The lowermost spike is in the axil of a slender, 2–4 cm long bract. It is generally rare and grows in wet forests, swamps, bogs and fens, scattered primarily east of the Rockies. • Two-seeded sedge was given the species name *disperma*, from the Greek *di*, 'two,' and *sperma*, 'seeds,' because the mature spikelets usually have 2 plump perigynia that look like 2 yellowish green seeds. Similarly, *trisperma* means '3-seeded,' in reference to the 3 rounded, 'seed-like' perigynia usually found in each spikelet.

C. trisperma

THIN-FLOWERED SEDGE • *Carex tenuiflora*
SPARSE-LEAVED SEDGE

GENERAL: Loosely tufted perennial from long, slender, yellowish brown rhizomes; stems very slender but firm, arching, 20–60 cm tall, roughened beneath head, brownish and clothed with old leaves at base.

LEAVES: 1–2 mm wide, soft, some almost as long as stem, **greyish green**.

FLOWER CLUSTER: 5–15 mm long, erect head of **2–4 closely bunched**, stalkless, **whitish green** spikes; female flowers at spike tips.

PERIGYNIA: Oval, 2–3.5 mm long, beakless, **pale green with a few small veins**, almost beakless; scales rounded, whitish, green midvein, almost hide perigynia; 2 stigmas.

WHERE FOUND: Fens, swamps and seepage areas; widely scattered across boreal forest, north to treeline; circumpolar.

NOTES: Rye-grass sedge (*C. loliacea*) resembles thin-flowered sedge, but its spikes are more widely spaced and its perigynia have many prominent veins. Rye-grass sedge grows in wet forests, swamps, seepage areas and on streambanks and is most common in the boreal forest from Lake Athabasca to interior Alaska. • Thin-flowered sedge can be recognized by the 3 (usually) small spikes bunched at the tip of its slender stem. • The species name is from the Latin *tenui*, 'slender' or 'thin,' and *flos*, 'a flower.' Although the flowers and flowering heads themselves do not appear to be particularly thin, the long stalks which support them are very slender.

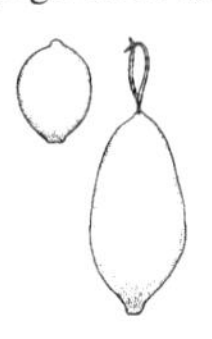

SHORT SEDGE • *Carex canescens*
GREY SEDGE

GENERAL: Densely tufted perennial from short, black rhizomes; stems erect to spreading, 10–60 cm tall, brownish at base and clothed with old leaves.

LEAVES: At stem base, usually shorter than stems, **soft, flat, 2–4 mm wide, bluish green** with whitish bloom.

FLOWER CLUSTER: 4–8 egg-shaped, greyish or silvery green spikes on erect or arching stalk; **lower spikes often separated**; **male flowers at base** of each spike and female above.

PERIGYNIA: Egg-shaped-oblong, 2–3 mm long, yellowish green to whitish brown with translucent dots or pits, **very short-beaked**, 15–30 per spike; scales straw-coloured, shorter than perigynia; 2 stigmas.

WHERE FOUND: Fens, swamps, wet meadows, streambanks and lakeshores; widespread across boreal forest; circumpolar.

C. canescens

NOTES: Also called *C. curta* or *C. lapponica*. • **Brownish sedge** (*C. brunnescens*) is similar, but it has smaller spikes with fewer, loosely spreading perigynia with sharp, 2-toothed beaks, and it does not have blue-green leaves. Brownish sedge is common across our region in peatlands and wet meadows. • *Canescens* means 'becoming greyish' and refers to the colour of the leaves. *Brunnescens* means 'becoming deep brown,' and probably refers to the perigynia of brownish sedge, though these are usually pale brown because of the pattern of translucent whitish dots or pits on their surfaces.

C. brunnescens

INLAND SEDGE • *Carex interior*

GENERAL: Densely tufted perennial from short, dark rhizomes; stems slender, wiry, 15–50 cm high.

LEAVES: Slender, 1–3 mm wide, thin, shorter than flowering stems.

FLOWER CLUSTER: 3–4 small, stalkless, few-flowered spikes; male flowers at base of uppermost spike(s).

PERIGYNIA: More or less pear-shaped, plump, firm, longer than yellowish brown scales, widely spreading at maturity, 2-sided; **beak less than $^1/_3$ length of perigynium**; 2 stigmas.

WHERE FOUND: Fens, seepage areas and wet meadows; widespread across southern boreal forest, north to southwestern N.W.T. and southern Yukon.

NOTES: Also called *C. muricata*. • While inland sedge has distinct, well-separated spikes, **narrow sedge** (*C. arcta*) has its spikes clumped in an oblong head. Narrow sedge is scattered across our region in marshes, fens and swamps. • The spreading perigynia of inland sedge give the spikes a distinctive star-like appearance. • Inland sedge was first described at Penn Yan, New York, but has since been found from coast to coast in Canada. The name *interior* is Latin for 'inner,' and refers to the inland distribution of this species, away from the coast.

DEWEY'S SEDGE • *Carex deweyana*

GENERAL: Loosely tufted perennial from short rhizomes; **stems weak, spreading**, 20–80 cm tall, roughened beneath head, clothed at base with old leaves.

LEAVES: Flat, soft, thin (2–4 mm wide), much shorter than flowering stems.

FLOWER CLUSTER: 2–7 stalkless spikes in **loose, crooked head, 2–6 cm long**; end spike has **female flowers at top**; side spikes usually all female, spaced apart from each other; lowest bract 3 cm long, slender.

PERIGYNIA: Pale green, thin, papery, 4–5 mm long, lance-elliptic to egg-shaped, **rounded and spongy at base**; taper to slender, **flattened** beak, with **sharp**, minutely saw-toothed **edges**; scales thin, whitish to yellowish brown; 2 stigmas.

WHERE FOUND: Moist woods, forest openings and meadows; widespread across southern boreal forest, north and west to southwestern N.W.T. and west-central Yukon.

NOTES: Sedges can be used for dying. They produce various shades of yellow and gold when different mordants are used: alum - golden tan; chrome - dark gold; copper - medium tan; tin - gold; iron - olive; no mordant - blah beige; with very slight fading by light. Young plants produce the brightest colours. • Dewey's sedge was named in honour of Rev. Chester Dewey, a professor and student of *Carex* at the University of Rochester, New York.

BRONZE SEDGE • *Carex aenea*
SILVERY-FLOWERED SEDGE

GENERAL: Densely tufted perennial from short rhizomes; stems **slender**, wiry, **40–100 cm tall**, smooth, often arching.

LEAVES: Shorter than stem, soft, flat, 2–4 mm wide.

FLOWER CLUSTER: 3–10 spikes, stalkless, 6–8 mm long, **well-separated (at least below)** in **curved**, 5–6 cm long cluster; male flowers at base of each spike.

PERIGYNIA: 4–5 mm long, dull green, **become bronze, egg-shaped**, flattened, **wing-edged**, taper to long, fine-toothed beak; scales brownish, pointed, **conceal perigynia**; 2 stigmas.

WHERE FOUND: Open woods, meadows, and clearings; widespread across boreal forest, north and west to Great Bear River and interior Alaska.

NOTES: Included in *C. argyrantha* by some; in *C. praticola* by others. • **Meadow sedge** (*C. praticola*) (inset photo) closely resembles bronze sedge, but it is usually smaller and more slender and has narrower perigynia with smooth, not flattened beaks in silvery-green spikes. Meadow sedge is common across our region in dry to moist meadows and clearings. • The genus *Carex* has more species than any other genus of plants in the boreal forest. Well over 500 *Carex* species grow in North America. All are wind-pollinated and many people are sensitive to the large amounts of pollen they produce. • The name *aenea* means 'brassy green' or 'bronze,' in reference to the colour of the fruiting heads of this sedge.

C. aenea

C. praticola

BEBB'S SEDGE • *Carex bebbii*

GENERAL: Densely tufted from very short rhizomes; stems erect, slender, 20–80 cm tall, **roughened near top**.

LEAVES: Light green to yellowish green, flat, soft, 2–4 mm wide, shorter than stem.

FLOWER CLUSTER: 1–2.5 cm long, compact head of 3–12 stalkless spikes, greenish to straw-coloured, become brownish; **male flowers at base of each spike**; lower 1 or 2 spikes have inconspicuous bracts.

C. bebbii

PERIGYNIA: Pale green to straw-coloured, **narrowly egg-shaped**, 2.5–3.5 mm long, **about half as wide as long**, narrowly wing-edged, taper to flat, toothed beak; **scales** straw-coloured to brownish, **nearly as wide as, but shorter than, perigynia**; 2 stigmas.

WHERE FOUND: Marshes, ditches and wet meadows; in boreal forest and northern parkland, north and west to southern N.W.T. and northwest B.C.; most common in eastern part of our region.

C. crawfordii

NOTES: Crawford's sedge (*C. crawfordii*) is similar to Bebb's sedge, but it has narrowly lance-shaped, not egg-shaped, perigynia. It is found in drier habitats over a similar range, but it is most common in the southern part of our region and in Alaska. • The name *bebbii* honours Michael Schuck Bebb, a 19th-century botanist, best known for his study of willows. *Crawfordii* commemorates Ethan Allan Crawford, an early settler in the White Mountains of New England.

GOLDEN SEDGE • *Carex aurea*

GENERAL: Loosely tufted perennial from long, slender rhizomes; stems slender, usually 5–20 cm tall, light brown at base.

LEAVES: From near base, 1–4 mm wide, more or less flat.

FLOWER CLUSTER: 5–10 cm long; 1 narrow, 5–15 mm long male spike at stem tip (sometimes with a few female flowers at tip); 3–5 female spikes below, oblong, 5–15 mm long, 4–20-flowered, lower ones spreading on long stalks; bracts long, leaf-like, overtop spikes.

PERIGYNIA: Rounded egg-shaped, **beakless**, 2–3 mm long, **coarsely ribbed**, plump, **golden to orange** when mature; scales pale to reddish brown, with wide, green midvein, narrower than perigynia; 2 stigmas.

WHERE FOUND: Streambanks and wet meadows; widespread across our region.

NOTES: Included in *C. bicolor* by some. • A closely related species, **Garber's sedge** (*C. garberi*), is found on calcium-rich meadows and shores across our region. It resembles golden sedge, but its uppermost spike usually has **female flowers at the tip** and male flowers below. Many taxonomists consider Garber's sedge and golden sedge to be the same species. • Sedge seeds are tiny achenes, each enclosed in a sac-like membrane of fused bracts, called a perigynium. The perigynia of golden sedge are very distinctive. They become fleshy and nearly spherical—golden sedge is sometimes called 'grape sedge'—and, most noticeably, their colour can range from blue-green to yellow to orange at maturity, rather than the different shades of brown produced by most sedges. The spikes often lie on the ground when ripe. • The species name *aurea* means 'golden.'

BEAUTIFUL SEDGE • *Carex concinna*

GENERAL: Small, loosely tufted perennial from slender, scaly rhizomes; stems slender, **smooth**, arching, 5–20 cm tall, with brownish, somewhat shreddy sheaths at base.

LEAVES: At stem base, wide-spreading, **1–3 mm wide**, soft, much shorter than stem.

FLOWER CLUSTER: Crowded, 1–3 cm long; 1 male spike at end, **3–6 mm long**; 2–3 female spikes, few-flowered, erect, 5–10 mm long; bracts reduced to green, bladeless sheaths.

PERIGYNIA: Hairy, plump, 3-sided, 2.5–3 mm long, greenish to yellowish white, short-beaked; scales broad, dark reddish brown with whitish edges, about 1.5 mm long; **3 short, dark stigmas**.

WHERE FOUND: Damp, open woods, meadows and riverbanks; widespread across boreal forest.

NOTES: Richardson's sedge (*C. richardsonii*) also has hairy perigynia but its flower cluster is relatively loose (not compact) and the single, stalked male spike is longer (1–2.5 cm long) than that of beautiful sedge. It grows in calcium-rich meadows and in dry open forests north to the southern Northwest Territories.

C. richardsonii

BENT SEDGE • *Carex deflexa*

GENERAL: Loosely tufted from short rhizomes, **purplish red at base**; stems slender, 10–20 cm tall.

LEAVES: Pale green, 1–3 mm wide, soft, usually flat, **often spreading**.

FLOWER CLUSTER: End spike male, 2–5 mm long, **inconspicuous**; **2–3 female side spikes**, close together, stalkless, **often extend beyond male spike**; lowest bract leaf-like, longer than flower cluster.

PERIGYNIA: Green, stalked, elliptic-egg-shaped, 2–3 mm long, **short-hairy**, with short, flat beak; **scales** brown, purplish brown or reddish, pointed, with green midvein and whitish edges, **shorter than perigynia**; 3 stigmas.

WHERE FOUND: Dry open woodland and clearings; often on acidic soils; widespread but rarely abundant across boreal forest, north and west to southern N.W.T. and interior Alaska; rare in Alberta.

NOTES: There are 3 other small dry-site sedges with hairy perigynia. **Peck's sedge** (*C. nigromarginata* var. *elliptica*, also called *C. peckii*) is generally taller than bent sedge and has 3–4 mm long perigynia and scales with white rather than purple sides. It grows scattered across our region in open woods, mainly in the southern boreal forest and parkland. Both **Ross' sedge** (*C. rossii*, also called *C. deflexa* var. *rossii*) and **umbellate sedge** (*C. umbellata*, also called *C. abdita* or *C. rugosperma*) usually have additional female spikes on long stalks from near the base of the stem. Ross' sedge has stouter rhizomes and 3–4.5 mm long, longer-beaked perigynia. It occurs throughout the boreal forest. Umbellate sedge has relatively long-stalked male spikes that often appear separate from the female spikes, and its lowermost bracts are scale-like. It ranges across the southern boreal forest of the prairie provinces. The unrelated **spreading arctic sedge** (*C. supina* ssp. *spaniocarpa*) occurs in dry, sandy-gravelly habitats from northern Manitoba to Alaska. It is only 10–15 cm tall, and has shiny, leathery, smooth (not hairy) perigynia.

C. rossii *C. peckii*

NORWAY SEDGE • *Carex media*

GENERAL: Loosely tufted from short rhizomes, 20–60 cm tall; **stems** slender, erect or arched, **sharply triangular, rough**.

LEAVES: Soft, flat, 1.5–3 mm wide, **rough**, bunched near base, much shorter than stem.

FLOWER CLUSTER: Usually 3 closely-spaced, stalkless or short-stalked, erect or ascending spikes about 1 cm long; **end spike has male flowers at base; side spikes wholly female**; leaf-like bract at base of lowest spike, shorter to slightly longer than flower cluster.

PERIGYNIA: Greenish, elliptic to egg-shaped, swollen, faintly veined, hairless, **granular roughened**, 2–3.5 mm long, short beak; **scales** broadly egg-shaped, purple-black, whitish edges, **much shorter than to almost as long as perigynia and about as wide**; 3 stigmas.

WHERE FOUND: Streambanks, lakeshores, seepage areas and moist woods; widespread across boreal forest; circumpolar.

NOTES: Also called *C. norvegica.* • **Buxbaum's sedge** or **brown sedge** (*C. buxbaumii*), also called club sedge, has essentially stalkless spikes that are so close together they often look like a single spike. The scales of the female flowers have a long point that often extends past the tip of the perigynia. Buxbaum's sedge is a more or less circumpolar species that favours fen habitats across the boreal forest, north to Great Bear Lake and south-central Alaska. • Dry sedge leaves have been used for centuries as stuffing for boots and mittens to protect hands and feet from the cold in winter and to cushion the feet on long journeys in summer.

HAIR-LIKE SEDGE • *Carex capillaris*

GENERAL: Densely **tufted** perennial from short rhizomes; stems very slender, usually 10–50 cm tall, with dark, somewhat shreddy sheaths at base.

LEAVES: Mostly at stem base, shorter than stems, flat, thin, 0.5–2 mm wide.

FLOWER CLUSTER: 5–15 cm long; male spike at tip, 4–8 mm long (sometimes with a few female flowers); 2–4 female spikes, 4–17 mm long, **drooping** on 1–4 cm **long, slender stalks**; lowest bract leaf-like, with **long, tubular sheath**.

PERIGYNIA: 2–4 mm long, greenish brown, glossy when mature, lance-egg-shaped, 3-sided, taper to 1 mm long beak, longer but narrower than light brown scales; 3 stigmas.

WHERE FOUND: Moist, open forests, wetlands, and stream and pond edges, usually on calcium-rich soils; widespread across boreal forest, north through arctic islands; circumpolar.

NOTES: Raymond's sedge (*C. atratiformis* ssp. *raymondii*, also called *C. raymondii*) is a robust plant that grows to 70 cm tall and has broader (4 mm) leaves and longer (to 2 cm), thicker female spikes. It occurs sporadically across our region, north and west to Great Bear Lake and interior Alaska. • Hair-like sedge and sheathed sedge (p. 240) have a similar distribution and habitat, but sheathed sedge has larger perigynia (more than 3 mm long) and its spikes are more erect. • The flowering heads produce many small fruits, and hair-like sedge is considered a reasonable source of seed for small birds. • The name *capillaris* means 'hair-like,' and it refers to the thread-like branches of the drooping flower cluster.

SHEATHED SEDGE • *Carex vaginata*

GENERAL: Stems single or few together from long, slender creeping rhizomes, 10–60 cm tall, **smooth, irregularly bent**.

LEAVES: Soft, flat, hairless, 2–4 mm wide, mostly at stem base.

FLOWER CLUSTER: End spike male, 1–2 cm long, long-stalked; **2–3 female side spikes** (sometimes with male flowers at tip), 1–2.5 cm long, ascending, 5–12-flowered, **widely spaced**, slender stalks; **bracts** at base of female spikes have **conspicuous sheaths**, 1–2 cm long, **enlarged upwards**, with flat blade, shorter than spike.

PERIGYNIA: Yellowish green to light brown, elliptic to elliptic-egg-shaped, 3–5 mm long, faintly veined, smooth, beak about 1 mm long; **scales** purplish brown, **shorter than perigynia**; 3 stigmas.

WHERE FOUND: Bogs and wet coniferous woods; often with *Sphagnum* moss; common and widespread across boreal forest, north to low Arctic; circumpolar.

NOTES: This common sedge is said to have moderate value as winter forage for wildlife. • The name *vaginata* is derived from the Latin *vagina*, 'a sheath,' and refers to the conspicuous 1–2 cm-long sheaths formed by the leafy bracts immediately below the spikes.

GREEN SEDGE • *Carex viridula*

GENERAL: More or less densely tufted, 10-40 cm tall; stems stiff, erect, smooth, **yellowish green**.

LEAVES: Yellowish green, flat to channelled, 1-3 mm wide, **equalling or longer than stem**.

FLOWER CLUSTER: End spike male (sometimes with a few female flowers at base), stalkless or short-stalked; **2-6 female side spikes, crowded**, pale green; upper spikes stalkless or nearly so, **barely shorter than male spike**; lower spikes remote, short-stalked; **leaf-like bracts much longer than flower cluster**.

PERYGYNIA: Yellowish green turning brown, egg-shaped, 2-3 mm long, **crowded, spreading, ribbed**, abruptly tapered to 2-toothed beak; **scales** egg-shaped, pale brown with whitish edges, **much shorter than perigynia and about as wide**; 3 stigmas.

WHERE FOUND: Wet areas, shores and springy places; particularly common on calcium-rich or slightly alkaline soils; across our region, north and west to Great Bear Lake and interior Alaska; circumpolar.

NOTES: Also called *C. oederi* ssp. *viridula*. • The species name *viridula*, from the Latin *viridis*, 'green,' refers to the distinctive yellowish green colour of this plant. • The seeds of many sedges are distributed by water; they are bouyant and able to floating great distances.

MUD SEDGE • *Carex limosa*
SHORE SEDGE

GENERAL: Perennial from long, slender, scaly rhizomes, **yellowish, felt-covered roots**; stems single or few together, slender, 20–40 cm tall, reddish at base.

LEAVES: Few, much shorter than stem, with bluish white bloom, deeply channelled, 1–3 mm wide.

FLOWER CLUSTER: Small group of 2–4 spikes on slender, nodding, 1.5–5 cm long stalks; male spike at tip, **1–3 cm long**; 1–3 **nodding** female spikes, short (1–2 cm long), cylindrical; lowest bract leaf-like, 2–7 cm long.

PERIGYNIA: Bluish green to straw-coloured, 2.5–4 mm long, 8–10-veined, broadly egg-shaped with minute beak; **scales almost hide perigynia**, brownish, purplish or straw-coloured, with **broad tip**; 3 stigmas.

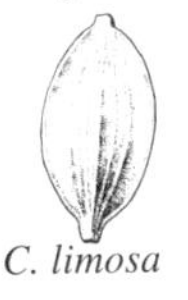
C. limosa

WHERE FOUND: Fens; widespread across boreal forest; circumpolar.

NOTES: Bog sedge (*C. paupercula*, also called *C. magellanica*) also has yellowish woolly roots, but it is densely tufted, and has shorter male spikes (less than 12 mm) and narrow, long-pointed scales. **Livid sedge** (*C. livida*), also called pale sedge, does not have a yellowish felt on its roots, and its female spikes are erect and loosely flowered. Both bog sedge and livid sedge have bluish green leaves and perigynia (livid sedge more so) and are common across our region. However, livid sedge is found mostly in open (rich) fens, while bog sedge is most common in treed (poor) fens. • Mud sedge has high nutritional value and is grazed by horses, cattle, sheep and wildlife.

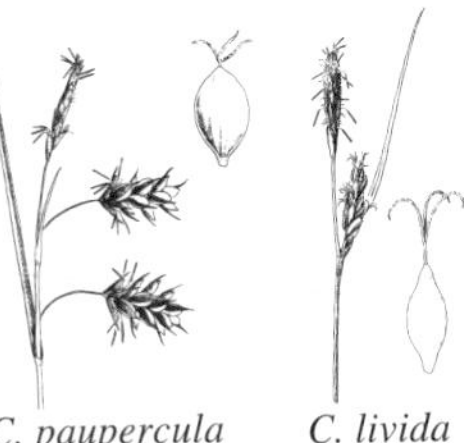
C. paupercula *C. livida*

HAIRY-FRUITED SEDGE • *Carex lasiocarpa*
SLENDER SEDGE

GENERAL: Loosely tufted perennial from long-creeping, tough, scaly rhizomes; stems **30–100 cm tall, dark reddish at base, with shredding lower sheaths**.

LEAVES: Very long and tapering, often longer than stems, **1–2 mm wide**, rolled inwards along edges, with minute, whitish, knot-like crosswalls between veins.

FLOWER CLUSTER: 8–15 cm long group of 3–6 **stalkless**, erect, widely spaced spikes; 1–3 (usually 2) male spikes at tip; 1–3 female spikes, 1–3 cm long; lowest bract 6–15 cm long, leaf-like, overtops spikes.

PERIGYNIA: Oblong-egg-shaped, 3–5 mm long, **inflated**, dull brown, **densely hairy**; beak 0.6–1 mm long, with **2 short, upright teeth**; scales reddish brown or purplish, narrower than perigynia, pointed; **3 stigmas**.

WHERE FOUND: Fens and edges of lakes and ponds; widespread across boreal forest, north and west to Great Bear Lake and disjunct in southern Alaska; circumpolar.

C. lasiocarpa *C. lanuginosa*

NOTES: Woolly sedge (*C. lanuginosa*, also called *C. pellita*) (inset photo) grows in marshes and along shores and also has hairy perigynia. However, it has wider (to 5 mm), flatter leaves, and shorter (about 3 mm long) globular perigynia with 2 long, somewhat spreading teeth. Woolly sedge has not been found north of about 56°N in our region. **Sand sedge** (*C. houghtoniana*) has larger (5–7 mm), finely hairy perigynia and grows in dry, sandy or gravelly habitats in the boreal forest of the prairie provinces. • Although woolly sedge has good forage value and is very palatable to livestock, it is seldom grazed by deer.

BEAKED SEDGE • *Carex utriculata*

GENERAL: Tall, coarse sedge, in clumps or mats, spreads by short rhizomes and long, creeping runners; **stems thick, spongy**, 0.5–1 m tall.

LEAVES: Yellowish green, rather thick, more or less **channelled** to flat, 4–10 mm wide, well-developed, often overtop flower clusters; sheaths have conspicuous crosswalls between parallel veins.

FLOWER CLUSTER: 10–30 cm long; 2–4 overlapping male spikes at tip; 2–5 well-separated, **cylindrical female spikes below**, 4–10 cm long, short-stalked or stalkless; lowermost bract leaf-like, overtops flower cluster.

PERIGYNIA: Egg-shaped, inflated, glossy, **yellowish green to straw-coloured**, veined, 3–8 mm long, ascending to spreading at maturity; **2 teeth**, 1–2 mm long, at tip; scales light to dark brown, narrower than perigynia; 3 stigmas.

WHERE FOUND: Common in fens, swamps, marshes and along shorelines; across our region; circumpolar.

C. utriculata *C. retrorsa*

NOTES: Beaked sedge was previously called ***C. rostrata***, but that name has recently been applied to a rare species of floating fens in the southern part of our region. *C. rostrata* is similar to beaked sedge but has narrower (1.5–4 mm wide) leaves that are strongly bluish green and roughened (hand lens) on the upper surface. • **Turned sedge** (*C. retrorsa*) resembles both beaked sedge and awned sedge (below), but it has closely packed female spikes and crowded perigynia that bend downwards. It grows in marshes north to the southern Northwest Territories. • Beaked sedge provides good forage in spring and it is often a major component of 'beaver hay'—fodder from pastures that are too wet for cultivation. At that time of year, it is often standing in water, but cattle will wade deep into the water to reach it. As the season progresses, the leaves become tough and unpalatable and by autumn elk are about the only animals to eat it.

AWNED SEDGE • *Carex atherodes*

GENERAL: Large, tufted perennial from long, slender rhizomes; stems stout, 30–120 cm tall, often reddish-tinged, with **shredding and eventually web-like sheaths** at base.

LEAVES: Few to several, **usually shorter than stems, dull green,** hairy, **flat**, 3–12 mm wide, with whitish, knot-like crosswalls between veins; **sheaths hairy**.

FLOWER CLUSTER: 15–25 cm long, with 2–10 erect, cylindrical spikes; 2–6 erect male spikes at tip; 2–4 female spikes, 2–12 cm long, widely separated, short-stalked or stalkless, erect; lowest bract leaf-like, sheathing, longer than spikes.

PERIGYNIA: 7–12 mm long, somewhat leathery, **yellowish green to light brown**, broadly lance-shaped, **many-ribbed**, taper to slender beak with **2 spreading, 1.2–3 mm long points**, ascending to spreading at maturity; scales bristle-tipped, reddish brown, narrower than, but about as long as, perigynia; 3 stigmas.

WHERE FOUND: Wet meadows, lakeshores, fens, ponds and marshes; widespread across our region, north and west to Great Slave Lake and interior Alaska; circumpolar.

NOTES: Lakeshore sedge (*C. lacustris*) is another large sedge with firm, many-ribbed perigynia. However, its perigynium-beak teeth are only 0.5 mm long and its leaves are hairless. Lakeshore sedge grows in marshes, treed fens and along lakeshores in the boreal forest of the prairie provinces. • Awned sedge is similar to beaked sedge (above) in appearance and habitat, but beaked sedge lacks the hairy leaf sheaths. • The young shoots of sedges are edible once they have been cooked. Young stems, especially the fleshy lower parts can be chewed raw, like celery.

WATER SEDGE • *Carex aquatilis*

GENERAL: Densely tufted perennial from **thick, scaly, cord-like rhizomes**; stems **rather slender**, 20–100 cm tall, reddish at base and surrounded by long leaf sheaths and old leaves.

LEAVES: Many, as long as or shorter than stems, often with **bluish white cast**, 2–5 mm wide.

FLOWER CLUSTER: 5–25 cm long, with 3–6 cylindrical spikes; 1–2 male spikes at tip; 2–4 **erect**, linear to oblong, female spikes below, 1–4 cm long, **short-stalked or stalkless**; lowest bract leaf-like, 7–35 cm long, extends **beyond uppermost spike**.

PERIGYNIA: 2.5–3 mm long, light green, often red-dotted, hairless, elliptic, **flattened, abruptly short-beaked**; scales blackish with yellowish to greenish midvein, narrower and **shorter than perigynia**; 2 stigmas.

WHERE FOUND: Fens, wet meadows, streambanks, shallow ponds and ditches; widespread and common across our region, north to Arctic coast; circumpolar.

NOTES: Water sedge is one of the most common and widespread plants of the boreal forest. • **Lens-fruited sedge** (*C. lenticularis*) looks much like water sedge, but it has finely veined perigynia. It grows on river flats, streambanks and muddy shores of ponds and lakes in the boreal forest, north and west to Great Bear Lake. • The Chipewyan used the roots of water sedge in a medicine taken by women whose menstrual period was delayed. Alaskan native peoples ate the succulent, sweet, pinkish stem bases raw. • In the Arctic, water sedge provides excellent forage for both domestic animals and wildlife. Its nutritional value is said to be equal to that of clover, with even higher levels of protein in some cases. It is very palatable to cattle and horses, and it is often completely grazed in wet meadows. • The seeds of many sedges are distributed by water; they are buoyant and able to float great distances.

WHITE BEAK-RUSH • *Rhynchospora alba*

GENERAL: Densely **tufted** perennial from short rhizomes; stems triangular, slender, usually 15–30 cm tall, with straw-coloured leaf sheaths at base.

LEAVES: Linear, 0.3–2.5 mm wide, scale-like, shorter than stems, channelled to flat, greyish green, often smaller low on stem.

FLOWERS: Several to many small, **2-flowered spikelets** in 1–4 stalked, head-like, 5–15 mm wide clusters; end cluster largest, with bract about as long as cluster; **scales whitish to pale brown**, egg-shaped.

FRUITS: Achenes, brownish green, lens-shaped, 1.5–2.5 mm long, taper to **narrowly triangular tubercle**; **10–12 stiff**, barbed **bristles** surround base.

WHERE FOUND: Fens and wet, peaty soil; rare and widely scattered across boreal forest, north and west to Great Slave Lake and southern Alaska; locally common in southeastern Manitoba; not yet found in Alberta; circumpolar, with gaps.

NOTES: The genus name *Rhynchospora* comes from the Greek *rhynchos*, 'a snout,' and *spora*, 'a seed,' and refers to the beaked achenes. The species name *alba* means 'white,' and refers to the whitish scales of the flowers.

Key to the Cotton-grasses (*Eriophorum*)

1a. 2 or more spikelets, with 1 or more leaf-like bracts below them 2

2a. Leaves narrow, long, channelled or triangular in cross-section .... ***Eriophorum gracile*** (see *E. angustifolium*)

2b. Leaves flattened, except perhaps at tips 3

3a. Midrib of scales extends to tip; sheaths green .... ***E. viridi-carinatum*** (see *E. angustifolium*)

3b. Midrib of scales not to tip; sheaths dark reddish-brown .... ***E. angustifolium***

1b. 1 spikelet, without leaf-like bracts below it 4

4a. Plants spread by stolons, stems usually single 5

5a. Bristles white; head round .... ***E. scheuchzeri*** (see *E. vaginatum* ssp. *vaginatum*)

5b. Bristles rust-coloured; head oblong .... ***E. chamissonis*** (see *E. vaginatum* ssp. *vaginatum*)

4b. Plants without stolons, densely tufted 6

6a. Scales have broad, pale edges; base scales bend downwards .... ***E. vaginatum* ssp. *vaginatum***

6b. Scales lack pale edges; base scales do not bend downwards 7

7a. Bristles bright white; plants to 30 cm tall .... ***E. callitrix*** (see *E. vaginatum* ssp. *vaginatum*)

7b. Bristles dull white to creamy; plants 30–60 cm tall .... ***E. brachyantherum*** (see *E. vaginatum* ssp. *vaginatum*)

TALL COTTON-GRASS • *Eriophorum angustifolium*
NARROW-LEAVED COTTON-GRASS

GENERAL: Perennial from stout, **spreading rhizomes**; **1 to few** stems, 20–80 cm tall; dark reddish brown sheaths clothe base.

LEAVES: Most at stem base, 2 or more on stem, flat below middle, triangular and channelled towards tip, **3–6 mm wide**, rough-edged with minute teeth.

FLOWER CLUSTER: 2–10 fluffy white spikelets with 2–3 **leafy**, often purplish **bracts** at base; spikelets drooping, with many, **white, silky, 2–3.5 cm long bristles**; scales broadly lance-shaped, dark with translucent edges, **slender midvein does not reach tip**; anthers 2.5–5 mm long.

FRUITS: Dark brown to black, 3-sided achenes, 2–2.5 mm long, each surrounded by white bristles from base.

WHERE FOUND: Fens, shallow ponds and lake edges; widespread across boreal forest, north past treeline to southern arctic islands; circumpolar.

NOTES: Also called *E. polystachion.* • **Slender cotton-grass** (*E. gracile*) has several smaller heads per stem, narrower (1–2 mm wide), 3-sided leaves and only 1 leaf-like bract below the flower cluster. **Thin-leaved cotton-grass** (*E. viridi-carinatum*) has flat, bright green leaves and 2 or 3 bracts, and its flower scales have a prominent green midvein that extends all the way to the tip. Both thin-leaved cotton-grass and slender cotton-grass are widely scattered across the boreal forest in fens, but they are seldom found north of Great Slave Lake. • At one time the fluffy seed heads of tall cotton-grass were collected as a substitute for cotton, known as 'arctic wool.' Unfortunately, the fibres were too brittle to withstand much twisting, and the industry was short-lived. • In the 1600s, cotton-grass was used in northern Europe to treat diarrhea and relieve coughing. However, it was taken with caution, as it was likely to cause a headache and sleepiness.

E. angustifolium

E. gracile

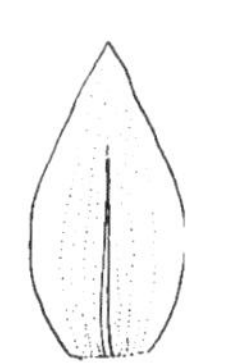

E. angustifolium scale

E. viridi-carinatum scale

SHEATHED COTTON-GRASS • *Eriophorum vaginatum* ssp. *vaginatum*

GENERAL: Densely tufted perennial in large, compact clumps; stems stiff, 3-sided, 20–60 cm tall.

LEAVES: Long, stiff, very slender at stem base; stem leaves consist of 1–2 **inflated, bladeless sheaths, usually below middle of stem**.

FLOWER CLUSTER: Single, **oblong spike**, 1–1.5 cm long; **anthers 2.5–3.5 mm long**.

FRUITS: Broadly egg-shaped, short-beaked achenes, 2.5–3.5 mm long, each surrounded by bright white, 2–2.5 cm long bristles; **scales** pointed, **lead grey to blackish, with pale edges, spreading or bent backwards at maturity**.

WHERE FOUND: Relatively dry bogs; frequently associated with *Sphagnum* moss; widespread across our region north to arctic islands; circumpolar.

NOTES: Our sheathed cotton-grass is replaced in parts of Manitoba and eastward by **ssp. *spissum***, which differs from ssp. *vaginatum* in having more inflated upper sheaths, spherical spikes with blackish scales, shorter anthers (1–2 mm) and narrower achenes. • There are 4 other common, single headed cotton-grasses in our region. **Close-sheathed cotton-grass** (*E. brachyantherum*) and **beautiful cotton-grass** (*E. callitrix*) are densely tufted like sheathed cotton-grass, but neither has pale edges on its scales, and only their base scales bend downwards. Close-sheathed cotton-grass has slender stems, is 30–60 cm tall, has dull-white to creamy bristles, and the uppermost sheath is above the middle of the stem. It is less common than sheathed cotton-grass, and grows in less acidic bogs and fens throughout our region. Beautiful cotton-grass has stout stems 20–30 cm tall, small anthers (0.6–1 mm long) and shining white bristles. It grows in fens and calcium-rich seeps and meadows across the northern part of our region. **Russet cotton-grass** (*E. chamissonis*, included in *E. russeolum* by some) has stolons, mostly single stems, 1.5–3 mm long anthers and mostly rust-coloured bristles in an oblong head. It is widespread throughout our region in bogs and poor fens. **Scheuchzer's cotton-grass** (*E. scheuchzeri*) is essentially a round-headed, white-bristled, tiny-anthered (0.5–1 mm) version of russet cotton-grass, and is found to the north and along the Rockies. • The seed fluff of cotton-grasses has been collected in great quantities in the past for use in stuffing pillows. It has also been made into paper and wicks for candles and lamps. • Sheep enjoy eating young cotton-grass plants, but generally they are considered of no agricultural value. • The genus name is derived from the Greek *erion*, 'wool,' and *phoros*, 'bearing,' in reference to the conspicuous cottony heads. The species name *vaginatum* refers to the inflated sheaths along the stem of sheathed cotton-grass.

E. brachyantherum *E. chamissonis*

E. vaginatum ssp. *vaginatum*

Key to the Spike-rushes (*Eleocharis*)

1a. 2 stigmas; achenes lens-shaped .... ***Eleocharis palustris***

1b. 3 stigmas; achenes 3-sided to nearly round 2

2a. Tubercle (bump at tip) distinct from achene, separated from it by narrow constriction; stems (culms) usually less than 10 cm tall .... ***E. acicularis***

2b. Tubercle blends into achene, does not form distinct cap; stems (culms) up to 30 cm tall .... ***E. quinqueflora*** (see *E. palustris*)

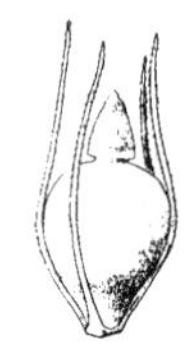

E. palustris

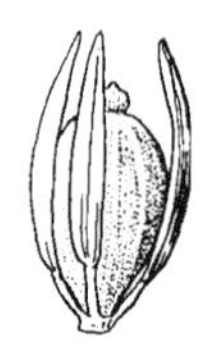

E. acicularis

E. quinqueflora

CREEPING SPIKE-RUSH • *Eleocharis palustris*

GENERAL: Perennial from **long, dark brown rhizomes**, forms carpets; stems round or oval in cross-section, **single or in clusters**, with reddish sheaths at base, 10–100 cm tall.

LEAVES: All at stem base, reduced to bladeless sheaths.

FLOWER CLUSTER: Single spikelet at stem tip, 0.5–2 cm long, brown, narrowly egg-shaped to lance-shaped, with **2–3 empty scales** at base; fertile scales arranged spirally, lance-shaped, usually have firm midvein to tip; **2 stigmas**.

FRUITS: Achenes, lens-shaped, yellow-brown, 1–1.5 mm long, **conical swelling** (like a nose cone) at tip, base usually has 4 barbed bristles, a bit longer than achene.

WHERE FOUND: Marshes, wet meadows, ditches, mudflats and lakeshores; widespread across our region, north to treeline; circumpolar.

NOTES: Few-flowered spike-rush (*E. quinqueflora*, also called *E. pauciflora*) differs in having smaller spikelets, flowers with 3 stigmas and triangular seeds with a beak. It is a circumpolar species found in calcium-rich wetlands across the boreal forest, and reaches Great Bear Lake and interior Alaska in the northwest. It is rare in the parkland and prairie. • Both the genus, *Eleocharis*, and species, *palustris*, names mean 'a plant that grows in marshes.' The common name 'spike-rush' refers to the appearance of the spike-like flower cluster.

E. quinqueflora

E. palustris

NEEDLE SPIKE-RUSH • *Eleocharis acicularis*

GENERAL: Loosely tufted from slender creeping rhizomes, **forms dense mats; stems slender, needle-like**, ridged and grooved, 3–12 cm tall.

LEAVES: Reduced to bladeless sheaths at base of stem.

FLOWER CLUSTER: Single, egg- to lance-shaped, **spike at stem tip**, 2–7 mm long, **somewhat flattened**, 3–15-flowered; scales reddish brown with green midvein and pale, thin edges; **3 stigmas**.

FRUITS: Whitish, egg-shaped achenes, widest above middle, 0.7–1.1 mm long, with several lengthwise ridges connected by many fine cross-ridges; conical swelling at tip; 3–4 bristles, as long as or slightly longer than achene (rarely absent).

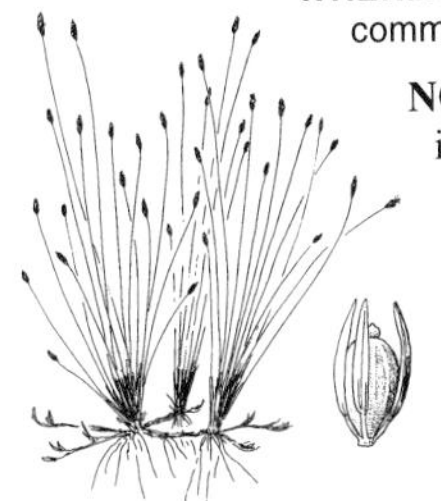

WHERE FOUND: Marshes, sloughs, shores, mudflats and other wet places; common and widespread across our region; circumpolar.

NOTES: Needle spike-rush is often sterile and submerged under water in the northernmost parts of its range. • The genus name *Eleocharis* is taken from the Greek *helos*, 'a marsh,' and *charis*, 'grace,' in reference to the habitat and graceful appearance of these plants. *Acicularis* means 'needle-like,' from the Latin *acus*, 'a needle.' The small, slender, stiff stems and leaves of these plants could be likened to a carpet of needles standing on end.

Key to the Bulrushes (*Scirpus*)

1a. 1 involucral bract (or none), not leaf-like; stem leaves much reduced in size 2

2a. 1 to many spikelets, appear to be on side of stem; bract grows as continuation of stem; stems (culms) stout, round 3

3a. Stems (culms) soft; flower cluster often droops; scales about as long as achenes, midrib strong .... ***Scirpus lacustris* ssp. *validus***
(see p. 215 'Aquatics' section)

3b. Stems (culms) firm; flower cluster erect, often compact; scales longer than achenes, with conspicuous reddish brown lines, midrib inconspicuous .... ***S. lacustris* ssp. *glaucus***
(see *S. lacustris* ssp. *validus*, in 'Aquatics' section)

2b. 1 spikelet at end of stem; bract small (when present), often shorter than spikelet 4

4a. Stems sharply 3-sided, rough towards top; bristles extend 1–2.5 cm beyond spikelets .... ***S. hudsonianus***

4b. Stems rounded, smooth; bristles only slightly, if at all, surpass spikelets .... ***S. caespitosus***

1b. 2 or more involucral bracts, leaf-like; stem leaves usually well developed 5

5a. Bristles not barbed, far longer than achenes and scales; 3 stigmas .... ***S. cyperinus***

5b. Bristles barbed nearly to base, equal to or slightly longer than achenes, not longer than scales; 2 stigmas .... ***S. microcarpus***

TUFTED BULRUSH • *Scirpus caespitosus*
TUFTED CLUBRUSH

GENERAL: Densely tufted perennial, often forms tough **tussocks**; stems very slender, wiry, **round**, 5–40 cm tall, base clothed with dense, drab, **tough, leaf sheaths**.

LEAVES: Most at stem base, reduced to light brown, scale-like blades on sheaths; **uppermost sheath green, with 5–8 mm long blade**.

FLOWER CLUSTER: Single, 2–4-flowered spikes at stem tips, 3.5–6 mm long; scales chestnut brown, pointed, lowermost has **blunt, slender bristle** about as long as spike.

FRUITS: Brown, narrowly egg-shaped, 1.5–2 mm long, 3-sided, smooth achenes, base surrounded by 6 delicate, white bristles about twice as long as achene.

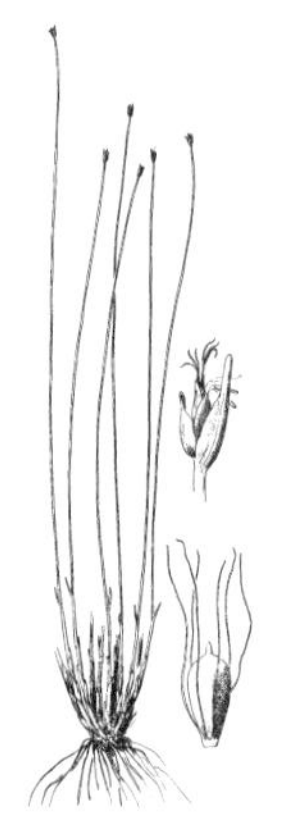

WHERE FOUND: Fens and peaty tundra; widespread across boreal forest, north past treeline to Arctic coast; circumpolar.

NOTES: Also called *Trichophorum caespitosum*. • *Scirpus* was the Latin name for some of the bulrush species. *Caespitosus* means 'tufted,' or 'growing in clumps or patches.'

SMALL-FRUITED BULRUSH • *Scirpus microcarpus*
SMALL-FLOWERED BULRUSH

GENERAL: Perennial from sturdy, spreading rhizomes; stems usually **clustered**, stout, triangular, 30–100 cm tall (sometimes to 150 cm).

LEAVES: Several, basal and along the stem, **flat, 5–15 mm wide; sheaths often purplish tinged** and with **whitish, knot-like crosswalls**.

FLOWER CLUSTER: Broad, spreading cluster of many short spikelets in small groups at end of spreading stalks; usually 3 leaf-like bracts at base, longest often longer than flower cluster; spikes usually 3–4 mm long; scales egg-shaped, green with purplish black markings.

FRUITS: Pale, lens-shaped, 1–1.5 mm long achenes, each surrounded by 4 (sometimes 6) whitish, barbed, slightly longer bristles.

WHERE FOUND: Marshes, streambanks, ponds and wet ditches; widespread across boreal forest and parkland, north and west to Great Slave Lake, disjunct in interior Alaska.

NOTES: The really big bulrushes have a single bract that appears as a continuation of the stem (see common great bulrush, p. 215), while small-fruited bulrush has leaf-like bracts. • Although all species of *Scirpus* are edible, some taste better than others. The young shoots can be eaten raw or cooked as a vegetable. The Cree ate the 10 cm of stem immediately above the root when it was freshly dug—it can be quite thirst quenching. The roots of many species are quite starchy. Young roots, crushed and boiled, yield a sweet syrup. They can also be roasted and eaten like potatoes, after removing the outer skin and fibres. The roots are also edible raw, or they can be dried, ground and made into a nutritious white flour. Pollen can be added to bread, mush or pancakes; it was also collected in large quantities, pressed into cakes and baked. The seeds can be parched and ground into meal, or they can be eaten whole, often as an addition to mush or breads.

WOOL-GRASS • *Scirpus cyperinus*

GENERAL: Densely tufted with fibrous roots; stems slender, stiff, leafy, slightly 3-sided, 0.5–1 m tall.

LEAVES: Most at stem base, 2–5 mm wide, up to 30 cm long, rough, upper ones often taller than stem.

FLOWER CLUSTER: Compound umbel, spreading or contracted; spikelets stalked, 3–5 mm long; bracts leaf-like, largest much longer than flower cluster; scales greenish black with whitish edges.

FRUITS: Sharply beaked, white or yellowish achenes, about 1 mm long, each surrounded by 6 white or **light brown bristles, much longer than achene**.

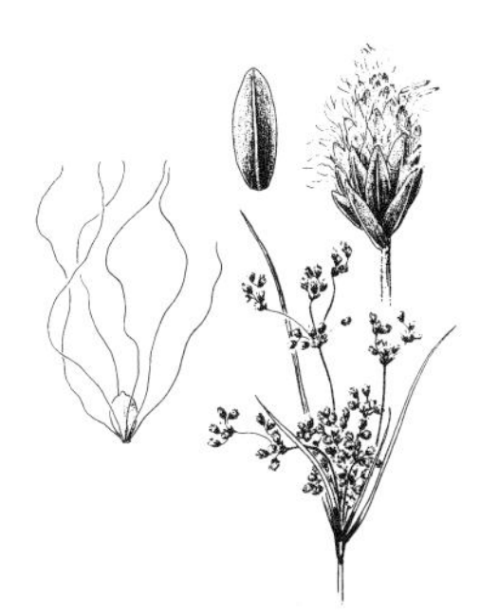

WHERE FOUND: Wet meadows, marshes and ditches; scattered across boreal forest of prairie provinces; rare in Alberta, more common to east.

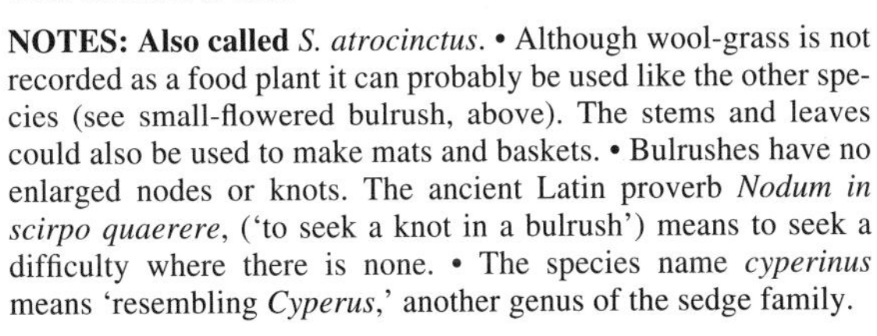

NOTES: Also called *S. atrocinctus*. • Although wool-grass is not recorded as a food plant it can probably be used like the other species (see small-flowered bulrush, above). The stems and leaves could also be used to make mats and baskets. • Bulrushes have no enlarged nodes or knots. The ancient Latin proverb *Nodum in scirpo quaerere*, ('to seek a knot in a bulrush') means to seek a difficulty where there is none. • The species name *cyperinus* means 'resembling *Cyperus*,' another genus of the sedge family.

HUDSON BAY BULRUSH • *Scirpus hudsonianus*

GENERAL: Loosely tufted from slender rhizomes; stems 3-sided, rough-edged, 10–40 cm tall.

LEAVES: Lower sheaths almost bladeless, upper ones have 1–2 cm long blades.

FLOWER CLUSTER: Single, egg-shaped spike at stem tip, 5–7 mm long; outer bract blunt-tipped, shorter than or as long as spikelet.

FRUITS: Brownish, 3-sided achenes, about 1.5 mm long, with pointed tip, each surrounded by 6 thin, flattened, **long** (2–2.5 cm) **white bristles**.

WHERE FOUND: Fens and springs; widespread and locally common across boreal forest, north and west to Great Slave Lake and south-central Alaska; circumpolar.

NOTES: Also called *Trichophorum alpinum*. • This little bulrush is easily identified by the tufts of soft, white hairs that crown each flowering spike, making it look like a small cotton-grass. • The common name and the species name *hudsonianus* are a reference to Hudson Bay, where plants of this species were first collected.

The Rush Family (Juncaceae)

The rush family has 2 genera—*Juncus* (rushes) and *Luzula* (wood-rushes)—and its members usually have 2 series of 3 scale-like floral bracts, like miniature brown lilies. Like lilies, but unlike sedges and grasses, rushes have fruits that are 3-chambered capsules with several to many seeds. Rush stems are usually round in cross-section. Remember: 'Sedges have edges and rushes are round.'

Like the sedges, rushes tend to be species of open, often wet habitats. The wood-rushes have similar tendencies, although some species (e.g., small-flowered wood rush, *Luzula parviflora*) also inhabit moist forests.

Key to the Rushes*

1a. Capsules 3-chambered; many seeds; leaves stiff, hairless (*Juncus*, the rushes) 2

- **2a.** Involucral bract (lowest bract of flower cluster) resembles extension and continuation of stem, flower cluster thus appears to be on side of stem 3
  - **3a.** Involucral bract channelled along upper side .... ***Juncus vaseyi*** (see *J. tenuis*)
  - **3b.** Involucral bract cylindrical, not channelled 4
    - **4a.** Stem finely grooved; sheaths dull, light-brown; flower cluster in middle of 'stem'; perianth whitish .... ***J. filiformis*** (see *J. balticus*)
    - **4b.** Stem smooth; sheaths shiny brown; flower cluster above middle, in upper third of 'stem'; perianth reddish brown .... ***J. balticus***
- **2b.** Involucral bract leafy or scale-like, does not form continuation of stem, flower cluster thus appears to be at stem tip 5
  - **5a.** Plants annual .... ***J. bufonius***
  - **5b.** Plants perennial 6
    - **6a.** Flowers borne singly in flower cluster, each with pair of bractlets 7
      - **7a.** Capsules longer than perianth; seeds have tail-like appendages 8
        - **8a.** Capsules blunt, only slightly longer than perianth; seed appendages short .... ***J. vaseyi*** (see *J. tenuis*)
        - **8b.** Capsules pointed, about 1.5 times as long as perianth; seed appendages long .... ***J. castaneus*** (see *J. nodosus*)
      - **7b.** Capsules shorter than, or as long as, perianth; seeds lack appendages 9
        - **9a.** Perianth 4–6 mm long; leaves mostly less than half as long as stem; auricles short, yellowish .... ***J. dudleyi*** (see *J. tenuis*)
        - **9b.** Perianth 3–5 mm long; leaves mostly more than half as long as stem; auricles long, whitish .... ***J. tenuis***
    - **6b.** Flowers borne in heads with several flowers, lacking bractlets 10
      - **10a.** Leaves have regularly-spaced cross walls 11
        - **11a.** Heads spherical, short-stalked, many-flowered; capsules lance-shaped .... ***J. nodosus***
        - **11b.** Heads more cone-like, long-stalked, few-flowered; capsules oblong .... ***J. alpinus***
      - **10b.** Leaves without regularly-spaced cross walls, or with cross walls only near tips .... ***J. longistylis*** (see *J. alpinus*)

1b. Capsules 1-chambered; 3 seeds; leaves flexible, usually hairy on edges near base (*Luzula*, the wood-rushes) 12

- **12a.** Flower cluster has flowers in groups of 2 or 3, not in small heads .... ***Luzula parviflora***
- **12b.** Flower cluster has flowers in small heads .... ***L. multiflora*** (see *L. parviflora*)

* The rush portion of this key is derived from a key to the genus *Juncus* by Adolf Ceska in Douglas et al. (1994)

WIRE RUSH • *Juncus balticus*
ARCTIC RUSH

GENERAL: Perennial from **strong rhizomes**; stems stout, single, often in dense rows (like teeth of comb), rounded, 10–60 cm tall, base 1.5–4 mm thick.

LEAVES: All at stem base, usually **reduced to bladeless**, pointy-tipped, shiny, **yellowish brown sheaths**.

FLOWER CLUSTER: Dense to diffuse cluster of several to many flowers, usually 2–4 cm long, **seems to grow from 1 side of stem** because erect, rounded, 2.5–25 cm **bract looks like continuation of stem**; 6 flower scales, greenish to purplish brown, **4–6 mm long**, lance-shaped; **6 stamens**; anthers longer than filaments.

FRUITS: Capsules, egg-shaped, abruptly sharp-pointed, firm, about as long as scales; seeds elliptic-egg-shaped, 0.8–1 mm long, finely net-veined, minute point at tip.

WHERE FOUND: Wet meadows, river banks, sandbars, lake and slough edges and wetlands; widespread across our region, north past treeline to southwestern arctic islands and northern Alaska; circumpolar.

NOTES: Also called *J. arcticus* ssp. *balticus*. • **Thread rush** (*J. filiformis*) has a greenish flower cluster above a slender bract that is as long as or longer than the stem. Thread rush grows in moist meadows and on sandy shores throughout our region. • The flowers of wire rush produce a pinkish dye when they are chopped, boiled slowly for 2 hours and allowed to cool. If alum or cream of tartar is added as a mordant, a green colour results. The stems produce brown to green colours. • Although rushes are of little economic importance, they can be very important in natural systems, where they colonize open sites. Wire rush often forms extensive networks of tough rhizomes that stabilize soil.

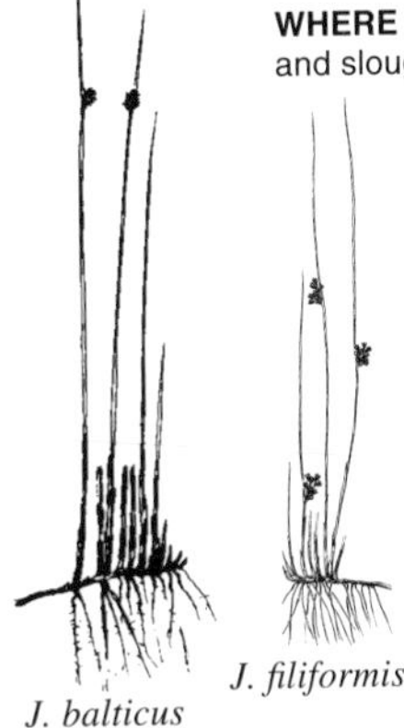

J. balticus *J. filiformis*

ALPINE RUSH • *Juncus alpinus*

GENERAL: Perennial, in small tufts from creeping rhizomes; stems 15–50 cm tall.

LEAVES: 1–2 small leaves along the stem, **rounded, with crosswalls**.

FLOWER CLUSTER: Few to several, 2–10-flowered heads on erect or ascending branches; some flowers often elevated on short stalks above others in head.

FRUITS: Pale brown capsules, slightly longer than scales.

WHERE FOUND: Wet meadows and shores; widespread across boreal forest; circumpolar.

NOTES: Also called *J. alpinoarticulatus*. • **Long-styled rush** (*J. longistylis*) is a similar species, but it has flat or channelled leaves that do not have crosswalls. It grows in wet meadows, marshes and fens, and on lakeshores and riverbanks across North America, but in our region it is found mostly east of the Rockies in the southern boreal forest, parkland and prairie. • According to Greek mythology, the rush was created when Polyphemus, the burly Cyclops in Homer's *Odyssey*, discovered the beautiful maiden Galatea in the arms of her lover, a shepherd named Acris. Polyphemus crushed poor Acris with a rock, and as Galatea wept over his body, she changed his flowing blood into water that would flow forever. When the water became clear, the body of the young shepherd appeared waist-deep in the water. Gradually the arms lengthened and the shoulders sprouted green blades, and soon the edge of the brook was lined with rushes. • The genus name is derived from the Latin *jungo*, 'to join' or 'to bind.' In the past, rushes were frequently used for tying things together.

J. alpinus *J. longistylis*

KNOTTED RUSH • *Juncus nodosus*

GENERAL: Perennial from slender **rhizomes with tuber-like thickenings** ; stems single, 10–40 cm tall.

LEAVES: Rounded, with **crosswalls**; upper stem leaves taller than flower cluster.

FLOWER CLUSTER: Few to several, 8–20-flowered, spherical heads; scales 3–4 mm long, narrowly-lance-shaped, long-pointed; bract longer than flower cluster.

FRUITS: Narrowly lance-shaped to awl-shaped capsule, 3-sided, longer than scales; seeds pointed.

WHERE FOUND: Moist ground, marshes and shores; found across boreal forest and parkland, north to southern N.W.T. and just reaching interior Alaska.

NOTES: Chestnut rush (*J. castaneus*) is common across the western part of our region and grows on wet shores and river flats, and in marshes and fens. It has broad leaves that lack crosswalls, 1–3 heads of 4–10 flowers each, large (6–9 mm long), chestnut to purplish brown capsules, and seeds with long, white tails. • The name 'rush,' is derived from the Norwegian *rusk*, which means 'to plait' or 'to twist.' Today this name usually refers to plants in the family Juncaceae, but in the past it encompassed a wide variety of similar plants, including the bulrushes (*Scirpus* spp.), sedges (*Carex* spp.), horsetails (*Equisetum* spp.) and cattails (*Typha* spp.). All of these plants were probably used in similar ways, depending on which species were available locally. • The species name *nodosus* refers to the distinctive node-like thickenings on the roots.

SLENDER RUSH • *Juncus tenuis*

GENERAL: Tufted perennial; **stems more or less leafless**, round, 10–60 cm tall.

LEAVES: 1–3 at stem base, flattened or inrolled, about 1 mm wide, usually at least half as long as flowering stems.

FLOWER CLUSTER: At stem tips, 3 to many, many-flowered heads, form 2–6 cm long, **compact to loose cluster**; bract usually longer than flower cluster; flower scales greenish to tan, **slender-tipped**, lance-shaped, 3–5 mm long; 6 stamens, anthers shorter than filaments.

FRUITS: Capsules, egg-shaped, rounded and dimpled at tip; seeds elliptic, about 0.5 mm long.

WHERE FOUND: Wet ground, often in disturbed areas and clearings; widespread across southern boreal forest and parkland, north and west to southern Yukon and interior Alaska; seldom north of 55°N in prairie provinces.

NOTES: Dudley's rush (*J. dudleyi*, also called *J. tenuis* var. *dudleyi*) resembles slender rush, but it has slightly longer flower scales (4–6 mm), its leaves are less than 1/2 as long as the stem, and it has short, yellowish lobes where the leaves join the stem. It grows in moist meadows, in clearings, on riverbanks and on disturbed soil, scattered throughout our region but most common east of the Rockies. • **Big-head rush** (*J. vaseyi*) has a compact flower cluster, capsules that are longer than the flower scales, seeds with tails, and somewhat channelled leaves. It grows in moist meadows and on river flats, shores and wetland edges, scattered across our region, north and west to Great Slave Lake. • Rushes quickly become tough and unpalatable as summer progresses, but in spring, when they are still young and tender, some species are occasionally eaten by muskrats, deer and waterfowl. Many birds eat rush seeds. Livestock seldom eat rushes, though they may consume a few in the spring. • The species name, *tenuis*, is Latin for 'slender,' in reference to the leaves.

J. tenuis

J. dudleyi

J. vaseyi

TOAD RUSH • *Juncus bufonius*

GENERAL: Annual from fibrous roots, no rhizomes; stems thread-like, usually tufted and branched from base, erect to sprawling, channelled, 5–20 cm tall.

LEAVES: Few to several at stem base and along stem, usually **less than 1 mm wide**, flat to rolled, 0.5–10 cm long, light green, brown with age.

FLOWER CLUSTER: At stem tips, commonly spreading and **much-branched**; flowers usually single in axils of transluscent bracts; flower scales **greenish**, narrowly lance-shaped, 3–8 mm long; 6 stamens, anthers shorter than filaments.

FRUITS: Capsules, oblong-elliptic, rounded at tip with tiny, abrupt point, chestnut brown, shorter than scales; seeds cylindrical-elliptic, 0.3–0.5 mm long, finely ridged, minute point at tip.

WHERE FOUND: Lakeshores, streambanks and moist sites in fields, clearings and disturbances; widespread across our region; circumpolar.

NOTES: This little plant is our only annual rush. A worldwide, variable weed, toad rush apparently consists of both native and introduced races in North America. • If you look closely at most rush flowers, you will see that they have both petals and sepals, but these have been reduced to tiny, green or brown scales. The female parts mature first, ready to catch wind-carried pollen from other plants. Later, the anthers appear and disperse their pollen. • Toad rush usually produces lots of flowers in the spring and early summer, but if these are not fertilized, another set of flowers appears later in the summer. These are relatively small and inconspicuous, and they never open. Instead the anthers and stigmas mature inside the closed flower and the plant pollinates itself, producing seed for the next year. • The species name *bufonius* translates from Latin as either 'of the toad' or 'living in damp places.'

SMALL-FLOWERED WOOD-RUSH • *Luzula parviflora*

GENERAL: Perennial from spreading rhizomes; stems single or in small tufts, usually 30–60 cm tall; basal sheaths chestnut or purplish brown.

LEAVES: Most at stem base, **3–6 on stem**, **4–10 mm wide**, **flat**, sometimes with long, white hairs along edges.

FLOWER CLUSTER: Open, loosely branching, cluster of small groups of flowers **nodding** on long, slender stalks; 6 flower scales, brown, pointed, **1.8–2.5 mm long**; bracts shorter than flower cluster, edges torn, **often with long hairs**.

FRUITS: Dark brown, egg-shaped capsules, 2–2.5 mm long, equal to or longer than flower scales; many seeds, brown, 1.1–1.5 mm long.

WHERE FOUND: Moist, open forests, thickets, meadows, shores and wetlands; widespread across boreal forest; circumpolar.

NOTES: Field wood-rush (*L. multiflora*, also called *L. campestris* ssp. *multiflora*) has flowers and fruits crowded in dense, rounded clusters. It grows scattered across our region in dry to moist meadows, clearings and forest openings, and on turfy shores and pond edges. • Livestock will eat wood-rushes in the spring, when they are tender and succulent, but the plants become tough and unpalatable as the season progresses. • The outer surfaces of wood-rush seeds produce a sticky mucilage when they become wet. This helps to establish the seedling, but it can also aid in seed dispersal by adhering to passing animals, which then carry the seed to new areas.

L. multiflora

Poaceae (Grass Family)

Grasses tend to be overlooked in favour of other, more glamorous flowering plants, but the grass family (Poaceae or Gramineae) is undoubtedly the most useful to humans. Grasses now provide us with 3 times more food than peas and beans, tubers, fruit, meat, milk and eggs combined. The cultivation of cereals is a matter of life or death for many people living today. Grasses include sugar cane, and cereals such as wheat, barley, oats, rye, corn and rice. Grains such as barley can be brewed into beers and whiskeys, rye into whiskey, rice into sake, and molasses (from sugar cane) into rum. Grasses are used for forage and turf, and they provide materials for weaving, thatch, adobe, or bamboo structures. Grasses in general were used widely by native peoples for lining steam-cooking pits, wiping fish, covering berries, stringing food for drying, spreading on floors, or as bedding. Many species of wildlife, from large grazing mammals to waterfowl, depend on grass and grassland or wetland habitats for food, shelter and the completion of their life cycles.

The grass family is very large and complex. In our region there are over 100 grass species, most of which are native—although introduced European grasses are an important component of the weedy boreal flora. Only the most common species encountered in the region are presented here.

To assist in identification, the grasses are picture-keyed first into tribes and then into genera. Detailed drawings of the flowering parts and leaf axils are provided with each species to illustrate key characteristics. Consult Hubbard (1969), Looman (1982) and the general floras mentioned in the introduction for more information about our grasses.

Grass flowers are small but complex, and they have acquired a peculiar but necessary terminology. The tiny flowers occur in the axils of a small scale-like inner bract (**palea**), and a larger outer bract (**lemma**); the flower, lemma and palea together are called a **floret**. Below the florets are 2 more bracts, the **glumes**; florets and glumes together are called a **spikelet**, which is the basic unit of the flower cluster. The leaves of most grasses sheath the stem before diverging as a narrow blade, and where the sheathing part joins the blade there are 2 distinctive organs useful for identification: the **ligule** (typically a membranous flap or appendage), and the **auricle** (an ear-like lobe or flange). Grass stems are usually round in cross-section and hollow between the nodes. If your 'grass' specimen is either triangular in cross-section or has a solid, round stem, refer to the 'Sedges' section (pp. 227–49) or the 'Rushes' section (pp. 250–53).

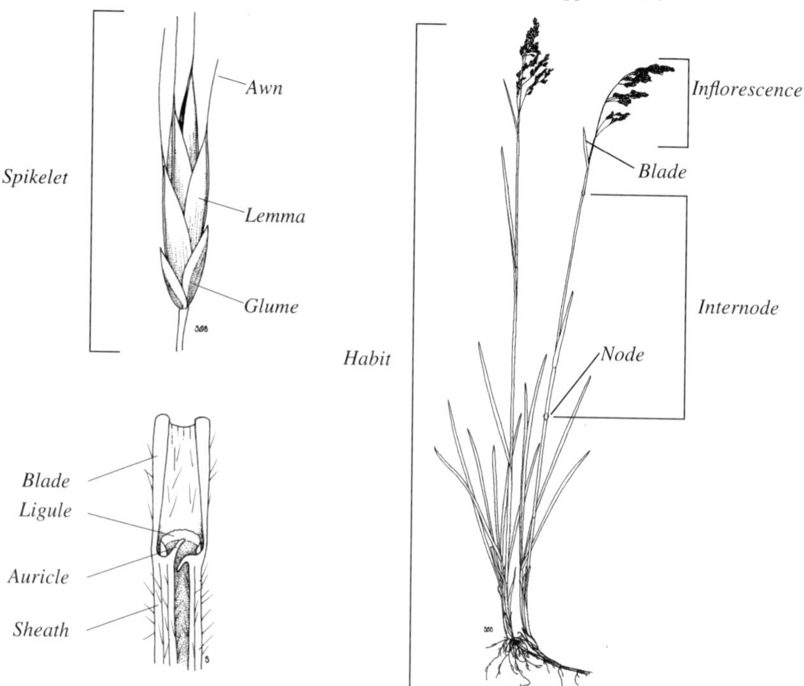

Picture-Key to Common Grass Tribes

1a. Flower cluster is a spike (spikelets without stalks) 2

2a. Spikes single, at stem tips; spikelets arranged singly, or in 2s or 3s on opposite sides of main axis .... **Hordeae** (Barley tribe)

2b. Spikes usually occur in groups; spikelets arranged in 2 rows on 1 side of main axis .... **Chloridae** (Chloris tribe)

1b. Flower cluster is a panicle or raceme (spikelets with stalks) 3

3a. Each spikelet has 1 floret; glumes small .... **Agrostideae** (Timothy tribe)

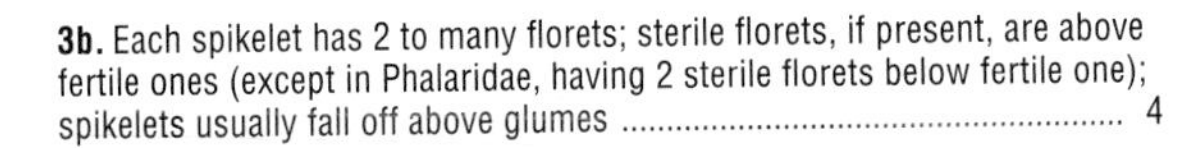

3b. Each spikelet has 2 to many florets; sterile florets, if present, are above fertile ones (except in Phalaridae, having 2 sterile florets below fertile one); spikelets usually fall off above glumes 4

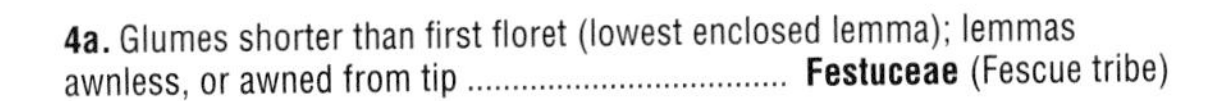

4a. Glumes shorter than first floret (lowest enclosed lemma); lemmas awnless, or awned from tip .... **Festuceae** (Fescue tribe)

4b. Glumes equal to, or longer than first floret (lowest enclosed lemma) .. 5

5a. Glumes narrow; lemmas awned from back, or awnless; sterile florets, if present, are above fertile ones .... **Aveneae** (Oat tribe)

5b. Glumes broad, boat-shaped; lemmas awned from notched tip, or awnless; spikelets have 1 fertile floret above 2 sterile ones .... **Phalarideae** (Canary grass tribe)

Grass Tribes, Keys to Genera

Hordeae (Barley tribe)

1 spikelet at each node

Agropyron
(wheat grass)

2 spikelets at each node, alike

Elymus
(wild rye)

3 spikelets at each node, side pair reduced to awns

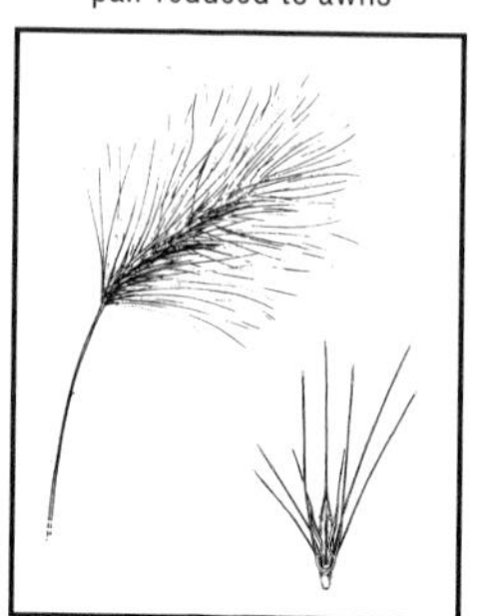

Hordeum
(barley)

Chloridae (Chloris tribe)

Spikelets arranged in 2 rows on 1 side of central stalk (rachis), form 1-sided spikes or racemes

Beckmannia
(slough grass)

Agrostideae (Timothy tribe)

Flower cluster spike-like

Glumes awned, with comb-like fringe of hairs on keel

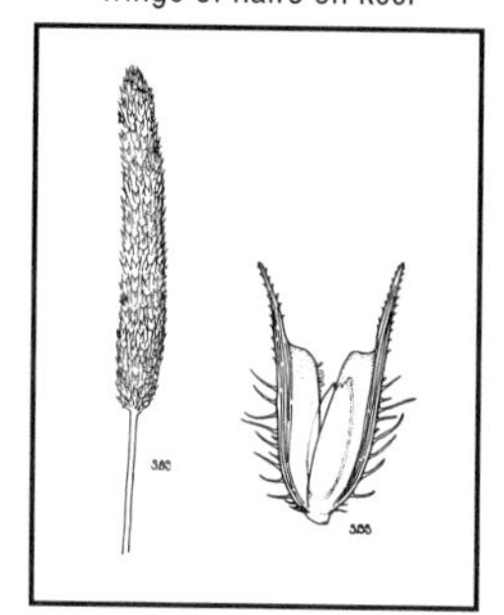

Phleum
(timothy)

Glumes awnless; lemma awned from near middle

Alopecurus
(foxtail)

Agrostideae (Timothy tribe) [continued]

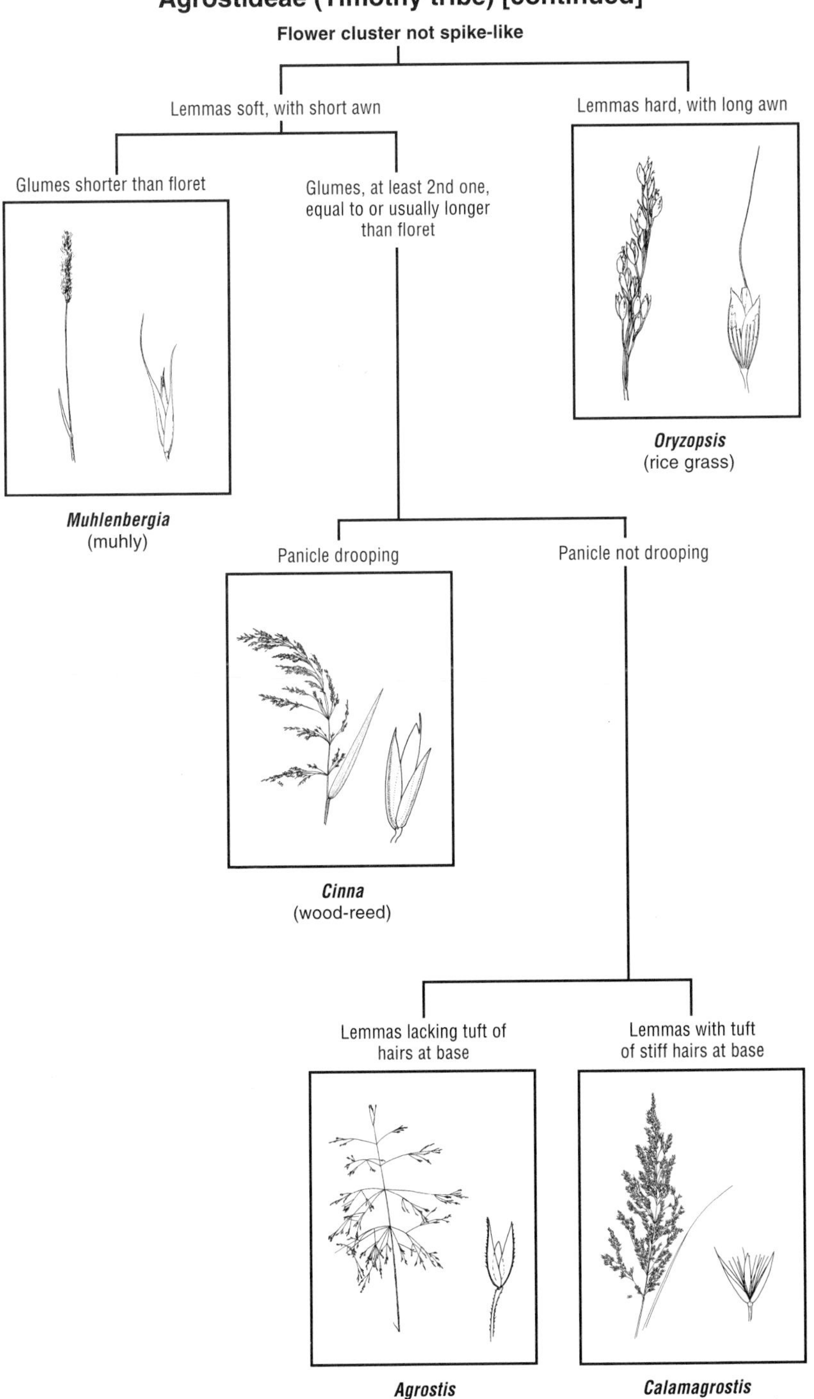

Festuceae (Fescue tribe)

Lemmas without awns

Lemmas with veins converging at tip, often cobwebby at base

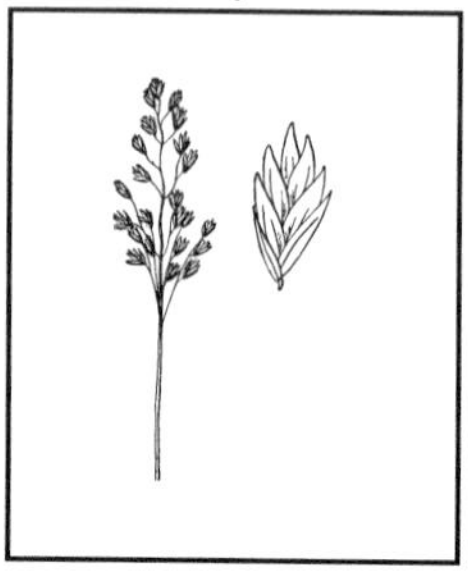

Poa
(bluegrass)

Lemmas with 7 prominent, parallel (not branching) veins

Glyceria
(manna grass)

Lemmas with awns

Flower cluster usually open, if narrow, spikelets not densely crowded; plants less than 2 m tall

Thickened base of lemma long-hairy

Schizachne
(false melic)

Thickened base of lemma not long-hairy

Spikelets large, in panicles; lemmas awned from between 2 teeth at tip

Bromus
(brome)

Spikelets small; lemmas awned from tip

Festuca
(fescue)

Flower cluster plume-like; plants tall (2-3 m), reed-like

Phragmites
(reed grass)

Aveneae (Oat tribe)

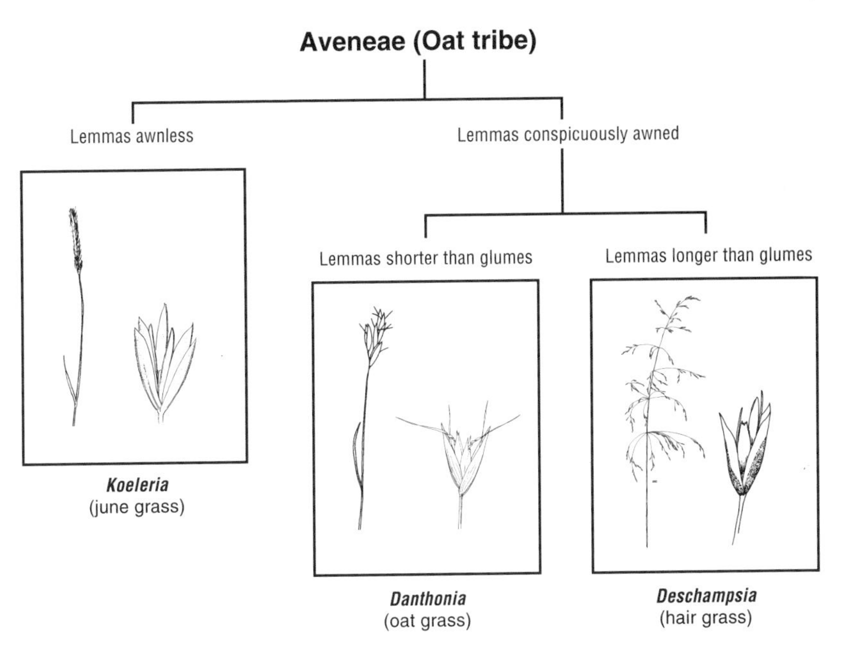

Koeleria
(june grass)

Danthonia
(oat grass)

Deschampsia
(hair grass)

Phalarideae (Canary grass tribe)

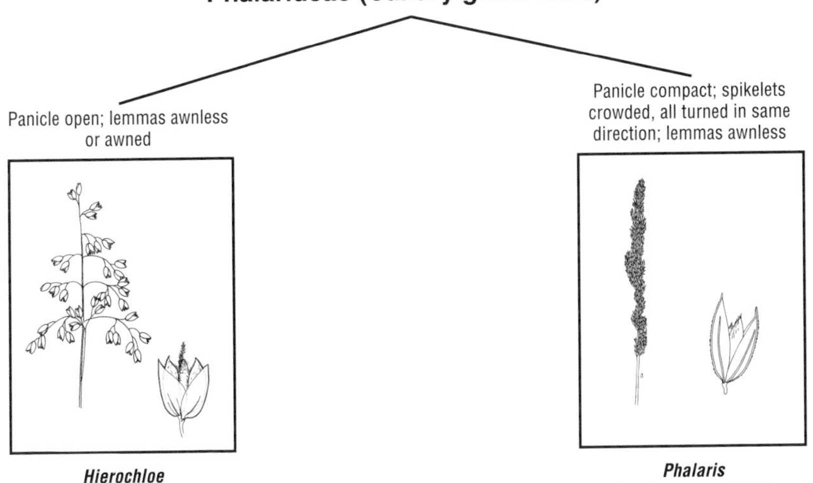

Hierochloe
(sweet grass)

Phalaris
(reed canary grass)

FOXTAIL BARLEY • *Hordeum jubatum*

GENERAL: Tufted perennial; stems usually 30–60 cm tall, erect or curved at base.

LEAVES: Greyish green, rough, hairy, flat, usually 2–6 mm wide; ligules less than 0.6 mm long.

FLOWER CLUSTER: Spike, usually nodding, 5–10 cm long including **long hair-like awns**, green to **purplish**; at maturity breaks into units of 3, 1-flowered spikelets; central spikelet bisexual; 2 side spikelets usually reduced to 1–3 spreading awns; glumes hair-like, usually 2–6 cm long; lemmas 5–8 mm long (sometimes to 10 mm), with **2–9 cm long awns**.

WHERE FOUND: Saline marshes, meadows, roadsides, and waste ground; widespread across North America; north past treeline to Arctic coast.

NOTES: When the spike matures, it breaks into pieces, and the long, wiry, barbed awns help to disperse the spikelets as they are carried away by the wind or by passing animals. The barbs hook onto the fur of passing animals or catch onto clothing, and the bristles work their way in as the carrier moves. This can be a nuisance for humans, but it is a real hazard for dogs, sheep and other animals, especially if the barbs become embedded in the soft tissue around their eyes, nose or mouth. • The Chipewyan call foxtail barley 'arrow grass' or 'enemy grass,' and they recognize it as a hazard for their dogs, similar to porcupine quills. To get rid of nuisance dogs, foxtail barley was mixed with their feed. As the barbs punctured the gastrointestinal tract, they would cause bleeding and infection and eventually the animal would die.

SLENDER WHEAT GRASS • *Agropyron trachycaulum*

GENERAL: Loosely tufted perennial, no rhizomes; stems slender, erect, 50–90 cm tall (sometimes to 150 cm).

LEAVES: Several on stem, rough-hairy, usually flat, 3–6 mm wide; auricles absent or short; ligules very short (to 0.5 mm).

FLOWER CLUSTER: Spike slender, green, erect or slightly curved, 5–25 cm long; spikelets 15–20 mm long, 1 per node, usually well-spaced (at least below), 5–8-flowered; glumes 10–12 mm long, 3–7-veined, awn-tipped, nearly as long as spikelet; lemmas 12–15 mm long, awnless or short-awned, rough-hairy; anthers 1–2 mm long.

WHERE FOUND: Open woods, meadows, shores and disturbed areas; widespread across boreal forest, parkland and prairie; reaches Great Bear Lake, Mackenzie delta and interior Alaska to north and west.

NOTES: Also called *A. caninum* and *Elymus trachycaulus.* • **Awned wheat grass** (*A. trachycaulum* var. *unilaterale*, also called *A. subsecundum*) has lemmas with 15–40 mm long awns and overlapping spikelets in a coarse spike 5–20 cm long. It grows in grassy meadows and open (often aspen) forests, and on hillsides and gravelly terraces across our region. • The introduced, weedy **quack grass** or **couchgrass** (*A. repens* also called *Elyrigia*) has long, yellowish, creeping rhizomes, fairly wide (to 1 cm) leaves and lemmas with 1–10 mm long awns. It grows in disturbed habitats across our region. • Slender wheat grass is easily destroyed by cultivation or prolonged grazing, but it can provide excellent forage for sheep, cattle and wildlife. • The grains are edible, but tedious to gather as they do not readily fall free from their surrounding bracts. Wheat grass grains can be ground or crushed and used to make flour or porridge. Grains infected with the **poisonous fungus** ergot (*Claviceps*) should never be eaten (see *Bromus ciliatus*, p. 272).

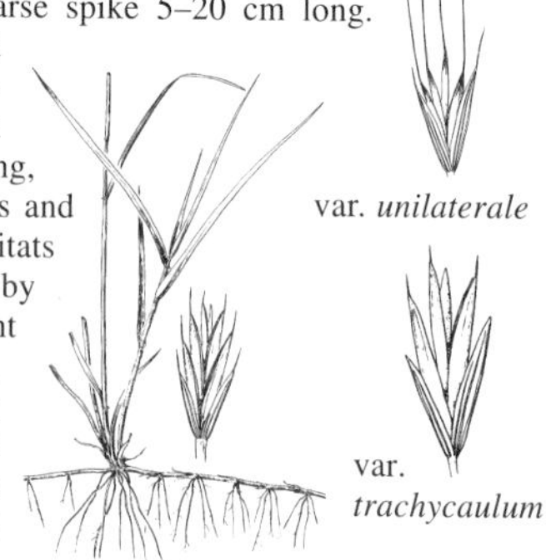

A. repens

HAIRY WILD RYE • *Elymus innovatus*
FUZZY-SPIKED WILD RYE

GENERAL: Tufted perennial, often sod-forming with creeping, scaly rhizomes and deep, spreading roots; stems erect, 0.5–1 m tall.

LEAVES: Thin, stiff, rough, 2–7 mm wide, flat or **inrolled**; auricles well-developed, claw-like; ligules very short (to 0.5 mm long).

FLOWER CLUSTER: Dense, erect, **very fuzzy** spike, 4–12 cm long; spikelets 2 per node, 3–5-flowered, usually densely purplish or greyish, hairy; glumes very narrow, densely hairy; lemmas coarsely hairy, 7–10 mm long, tipped with **1–4 mm long awn**.

E. innovatus

E. canadensis

WHERE FOUND: Open woods, clearings, slopes, meadows and floodplains; widespread across western boreal forest.

NOTES: Canada wild rye (*E. canadensis*) has 8–14 mm lemmas with curved, 20–40 mm awns. It grows on riverbanks, sandy shores and gravelly terraces, east from the Rockies. **Smooth** or **blue wild rye** (*E. glaucus*) has 10 mm long lemmas with more or less straight, 10–20 mm long awns. It grows in moist meadows and open forests, and is scattered southeast from the southern Yukon, mostly in the southern parts of our region. • Several species of wild rye have been used for food. The grains can be gathered, ground into flour and made into meal for cakes and porridge. Wild rye is susceptible to infection by ergot, a **poisonous fungus**. • Hairy wild rye is considered of low palatability, but it provides forage for domestic stock in boreal regions and is an important native forage plant. It is also an important winter forage for mountain sheep in the northern Rockies.

TIMBER OAT GRASS • *Danthonia intermedia*

GENERAL: Tufted perennial; stems 5–50 cm tall.

LEAVES: Mainly at stem base, flat or inrolled, 1–3 mm wide; long hairs on sheaths and blades; **ligules** to 1 mm long, **have hairy fringe**.

FLOWER CLUSTER: Dense, tuft-like panicle, 2–9 cm long with erect to ascending branches; **1 spikelet per branch**, 12–15 mm long, 3–8-flowered, break into single florets; glumes similar, 10–15 mm long; lemmas 7–8 mm long (sometimes to 12 mm), with **2 slender-pointed teeth and 5–15 mm long, twisted awn at tip, hairy at base and edges only**.

WHERE FOUND: Grassland, meadows, and open woods; widespread across southern boreal forest, parkland and prairie from southwestern Manitoba (where it is rare) to southern Yukon.

NOTES: Included in *D. californica* by some. • Timber oat grass is one of the first grasses to appear in the spring. It is a distinctive grass with large, oat-like spikelets that often turn purplish. The small florets never open. Instead, they produce seed from self-fertilization, often within the lower sheaths. • Oat grasses are fairly palatable for livestock, but they are seldom abundant enough to be important as fodder. • The genus *Danthonia* was named in honour of Etienne Danthione, a French botanist who lived in the early 1800s. The species name *intermedia* is from the Latin *inter*, 'between' and *medius*, 'middle,' suggesting that this oat grass is intermediate between other species of the genus.

TIMOTHY • *Phleum pratense*

GENERAL: Short-lived, tufted perennial, often forms large clumps with fibrous roots; stems stiffly erect, 40–90 cm tall; **bulbous** (corm-like) at **base**.

LEAVES: 6–12 mm wide, 5–20 cm long, light green to greyish green, flat, tapered; ligules 2–4 mm long.

FLOWER CLUSTER: Very dense, spike-like, short-branched cylindrical panicle, usually **5–10 cm long, to 1 cm thick**; spikelets crowded, flattened; glumes 3–5 mm long, with **comb-like, hairy fringe on ridges** and stout, curved awns 1–2 mm long; lemmas shorter than glumes.

WHERE FOUND: Roadsides, pastures, clearings; introduced (from Eurasia) and partly naturalized near settled areas; widespread across most of North America; north and west to southern N.W.T. and interior Alaska; circumpolar.

NOTES: Like other grasses, timothy is wind-pollinated. Its pollen grains are extremely small and light—1 g of pollen contains about 210 million pollen grains. • The pollen produced by timothy is very allergenic, and this grass is one of the species most frequently used in preparing vaccines and allergens for immunization against hay fever. • Although timothy is a rather coarse grass generally avoided by livestock, it is considered a good hay plant if it is cut just as the seeds are ripening. Although timothy does not do well with intensive grazing or mowing, it is probably the most important perennial pasture grass in temperate North America. Timothy is often the main grass sown with clover in hay meadows. Because it is relatively tall, it is a good hay species. Timothy is also drought resistant, capable of growing in upland sites and on poor sandy soils. Although the grains are small and delicate-looking, they are not easily digested, and large herbivores often carry them to distant sites.

SHORT-AWNED FOXTAIL • *Alopecurus aequalis*
LITTLE MEADOW-FOXTAIL

GENERAL: Tufted perennial; stems erect to prostrate, root at lower nodes, 15–60 cm tall.

LEAVES: Flat, 1–4 mm wide, grey-green, limp when submerged; ligules 4–8 mm long, pointed.

FLOWER CLUSTER: Slender, spike-like, short-branched cylindrical panicle, 2–7 cm long, **3–5 mm thick**, pale green; **glumes 2–3 mm long**, blunt-tipped, long hairy; lemmas 2.5 mm long with **0.5–3 mm long awn** attached near middle.

WHERE FOUND: Marshes, ditches, shallow water and muddy shores; widespread across our region; circumpolar.

NOTES: The blunt, translucent-tipped glumes of short-awned foxtail spikelets are opposite and joined near their base. • Short-awned foxtail has also been called 'orange foxtail' because of its bright orange to golden-yellow anthers. • Although short-awned foxtail is not harvested as a crop, the seeds have been eaten in times of famine in many Asian countries and it is considered a good forage grass for livestock. • Short-awned foxtail flowers for most of the growing season (from June to September) and spikes at all stages of development may be found on the same plant. The spikes mature from the top downwards; the stigmas protrude from the spikelets a day or more before the anthers appear. This favours cross-pollination. When the spikelets are ripe, they fall from the panicle as a unit, leaving only the bare axis of the spike behind. The mature grains are soft and presumably rich in oils, like those of other northern grasses.

ROCKY MOUNTAIN FESCUE • *Festuca ovina* var. *saximontana*

GENERAL: Small, **densely tufted** perennial; stems 10–50 cm tall, smooth.

LEAVES: At stem base, very slender (about 1 mm wide), tightly rolled, **erect**, rough; ligules short (to 0.4 mm), finely fringed.

FLOWER CLUSTER: Linear, interrupted, bluish to yellowish green (rarely purple) panicle, usually 4–8 cm long (sometimes to 12 cm); spikelets 2–5-flowered; glumes lance-shaped, 2–3 mm long; lemmas lance-shaped, 2–4 mm long, taper to **1–3 mm long awn**.

WHERE FOUND: Open woods, clearings, rocky slopes and grasslands; widespread across boreal forest north and west to southern N.W.T. and interior Alaska.

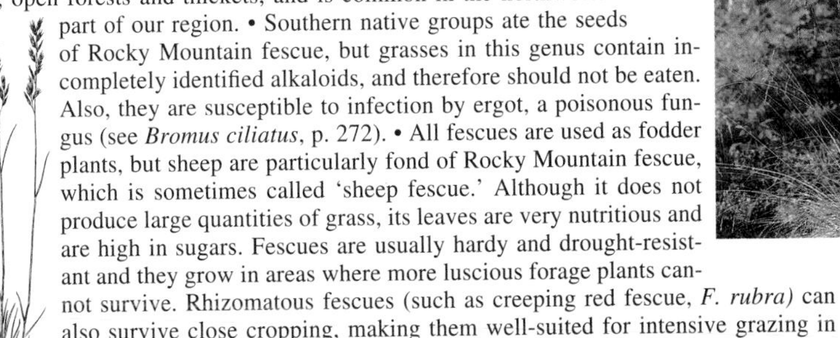

NOTES: Also called *F. saximontana.* • **Northern rough fescue** (*F. altaica* or *F. hallii*, included in *F. scabrella* by some) (inset photo) is a big bunchgrass, 50–100 cm tall (or taller), with long (to 15 mm), glossy spikelets and pointy, but unawned, lemmas. It grows in grasslands, rocky meadows, open forests and thickets, and is common in the northwestern part of our region. • Southern native groups ate the seeds of Rocky Mountain fescue, but grasses in this genus contain incompletely identified alkaloids, and therefore should not be eaten. Also, they are susceptible to infection by ergot, a poisonous fungus (see *Bromus ciliatus*, p. 272). • All fescues are used as fodder plants, but sheep are particularly fond of Rocky Mountain fescue, which is sometimes called 'sheep fescue.' Although it does not produce large quantities of grass, its leaves are very nutritious and are high in sugars. Fescues are usually hardy and drought-resistant and they grow in areas where more luscious forage plants cannot survive. Rhizomatous fescues (such as creeping red fescue, *F. rubra)* can also survive close cropping, making them well-suited for intensive grazing in pastures and for repeated cutting in lawns. As with the other tufted or with bunchgrass fescues, northern rough fescue decreases in abundance with intensive grazing, and eventually disappears from over-grazed sites.

JUNE GRASS • *Koeleria cristata*

GENERAL: Densely tufted perennial, seldom has rhizomes; stems smooth, erect, 20–50 cm tall, hairy towards top; lower sheaths often hairy.

LEAVES: Most at stem base, flat, or slightly inrolled, 1–3 mm wide, bluish green, stiff; ligules 0.5–1 mm long, usually fringed with short hairs.

FLOWER CLUSTER: Dense, spike-like panicle, 2–8 cm long, shiny, pale green to purplish; spikelets 4–5 mm long, 2–4-flowered, flattened; upper glume 3–4 mm long, 3–5-veined; lower glume smaller, 1-veined; lemmas rounded, shiny, somewhat thin and translucent, pointed or with **short awn** from just below tip.

WHERE FOUND: Grassland, rocky ridges and open forests, on well-drained soils; widespread across southern boreal forest, parkland, and prairie north and west to Great Slave Lake and southern Yukon; circumpolar.

NOTES: Also called *K. macrantha* and *K. gracilis.* • June grass is a good forage grass, but the plants are usually widely spaced. • Many grasses are known for their ability to withstand repeated grazing and mowing. They are able to do this because their leaves can grow from their bases. If the leaf is cut or bitten off, it can grow again to full size. In fact, grasses that are often closely clipped usually become leafier. • The genus *Koeleria* was named in honour of German botanist George Wilhelm Koeler (1765–1807).

PURPLE REED GRASS • *Calamagrostis purpurascens*

GENERAL: Tufted perennial with very short rhizomes; stems 30–70 cm tall.

LEAVES: Flat or slightly inrolled, **rather thick**, rough, 2–4 mm wide; ligules 2–4 mm long.

FLOWER CLUSTER: Dense, usually pinkish or purplish panicle, 4–15 cm long; spikelets 1-flowered, 6–8 mm long; glumes longer than lemmas, 5–8 mm long, rough; lemmas have 4 short, bristle-like teeth at tip, a **bent awn sticking out** about 2 mm and many lon2g **callus hairs, about 1/3 as long as lemma**.

WHERE FOUND: Dry, open woodlands, clearings, hillsides and ridges; widespread across northern boreal forest from northern Manitoba and Lake Athabasca, north past treeline through arctic islands.

NOTES: Purple reed grass plants are usually well-spaced. The large, purplish heads are conspicuous late in the season. • The common name 'reed grass' is a direct translation of the genus name, which is taken from the Greek *calamos*, 'reed' and *agrostis*, 'grass.' The species name *purpurascens* means 'purplish' or 'becoming purple,' and refers to the reddish to purplish flower clusters and seed heads of this grass.

NORTHERN REED GRASS • *Calamagrostis inexpansa*

GENERAL: Tufted, from slender rhizomes; stems 40–100 cm tall, **harsh, rough**.

LEAVES: Firm, flat or rolled inwards, 2–4 mm wide, very rough; **ligules large, 4–7 mm long, often torn along edge**.

FLOWER CLUSTER: Dense, firm, narrow panicle, 5–15 cm long; lemmas rough with straight awn and many **callus hairs 1/2 to 3/4 as long as lemma**; glumes 3–4 mm long, slender-pointed.

WHERE FOUND: Marshes, fens, shores and low meadows; common and widespread across our region.

NOTES: Narrow reed grass (*C. neglecta*, also called *C. stricta*) is a similar species, but has smooth stems and leaves, limp leaf blades, shorter (1–3 mm), smooth-edged ligules and thinner glumes. It occurs across our region in wet meadows and marshes and on lakeshores. Narrow reed grass is not found in the prairies and it occurs farther northwards into the Arctic than northern reed grass. • Reed grasses are mostly palatable and nutritious to wildlife and livestock, but they are considered poor or low quality fodder plants because their leaves and stems become very rough as they mature. In the Arctic, however, reed grasses become more important as fodder plants.

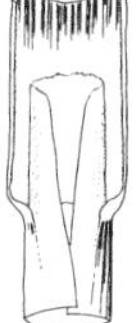

C. neglecta

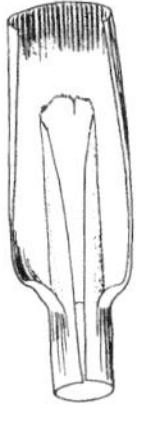

C. inexpansa

BLUEJOINT • *Calamagrostis canadensis*
MARSH REED GRASS

GENERAL: Tussock-forming perennial from creeping rhizomes; stems 60–120 cm tall.

LEAVES: Many, **4–10 mm wide**, flat, **rather limp**; no auricles; ligules 3–6 mm long.

FLOWER CLUSTER: Drooping panicle, 10–20 cm long, narrow and rather dense to loose and relatively open; spikelets 1-flowered; glumes 3–4 mm long, pointed; lemma 2–3 mm long with **straight awn** from just below middle, and **many callus hairs nearly as long as lemma**.

WHERE FOUND: Moist woods, meadows, wetlands, lakeshores and clearings; widespread across our region; north past treeline to Arctic coast; circumpolar.

NOTES: Polar grass (*Arctagrostis latifolia*) resembles bluejoint, but its lemmas are not awned and do not have hairy tufts at their bases. It has runners, stems to 1 m tall, short, broad, flat leaves and a compact, lance- to pyramid-shaped cluster of dark purple spikelets. Polar grass is widespread and circumpolar, and grows in wet meadows and on moist shores, river flats and peaty barrens across the northern part of our region. • Bluejoint is the most common reed grass in North America. In the north it is an important food for bison. • The Cree used bluejoint to make mattresses when they did not have better materials. They also lined winter storage pits with bluejoint and covered the vegetables in these pits with a thick layer of bluejoint to protect them from the frost. • Together with sedges, bluejoint often forms the bulk of the 'beaver-hay' that grows naturally in meadows too wet for cultivation. • Because of its dense leafiness, wide range of habitats, and high genetic diversity, bluejoint would seem to have a high potential for exploitation. However, its light, hair-surrounded seeds are difficult to collect and sow, and this has limited testing and cultivation.

TUFTED HAIR GRASS • *Deschampsia caespitosa*

GENERAL: Densely tufted perennial from fibrous roots; many stems, 20–120 cm tall.

LEAVES: At stem base, **rather stiff**, 1.5–5 mm wide, flat or folded; **ligules prominent, pointed, up to 10 mm long**.

FLOWER CLUSTER: Open, loose panicle, 10–25 cm long, often drooping, with spreading thread-like branches; spikelets near branch tips, 3–5 mm long, usually 2-flowered, **bronze** or purplish, **glistening**; lemmas thin, smooth, **2.5–3.5 mm long**, with fragile awn from near base and tuft of short hairs at base.

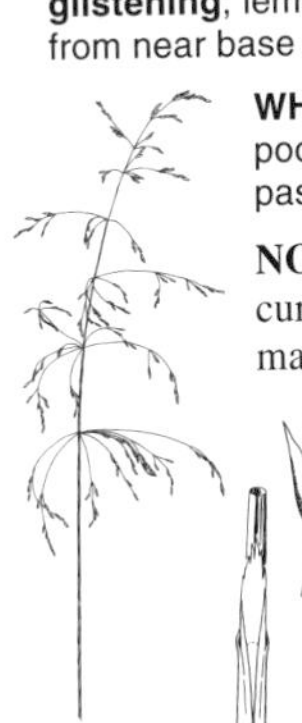

WHERE FOUND: Wet meadows, shores, riverbanks and poorly drained soils; widespread across our region; north past treeline to Arctic coast; circumpolar.

NOTES: Tufted hair grass is a variable species that occurs in a wide range of habitats, from coastal salt marshes to alpine tundra, and ranges from the Arctic to high altitudes in the tropics. • Tufted hair grass is so tough and rigid that horses and cows usually refuse to eat it. Under natural conditions it is generally considered an agriculturally worthless grass of wet soils. However, its vigorous growth under cultivation suggests it may have potential as a forage grass, but its dense tussocks make it difficult to cut into hay. • The attractive flower clusters can be used fresh or dried in flower arrangements.

ROUGH HAIR GRASS • *Agrostis scabra*
TICKLE GRASS, HAIR BENTGRASS

GENERAL: Small, **densely tufted** perennial from fibrous roots; stems slender, **rough, 20–70 cm tall**.

LEAVES: Many, most at stem base, **narrow** (1–3 mm wide or less), **short**, rough; no auricles; ligules thin, translucent, 2–5 mm long.

FLOWER CLUSTER: Very **diffuse panicle**, 10–25 cm long, often purplish, with hair-like, **rough**, brittle, **spreading branches**; spikelets 1-flowered, 2–2.7 mm long, **near branch tips**; glumes 2–2.5 mm long, pointed; lemmas delicate, 1.5–1.7 mm long.

WHERE FOUND: Open woods and moist meadows, shores and beaver dams; common in disturbed areas; widespread across our region.

NOTES: Redtop (*A. stolonifera*; *A. alba* in part) is an introduced species used extensively as a lawn turf grass. It is often abundant on moist disturbances in the southern boreal forest and throughout much of North America. Redtop stems are tufted but usually bent at the base, and it can be distinguished by its well-developed rhizomes or runners. It also has a large, stiff, less diffuse flower cluster with smooth branches bearing spikelets almost to their base, and its leaves are 3–8 mm wide—broader than those of rough hair grass. • The grains of the hair grasses are tiny and these species have no particular agricultural value. However, many are used as fodder plants in haylands and pasture. • The delicate, loose flower clusters become brittle when they dry at maturity, so they easily break away from the stalk to be blown about by the wind, scattering their seeds as they tumble along. • Rough hair grass often is so abundant that it casts a pinkish haze over the ground in middle to late summer; some people call it 'fire grass.' Another common name is 'tickle grass.' You'll understand this name best by walking bare-legged through a patch.

BOG MUHLY • *Muhlenbergia glomerata*

GENERAL: Perennial, from scaly, creeping rhizomes; stems erect, slender, 30–60 cm tall, simple or with a few erect branches at base; nodes and internodes minutely hairy or internodes hairless but rough.

LEAVES: Erect, firm, flat, rough, 2–6 mm wide; minute ligules.

FLOWER CLUSTER: Greenish or purplish panicle, 4–7 cm long, 5–10 mm wide, lower branches somewhat widely spaced; spikelets 5–6 mm long, 1-flowered; **glumes** narrowly lance-shaped, **taper to long, rough awns**; lemmas hairy below middle.

WHERE FOUND: Fens, peaty meadows and wet shores; widespread, but rarely abundant, across boreal forest of prairie provinces north to Great Slave Lake and southeastern Yukon Territory.

NOTES: Mat muhly (*M. richardsonis*) is a densely tufted or matted grass with thick, scaly rhizomes; wiry stems; short, hair-like leaves; and an interrupted, spike-like flower cluster with lead-coloured spikelets with short, blunt-tipped glumes. It grows on grassy, alkaline flats and wet, gravelly, usually calcium-rich shores and riverbanks, eastwards from the southern Yukon. • Bog muhly could be called an ant hill indicator species as it seems to particularly favour growing on the loose soil of elevated ant mounds in fen habitats. • Most species of muhly are fairly palatable to animals, but they are rarely common enough to be considered important forage grasses. • The genus was named in honour of G.H.E. Muhlenberg (1753–1815), a Lutheran minister from Pennsylvania who enjoyed studying the grasses. The species name *glomerata* is from the Latin *glomus*, 'a ball' or 'a sphere,' in reference to the densely clustered spikelets characteristic of this species.

DROOPING WOOD-REED • *Cinna latifolia*
WOOD REEDGRASS

GENERAL: Loosely tufted, sweet-scented perennial; stems slender, 60–150 cm tall.

LEAVES: Limp, flat, **5–15 mm wide**, narrow quickly to sharp tip, rough, **spread at right angles to stem**; **ligules hairy, 5–10 mm long**.

FLOWER CLUSTER: Open, nodding, green or yellowish panicle, 10–30 cm long, with clusters of slender branches; spikelets 1-flowered, 3–4 mm long, detach as unit at maturity; glumes narrow, slender-pointed, 3–3.5 mm long; lemmas similar to glumes, with tiny awn or awnless.

WHERE FOUND: Moist woods, meadows, streambanks, and disturbed areas; widespread across boreal forest; north and west to southern N.W.T. and southern Alaska; circumpolar.

NOTES: Drooping wood-reed increases tremendously on disturbed sites. • Fresh wood-reed leaves can be burned slowly to produce a mosquito-repelling smoke. • The name *Cinna* was taken from the Greek *kinni*, which referred to an unknown grass. The species name *latifolia* is derived from the Latin *latus*, 'broad,' and *folium*, 'leaf,' in reference to the wide leaves of this grass.

ROUGH-LEAVED RICE GRASS • *Oryzopsis asperifolia*
WHITE-GRAINED MOUNTAIN RICE GRASS

GENERAL: Loosely tufted perennial; stems erect to wide-spreading, 20–70 cm tall.

LEAVES: Most at stem base, erect, **firm**, later limp and laying on ground, **flat** or slightly rolled, **3–10 mm wide**, rough, taper at both ends; ligules short, to 1 mm long, hairy-fringed.

FLOWER CLUSTER: Simple, few-flowered, spike-like panicle, 5–10 cm long, with erect branches; spikelets 1-flowered, 6–9 mm long; **glumes 6–8 mm long**, green with translucent edges, abruptly pointed; lemmas hard, rounded, 6–9 mm long, whitish, densely hairy at base, with 5–10 mm long awn at tip.

WHERE FOUND: Dry to moist open woods and clearings; widespread across southern boreal forest and parkland; north and west to southern N.W.T. and northern British Columbia.

O. pungens

O. asperifolia

NOTES: Northern rice grass (*O. pungens*) is short (to 40 cm tall) and densely tufted with short, wiry, hair-like leaves, and it has small spikelets (3–4 mm long). Except for the shape of its spikelets, it does not look much like white-grained mountain rice grass and it tends to grow in drier habitats—sandy-gravelly soils in open forests and on terraces and outcrops—across the boreal forest, north and west to Great Bear Lake and the southern Yukon. • The rice grasses can dominate dry forested sites in the southern parts of our region. Their clusters of reclining leaves form characteristic circles on areas of forest floor without many herbs. • The genus is named *Oryzopsis*, from the Greek *oruza*, 'rice,' and *opsis*, 'like,' because the swollen spikelets of these grasses are very similar to those of cultivated rice (*Oryza sativa*). The species name *asperifolia* is from the Latin *asper*, 'rough,' and *folium*, 'leaf,' in reference to the relatively large, rough leaves.

SLOUGH GRASS • *Beckmannia syzigachne*

GENERAL: Annual, tufted; stems stout, light green, 30–100 cm tall.

LEAVES: Flat, firm, rough, 3–10 mm wide.

FLOWER CLUSTER: Narrow, crowded panicle, composed of **many spikes**; spikelets **1-flowered**, about 3 mm long, **circular, flat**, nearly stalkless in **2 rows** along 1 side of slender axis; glumes as long as spikelet, cross-wrinkled, ridged, abruptly pointed; lemma narrow, pointed, 5-veined, tip barely protrudes beyond glumes.

WHERE FOUND: Shores, marshes, wet meadows and ditches; common across our region.

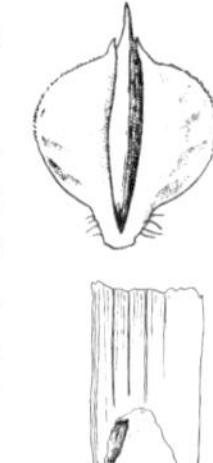

NOTES: Slough grass is a very distinctive and attractive grass. • The seeds of slough grass have been boiled to make meal or porridge in many places around the world, including Japan, Korea, China, Siberia and North America. In Alberta, native groups parched the grains and then used them whole or ground them into flour. • Slough grass provides good hay and pasture until it flowers, after which it becomes too rough and fibrous. It can be a valuable forage grass in wet sites, and it is also able to tolerate relatively high levels of salt in the soil. • The genus is named in honour of Johann Beckmann, a German botanist.

COMMON SWEET GRASS • *Hierochloe odorata*

GENERAL: Sweet-smelling perennial with many, leafy shoots spreading from **long rhizomes**; stems erect, **often single, 30–60 cm tall, base usually purplish**.

LEAVES: 2–3 stem leaves, short, 2–8 mm wide; non-flowering shoots often have long blades; ligules 1–3 mm long.

FLOWER CLUSTER: 4–10 cm long, **open**, pyramidal **panicle**, spreading to backwards-bending branches; spikelets 4–7 mm long, shiny bronze to golden yellow or purplish, tulip-shaped; glumes broad, 4–6 mm long; lemmas hairy, **awnless**.

WHERE FOUND: Moist meadows, lakeshores, streambanks and disturbed areas; less commonly in dry open areas; widespread across our region; circumpolar.

NOTES: Sweet grass was widely used by native peoples, especially the groups of the great plains, as incense in holy ceremonies. It was braided and lit or placed on hot coals and burned as an offering during many gatherings and activities. When participants in a ceremony undertook a prolonged fast, sweet grass could be chewed to extend their endurance. Sweet grass was mixed with tobacco and smoked during some ceremonies. • Sweet grass was also widely used as medicine, perhaps because it was considered holy. Sweet grass tea was used to relieve coughing, vomiting, sore throats and bleeding, and to help a woman expel afterbirth. Externally it was used to treat sore eyes, saddle sores and chapped skin, to keep hair from falling out, and to purify a woman after she had given birth. The smoke of burning sweet grass was inhaled to stop nosebleeds and to relieve colds. • The sweet smell of this grass comes from coumarin, a chemical that also acts as a potent anticoagulant. • In the home, sweet grass was sewn into clothing or placed with stored items as a natural sachet. The bronze-coloured flower clusters are attractive and sweet-smelling in dried flower arrangements. • Because of its long spreading rhizomes, sweet grass stabilizes loose soil on slopes. Often it is found growing on roadsides and in other disturbed areas.

REED CANARY GRASS • *Phalaris arundinacea*

GENERAL: Robust perennial from **long, thick, pinkish scaly rhizomes**; stems blue-green, 60–200 cm tall.

LEAVES: Up to 10 on stem, **flat, 6–20 mm wide**, 10–30 cm long, rough; ligules 2–5 mm long, usually tattered and turned backwards, slightly short-hairy; no auricles.

FLOWER CLUSTER: Panicle, compact (at least initially), 6–18 cm long; **spikelets 3-flowered**, 5–6 mm long, pale or purplish, **crowded on side branches**; glumes 4–5 mm long, narrow, ridged, with minute hairs; **1 fertile lemma**, at tip of spikelet, **3–4 mm long, shiny** when mature; 2 sterile lemmas below fertile lemma, reduced to tiny, brownish, hairy scales.

WHERE FOUND: Marshes, streambanks, shores, and wet, disturbed sites; introduced for forage in some hay fields; widespread across our region, north and west to southern N.W.T and southern Yukon; more or less circumpolar.

NOTES: Dried reed canary grass was first rubbed (to soften it) and then stuffed into clothing and footgear for insulation. It was also widely used for mats, either woven for repeated use, or simply spread in a layer on the ground. At one time, reed canary grass was fermented in urine, producing a sudsy solution that was used for washing. The dried roots were used as scrubbers in steambaths. • 'Canary grass' was the name given to a close relative, *Phalaris canariensis*, because it was cultivated in southern Europe to produce seed for caged songbirds or because it was first described from the Canary Islands. • *Phalaris* was an ancient Greek name for an unknown grass. The species name *arundinacea* means 'resembling a reed'; it is from the Latin *(h)arundo*, 'a reed' or 'a cane.'

COMMON REED GRASS • *Phragmites australis*
GIANT REED GRASS

GENERAL: Stout, tall, **reed-like** perennial, from stout, extensively spreading rhizomes and stolons; stems leafy, erect, **1–3 m tall**.

LEAVES: Flat, 1–4 cm wide, 20–40 cm long, stiff, with **loose overlapping sheaths that twist in wind, aligning leaves on 1 side of stem**; **ligules 1.5–3 mm long, fringed with hairs**.

FLOWER CLUSTER: Dense, feathery panicle, 10–40 cm long, purplish in flower, straw-coloured to greyish in fruit; spikelets 3–6-flowered, 6–15 mm long, with tuft of **long, silky hairs** from near base longer than florets.

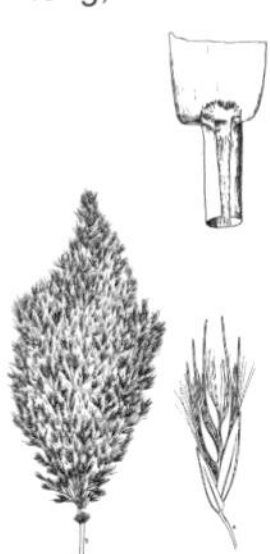

WHERE FOUND: Marshes, ditches and shores; common across southern boreal forest and parkland; north and west to southwestern N.W.T.; circumpolar.

NOTES: Also called *P. communis*. • The Chipewyan occasionally ate the inside of the base of the stem in spring. They also used hollow sections of the stems as pipe stems. There are some reports of native people eating the roots of common reed grass, but mostly they used the sugar these tall grasses produce. When the stems are punctured, they exude a pasty substance that hardens into a gum. This was collected and compressed into balls that were eaten like candy. Some native groups put bundles of cut, dried reeds on blankets and shook off the sugar crystals. This was then dissolved in water to make a sweet, nourishing drink. Native groups to the south collected, dried and ground the stalks, and sifted out the flour. This flour contained so much sugar that when it was placed near a fire it would swell and turn brown. It was then eaten like taffy—an early version of the roasted marshmallow. • Common reed grass is common all over the world. It is often important for binding the soil on river banks. • Moose, caribou and muskrats all feed on the thick, sweet, nutritious root stocks, which have been suggested as a source of fodder for livestock.

TALL MANNA GRASS • *Glyceria grandis*

GENERAL: Tufted perennial with creeping rhizomes, sod-forming; stems stout, tufted, 1–1.5 m tall.

LEAVES: Flat, 15–40 cm long, 6–12 mm wide; sheaths closed to near top.

FLOWER CLUSTER: Large, open, somewhat nodding **panicle, 20–40 cm long**; spikelets 5–6 mm long; **glumes whitish, pointed; lemmas purplish**, narrowly egg-shaped, about 2.5 mm long.

WHERE FOUND: Sloughs, marshes, low meadows, ditches, shores and damp ground; common and widespread across our region, north to central Yukon and interior Alaska.

NOTES: Graceful manna grass (*G. pulchella*) resembles tall manna grass but it is generally smaller (40–60 cm tall). Also, it has narrower leaves (2–6 mm wide), a shorter (less than 20 cm), few-flowered panicle with bronze-purple spikelets, blunt glumes and lemmas with broad (rather than narrow), translucent edges. It is a less common species, but its habitat and range are similar to that of tall manna grass. • The grains of many species of *Glyceria* are edible—hence the name manna grass. Manna was the food miraculously supplied to the Israelites in the wilderness. The seeds of some European manna grasses were considered a delicacy, used to thicken soups and gruels and to make good quality bread. In the 1700s, these seeds were sold to high-ranking Germans and Swedes for use in food, because of their excellent flavour and nourishing attributes. Linnaeus reported that geese fattened more quickly on manna seeds than on any other feed. The small grains are most easily gathered by beating the seed heads over a sheet or other container, much the same way native people collected wild rice. • Manna grasses are all tender and readily eaten by many forms of wildlife but furnish comparatively little forage since they occur in wet areas, usually in limited quantity. Tall manna grass provides excellent cover for waterfowl.

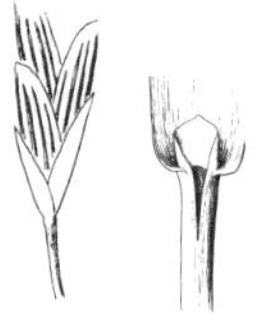

FOWL MANNA GRASS • *Glyceria striata*

GENERAL: Perennial, from rhizomes, often in large clusters; stems slender, pale green, 30–80 cm tall.

LEAVES: Flat or folded, 2–4 mm wide; sheaths closed to near top.

FLOWER CLUSTER: Open, egg-shaped panicle, 10–20 cm long, erect or drooping at tip; **spikelets** 3–4 mm long, **often purplish**; glumes blunt; lemmas oblong, blunt, about 2 mm long, **prominently 7-veined**, more or less pale at tip.

G. striata

WHERE FOUND: Shallow water and wet meadows; common across our region, north and west to upper Mackenzie valley; disjunct in southern and interior Alaska.

NOTES: Northern manna grass (*G. borealis*) is another species with similar habitat requirements and range. It is distinctive in the genus in that it has upright, narrow panicles with long (7–15 mm), narrow, cylindrical spikelets. It also has flattened leaf sheaths. • Manna grasses are rather coarse, but they are eaten by cattle up until the time of flowering. Unfortunately, they are often infected with smut (a fungal disease) at the time of flowering and fruiting, and this can form strong acids that are harmful to cattle. Manna grass plants contain cyanogenic glycosides, which can release hydrocyanic acid, a violent **poison**. Fowl manna grass may have caused the death of calves in Maryland. The hay, however, is said to be harmless. • The species name, from the Latin *stria*, 'furrow' or 'channel,' refers to the prominent parallel veins on the lemmas.

G. borealis

KENTUCKY BLUEGRASS • *Poa pratensis*

GENERAL: Sod-forming perennial with **long creeping rhizomes**; stems erect, 30–100 cm tall, often tufted.

LEAVES: Many, soft, flat or folded, 2–5 mm wide, dark green, distinctly boat-shaped at tip; **ligules 1–3 mm long**; no auricles.

FLOWER CLUSTER: Rather dense to open, **pyramid-** or egg-shaped, 5–15 cm long panicle, tends to curve to 1 side when mature, **usually 3–5 branches at each joint**; spikelets crowded, 3–5-flowered, 3–6 mm long, green or purplish-tinged; **lemmas 3.5–4 mm long**, distinctly 5-veined, hairy, **with cobwebby hairs at base**.

WHERE FOUND: Open woods and prairie; widely introduced and possibly native; widespread across our region, north and west to Great Bear Lake and western Alaska; circumpolar.

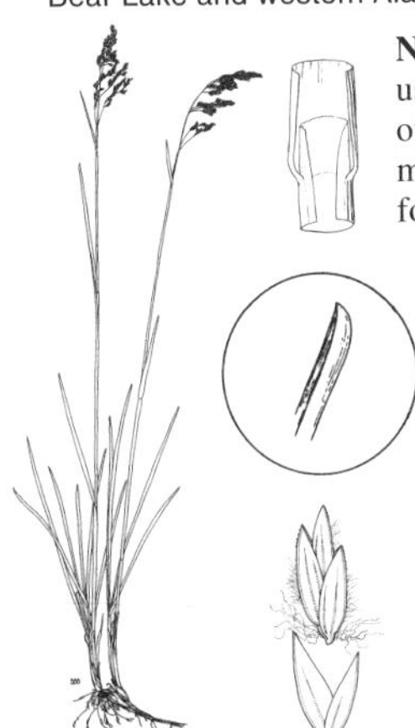

NOTES: Kentucky bluegrass, an extensively used lawn and pasture grass, is highly tolerant of close grazing and mowing. In the western mountain ranges, bluegrasses rank high as forage species and as cultivated pasture grasses. Although the grains of the bluegrasses are edible, they are much too small to warrant collecting. Also, bluegrasses can be heavily infected with the **poisonous** fungus, ergot (see *Bromus ciliatus*, p. 272). • Kentucky bluegrass is the state flower of Kentucky, 'the bluegrass state.' Mandolin player Bill Monroe, a Kentucky native, named his band the 'Bluegrass Boys,' and the fast and furious country-tinged music they played came to be called 'bluegrass' music.

FOWL BLUEGRASS • *Poa palustris*

GENERAL: Tall, loosely tufted perennial from fibrous roots; stems 30–100 cm tall, usually curved and purplish at base.

LEAVES: Soft, flat or folded, boat-shaped at tip, 1–4 mm wide, rough; ligules 3–5 mm long.

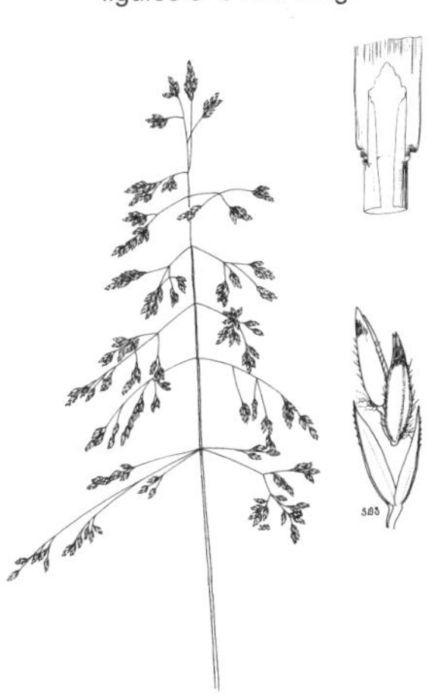

FLOWER CLUSTER: Open, loose panicle, 10–30 cm long, pyramidal or oblong, **fine, spreading branches in whorls of 4 or 5**; spikelets 2–4-flowered, 3–4 mm long; lemmas usually bronze at tip, **cobwebby at base, 2–3 mm long**.

WHERE FOUND: Moist meadows, woods and clearings, wet ditches and shores; widespread across our region, north and west to Great Bear River and southern Alaska; circumpolar.

NOTES: The Chipewyan boiled the panicles of fowl bluegrass to make a hair rinse that was said to make a woman's hair grow longer and thicker. • Bluegrasses are often used to make immunizing vaccines or allergens to treat people suffering from hay fever.

TIMBERLINE BLUEGRASS • *Poa glauca*
GLAUCOUS BLUEGRASS

GENERAL: Densely tufted perennial, often blue-green **with whitish bloom**; stems stiff, 10–30 cm tall (sometimes to 60 cm).

LEAVES: Most at stem base, 1–2 mm wide, flat, but boat-shaped at tip; ligules 0.5–2 mm long, hairy; no auricles.

FLOWER CLUSTER: Compact to open, **narrow**, 3–10 cm long panicle, ascending branches; few spikelets, 2–3-flowered, 5–6 mm long, bluish green to purplish; lemmas 2.5–4 mm long, with short hairs, but **no basal tuft of cobwebby hairs**.

WHERE FOUND: Rocky slopes and sandy soils; widespread across boreal forest, north past treeline through arctic islands; circumpolar.

NOTES: Both the common name 'bluegrass' and the species name *glauca* refer to the bluish white bloom that sometimes covers the leaves, stems and flowering heads of timberline bluegrass. Some particularly 'blue' varieties of bluegrass are cultivated for use as decorative plants.

FRINGED BROME • *Bromus ciliatus*

GENERAL: Coarse, loosely tufted perennial; stems slender, 60–100 cm tall; nodes hairless or sometimes hairy; sheaths hairless or sometimes short-hairy.

LEAVES: Flat, 4–8 mm wide, hairless to sparsely soft hairy.

FLOWER CLUSTER: Loose, open, nodding panicle 10–20 cm long; upper glume lance-shaped, 7–10 mm long, 3-veined; lower glume narrower, 5–7 mm long, 1-veined; spikelets greenish 1–2 cm long, 4–10-flowered; **lemmas** 8–10 mm long, **long-hairy near edges**, hairless or sparingly hairy on back, rarely hairless throughout, with straight, 3–5 mm long awn.

WHERE FOUND: Open woods and meadows; found across our region, north and west to Great Bear Lake and interior Alaska.

NOTES: • Whenever grains are eaten, care must be taken to avoid any that have become infected with ergot (*Claviceps*), a **poisonous fungus**. In infected plants the grain is replaced by a black mass of spores, which, if eaten, causes **severe illness or death**, in both livestock and humans. Chronic poisoning from eating small amounts of ergot for long periods of time is characterized by loss of the nails in mild cases and loss of hands or feet plus occasional gangrene of internal organs in extreme cases. 'Convulsive ergotism,' a dreaded disease reported in many historical accounts from Europe in the 1700s, appears to result from eating large amounts of ergot in a short time. Medicinally, some alkaloids extracted from ergot are used to induce uterine contractions in labour. • Grasses are wind-pollinated. Their anthers, dangling by hair-like filaments, are easily shaken by the slightest breeze. Many people with allergies react to the large amounts of pollen produced by many grasses. • *Bromus* is from the Greek *bromos*, 'an oat'; *ciliatus* is Latin for 'fringed with hairs,' and refers to the rows of long hairs that fringe the lemmas of the spikelets.

NORTHERN BROME • *Bromus inermis* PUMPELLY BROME ssp. *pumpellianus*

GENERAL: Perennial from rhizomes, often sod-forming; stems erect, 20–100 cm tall, with **hairy nodes**.

LEAVES: Flat, 5–15 mm wide,prominent veins; auricles usually present on some leaves; ligules 1–2 mm long.

FLOWER CLUSTER: Ascending to erect, narrow to open panicle, 5–20 cm long, 1–4 erect or arching branches at each node; spikelets 15–35 mm long, 7–10-flowered, relatively narrow, often purplish-tinged; upper glume 6–11 mm long, 3-veined; lower glume smaller, usually 1-veined; lemmas 9–12 mm long, **fuzzy-hairy**, with straight, **0–3 mm long awn**.

WHERE FOUND: Meadows, open woods, fields and waste ground; widespread across boreal forest; extends north past treeline to Arctic coast; circumpolar.

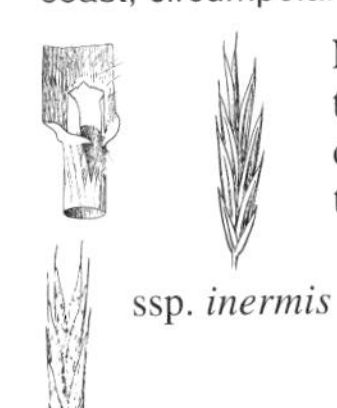

ssp. *inermis*

ssp. *pumpellianus*

NOTES: Also called *B. pumpellianus*. • This native species can be distinguished from the introduced **smooth brome** (*B. inermis* ssp. *inermis*) by the presence of hairs on the lemmas and at the stem nodes. Smooth brome is widespread across our region and well established on roadsides and other disturbed areas, and in old fields and meadows. It is an important component of hayfields and pastures. • Smooth brome is a fairly drought-resistant grass capable of growing on sandy and stony soils. It has been cultivated as a hay and fodder plant in Europe and North America for many years and it is still frequently seeded in western rangelands. Today it is sown extensively along new highways and on earthen dams because of its rapid growth and its soil-binding abilities once it has become established. • Smooth brome is highly invasive of natural habitats, posing a considerable danger to natural areas.

FALSE MELIC • *Schizachne purpurascens* PURPLE OAT GRASS

GENERAL: Slender, loosely tufted perennial from rhizomes; stems erect, with curved base, hairless, 40–80 cm tall.

LEAVES: Flat, 1–6 mm wide, limp, narrow at base; ligules 0.5–1.5 mm long; sheaths closed at first but splitting with maturity.

FLOWER CLUSTER: Open, drooping panicle, 6–15 cm long, **often purplish**, few branches (1–2 per node); 1–2 spikelets per branch, 3–5-flowered, 20–25 mm long, purplish, bronze or pale; glumes of different lengths, much shorter than lemmas; lemmas lance-shaped, 8–10 mm long, with **10–15 mm long bent awn from just below 2 teeth at tip**, and tuft of silky hairs at base.

WHERE FOUND: Open woods and grassland; widespread across our region, north and west to Great Bear Lake; disjunct in southern Yukon and south-central Alaska.

NOTES: The Chipewyan made wicks for grease lamps by twisting together the soft grasses collected from mouse nests. The strand was then laid across the grease in a birch-wood bowl with one end sticking out to be lit. • The genus name *Schizachne* is derived from the Greek *schizo*, 'split,' and *achne*, 'chaff,' in reference to the 2-cleft tip of the lemma. The species name means 'purplish' or 'becoming purple'; it refers to the reddish to purplish flower clusters and seed heads of this grass.

Ferns and Allies

Ferns and their 'allies' (horsetails and club-mosses) are vascular plants—that is, they have internal tubes for conducting fluids, like most plants in this guide (except lichens and most bryophytes). However, they reproduce not by seeds but by spores (as do bryophytes and lichens). Ferns, horsetails and club-mosses tend to be most abundant in moist habitats, although some have adapted to dry sites.

Our region is home to approximately 20 species of 'true' ferns, 6 species of grape ferns and rattlesnake ferns, 10 species of horsetails, and 6 species of club-mosses. More information about ferns and allies is available in Cody and Britton (1989), Lellinger (1985), Taylor (1970), Vitt et al. (1988), and in the general floras mentioned in the introduction to this guide.

Ferns

The grape or rattlesnake ferns (*Botrychium*) are relatively small, with fleshy roots and short, vertical, buried stems (**rhizomes**). Each rhizome bears a single leaf divided into a sterile expanded blade and a fertile, spike-like or panicle-like portion with clusters of spore sacs (**sporangia**) arranged in double rows. The spore sacs are spherical, and split in 2 to release numerous yellow spores.

The 'true' ferns have creeping or erect rhizomes (often very scaly) and stalked, erect or spreading, often large leaves (**fronds**). The leaf blades are curled as buds (called '**fiddleheads**'), and they are usually lobed, divided, or variously compound. Fertile and sterile leaves are most often alike, but they are dissimilar in some genera. The spore sacs (**sporangia**) are grouped in spore clusters (**sori**), which are sometimes covered by a membrane called the **indusium**.

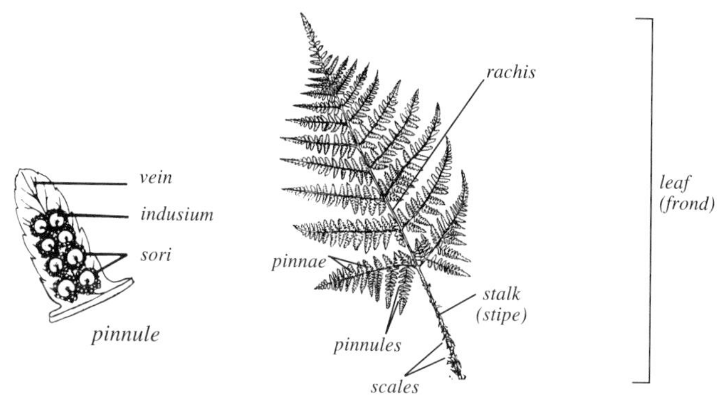

Pinnate Fern Leaf (Frond)

Key to Ferns

1a. Spore sacs in clusters on leafless stalk that projects from upper side of vegetative leaf 2

2a. Plants usually less than 20 cm tall; sterile leaves longer than broad, pinnate, divisions broadly club- to fan-shaped .... ***Botrychium lunaria*** (see *B. virginianum*)

2b. Plants usually more than 20 cm tall; sterile leaves not much longer than broad, often shorter, 1–3 times pinnate, divisions not club- or fan-shaped .... ***Botrychium virginianum***

1b. Spore sacs on underside of vegetative leaves or on separate, modified fertile leaves 3

3a. Sterile and fertile leaves dissimilar; ultimate segments of fertile leaves long and narrow, with edges strongly rolled in towards lower surface 4

4a. Sterile and fertile leaves 2–3 times pinnate, parsely-like; plants up to 30 cm tall; grow on rocky, dry sites .... ***Cryptogramma crispa***

4b. Sterile and fertile leaves once pinnate; plants 0.5–1.5 m tall; grow on rich, alluvial sites .... ***Matteuccia struthiopteris***

3b. Sterile and fertile leaves similar 5

5a. No indusium (sometimes present but inconspicuous in *Woodsia*) 6

6a. Leaves evergreen, pinnately lobed, not divided into leaflets .... ***Polypodium virginianum***

6b. Leaves deciduous, 2–3 times pinnate 7

7a. Leaves delicate, smooth, lower 2 divisions nearly as long as end one; ultimate leaflets pointed .... ***Gymnocarpium dryopteris***

7b. Leaves firm, gland-dotted, lower 2 divisions about half as long as end one; ultimate leaflets blunt .... ***Gymnocarpium robertianum*** (see *G. dryopteris*)

5b. Indusium present 8

8a. Indusium radiates in hair-like or deeply torn segments; veins do not reach edge of leaf 9

9a. Main leaf axis and leaflets lack chaffy scales; leaf stalk and main axis yellowish, dark brown only at base .... ***Woodsia glabella*** (see *W. ilvensis*)

9b. Main leaf axis and leaflets have chaffy scales; leaf stalk and main axis reddish brown to dark purple .... ***Woodsia ilvensis***

8b. Indusium flap- or hood-like, or rounded to kidney-shaped 10

10a. Indusium flap- or hood-like 11

11a. Indusium flap-like, elongated, attached along 1 edge, often fringed with long, coarse hairs; plants to 1 m tall; found in moist woods .... ***Athyrium filix-femina***

11b. Indusium hood-like, attached at a point; plants to 30 cm tall; often grow among rocks .... ***Cystopteris fragilis***

10b. Indusium round to kidney-shaped 12

12a. Plants to 1 m tall; leaves usually 3 times divided; often found in moist woods .... ***Dryopteris austriaca***

12b. Plants to 30 cm tall; leaves twice divided, rust-coloured below; dry leaves have spicy aroma; often grows among rocks .... ***Dryopteris fragrans*** (see *D. austriaca*)

VIRGINIA GRAPE FERN • *Botrychium virginianum*
RATTLESNAKE FERN

GENERAL: Erect perennial from short, soft rhizomes and cluster of fleshy roots; stems single, **15–80 cm tall**.

LEAVES: Single, stalkless blade borne near middle of stem, thin, broadly triangular, 20–50 cm long, 3 main lobes, **1–3 times pinnate**, **sterile**; fertile leaf single (see spore clusters, below).

SPORE CLUSTERS: Many, spherical, yellow, stalkless, on spike-like, pinnately compound, fertile leaf, 2–15 cm long, with 3–20 cm long stalk from base of sterile leaf.

WHERE FOUND: Moist woodlands, thickets, and meadows; widespread across southern boreal forest north and west to southwestern N.W.T. and northern British Columbia; circumpolar.

NOTES: Moonwort (*B. lunaria*) (inset photo) is a much smaller grape fern (usually 7-15 cm tall) that is scattered across boreal North America and around the world in open grassy sites, sometimes in thickets and open deciduous forest. Its single, fleshy, yellowish green leaf, with 2–5 pairs of fan-shaped 'leaflets,' is distinctive. • Native peoples used the roots of Virginia grape fern to make poultices or lotions for snakebites, bruises, cuts and sores. However, these plants are usually rare or uncommon, and they should not be collected. • Virginia grape fern is also called 'rattlesnake fern,' presumably because the fertile spike emerging from the sterile leaf somewhat resembles the tail of a rattlesnake. • The species name, *virginianum*, indicates that this species was first identified in Virginia. Many plants are named after Virginia, because many North American botanists began collecting there.

OSTRICH FERN • *Matteuccia struthiopteris*

GENERAL: Large, **erect, spreading** perennial, **0.5–1.5 m tall**, from erect, woody base covered in old leaf stalk bases; spreads by black underground stolons and coarse, scaly rhizomes.

LEAVES: Sterile leaves green, **many**, in whorls, **lance-shaped** in outline, **tapered at both ends, once pinnate**; 20–30 leaflet pairs, alternate, smaller towards base, deeply cut into many, oblong, blunt lobes, with edges curved under; stalks short, dark green to black at base; olive-green to blackish fertile leaves markedly different (see spore clusters, below).

SPORE CLUSTERS: On many, rigid, erect, 20–60 cm tall fertile leaves which **turn dark brown**, on veins, long, covered by rolled-under edges of leaflets.

WHERE FOUND: Moist to wet forests, along streambanks and in swamps; widespread across boreal forest, north and west to southwestern N.W.T.; disjunct in Alaska; circumpolar.

NOTES: The Chipewyan mixed ostrich fern with other herbs in compound medicines, but never used it alone. It was also considered a heart medicine, a use learned through a dream. The Woods Cree used the base of the stalks of green leaves as an ingredient in a medicinal tea to treat back pain and to speed the expulsion of the afterbirth following delivery. • Of all our ferns, the fiddleheads of this species are considered the largest, tastiest, and safest to eat. Fiddleheads in markets are usually ostrich fern. Farther east, some native groups cooked, boiled, or roasted the scaly rootstocks for food. • The name 'ostrich fern' refers to the feather-like arrangement of the leaf segments. The genus *Matteuccia* is named for Italian physicist Carlo Matteucci (1800–68). This species often forms large clonal populations which can reach 1/4 ha in size in open woods.

LADY FERN • *Athyrium filix-femina*

GENERAL: Large, clustered, erect to spreading perennial from stout, densely scaly, short rhizomes.

LEAVES: Narrowly to broadly **lance-shaped** in outline, 40–100 cm long (sometimes to 2 m), **tapered at both ends**, 2–3 times pinnate; 20–30 leaflet pairs, alternate, linear-lance-shaped, up to 15 cm long and 4 cm wide; 30–60 smaller divisions per leaflet, 1–2 cm long, alternate, toothed or again divided; stalks stout, fragile, scaly at base, about 1/3 as long as blade.

SPORE CLUSTERS: Long and curved, with crescent- to kidney-shaped indusium, on all leaves.

WHERE FOUND: In moist to wet forests and thickets and along streambanks; sometimes in swamps; scattered across boreal forest north and west to central Yukon and western Alaska; generally south of 55°N in Manitoba; circumpolar.

NOTES: The large leaves of lady fern were used by native peoples for laying out or covering food, especially drying berries. The fiddleheads were eaten in the early spring when they were 7–15 cm tall. They were boiled, baked, or eaten raw with grease. • Native peoples used tea made from lady fern roots to stimulate urination, to stop breast pains associated with childbirth, and to stimulate milk production in caked breasts. Tea made from the stems was taken to ease labour. The roots were dried and ground to make a powder that was dusted on sores to aid healing. • Oil from from the rootstocks of lady fern or male fern (*Dryopteris filix-mas*) has been used since the times of Theophrastus and Dioscorides (1st century A.D.) to expel worms from both humans and livestock. A single, strong dose was often sufficient, but if the dose was too large, it could cause muscular weakness, coma and, most frequently, blindness.

SPINULOSE SHIELD FERN • *Dryopteris austriaca*
SPINY WOOD FERN

GENERAL: Large, **clustered, erect, spreading** perennial, **usually 30–90 cm tall**; from stout rhizomes covered with brown scales and remains of old stalk bases.

LEAVES: Triangular to oblong-lance-shaped in outline, 10–60 cm long, 2–3 times pinnate; 5–15 (sometimes 20) leaflet pairs, slightly alternate, **lowest pair not symmetrical**; 12–20 pairs of smaller divisions, slightly alternate; smallest segments softly spiny-toothed; stalks dark, usually shorter than leaves, with many brown scales, especially near base.

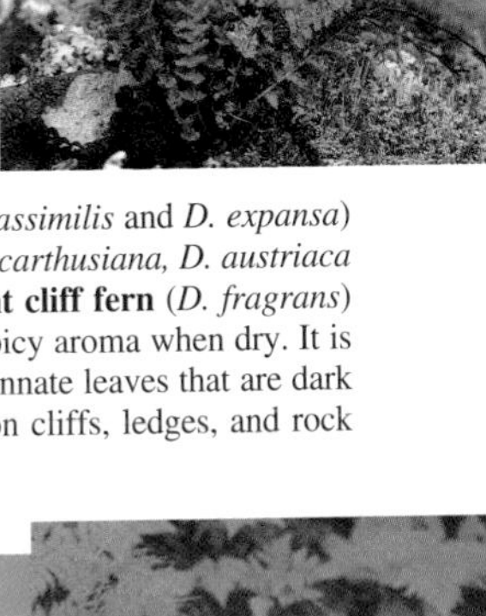

SPORE CLUSTERS: Rounded dots on veins, partly covered by rounded to kidney-shaped indusium.

WHERE FOUND: Moist forests, thickets and openings; sometimes in swamps; widespread across boreal forest; rare in parkland; north and west to southwestern N.W.T. and western Alaska; more or less circumpolar.

NOTES: Also called *D. carthusiana*, *D. spinulosa*, *D. dilatata*, *D. assimilis* and *D. expansa*. • Spinulose shield fern includes a complex group of species that have undergone a taxonomic revision in recent years. Basically, it includes 2 major groups, **broad spinulose shield fern** (which includes *D. dilatata*, *D. assimilis* and *D. expansa*) and **narrow spinulose shield fern** (which includes *D. carthusiana*, *D. austriaca* and *D. spinulosa*). • **Fragrant shield fern** or **fragrant cliff fern** (*D. fragrans*) (inset photo) is so named because its leathery, evergreen leaves have a spicy aroma when dry. It is much smaller (to 30 cm tall), with dense tufts of narrow, usually twice pinnate leaves that are dark green above and rust-coloured below from dense, dry scales. It grows on cliffs, ledges, and rock slopes across our region, primarily north of 57°N.

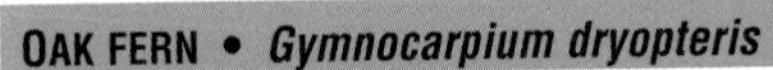

OAK FERN • *Gymnocarpium dryopteris*

GENERAL: Delicate perennial from slender, spreading, blackish rhizomes with brown, fibrous scales; stalks **usually single, but often in masses, erect**, 8–35 cm tall.

LEAVES: Broadly triangular in outline, 5–18 cm long, 5–25 cm wide, **2–3 times pinnate, hairless**; 3 major, approximately equal divisions, each with 4–7 pairs of lobes or leaflets, smallest segments round-toothed; stalks very slender, shiny, straw-coloured, sparsely scaly at base, usually longer than blades.

SPORE CLUSTERS: Small, **circular** dots on veins, near leaflet edges, **no indusium**.

WHERE FOUND: Moist forests, thickets and rocky slopes; scattered across boreal forest north to Great Slave Lake and approximately 65°N in Yukon and Alaska; circumpolar.

NOTES: Also called *Dryopteris disjuncta*. • **Northern oak fern** (*G. jessoense*, also called *G. robertianum* or *Dryopteris robertiana*) is found on limestone and shale ledges and rock slopes across northern North America and around the world. Because of its specific habitat preferences, it is usually uncommon. Northern oak fern resembles oak fern, but its leaves are more narrowly triangular and they have dense, minute glands. • The Cree used crushed oak fern leaves to repel mosquitoes and to soothe mosquito bites. • These delicate little ferns can provide ground cover in shady locations. They often need soil that has been treated with lime. • Oak fern is a characteristic 'indicator species' of moist sites. As moisture levels increase other species such as horsetails (*Equisetum* spp.) become more abundant. • The genus is named *Gymnocarpium*, meaning 'naked fruit,' because the spore clusters are not covered by an indusium.

ROCK POLYPODY • *Polypodium virginianum*

GENERAL: Evergreen, fronds arise singly along scaly, branching rhizome.

LEAVES: Oblong to lance-shaped in outline, 5–20 cm long, 2–7 cm wide; 10–20 leaflet pairs, alternate to nearly opposite.

SPORE CLUSTERS: Large, round, at leaflet edges; no indusium.

WHERE FOUND: Rock outcrops, mossy ledges and cliffs; common in wooded parts of Precambrian Shield across prairie provinces north to Great Bear Lake; circumpolar.

NOTES: Also called *P. vulgare* var. *virginianum*. • Includes *P. sibiricum*. • The Woods Cree made a medicinal tea from rock polypody to treat tuberculosis. Ancient Greco-Roman physicians and, more recently, herbalists prescribed it as a remedy for coughs and chest problems. They also used rock polypody as an appetite stimulant and as a mild to strong laxative, depending on the strength of the preparation. The powdered rhizomes were taken internally to expel tapeworms and applied externally as a liniment. Animal experiments have verified these effects. • Because the rhizomes of these ferns have a sweetish taste, rock polypody has also been called 'angel-sweet' or 'wild licorice.' The licorice-flavoured rhizomes contain sucrose and fructose, as well as osladin, a compound that is said to be 300 times sweeter than sugar. Although polypody rhizomes have seldom been used as food, they are occasionally used for flavouring, and were often chewed as an appetizer. However, care should be taken when using ferns, as the **toxicity** of most species is not known. Rock polypody can cause a **rash**, but this soon disappears. • The genus name comes from the Greek *polys*, 'many,' and *podos*, 'a foot,' in reference to the branching rhizomes.

RUSTY WOODSIA • *Woodsia ilvensis*

GENERAL: Densely tufted perennial from branching, scaly rhizomes; scales brown, 3–4 mm long, toothed, with sparse, hairy fringe.

LEAVES: Oblong-lance-shaped in outline, 5–20 cm long, 10–35 mm wide, firm, dark green, **hairy and usually scaly below, commonly aromatic**; mostly 10–20 leaflet pairs, oblong, deeply notched; edges often rolled under and lobed; stalks **jointed near base, brown, firm, scaly, hairy**.

SPORE CLUSTERS: Rounded, separate but often merge with age; **minute, disc-shaped indusium with 10–20 large, dark hairs**, curved up around spore clusters.

WHERE FOUND: Rock crevices; usually on Precambrian or acid rocks; scattered across boreal forest; circumpolar.

W. ilvensis

W. glabella

NOTES: Smooth woodsia (*W. glabella*) (inset photo) is similar, but it has straw-yellow or greenish stalks that are not scaly above the joint, narrower, hairless leaves (6–14 mm wide), and an indusium with 5–8 hairs that are only slightly longer than the spore clusters. Smooth woodsia grows scattered across the boreal forest and around the world in moist, calcium-rich, often shaded rock crevices, ledges and rock slopes. It is rare in the prairie provinces but becomes more common in the north. • Although both of these little woodsia ferns are too widespread to be considered rare on a national or worldwide basis, they have scattered distributions and are classified as rare in many parts of their ranges. Rusty woodsia is rare in the Yukon Territory, Alberta, British Columbia, Illinois and Iowa; endangered in North Carolina; and possibly extirpated in Rhode Island. Smooth woodsia is rare in British Columbia, Alberta, Saskatchewan, Manitoba, Ontario, Nova Scotia, New Hampshire and New York; and endangered in Maine, Minnesota and Vermont. • The genus *Woodsia* was named in honour of Joseph Woods (1776–1864), an English architect and botanist who studied the genus *Rosa*.

FRAGILE BLADDER FERN • *Cystopteris fragilis*

GENERAL: Delicate perennial from slender, densely scaly, creeping rhizomes, 4–35 cm tall; usually in small, erect clusters.

LEAVES: Lance-shaped in outline, 7–18 cm long, tapered at both ends, 2–3 times pinnate, essentially hairless; 8–15 leaflet pairs; 3–8 pairs of smaller divisions, irregularly toothed; stalks **straw-coloured**, 3–12 cm long, brittle, **hairless**, as long as or shorter than blades.

SPORE CLUSTERS: Small, **roundish** dots on veins, partially covered by **hood-like**, toothed or lobed **indusium**, soon withers and curls back.

WHERE FOUND: Moist to dry, rocky woods, openings, slopes and cliffs; widespread across our region, north through arctic islands; circumpolar

NOTES: This small fern was not widely used in the North, but native groups farther south used a tea of fragile bladder fern to relieve stomach disorders and chills and to bathe injuries. • Fragile bladder fern is one of our most widely distributed ferns. Its delicate leaves grow best among rocks in moist, shady sites. They are very sensitive to frost and die back quickly with the onset of winter. • This species is sometimes confused with rusty woodsia, but that fern has hairy fronds and its indusium is composed of hair-like segments radiating from the base of the spore cluster and is never flap-like. • *Cystopteris* is from the Greek *kystos* 'bladder,' and *pteris*, 'fern,' in reference to the hood-like indusium; *fragilis* means 'brittle,' in reference to the stems.

PARSLEY FERN • *Cryptogramma crispa*

GENERAL: Small, slender, **densely clustered** evergreen perennial with short, scaly rhizomes and many fibrous roots; sterile and fertile **leaves markedly different**.

LEAVES: Sterile leaf blades egg-shaped in outline, 5–7 cm long, crisply firm, hairless, usually **3 times pinnate**; 3–6 leaflet pairs, largest at base; 2–4 pairs of smaller divisions, lobed or finely toothed; stalks **straw-coloured** to greenish, as long as or longer than blades; fertile leaves erect, taller than sterile leaves, to 20 cm tall.

SPORE CLUSTERS: Borne along length of leaflets, **covered by thin, yellowish, rolled-under leaflet edges**.

WHERE FOUND: Rocky outcrops and crevices, usually on acidic rock; widely scattered across boreal forest, north and west to Great Bear Lake and southern Alaska.

NOTES: Also called *C. acrostichoides*. • Parsley fern avoids lime-rich soils and only grows on acidic soils. • The genus name *Cryptogramma* is derived from the Greek *cryptos*, 'hidden,' and *gramma*, 'line,' in reference to the lines of spore clusters hidden under the rolled-under leaflet edges. *Crispus* means 'curled' or 'wrinkled,' again in reference to the leaves. This species is called 'parsley fern' because the fronds look somewhat like parsley.

Horsetails

The horsetails are non-woody plants with rhizomes and erect, aerial, usually hollow, grooved, regularly jointed stems containing silica, which makes them harsh to the touch. Slender, green branches (which could be mistaken for leaves) grow from the conspicuous nodes of some species. The **leaves** are **reduced** to tiny whorls, fused to form a sheath around the stem, with only their tips free. They usually lack chlorophyll (the stems and branches are green and photosynthetic). Spore-bearing cones (**strobili**) are produced atop the stems. The horsetail family (Equisetaceae) has a single genus, *Equisetum*, with about 20 species worldwide.

Key to the Horsetails (*Equisetum*)

1a. Stems evergreen, perennial, usually unbranched; cones sharp-pointed at tip (the scouring-rushes) 2

2a. Stems low, thin, zigzag, rather curly and twisted, not hollow; sheaths have 3 teeth .... ***Equisetum scirpoides***

2b. Stems erect, stiff, hollow; sheaths of main stem have more than 3 teeth 3

3a. Stems slender (1–3 mm thick), with 5–12 ridges, up to 1/3 of stem hollow .... ***E. variegatum*** (see *E. scirpoides*)

3b. Stems stout (4–10 mm thick), 3/4 or more of stem hollow .... ***E. hyemale***

1b. Stems annual, usually with regularly whorled branches; cones blunt (the horsetails) 4

4a. Fertile (cone-bearing) and sterile stems similar; cones produced in summer; often grows in shallow water 5

5a. Stems have 10–30 shallow ridges; central cavity 1/2 to 4/5 diameter of main stem .... ***E. fluviatile***

5b. Stems have 5–10 deep ridges; central cavity less than 1/3 diameter of main stem .... ***E. palustre*** (see *E. fluviatile*)

4b. Fertile (cone-bearing) and sterile stems dissimilar; cones produced in spring; usually not growing in aquatic habitats 6

6a. Twice branched; sterile stems have 2 rows of spines on each ridge; stem sheaths brownish; teeth grouped in several broad lobes .... ***E. sylvaticum***

6b. Usually only once branched; sterile stems have blunt bumps or cross-ridges on their ridges; stem sheaths greenish; teeth separate or nearly so 7

7a. Teeth of stem sheaths dark brown throughout; fertile stems permanently whitish or brownish, unbranched, soon wither; sterile stems have smooth or inconspicuous low bumps; first segment of primary branches much longer than stem sheath .... ***E. arvense***

7b. Teeth of stem sheaths brown with conspicuous white edges; fertile stems do not wither, become green,branched; sterile stems have conspicuous bumps; first segment of primary branches shorter than or equal to length of stem sheath .... ***E. pratense***

DWARF SCOURING-RUSH • *Equisetum scirpoides*

GENERAL: Perennial from brown, creeping, widely branched, 1–1.5 mm thick rhizomes; many stems ascending to prostrate, **6–16 cm long**, 0.5–1 mm thick, crooked, clustered, unbranched, **evergreen, solid (not hollow), 6-ridged**.

LEAVES: Small scales fused into loose sheaths, with black band above green base, with **3 teeth**, slender, white-edged.

SPORE CLUSTERS: In 2–5 mm long, short-stalked, **pointed, persistent** cones at stem tips.

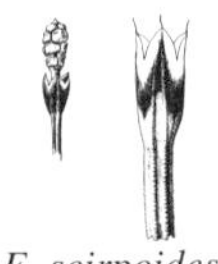

E. scirpoides

WHERE FOUND: Moist woods and thicket; widespread and common (though usually inconspicuous) across boreal forest, north to southern arctic islands; circumpolar.

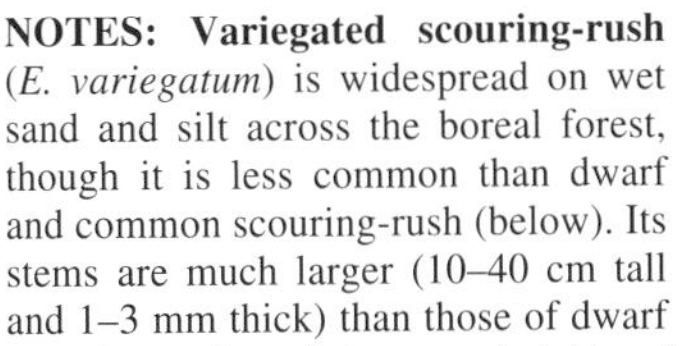

E. variegatum

NOTES: **Variegated scouring-rush** (*E. variegatum*) is widespread on wet sand and silt across the boreal forest, though it is less common than dwarf and common scouring-rush (below). Its stems are much larger (10–40 cm tall and 1–3 mm thick) than those of dwarf scouring-rush, and they are straight and erect rather than irregularly twisted and prostrate. Its sheaths have 6-8 teeth. • Dwarf scouring-rush is one of the smallest horsetails and easily overlooked, but once you start noticing it, you will discover that it grows almost everywhere in the boreal forest. • The horsetail family is over 400 million years old. During the Paleozoic era, forests of giant horsetails, with stems up to 1 m thick and 25 m in height, covered large parts of the world. Today's tallest horsetail, *E. giganteum*, grows in South America and reaches 10 m in height. Even our largest northern species are rarely over 1 m tall. • The species name *scirpoides* means 'rush-like,' echoing the common name 'scouring-rush.' This species is also called 'goosegrass' in some areas.

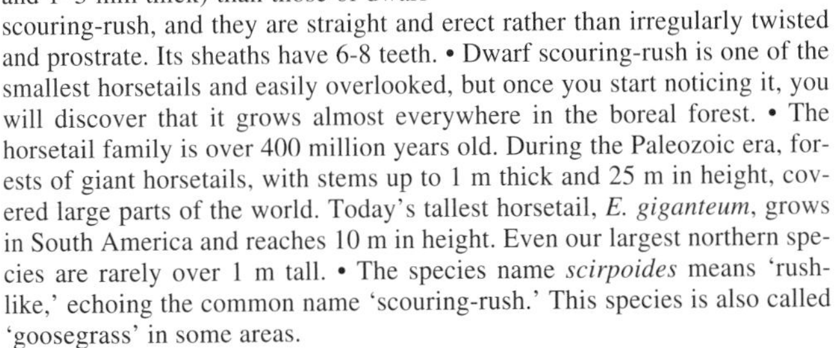

COMMON SCOURING-RUSH • *Equisetum hyemale*

GENERAL: Perennial from slender, blackish, creeping rhizomes; stems single (usually) or clustered, **unbranched, whitish green, evergreen, hollow, 4–10 mm thick, 30–100 cm tall, 18–40 roughened ridges**.

LEAVES: Small scales fused into sheaths 3–10 mm long, green to ashy grey, with brown to black teeth, usually with **black band** at tip and near base.

SPORE CLUSTERS: In short-stalked, **pointed, persistent** cones at stem tips.

WHERE FOUND: Moist sandy sites, in open woods, along streambanks and lakeshores; widespread across boreal forest north to 65°N in west and 55°N in Manitoba; circumpolar.

NOTES: The stiff stem segments of common scouring-rush have been used to make whistles. Because of the tiny silica crystals embedded in their tissues, horsetails have been used like sandpaper to hone, sharpen or polish everything from mussel shells and bone to pewter and fine woodwork—hence the name 'scouring-rush.' In music stores, pieces of common scouring rush are sold for shaping clarinet and oboe reeds. European peasants once used horsetails for scouring floors, but today they are more commonly used by campers for scouring pots, especially if the cook has let the soup, or worse yet the porridge, simmer a bit too long.

SWAMP HORSETAIL • *Equisetum fluviatile*

GENERAL: Perennial from wide-spreading, shiny, often reddish rhizomes; stems die back each year, single or clustered, erect, 10–100 cm tall, 3–8 mm thick, **simple or with few (usually) to many branches**; **large central cavity**; **10–30 ridges**, smooth or minutely cross-wrinkled; branches simple, **whorled**.

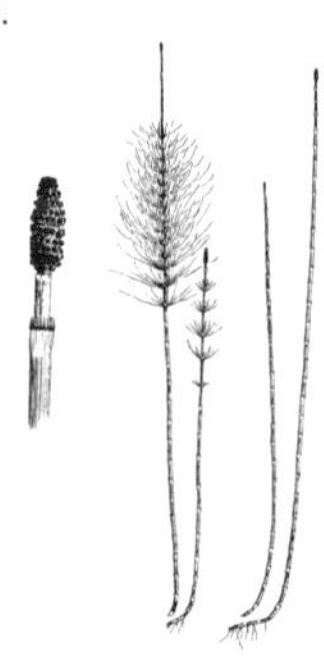

LEAVES: Small scales fused into green sheaths, 6–10 mm long, with 15–20 fine, dark brown to blackish teeth.

SPORE CLUSTERS: In stalked, blunt-tipped cones at stem tips, **fall off** after spores mature.

WHERE FOUND: Lake edges, wet river meadows, ditches, fens and marshes; widespread across boreal forest; circumpolar.

NOTES: Marsh horsetail (*E. palustre*) also inhabits wet meadows, fens and muddy river flats across our region, but its stems have a small, central cavity (less than 1/3 the diameter of the stem versus 2/3 in swamp horsetail) and stem sheaths with 8–10 brown or black teeth with whitish edges. • Because of their high silica content, horsetails have been used to treat many skin and eye problems. Horsetail tea can be used as a wash to relieve skin disorders and offensive body odour (including foot odour). It has also been used to strengthen hair, nails and teeth. Some native groups spread the ashes of these plants on skin sores, cankers and sore gums. • *Equisetum* is from the Latin *equus*, 'horse,' and *saeta* 'a bristle' or 'a coarse hair'; *fluviatile* means 'growing in a stream'—a fair description of its habitat.

WOODLAND HORSETAIL • *Equisetum sylvaticum*
WOOD HORSETAIL

GENERAL: Slender perennial from deep, creeping rhizomes; stems die back each year, erect, hollow, 15–60 cm tall; **2 markedly different stem types**; **fertile stems** unbranched at first, later have mostly **compound branches**, green to tan; **sterile stems much branched**, usually single, green, with 10–18 minutely spiny ridges; **many branches, again branched**.

LEAVES: Scales fused into sheaths around stems and branches; sheaths of sterile stems 3–12 mm long, **green below, brown upwards**, with 3–5 brown teeth, joined into several broad lobes; sheaths of fertile stems similar to but much larger and looser than those of common or meadow horsetail (p. 283).

SPORE CLUSTERS: In single, long-stalked, blunt-tipped, cones at fertile stem tips, appear in spring, **soon fall off**.

WHERE FOUND: Moist, open woods, thickets and meadows; widespread across boreal forest and northern parkland, north to treeline; circumpolar.

NOTES: Wood horsetail is our only horsetail with compound branches. • At one time, woodland horsetail tea was taken to treat kidney problems and fluid retention. The plant was also ground and placed in wounds to stop bleeding and promote healing. • Linnaeus apparently reported that woodland horsetail was an important food for horses in some parts of Europe. However, more recent reports of horsetail poisoning of livestock make this observation questionable. • Horsetails produce their food in their stems and branches, since their leaves have been reduced to rings of small, pointed scales. • The species name *sylvaticum* means 'of the forests,' emphasizing that woodland horsetail is most commonly found in forested habitats.

COMMON HORSETAIL • *Equisetum arvense*

GENERAL: Perennial, spreads extensively by dark-felted rhizomes with tubers; stems hollow, usually erect, die back each year, **2 markedly different types**; **fertile stems unbranched**, appear in early spring, usually thick and succulent, **brownish to whitish**, 10–30 cm tall; **sterile stems bottlebrush-like** (many whorls of slender **branches**), appear as fertile stalks **wither**, 1-several in clusters, 10–50 cm tall, slender, green, **10–12 ridged, minutely-roughened**; branches simple, **first branch segment longer than adjacent stem sheath**.

LEAVES: Reduced to small scales, usually fused into sheaths around stems and branches; sheaths of fertile stems have 8–12 large, pointed teeth; sheaths of sterile stems green, with 10–12 brownish or blackish teeth.

SPORE CLUSTERS: In long-stalked, blunt-tipped **cones** at tip of fertile stems.

WHERE FOUND: Moist to wet woods, meadows, swamps, fens, roadsides and other disturbed sites; widespread across our region, north through arctic islands to northern Ellesmere Island; circumpolar.

NOTES: Like all horsetails, common horsetail reproduces by spores. However, this species grows a special shoot to produce its spores—a small, pale brown, branchless plant with a spore cone at its tip. After the spore-bearing shoot matures and starts to die back, the horsetail sends up its characteristic green, bushy shoots. The small, fertile shoots are easily overlooked, and many people think they have found a new plant when they discover them growing along roads in the spring, but both forms are part of the same plant, the common horsetail. • Caribou, moose, sheep and bears eat this plant. In some parts of the Mackenzie Mountains, it is the main food of grizzlies in June and July.

MEADOW HORSETAIL • *Equisetum pratense*

GENERAL: Green, bottlebrush-like perennial, from spreading rhizomes; stems annual, erect, hollow, slender, 10-50 cm tall, 1-4 mm thick, die back each year; **fertile and sterile stems dissimilar**; fertile stems unbranched at first, later develop many whorls of branches; **sterile stems** mostly single, whitish green, **with 10-18 minutely roughened ridges**, many whorled branches, **fine**, 3-sided; **first branch segment not longer than adjacent stem sheath**.

LEAVES: Small scales fused into pale sheaths; sheaths 2-6 mm long, with 8-10 brown, **white-edged** teeth.

SPORE CLUSTERS: In 1-2 cm long cone, on long stalk at tip of bottlebrush-like shoot (whorled branches may be absent at first), **soon fall off**.

WHERE FOUND: Moist woods, thickets and meadows; widespread across our region; circumpolar.

NOTES: Meadow horsetail is often confused with common horsetail (above); however, all of meadow horsetail's shoots are green and have whorls of branches. Only common horsetail has small, brown, unbranched, fertile stems. The sterile stems of meadow horsetail are generally more slender and fragile looking than those of common horsetail, but the most reliable characteristic separating the 2 species is the length of the first branch segment relative to the length of the adjacent stem sheath. The branch segment is shorter than or equal to the stem sheath in meadow horsetail, but longer in common horsetail. • Horsetails contain an enzyme (thiaminase) that destroys vitamin B_1 (thiamine). In large quantities, they have caused deaths in livestock, though the poisoning is quickly reversed by removing horsetails from the diet. Their effect on humans is not completely understood, but raw horsetails can act as a **poison**. Cooking destroys the thiaminase. Only very small quantities should be taken internally, and people with **high blood pressure** or other cardiovascular problems are warned against using horsetail.

Club-mosses

The club-mosses (Lycopodiaceae) are low, evergreen, non-woody plants with branched, creeping or tufted stems covered by rows or spirals of small, narrow, scale-like, green leaves. Their spore cases (**sporangia**) are in cones at the tips of stems or in leaf axils.

L . annotinum

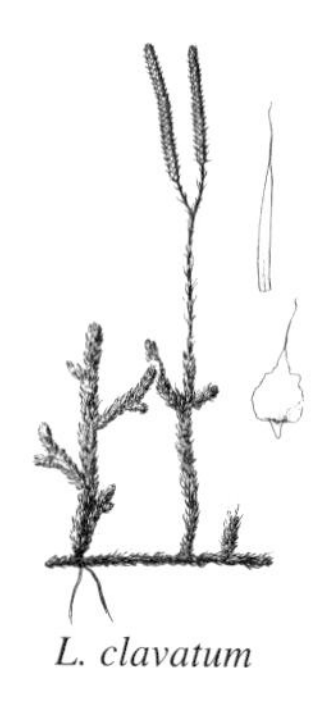

L. clavatum

L. obscurum

L. complanatum

STIFF CLUB-MOSS • *Lycopodium annotinum*

GENERAL: Perennial with horizontal, leafy, rooting stems creeping on or near ground surface, up to 1 m long; erect stems **simple to twice forked**, 5–30 cm tall, **bottlebrush-like**, with numerous bristly leaves.

LEAVES: Many, needle-like, 3–10 mm long, **usually spreading, firm, sharp-pointed**, in 8 vertical rows, **appear whorled**.

SPORE CLUSTERS: In **stalkless, single**, 12–35 mm long cones; spore clusters in axils of yellowish to greenish, egg-shaped, slender-pointed bracts, tightly clustered in cone.

WHERE FOUND: Moist forest, thickets, and heathland; widespread across boreal forest, north past treeline to Arctic coast; circumpolar.

NOTES: Running club-moss (*L. clavatum*) is found in dry, mossy sites across the boreal forest and around the world. It is similar to stiff club-moss, but running club-moss has a slender hair at the tip of each leaf (look closely), and its spore-bearing cones are on long stalks. • The Woods Cree used club-mosses to separate raw fish eggs from the membranous sacs in which they are produced. This was done by wriggling the egg mass and a bunch of stiff club moss together with the hands. The separated eggs were used to make fish-egg bread. • Club-moss spores have been used as a dusting powder in surgery, as baby powder and to treat various skin problems, including eczema and chaffed skin. The spores repel water so strongly that a hand dusted with them can be dipped into water without becoming wet. However, their use as an anti-absorbent is limited as they are known to irritate mucous membranes. The Carrier used to put club-moss spores into the nose to cause bleeding and thus cure headache. • The Cree used club-moss spores to divine the future of sick people. The spores were dropped into a container of water and if they moved towards the sun, the patient would survive. • Club-moss spores are very rich in oil, and they are highly flammable. At one time they were used by photographers and theatre performers as flash powder, giving the effect of lightning on the stage. Because they ignite explosively, club-moss spores were called 'witch's flour.'

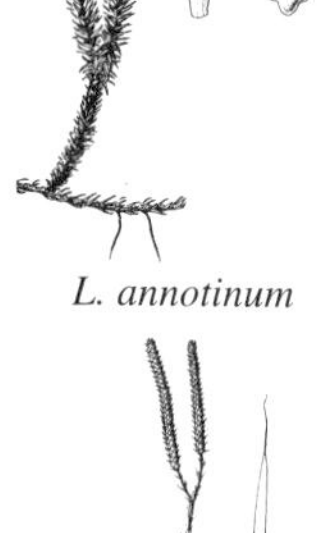

L. annotinum

L. clavatum

GROUND-PINE • *Lycopodium obscurum*

GENERAL: Perennial from extensively rooting, deep creeping rhizomes; stems erect, 7–30 cm tall, unbranched towards base, densely branched above, **bushy**, **tree-like**.

LEAVES: Many, dense, in 6 (sometimes to 8) vertical rows, sharp pointed, linear-lance-shaped.

SPORE CLUSTERS: In axils of greenish to yellowish brown, pointed, broadly egg-shaped bracts densely clustered in **single, stalkless** cones, 1–3 cm long, at stem tips.

WHERE FOUND: Moist sites in woods, thickets and clearings; widespread across boreal forest north and west to Great Bear River and west-interior Alaska.

NOTES: Also called *L. dendroideum* and *L. lagopus*. • Club-mosses have been boiled in water to make a medicinal tea that was cooled and used as an eye wash. At one time, fresh plants were put on the head to cure headaches and worn on clothing to ward off illness. • Ground-pines stay green all winter, and they are often used in Christmas decorations. • In the Paleozoic era (about 300,000 million years ago), the ancestors of modern club-mosses reached gigantic proportions. These huge plants, some up to 30 m tall, formed a major part of the plant material that developed into coal beds. • The 'ground-pine' gets its name from its resemblance to a miniature coniferous tree. The genus name *Lycopodium* is from the Greek *lycos*, 'wolf,' and *podus*, 'foot,' after a fancied resemblance of club-moss leaves to a wolf's paw.

GROUND-CEDAR • *Lycopodium complanatum*

GENERAL: Perennial with rooting, horizontal stems creeping on or near ground surface, up to 1 m long; erect-ascending stems, whitish green, much branched, **5–30 cm tall**, **flattened** (cedar-like).

LEAVES: In 4 vertical rows, sharp-pointed, **partially fused to stem**, in 3 forms: largest on upper side, lie flat against stem, with elevated base; smaller on lower side, awl-shaped; side leaves egg- to awl-shaped with short, free tips.

SPORE CLUSTERS: In axils of yellowish to greenish, egg-shaped bracts, densely clustered in **1–3**, 6–30 mm long cones on forked, 10–65 mm long stalks at stem tips.

WHERE FOUND: Open woodlands, thickets, heathland and rocky slopes; widespread across boreal forest, north to treeline; circumpolar.

NOTES: In Europe and North America, ground-cedar was dried, powdered and used to make a medicinal tea to increase urine production, stimulate menstrual flow and relieve spasms. This tea was considered useful for correcting 'female complaints' and was said to stimulate sexual desire. The spore powder was forced into the noses of people who needed reviving. It was also said that if the spores were boiled, the decoction would both kill lice and improve bad wine. The spores were used as a dusting powder in the drug trade, protecting abrading surfaces and preventing pills from sticking together.

Bryophytes (Mosses and Liverworts)

'Bryophytes' is a general term used to describe 2 large groups of small plants—the mosses and the liverworts.

All of the plants described previously in this guide contain internal tubes for transporting food and water within them—they are vascular plants. Bryophytes have poorly developed tubes—they are non-vascular plants. As well, most of the plants described previously reproduce themselves with seeds. Bryophytes (along with ferns, club-mosses and horsetails) lack seeds; they reproduce with spores instead.

Because they do a poor job transporting food and water internally, and because they require water for reproduction, bryophytes are small plants, and are often most abundant in wet places. Almost all of the mosses and liverworts in this region are common throughout the northern hemisphere, from boreal Canada and United States to Scandinavia, Russia, China and Japan.

Bryophytes consist of 2 generations: the **gametophyte**, the mosses and liverworts we see carpeting the ground throughout our region; and the **sporophyte**, a **capsule** raised above the gametophyte by a stalk (**seta**). Sporophytes are only present when conditions are right. In mosses they are relatively long-lived, but in most liverworts they usually last only a few days.

Moss and liverwort leaves are usually only 1 cell thick. Some mosses have leaves with a midrib (**costa**), and here the leaves are often more than 1 cell thick. All bryophytes lack true roots; instead they are attached to the substrate by **rhizoids**.

Most native peoples did not distinguish among the many mosses and liverworts that they used, apart from the sphagnum mosses. Mosses, in particular, were used throughout our region for many domestic tasks, including lining pits and as a source of moisture for both cooking and molding wood, wiping the slime off fish, stuffing mattresses and pillows, lining babies' cradles and bags, and covering floors. Mixed with pitch, mosses were used to caulk canoes and in more modern times they were mixed with mud to chink log cabins. Most groups preferred sphagnum mosses, but would use other species if sphagnum was not available.

As with other sections of the guide, this section describes only the most common bryophytes in our region. More information can be obtained from Conard and Redfearn (1979), Schofield (1992) and Vitt et al. (1988).

'If it were not for the mosses, it is difficult to say how barren the woods would be or how much beauty would be lost to nature.'

—E.M. Dunham, *How to Know the Mosses*, 1916.

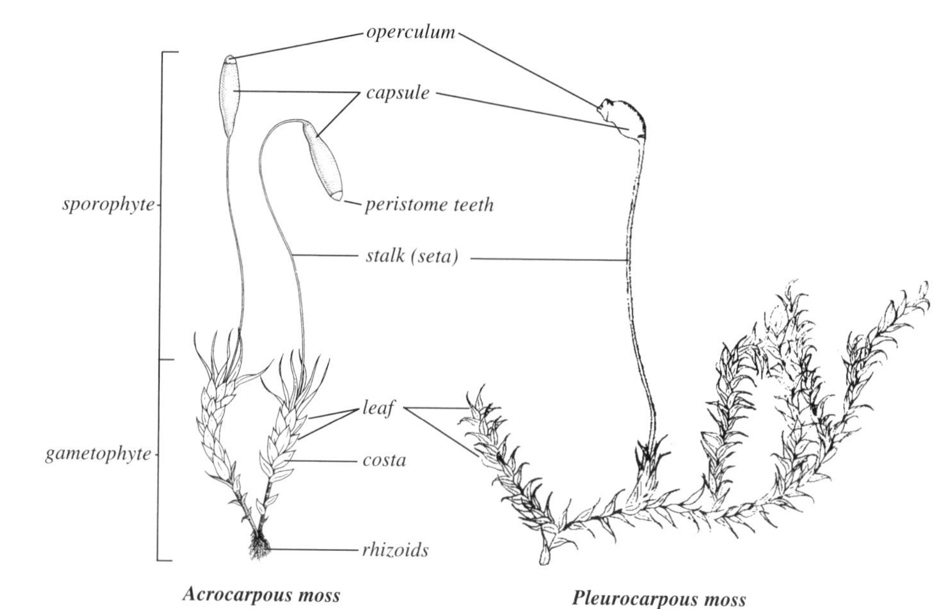

Acrocarpous moss ***Pleurocarpous moss***

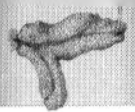

(Liverworts) Hepatics

Most liverworts are **leafy** (like mosses) but a few are **thalloid** (composed of a flattened body (**thallus**) that is not differentiated into stem and leaves).

Leafy liverworts have 2 or 3 rows of leaves on their stems. The leaves never have a midrib and often they are rounded or have 2-4 points at their tips, and are folded. If present, the leaves on the underside of the stem (the **underleaves**) are often reduced and usually pressed flat to the stem. Because the underleaves are reduced or absent, the larger stem leaves are nearly opposite each other on the stem. All the leaves are often flattened in the same plane, an arrangement that is rare in mosses.

The rows of leaves can either angle downwards from the point of attachment so that they overlap one another like shingles on a roof (**succubous**), or they can angle upwards and overlap like reversed shingles (**incubous**) when viewed from the upper surface.

The leafy liverwort sporophyte consists of a **foot**, an often colourless short **stalk**, and a **capsule**. The capsule usually develops in a sheath of fused leaves (the **perianth**). When the spores are mature, the stalk lengthens and the capsule splits lengthwise into 4 sections, releasing the spores. In our thalloid liverworts, the capsule stalk does not lengthen. Instead, the entire sporophyte grows on top of a stalk produced by the plant thallus (the gametophyte).

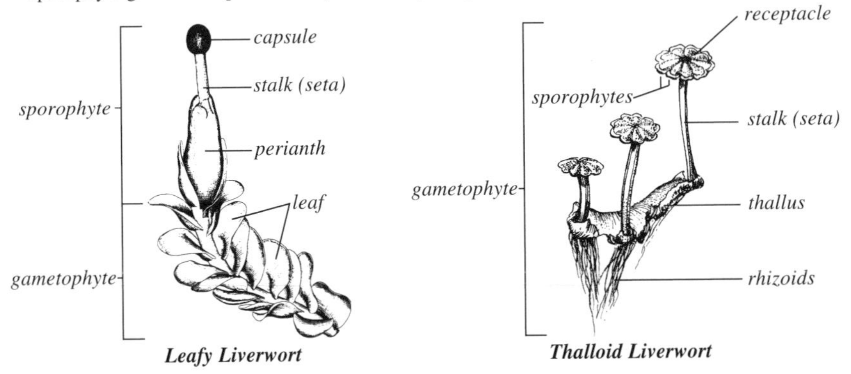

Leafy Liverwort *Thalloid Liverwort*

NORTHERN NAUGEHYDE LIVERWORT • *Ptilidium ciliare*

GENERAL: Red-brown to copper-red; stems stiff, upright, to 5 mm long, **form mats**, branches single to twice-divided and feathery; **easily removed from substrate**.

LEAVES: Unequally divided into 4–5 lobes, 'hand-shaped,' partly wrapped around stem, to 2 mm long by 2.5 mm wide; underleaves to 1 mm long by 1.3 mm wide, 2-lobed, with many tiny hairs.

SPOROPHYTES: Dark brown, nearly spherical capsule on long, yellowish stalk.

WHERE FOUND: Dry sites, with lichens (esp. *Cladonia*), usually on rock or on thin layer of soil over rock; sometimes mixed with other bryophytes; common across northern boreal forest; circumpolar.

NOTES: This species could be confused with naugehyde liverwort (p. 288), but that species usually grows on logs and tree bases, rather than rock or soil, and it is not as easily removed from its substrate. As the common name suggests, northern naugehyde liverwort is found farther north than naugehyde liverwort. • The name *ciliare* is from the Latin *cilium*, 'eyelash,' and refers to the small, eyelash-like hairs along the edges of the leaves.

NAUGEHYDE LIVERWORT • *Ptilidium pulcherrimum*

GENERAL: In very **dense patches, yellowish green to reddish brown**; branches feathery, **firmly attached to substrate**.

LEAVES: Alternate, unequally divided into 4–5 lobes, wrapped around stem; **about 1/2 size of northern naugehyde liverwort**.

SPOROPHYTES: Egg-shaped capsule on fleshy, 5–7 mm long stalk.

WHERE FOUND: Tree bases and rotting wood; occasionally on rocks or soil; very common and widespread across boreal forest northwestward to east-interior Alaska; circumpolar.

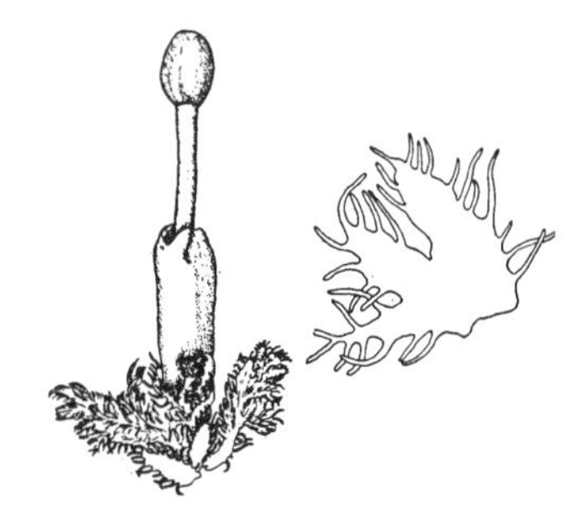

NOTES: Hand shaped leaves, which are divided into 3–5 lobes, are distinctive of this genus in our area. • The species name *pulcherrimum* is the diminutive form of the Latin *pulcher*, which means 'beautiful' or 'handsome,' so this name means 'little pretty.'

LEAFY LIVERWORT • *Lophozia ventricosa*

GENERAL: Stems 1–4 cm long, lie close to ground, erect at tip, greenish above, usually brownish to violet below, simple or little branched, with many rhizoids.

LEAVES: Alternate, **2-lobed**, not strongly overlapping, attached at angle to stem, wide-spreading to erect-spreading, about 1/2 embracing stem, flat to slightly ridged, smooth-edged except for triangular, pointed lobes at tip; generally no underleaves; **yellowish gemmae present in spherical groups at tips of upper leaf lobes**.

SPOROPHYTES: Oblong-egg-shaped, dark reddish brown capsule.

WHERE FOUND: Usually scattered amongst other bryophytes on tree bases, rotting wood and humus; common and widespread across boreal forest; circumpolar.

NOTES: *Lophozia* is a highly variable and taxonomically difficult genus with about 24 species occurring across our region. ***L. longidens*** has dark reddish brown gemmae; ***L. excisa*** is green, often has perianths and rarely produces gemmae; ***L. porphyroleuca*** has dark reddish stems and commonly has perianths and no gemmae. ***Tritomaria exsectiformis*** produces similar dark red gemmae, but it has asymmetrical, 3-lobed leaves. • The genus name comes from the Greek *lophos*, 'a point,' and refers to the pointed, lobed leaves. The species name *ventricosa*, meaning 'having a swelling on 1 side,' probably refers to the rather enlarged perianth in this species.

LITTLE HANDS LIVERWORT • *Lepidozia reptans*

GENERAL: Tiny, light to olive-green, 1.5–3 cm long; **branches widely spaced, pinnate, spreading**, usually prostrate or trailing; **forms loose mats** over other bryophytes.

LEAVES: Less than 1 mm long, widely spaced, **3–4-lobed to about half their length**; upper leaves curve downwards, resemble club-shaped mitts; underleaves somewhat smaller, erect, otherwise similar; cells rounded.

SPOROPHYTES: Fairly common; arise from long, prominent, slenderly spindle-shaped, toothed perianths.

WHERE FOUND: On decaying wood, especially rotten stumps and logs, across boreal forest; circumpolar.

NOTES: ***Blepharostoma trichophyllum*** has 3-lobed leaves, but its lobes are thread-like and extend almost to the base of the leaf. • *Lepidozia* liverworts prefer to grow on well-rotted wood. A simple kick test is a handy tool when identifying mosses on stumps. If your toe sinks into a moss-covered stump with a pleasant soft sound, you are probably looking at *Lepidozia* liverworts or common four-tooth moss (*Tetraphis pellucida*). If pain shoots through your foot, the stump is still in the early stages of decay and you will probably find a leafy liverwort (*Lophozia* spp.), copper wire moss (p. 294), and ragged moss (*Brachythecium* spp.) growing on its surface.

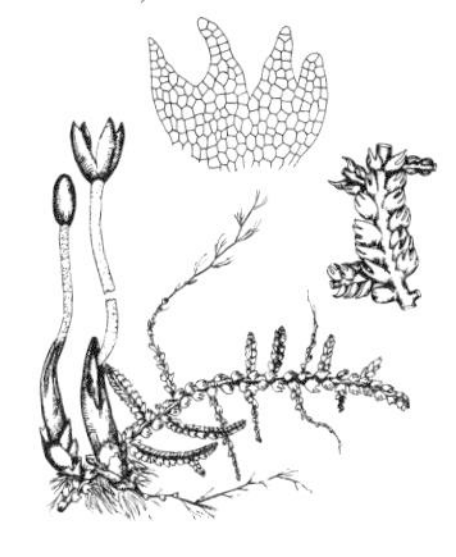

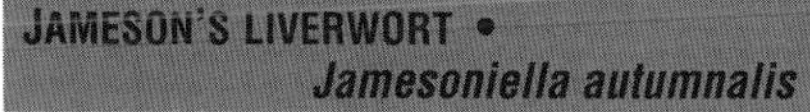

JAMESON'S LIVERWORT • *Jamesoniella autumnalis*

GENERAL: Olive green, **forms dense patches**; stems stout, 1–4 cm long, **lie flat on ground**, simple or branched, many rhizoids near stem tip.

LEAVES: Alternate, overlapping, attached at angle across stem, oblong-oval to round, unlobed, smooth-edged; underleaves present only on younger parts, inconspicuous, narrowly lance-shaped to awl-shaped.

SPOROPHYTES: Egg-shaped, reddish purple capsule on long, weak stalk.

WHERE FOUND: Mixed with other bryophytes on rotting wood or humus; inconspicuous but fairly common across boreal forest; circumpolar.

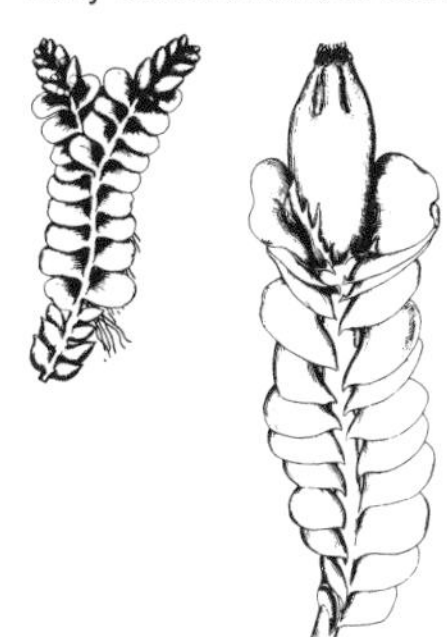

NOTES: Jameson's liverwort is tolerant of a wide range of environmental conditions. Its preferred habitat varies somewhat over its geographic range. • The species name *autumnalis* means 'characteristic of autumn.' The genus name is the diminutive of *Jamesonii*, a fern genus named in honour of Dr. William Jameson of Quito, Ecuador.

HARD SCALE LIVERWORT • *Mylia anomala*

GENERAL: Yellowish green to yellowish brown; leafy shoots about 3 mm wide; stems stout, irregularly twisted, 2–4 cm long, trailing (**erect when between mosses**), **usually simple** or with a few branches; many rhizoids.

LEAVES: Alternate, well-spaced to overlapping, spreading when wet, erect and pressed to stem when dry, egg-lance-shaped to rounded-oblong, unlobed, concave, **with distinctive border of larger cells**; leaves near stem tip usually different from those farther down; underleaves very small, mostly hidden by rhizoids; **yellowish green gemmae produced in clusters or scattered on edge near tips of upper leaves**.

SPOROPHYTES: Dark brown, egg-shaped-spherical capsule.

WHERE FOUND: Commonly found growing with rusty peat moss (p. 319) on hummocks; occasional on damp, peaty soil or on decaying wood; common and widespread across boreal forest, especially in peatlands; circumpolar.

NOTES: This is the only leafy liverwort in our region that has rounded (non-lobed) leaves that overlap from the top down, like shingles on a roof. • The species name *anomala*, from the Greek *ano*, 'not,' and *homalos*, 'the same,' means 'anomalous' or 'abnormal.'

CEDAR-SHAKE LIVERWORT • *Plagiochila asplenioides*

GENERAL: Light to olive green, dull; **usually in mats or patches**; stems to 10 cm long, unbranched to sparingly branched.

LEAVES: Roundish to elliptic, to 5 mm wide and 10 mm long (quite variable in size), edges bent backwards, **usually toothed**; underleaves tiny, inconspicuous, like hairs.

SPOROPHYTES: Stalk colourless, weak; capsule elliptic, up to 2 mm long.

WHERE FOUND: On rocks, soil, streambanks, forest floor and matted around tree bases; common across boreal forest north to central Yukon; circumpolar.

NOTES: *Plagiochila* is the largest genus of liverworts in the world. Cedar-shake liverwort is one of the few liverworts with toothed leaves. • The genus name comes from the Greek *plagios*, 'sloping,' and *cheilos*, 'lip,' and refers to the sloping mouth of the perianth. The species name *asplenioides* refers to the supposed resemblance of the leaves to those of spleenwort fern (genus *Asplenium*).

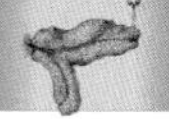

SNAKE LIVERWORT • *Conocephalum conicum*

GENERAL: Ribbon-like; **coarse, 6-sided markings** on upper surface, each with **dot or pore** at its centre; thallus 1–2.2 cm wide, sometimes with forked lobes, **strongly aromatic**; **no gemmae cups**; many rhizoids on lower surface.

SPOROPHYTES: Capsules occasionally produced beneath conical heads atop long, erect stalks; male plants produce stalkless pads with antheridia.

WHERE FOUND: On **moist, inorganic soil** in moist areas, as behind waterfalls, on sandy banks, and moist rock faces; scattered across southern boreal forest west, to northern B.C.

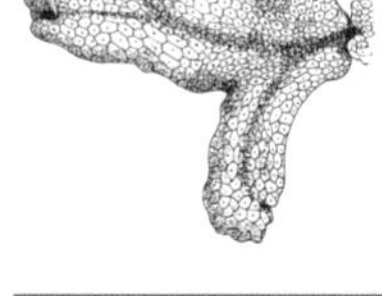

NOTES: This is our largest liverwort. It is much less common than green-tongue liverwort (below) and easily distinguished by its pronounced 6-sided markings. When it is crushed, snake liverwort gives off a pleasant odour. • The large, ribbon-like liverworts (*Conocephalum* and *Marchantia* spp.) have a complex internal structure, with tiers of green cells (looking like miniature cacti) that photosynthesize to produce food for the plant. • In some native languages, snake liverwort is known as 'seal's tongue.'

GREEN-TONGUE LIVERWORT • *Marchantia polymorpha*

GENERAL: Large, flattened, ribbon-like, 7–13 mm wide; branches forked, coarse, dull, with **indistinct pores** and **black mottles** across upper surface; usually has flaring, circular **cups containing several disc-shaped gemmae**; midrib well developed, present as depressed, darkened area on upper surface; lower surface black-purple, covered with triangular scales; single-sexed plants, sexes usually in separate colonies; male plants have sex organs in sacs on flattened, **lobed disc** at top of long (1–3 cm) stalk; female plants have sex organs beneath **finger-like lobes** that radiate outwards from tops of long stalks.

SPOROPHYTES: Develope from sex organs on female plants; stalks very short, normally not visible; capsules elliptic, open by fragmented tip.

WHERE FOUND: Streambanks, wet soil, roadside ditches, rock walls, often on burned ground; widespread in appropriate habitats across parkland and boreal forest; cosmopolitan.

NOTES: Green-tongue liverwort could be confused with snake liverwort (above), but that species is larger and has distinct 6-sided markings on its upper surface. • Species of a similar genus, ***Preissia,*** are smaller, have unstalked male sex organs and never have gemmae cups. • Early herbalists used green-tongue liverwort to treat tuberculosis and liver problems. In 1714, Culpepper reported that 'it is a singular good herb for all the diseases of the liver, both to cool and cleanse it, and helps the inflammations in any part, and the yellow jaundice likewise.' This remedy was based on the Doctrine of Signatures, in which plants were used to treat the part of the body they resembled. • Large colonies of green-tongue liverwort often indicate a recent fire. • Green-tongue was one of the first liverworts ever described, and today it is one of the best known and most commonly studied. It can be a troublesome weed in gardens, greenhouses, and tree nurseries, where it kills seedlings.

Mosses

The sporophyte of a moss is always attached to the gametophyte, which is usually conspicuous and leafy. The sporophyte is composed of a **foot**, a **stalk** (seta) and a **capsule** with an urn, a lid (**operculum**) and a circle of teeth (**peristome**) inside the mouth of the capsule. The sporophytes may be well developed with long stalks and inclined capsules with 1 or 2 rings of peristome teeth, or they may be smaller and simpler, on short stalks (sometimes completely covered by the leaves), and have upright capsules with no peristome. The capsule is often initially covered by a membranous hood (**calyptra**), which protects the developing sporophyte and is actually part of the gametophyte. Sporophytes grow from female sex organs, located either at the tips (**acrocarpous**), or along the sides (**pleurocarpous**) of branches. Pleurocarpous mosses are usually branched and creeping, while acrocarpous mosses are less frequently branched.

Moss stems sometimes have green, filamentous, branched structures (**paraphyllia**), or scale-like structures that cover branch buds (**pseudoparaphyllia**). Some species have stems covered with many-celled, root-like filaments that are never green. Any of these structures can be sufficiently abundant to give the stems a woolly appearance.

Moss leaves may or may not have a midrib, and in the majority of species the leaf blades are 1 cell thick. Leaf cells may be almost any shape, but the cells that attach the leaf to the stem (the **alar cells**) are often different in appearance (often enlarged or coloured) from the other cells of the leaf. Some mosses have very shiny leaves; in others the leaves are quite dull. Moss leaves can differ greatly in appearance between when they are moist and when they are dry.

What to Look For When Identifying Mosses

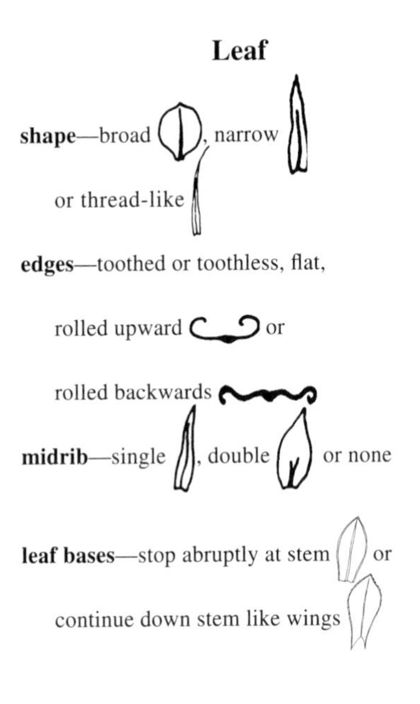

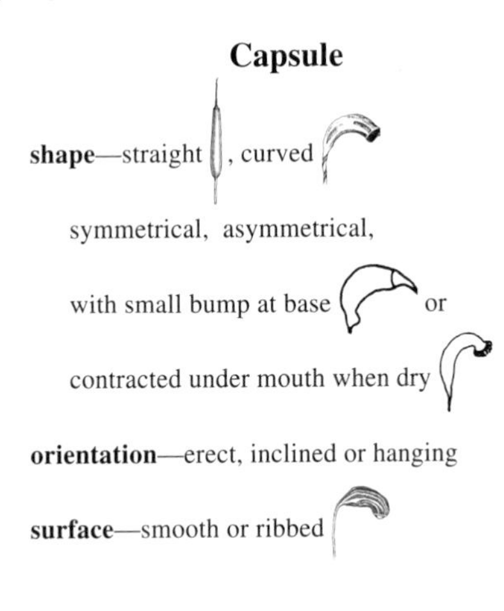

Capsule stalk

shape—long or short;

surface—smooth or rough

colour—red, straw, green, etc.

Capsule covering (calyptra)

PURPLE HORN-TOOTHED MOSS • *Ceratodon purpureus*
FIRE MOSS

GENERAL: Green, yellow-brown, or reddish, unbranched, to 1 cm tall; **form small, dense tufts**.

LEAVES: About 2 mm long, sharp-pointed, edges curled up or down nearly their entire length.

SPOROPHYTES: Stalk dark-red to brown, 2–3 cm long; **capsule dark red**, 2–4 mm long, cylindrical, inclined to horizontal, **deeply furrowed**.

WHERE FOUND: Soil, rock, dead wood, burned or logged-over areas, roadsides and roofs; **weedy species** in disturbed sites across our region; cosmopolitan, from sidewalks of New York to rocks of Antarctica.

Notes: Purple horn-toothed moss is possibly the most common moss in the world. It is common in burned areas, growing with copper wire moss (*Pohlia nutans*) (p. 294) and other pioneers such as cord moss (*Funaria hygrometrica*) (below), and green-tongue liverwort (*Marchantia polymorpha*) (p. 291). Its many, dark reddish stalks and horizontal, deeply furrowed, dark capsules are good identifying features. • The genus name *Ceratodon*, from the Latin *cerato*, 'horn-shaped,' refers to the forked teeth around the mouth of the capsule. The species name *purpureus* is Latin for 'purple,' and refers to the reddish purple colour of the capsules and their stalks.

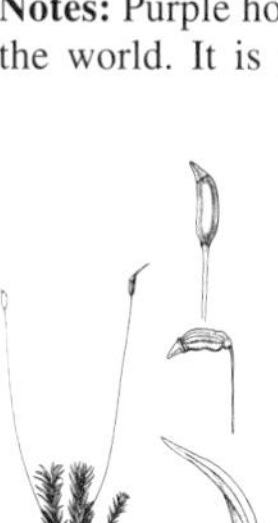

CORD MOSS • *Funaria hygrometrica*

GENERAL: Small, **light-green, almost bulb-shaped**; stems simple, short, less than 1 cm high; almost always annual.

LEAVES: 2–5 mm long, oval to egg-shaped, sharp-pointed, **erect**, concave, shiny, **contorted when dry**; midrib single, ends just before tip; more or less smooth-edged; leaf cells rectangular, large, thin-walled, smooth, similar throughout leaf.

SPOROPHYTES: Common, nearly always present, at branch tips; stalk **contorted** and twisted, **yellow**, position alters with changes in humidity; capsules asymmetrical, **pear-shaped**, 16-ribbed when mature, **hanging, bright yellow** when mature, green otherwise; capsule teeth double (superficially appear single); capsule hood large.

WHERE FOUND: Weed; found commonly **on disturbed, compacted soil** (rich in nitrogen), in greenhouses, and in moist depressions after fires; common across boreal forest, parkland and prairie; cosmopolitan, from Arctic to tropics.

NOTES: There are many small, erect, weedy mosses that produce spore capsules from the tips of their stems. Among the most common are cord moss, with broad leaves, contorted stalks and fat capsules; long-necked bryum (*Leptobryum pyriforme*) (p. 294) with linear leaves and pear-shaped capsules; **silvery bryum** (*Bryum argenteum*) with bulb-shaped, distinctively silvery plants; and purple horn-toothed moss (*Ceratadon purpureus,* above) with lance-shaped leaves, rolled leaf edges, and purple sporophyte stalks and capsules. • In some areas, cord moss is always found at the sites of abandoned bonfires. • The genus name *Funaria*, from the Latin *funale*, 'a cord,' and the common name 'cord moss' refer to the dry, twisted, cord-like capsule stalks. The species name *hygrometrica*, from the Greek *hygros*, 'wet' or 'moist,' and *metria*, 'measure,' refers to the twisting and curling of the stalk in response to changing moisture in the air.

LONG-NECKED BRYUM • *Leptobryum pyriforme*

GENERAL: Forms loose to dense, light green tufts; stems erect, 5–15 mm tall, **sometimes has dark red, spherical brood bodies in leaf axils**.

LEAVES: Lower leaves small and widely spaced; **upper leaves** larger, 4–6 mm long, **crowded at stem tips**, erect-spreading to wide-spreading, only slightly contorted when dry, **linear-lance-shaped to almost bristle-like**, smooth-edged or sometimes toothed near tip.

SPOROPHYTES: Single; stalk straight, yellowish red or orange-brown, 1–3 cm long, curved below capsule; **capsule horizontal to hanging**, brown or yellow-brown, 1.5–2.5 mm long, **pear-shaped with long, narrow neck** as long as rest of capsule, neck **strongly wrinkled when dry**.

WHERE FOUND: Common weedy species on soil, rotten wood, humus or occasionally rock; often in burned-over or otherwise disturbed habitats; widespread across boreal forest, parkland and prairie; cosmopolitan, on all continents except Antarctica.

NOTES: The slender, bristle-like leaves, spreading from a clasping base, combined with the shiny, golden-brown, pear-shaped, nodding capsules are distinctive of this species. Long-necked bryum is a common weedy moss in greenhouses. It looks more like a *Dicranella* than like its closest relatives, the *Bryums*. • The genus name *Leptobryum*, from the Greek *leptos*, 'slender,' means 'slender *Bryum*' and refers to the long slender leaves and sporophyte stalks of these mosses. The species name *pyriforme,* from the Latin *pira*, 'a pear,' and *forma*, 'form,' describes the pear-shaped capsules.

COPPER WIRE MOSS • *Pohlia nutans*
NODDING POHLIA

GENERAL: In green or yellowish green tufts; stems simple, red, 1–4 cm tall.

LEAVES: Overlapping to erect-spreading, **dull green**, 1.5–3.5 mm long; **lower leaves short**, egg-shaped to egg-lance-shaped; **upper leaves longer and more crowded**, lance-shaped to linear-lance-shaped, pointed, **toothed in upper part, often twisted around stem when dry**.

SPOROPHYTES: Single; stalk 15–30 mm long, yellow or red-brown; capsule 2–4 mm long, orange-brown, horizontal to hanging, **neck about 1/3 length of capsule**.

WHERE FOUND: Soil, humus, bases of trees, rotting and charred wood, tops of old stumps, banks and rock crevices, from open exposed habitats to shaded closed forests; common and widespread across boreal forest, parkland and prairie; circumpolar.

NOTES: Copper wire moss is our most common *Pohlia*, but it is so notably undistinguished that there are no reliable field characteristics with which to separate it from many other species of *Pohlia* and *Bryum*. Another common *Pohlia*, **glaucous thread moss** (*P. cruda*), is identified by its glossy, broad, toothed, yellow-green leaves. • The common name 'copper wire moss' refers to the stiff copper-coloured stalks of the sporophytes. The *Pohlia* genus was named in honour of J.E. Pohl, a German physician. The species name *nutans*, from the Latin *nutare*, 'to nod,' refers to the nodding capsule.

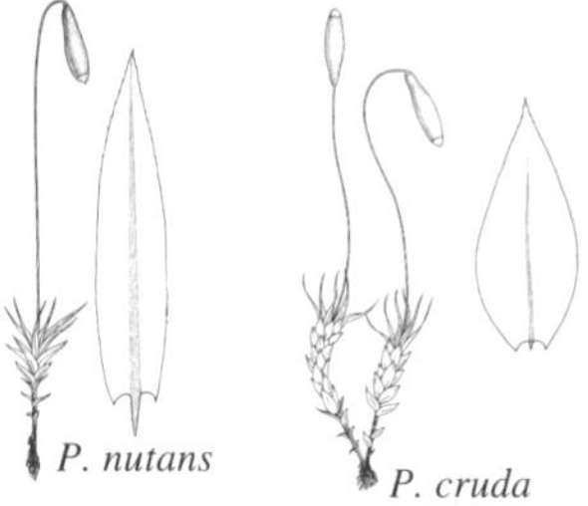

ERECT-FRUITED IRIS MOSS • *Distichium capillaceum*

GENERAL: In **erect, dense tufts**; stems slender, 1–6 cm tall, green and glossy towards top, brown below.

LEAVES: In 2 rows, oblong-lance-shaped, 2–4 mm long, abruptly narrowed to long, roughened point; bases shiny, white, **clasp stem**.

SPOROPHYTES: Single; stalk 8–18 mm long, red or reddish brown, twisted when dry; **capsule cylindrical**, brown, 1–2 mm long, **straight** to slightly curved, **erect to inclined, regular**, smooth or slightly wrinkled when dry and empty.

WHERE FOUND: Most often found in rock crevices, particularly on calcium-rich substrates, but not uncommon on soil, and occasionally occurs on rotting wood; common and widespread across boreal forest, parkland and prairie; circumpolar.

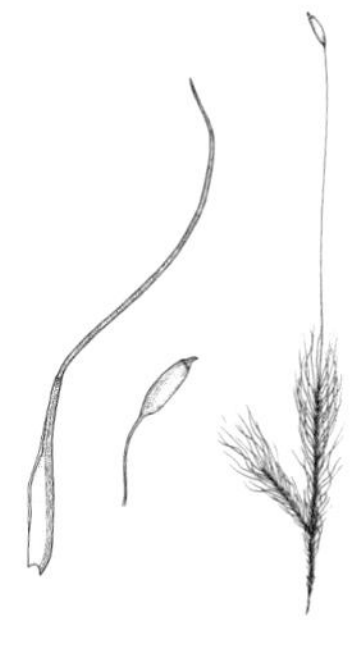

NOTES: Erect-fruited iris moss is a strongly flattened, leafy plant with glossy, overlapping, sheathing leaf-bases that look like miniature ears of barley. • A less common species, **inclined-fruited iris moss** (*D. inclinatum*), usually has shorter stems (up to 3 cm tall), more crowded leaves and shorter, egg-shaped, horizontal, irregular capsules on longer stalks. • The genus name, from the Greek *di*, 'twice,' and *stichos*, 'a row,' refers to the arrangement of the leaves. The species name *capillaceum*, from the Latin *capillus*, 'hair,' refers to the long slender, hair-like leaves.

SLENDER-STEMMED HAIR MOSS • *Ditrichum flexicaule*

GENERAL: In erect, loose tufts; **stems** yellow-green to brownish, **densely matted with rhizoids below**, 1.5–6 cm tall.

LEAVES: Erect-spreading, irregularly twisted, 2–7 mm long, lance-shaped, gradually long-pointed, channelled above, irregularly toothed at tip.

SPOROPHYTES: Single; stalk dark red, 8–25 mm long; **capsule egg-shaped to cylindrical**, 1.5–2 mm long, chestnut-brown, **erect** or nearly so, **smooth** to slightly wrinkled or furrowed when dry and empty.

WHERE FOUND: Strongly associated with calcium-rich soil and rock, usually in dry and exposed, often disturbed, places; also often occurs in rock crevices, prairies, fens, meadows and on humus; common and widespread across boreal forest, parkland and prairie; circumpolar.

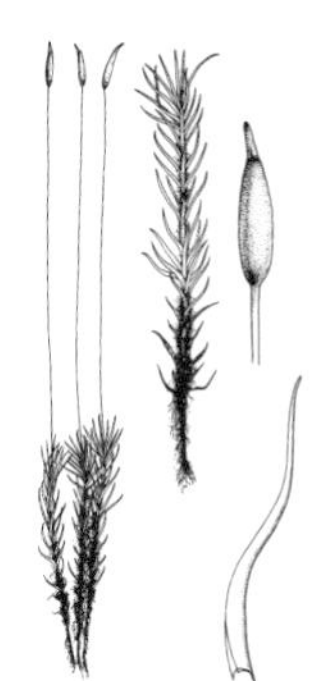

NOTES: Slender-stemmed hair moss often grows with iris mosses (*Distichium* spp.), but is readily distinguished from them in that its leaves are attached all around the stem rather than in 2 rows. • This highly variable moss is perhaps the most common moss on calcium-rich rock outcrops in western Canada. • Slender-stemmed hair moss often has brittle branches at its stem tips. These break off and aid in dispersal. • The genus name *Ditrichum*, from the Greek *di*, 'two' and *trichos*, 'hair,' refers to the capsule teeth, which are split into 2 hair-like divisions. The species name *flexicaule*, from the Latin *flexus*, 'to curve' or 'to bend,' and *caulis*, 'a stalk,' refers to the long, irregularly twisted stems.

FRAGILE SCREW MOSS • *Tortella fragilis*

GENERAL: In dense tufts; stems erect, 1–5 cm tall, dull yellowish green to dark green, turn brown with age.

LEAVES: Stiff, linear-lance-shaped, 4–5 mm long, narrow abruptly from sheathing base to awl-shaped tip, erect-spreading when moist, **slightly contorted when dry, tips of upper leaves often broken off**.

SPOROPHYTES: Rare; single; stalk 10–20 mm long, straight, smooth, yellowish to reddish, twisted when dry; capsule yellowish to reddish brown, 2–3 mm long, cylindrical, straight, erect, smooth, often wrinkled when dry and empty; **teeth around capsule mouth twisted into spiral**.

WHERE FOUND: On calcium-rich rock or soil, often on ledges of cliffs or in rock crevices; less frequently on logs or peaty humus in wet, calcium-rich habitats; widespread, but rarely abundant across boreal forest, parkland and prairie; circumpolar.

NOTES: A closely related species, **twisted moss** (*T. tortuosa*) has strongly rippled and twisted leaves (when dry) that do not break off at the tip. It occasionally grows with fragile screw moss. It is rare in the boreal forest, where it may be found on sand at the edges of streams and lakes. • The broad 'V' on each leaf (made by a wedge of green cells from the upper leaf meeting the clear cells of the leaf base) is characteristic of this genus. • The broken pieces from the leaf tips of fragile screw moss could be an important means of vegetative reproduction. This species rarely produces spore capsules. • The genus name *Tortella* is a diminutive form of the Latin *tortus*, 'twisted' or 'crooked,' and refers to the corkscrew twist of the capsule teeth. The species name *fragilis* refers to the fragile leaf tips.

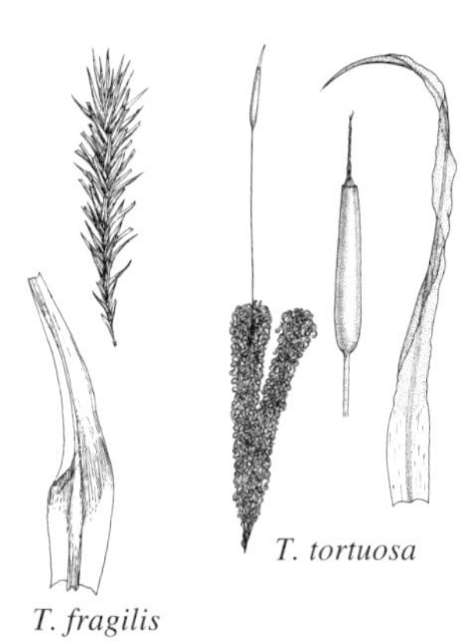

HAIRY SCREW MOSS • *Tortula ruralis*
SIDEWALK MOSS

GENERAL: Overall colour **brick red when wet** (base reddish, top greener)**, blackish when dry**; 2–3 cm tall, **forms upright tufts with whitish appearance**.

LEAVES: Folded and twisted around stem when dry; spreading or curved outwards and downwards at right angles when moist; **tip blunt, with long hairpoint**, sometimes as long as leaf.

SPOROPHYTES: Stalk 1–2 cm long; capsules 2–3 mm long, straight, with long, twisted teeth encircling mouth.

WHERE FOUND: Sandy soil, rock and dry, sunny, often calcium-rich sites across boreal forest, parkland and prairie; circumpolar.

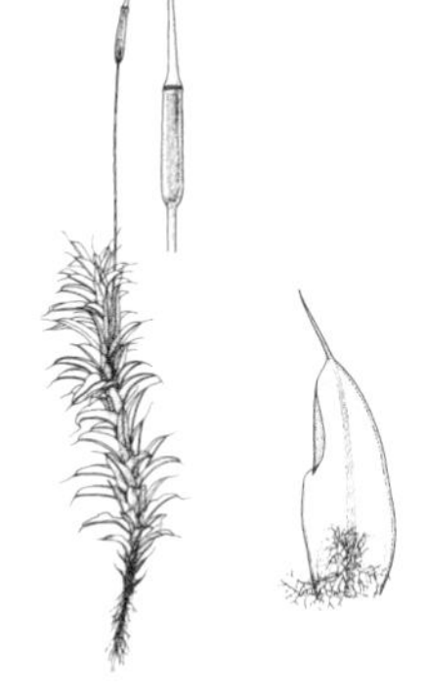

NOTES: The dull, broad, widely spreading leaves of hairy screw moss, with their clear bristle tips, are unmistakable. • The genus name *Tortula*, from the Latin *tortus*, 'twisted' or 'crooked,' refers to the twisted leaves and capsule teeth. The species name *ruralis* is Latin for 'of the countryside,' where this moss is widespread.

JUNIPER HAIR-CAP • *Polytrichum juniperinum*
JUNIPER MOSS

GENERAL: Bluish green, shiny, short to long, slender to stout, upright, unbranched; grows in **mats** or more rarely as **closely associated individuals**; stems 1–10 cm tall.

LEAVES: 4–8 mm long, upright-spreading when dry, wide-spreading when moist, thick, edges toothless, tip extends into **short, toothed, reddish bristle-point**; leaves similar to those of awned hair-cap (below).

SPOROPHYTES: Common; stalk upright, wiry, 2–6 cm long, reddish; capsule reddish brown, 2.5–5 mm long, **4-sided**, vertical, becomes horizontal with age, puckered at base; 64 short, blunt teeth around capsule mouth; capsule hood long, with **long hairs**, covers entire capsule.

WHERE FOUND: Soil, humus and rock, stumps, banks, trailsides, dry open woods; frequent after fire or logging; commonest on dry, exposed, acidic sites; common and widespread across boreal forest; cosmopolitan.

NOTES: Hair-caps are easily recognized by their thick, stiff leaves with characteristic dark green ridges running lengthwise along the upper surface of the midrib. Unlike most other mosses, hair-caps have leaves with a well-developed system of tiny tubes for carrying water. • Herbalists considered juniper hair-cap a powerful diuretic, and the tea made from this moss was used to treat urinary obstructions, dropsy and the like. Because it caused no nausea, it was considered an excellent remedy when it was necessary to continue treatment indefinitely. • Juniper hair-cap is an important pioneer on many types of soils and a stabilizer of sand. It is said that a colony of hair-cap can hold a vertical 20 cm bank of gravel in place. • The name *juniperinum* means 'like juniper'; the leaves of this hair-cap are similar in shape and colour to those of the common juniper.

AWNED HAIR-CAP • *Polytrichum piliferum*

GENERAL: Small, 1–4 cm tall, powdery green to brown, stem bases leafless, tops crowded with leaves; forms dense tufts.

LEAVES: 4–7 mm long, point up when dry, point out when wet; entire edge folds in over ridge on leaf; **leaf tip ends in colourless point** visible from several metres away as whitish coating around each plant.

SPOROPHYTES: Stalk brown, 3–4 cm tall, thick, wavy; capsules reddish brown, 2–3 mm long, **4-sided**.

WHERE FOUND: Dry, sterile, sandy or gravelly soil, exposed rocks, roadsides, old fields; common in suitable habitats across boreal forest; occasionally found in prairie and parkland; cosmopolitan.

NOTES: Awned hair-cap could be confused with juniper hair-cap (above). Both grow on dry sites and both have leaf edges that roll over the upper surface, but long white 'hair-points' on the leaves are characteristic of awned hair-cap. Slender hair-cap (p. 320) also resembles these 2 species, but it has fuzzy, whitish stems and grows on organic material in peatlands. • Hair-caps found on drier sites usually have their leaf edges rolled in to cover the leaf surface. This adaptation reduces moisture loss. • Awned hair-cap is an important pioneer species that colonizes acidic soils and sand. It has adapted to growing on wind-blown sand, and produces many rhizoids and branches when it is covered by sand. • Hair-caps are almost always considered an indicator of acidic conditions. • The name *Polytrichum*, from the Greek *poly*, 'many,' and *trichos*, 'hair,' refers to the hairy cap covering the spore capsule. The species name *piliferum*, from the Latin *pilus*, 'a hair,' refers to the white hair at the tip of each leaf.

PIPECLEANER MOSS • *Rhytidium rugosum*
CRUMPLED-LEAF MOSS

GENERAL: Yellow, greenish yellow or golden brown, stems more or less erect, 2–8 cm high, **fuzzless**; branched on 1 side or with symmetrical, feathery branches with hooked tips; forms **loose mats**.

LEAVES: Crowded, curved 1 way, **furrowed and rumpled**, narrow to long, thin tip; edges rolled under.

SPOROPHYTES: Extremely rare; stalk red, 2–2.5 cm long; capsule brown.

WHERE FOUND: Dry, well-drained, thin soil; exposed calcium-rich areas; dry tundra; widespread, but only locally abundant, in suitable habitats across boreal forest, parkland and prairie; circumpolar.

NOTES: Pipecleaner moss often grows with wiry fern moss (*Thuidium abietinum*) (p. 310). • Pipecleaner moss is one of the few terrestrial mosses in our region that has crumpled (rugose) leaves. Electric eels (*Dicranum polysetum*, p. 307) and wavy dicranum (*D. undulatum*, p. 320) also share this characteristic. • *Rhytidium* comes from the Latin *rhytidio*, meaning 'wrinkled,' and refers to the rumpled leaves. The species name *rugosum*, from the Latin *ruga*, 'a wrinkle,' also emphasizes the crumpled appearance of the leaves.

VELVET FEATHER MOSS • *Brachythecium velutinum*

GENERAL: Small, dark- or yellow-green, irregularly branched, tips of branches often curved and pointing in 1 direction; **forms soft, silky mats**.

LEAVES: Widely spaced, 1–1.5 mm long, loosely erect to widely spreading, narrowly to broadly lance-shaped, straight or curved and pointing in 1 direction, gradually pointed, **usually twisted at tip**, smooth, **edges toothed nearly to base**, midrib 1/2 to 2/3 length of leaf, **usually with 1 or more spines at tip.**

SPOROPHYTES: Single; stalk brown or reddish brown, **rough throughout** (use hand lens), 7–15 mm long; capsules dark brown, 1.5–2 mm long, oblong-cylindrical curved, asymmetrical, horizontal to somewhat hanging.

WHERE FOUND: On soil and rock, rotting wood, or bark at base of trees, in dry to moist woods; prefers calcium-rich habitat; widely scattered across southern boreal forest north to southern Yukon.

NOTES: This is one of the smallest of the *Brachythecium* species. *Brachythecium* is a large, difficult genus in need of study. Of the 15 or so species in our region, most have yellow-green, lance-shaped, sharply pointed leaves and sprawling or creeping stems that branch frequently but irregularly. The genus name *Brachythecium*, derived from the Greek *brachys*, 'short,' and *theca*, 'a case,' refers to the relatively short, thick spore capsules of the mosses in this genus. *Velutinum* means 'velvet-like'; mats of this small, soft moss, with the tips all swept in one direction, were thought to resemble velvet.

YELLOW COLLAR MOSS • *Splachnum luteum*
FAIRY PARASOLS

GENERAL: Translucent green; **in small patches with yellow 'umbrellas' or 'skirts'** above them.

LEAVES: Lightly crisped, pointed, with large teeth along edges.

SPOROPHYTES: Stalks red, 2–15 cm long; capsule brownish, urn-shaped, 1–1.5 mm long, with broad, bright yellow 'umbrella,' 4.5–11 mm wide below.

WHERE FOUND: Occasional on moose (or other herbivore) dung, usually in peatlands, most common in coniferous forest zones, across central and northern boreal forest; circumpolar.

NOTES: A similar but less common species, **red collar moss** (*S. rubrum*) has red umbrellas. **Globe-fruited splachnum** (*S. sphaericum*) is the most common and widespread collar moss in the boreal forest, but it is much less noticeable and is often overlooked. It has a small, purplish umbrella, not much wider than the body of the capsule. • The sporophytes with their expanded umbrellas closely resemble small mushrooms. • The umbrella (technically called a hypophysis) has thick, spongy tissue, seldom found in mosses. Collar mosses almost always grow on dung, and it is not surprising that most of our species have adapted to using flies to disperse their spores. The well-decomposed dung on which these mosses usually grow smells like fresh earth, but the moss's umbrellas give off sour or musty odours to attract insects in search of fresh dung. The umbrella provides a landing platform for flying insects and the sticky clumps of spores adhere easily to hairy visitors. After a fly has picked up spores it moves on to the next patch of fresh droppings where it lays its eggs and deposits the moss spores.

S. rubrum

S. sphaericum

BROWN TAPERING SPLACHNUM • *Tetraplodon mnioides*

GENERAL: In **dense tufts**; stems erect, often branched, 1–6 cm tall, green towards top, yellow-brown below, with many rhizoids.

LEAVES: Egg-lance-shaped to lance-shaped and widest above the middle, 2–5 mm long, somewhat contorted when dry, narrow abruptly to long, awl-shaped tip, **edges smooth**, inrolled.

SPOROPHYTES: Single; **stalk stiff**, dark red, **1–3 cm long**; capsule erect, cylindrical, **projects well above leaves, dark red**, turns black with age, 3–5 mm long, **lower part only slightly wider than upper part and usually slightly longer**.

WHERE FOUND: On dung and decaying animal remains in peatlands and moist forests; widespread but uncommon across boreal forest; circumpolar.

T. angustatus

T. mnioides

NOTES: Brown tapering splachnum is probably the most common dung moss in the boreal forest. **Narrow-leaved splachnum** (*T. angustatus*) is similar, but it has irregularly toothed leaf edges, shorter stalks (2–6 mm) and capsules that are yellowish or greenish when young and turn black when they mature. Both species have a similar distribution, but narrow-leaved splachnum is less common. • *Tetraplodon* species usually grow in drier forest sites than *Splachnum* species. • The genus name *Tetraplodon*, from the Greek *tetra*, 'four,' refers to the capsule teeth, which are often fused in 4s when young.

COMMON FOUR-TOOTH MOSS • *Tetraphis pellucida*

GENERAL: Small, dull red-brown, 8–15 mm high.

LEAVES: Erect-spreading, **almost appear to be in 3 rows**, flat, toothless; **lower leaves small, widely spaced**, 1–2 mm long, egg-shaped, pointed; **upper leaves** longer, narrower, more **crowded**, up to 3 mm long, midrib ends below tip.

SPOROPHYTES: Single; stalk usually erect, smooth, twisted when dry, 6–14 mm long; capsule 2–3 mm long, narrowly cylindrical, smooth, erect, symmetrical, **with 4 teeth at capsule mouth**.

WHERE FOUND: Most common on old, soft, rotting wood, rarely on peat; prefers acidic substrate; widespread and relatively common across boreal forest; circumpolar.

NOTES: The narrow, cylindrical, smooth, upright capsules of this moss, with 4 teeth at the mouth, are unmistakable. • Common four-tooth moss often grows with copper wire moss (p. 294) in moist coniferous forests. It is common on soft, rotting wood. 'Kickable' stumps usually have common four-tooth moss growing on them; if a swift kick results in a sore toe, you probably won't find this moss. • The name *Tetraphis*, from the Greek *tetra*, 'four,' refers to the 4 teeth at the mouth of the capsule, a dead giveaway to the identity of this genus. The species name *pellucida*, from the Latin *pellucere*, 'to shine through,' refers to the near transparency of the wet leaves.

MOUNTAIN CURVED-BACK MOSS • *Oncophorus wahlenbergii*

GENERAL: In tufts, green or yellow-green towards top, brown below; stems 1–2 cm tall, often branched, with some rhizoids below.

LEAVES: Wavy and contorted when dry, 2–5 mm long, **narrow abruptly from egg-shaped base to lance-awl-shaped tip; edges flat**, smooth or toothed near tip.

SPOROPHYTES: Stalk yellow, 1–1.5 cm long; capsule yellow-brown, 0.8–1.4 mm long; **small bump just above base on lower side**.

WHERE FOUND: Common on logs and rotting wood across boreal forest; circumpolar.

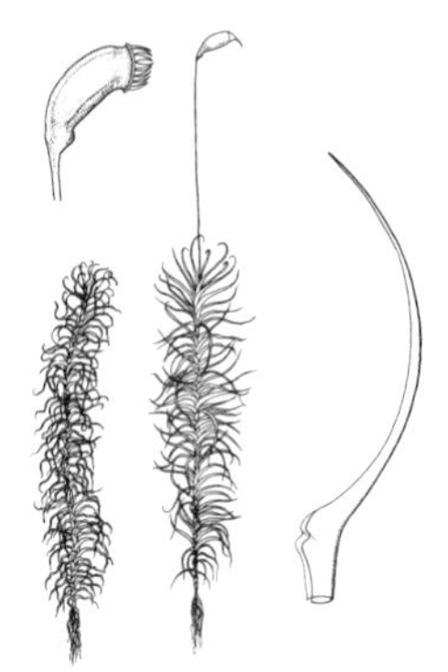

NOTES: Green spur-fruited fork moss (*O. virens*) is similar, but its leaves narrow gradually from an egg-shaped or oblong base and the leaf edges curve downward, rather than lying flat. It occasionally grows on rotten wood, but it is usually found along streams and beside fens. Both species have similar distributions. • The genus name, *Oncophorus*, from the Greek *onchos*, 'a mass,' and *phoros*, 'a bearer' or 'a producer,' refers to the Adam's apple-like swelling at the base of the capsule. The species is named in honour if G. Wahlenberg, a 19th-century Swedish botanist, who was the first to collect the species.

WHIP FORK MOSS • *Dicranum flagellare*

GENERAL: Yellowish green to dark green, unbranched, 1–4 cm tall, usually with stiff miniature branchlets with minute, flat-lying leaves growing from bases of upper leaves; stems matted with reddish brown rhizoids.

LEAVES: Irregularly curled and wavy when dry, not wrinkled, lance-shaped, 3–5 mm long, pointed, concave below, tube-shaped above; smooth-edged or toothed near tip.

SPOROPHYTES: Stalk yellow to brown, single, 1–2 cm long; capsule yellowish brown to brown, **erect, straight**, 2–2.5 mm long, furrowed lengthwise when dry.

WHERE FOUND: Rotten wood or bases of trees; occasionally on humus; fairly common across boreal forest north to N.W.T.; circumpolar.

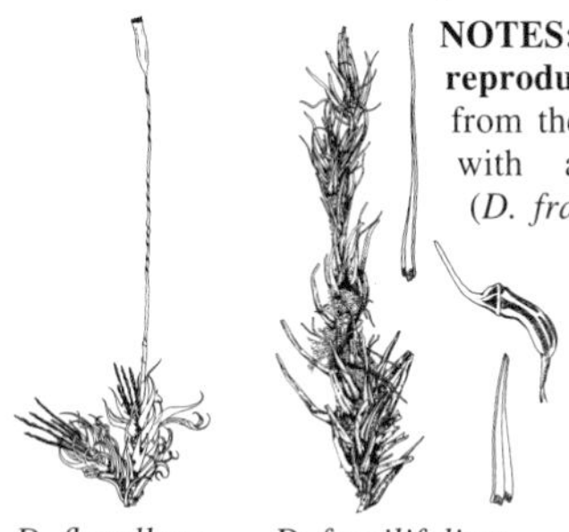

D. flagellare *D. fragilifolium*

NOTES: Whip fork moss **reproduces asexually by dropping its tiny whip-like branchlets** from the upper leaf axils. • Whip fork moss could be confused with a less common species, **fragile cushion moss** (*D. fragilifolium*). However, fragile cushion moss has no stiff branchlets and its straight leaves usually have their tips broken off. Also, the capsules of fragile cushion moss are inclined and curved rather than erect and straight. • The species name *flagellare*, from the Latin *flagellum*, 'a whip,' refers to the stiff, whip-like branchlets.

CURLY HERON'S-BILL MOSS • *Dicranum fuscescens*
DUSKY FORK MOSS

GENERAL: Green to dark green, unbranched, 1–4 cm tall; stems reddish brown; forms small to sizable patches and cushions.

LEAVES: Curve to 1 side, 4–7 mm long, narrow, pointed; contorted when dry, ridged in upper part (V-shaped).

SPOROPHYTES: Stalk yellow to dark, 1–2 cm long; capsule light or dark brown, inclined, to 4 mm long, furrowed lengthwise when dry, asymmetrical; may have small bump just above base on lower side.

WHERE FOUND: Usually on rotting wood or at base of living trees; common across boreal forest; circumpolar.

NOTES: Curly heron's-bill moss could be confused with broom moss (p. 308), but broom moss leaves are glossy rather than dull, and have flat rather than wavy edges when dry. At a more detailed level, broom moss has long rather than short cells and is ridged along the back of its midrib. Broom moss also tends to grow in looser mats. • The common name 'curly heron's-bill moss' refers to the curved, long-pointed leaves, like a curled heron's bill. The species name *fuscescens* is from the Latin *fuscus*, 'dark,' and refers to the brownish colour that this moss often becomes with age.

RED-MOUTHED MNIUM • *Mnium spinulosum*

GENERAL: Lush-green; stems about 1–2 cm high, erect, simple, usually with many sporophytes; often forms dark-green, glistening mats.

LEAVES: 3–5 mm long, erect-spreading to spreading when moist, shrivelled and pressed flat when dry, oval lance-shaped, **gradually and sharply pointed**, bordered by linear cells in several rows; **edges have large, double teeth in upper half of leaf**; leaf cells 6-sided, smooth, thin-walled throughout.

SPOROPHYTES: Common, at plant tips; stalks long, slender, **1–4** per plant; capsules cylindrical, hanging, smooth; capsule teeth **deep red**, double, well developed.

WHERE FOUND: On soil and needles, or tree bases, in moist, heavily shaded habitats under coniferous trees; found across boreal fores; circumpolar.

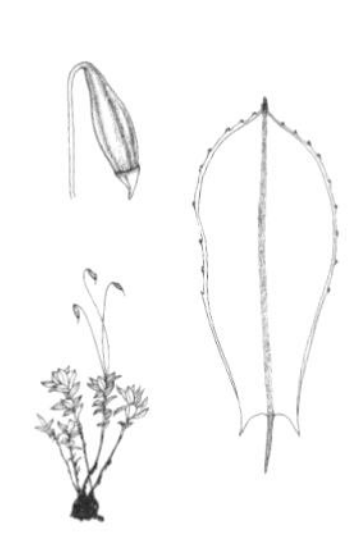

NOTES: Edged lantern moss (*M. marginatum*) could be confused with red-mouthed mnium, but it has yellowish (rather than red) capsule teeth and the teeth along the edges of its leaves are smaller. Red-mouthed mnium can always be identified microscopically by its leaf border, which is at least 3 cells thick (other *Mniums* have borders 1–2 cells thick), and by the deep red teeth around the mouth of its capsule. • The species name *spinulosum,* from the Latin *spina*, 'a spine,' refers to the distinctive spiny teeth of the leaf edges.

WOODSY LEAFY MOSS • *Plagiomnium cuspidatum*
WOODSY MNIUM

GENERAL: Yellow-green to dark green, 2–3.5 cm tall; **forms loose, often extensive tufts**; stems simple, fertile stems erect, sterile stems spreading or horizontal.

LEAVES: Small towards base, larger and more crowded towards top, 3–4.5 mm long, erect to erect-spreading, **strongly wavy and contorted when dry, obovate**, with tooth-like point, leaf-base extends down stem, bordered by single layer of linear cells, with **single, sharp, 1-celled teeth in upper half**.

SPOROPHYTES: Stalk **single**, yellowish or brownish, 1–3 cm long; capsule yellowish or brownish, hanging, 2–3 mm long.

WHERE FOUND: On moist soil or humus, rotting wood and bark at base of trees; common and widespread across boreal forest and northern parkland; circumpolar.

NOTES: Also called *Mnium cuspidatum*. • Woodsy leafy moss, probably the most common of the 'leafy mosses,' is easily recognized by its obovate leaves, which are strongly twisted and wavy when dry and toothed only above the middle. It could be confused with **toothed mnium** (*P. ciliare,* also called *P. affine* var. *ciliare*), but the leaves of that species are toothed to the base with teeth formed by 1–4 cells. • The genus name *Plagiomnium*, from the Greek *plagios*, 'oblique,' probably refers to the obliquely arching shoots of many species. The species name *cuspidatum*, from the Latin *cuspis*, 'a point,' refers to the sharp-pointed teeth along the leaf edges.

P ciliare *P. cuspidatum*

DRUMMOND'S LEAFY MOSS • *Plagiomnium drummondii*

GENERAL: Light, **clear green**, 2–6 cm high; stems unbranched, erect, with rhizoids on lower part; forms tufts.

LEAVES: More or less in 2 rows on sterile plants, smaller at ends; erect on fertile plants, larger at top; small point on tip of each leaf; **leaves not much changed in drying**, lightly crisped, very clean and crinkly, **edges single-toothed in upper half**.

SPOROPHYTES: Stalk yellow or orangish, 1–3 cm long, groups of **2–4 per plant**; capsule yellow, hanging, 2–3 mm long.

WHERE FOUND: Soil, humus, bark, tree bases and rocks in moist forest; scattered across southern boreal forest and northern parkland.

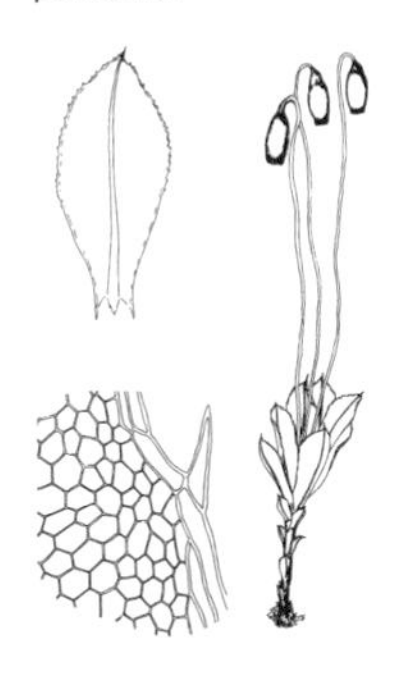

NOTES: Also called *Mnium drummondii*. • Both Drummond's leafy moss and woodsy leafy moss (p. 302) have leaf edges that are singly toothed on the upper half only. However, Drummond's leafy moss produces spore capsules in clusters of 2–4, whereas the sporophytes of woodsy leafy moss are single. Drummond's leafy moss also has shiny leaves (like they have been varnished), whereas the leaves of woodsy leafy moss are dull. • At one time, all of the 'leafy mosses' were included in the genus *Mnium*. Recently, they have been divided into 3 genera: *Mnium*, which has leaves with double teeth; *Plagiomnium*, leaves with single teeth; and *Rhizomnium*, leaves with no teeth.

BLUNT-LEAVED BRISTLE MOSS • *Orthotrichum obtusifolium*

GENERAL: Dark green to brown; forms dense tufts; stems erect, often forked, 5–10 mm tall.

LEAVES: Spreading when moist, pressed flat when dry, shape changes little in drying, 1.5–2 mm long, egg-shaped, **blunt or rounded**, midrib ends well below tip, **often has many**, small, green or red-brown, **brood bodies on upper surface** (sometimes on both surfaces).

SPOROPHYTES: Single; **stalk very short**, less than 0.5 mm long; **capsule hidden by upper leaves or barely poking above them**, dark brown, cylindrical, about 2 mm long, **8-ribbed; capsule covering hairless**.

WHERE FOUND: On the bark of hardwoods, particularly aspen and balsam poplar, in open places; rarely on wooden fences; common and widespread in boreal forest and northern parkland.

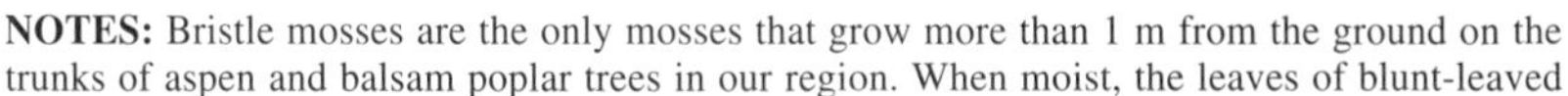

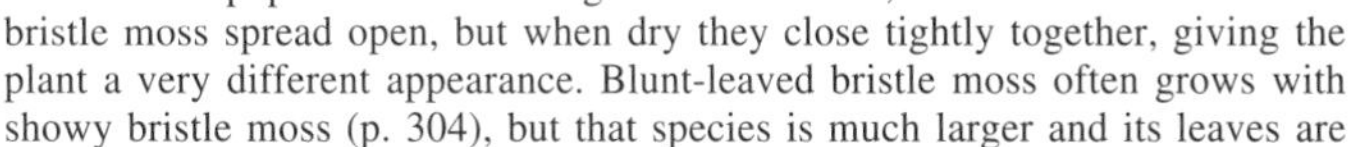

NOTES: Bristle mosses are the only mosses that grow more than 1 m from the ground on the trunks of aspen and balsam poplar trees in our region. When moist, the leaves of blunt-leaved bristle moss spread open, but when dry they close tightly together, giving the plant a very different appearance. Blunt-leaved bristle moss often grows with showy bristle moss (p. 304), but that species is much larger and its leaves are pointed rather than rounded at the tip. • The genus name *Orthotrichum*, from the Greek *orthos*, 'straight,' and *thrichos*, 'hair,' refers to the straight, erect hairs on the capsule covering of many species. The species name *obtusifolium*, from the Latin *obtusus*, 'blunted' or 'dull,' and *folium*, 'a leaf,' describes the blunt or rounded appearance of the leaves.

SHOWY BRISTLE MOSS • *Orthotrichum speciosum*

GENERAL: Dark green to brownish; forms loose to dense tufts; stems coarse and stiff, 1–5 cm tall.

LEAVES: Loosely erect when dry, erect-spreading with tips curved backwards when moist, 3.5–4 mm long, lance-shaped, **narrow gradually to pointed tip**; edges curled backwards nearly to tip, smooth; **midrib ends at or near tip**.

SPOROPHYTES: Single; stalk 1.5–2 mm long; **capsule just pokes above leaves or is completely above them**, pale brown, 2–2.5 mm long, oblong-cylindrical, tapers to long neck, **indistinctly 8-ribbed or smooth; capsule covering has many hairs**.

WHERE FOUND: On the trunks and branches of trees (particularly poplars), soil and rock outcrops (particularly in northern areas), occasionally on logs; widespread in parkland and boreal forest.

NOTES: In the north, showy bristle moss is often common around bird perches. It commonly grows with blunt-leaved bristle moss (p. 303) on the trunks of aspen and poplar trees, more than 1 m from the ground, but it can also form extensive mats on rock outcrops. • The species name *speciosum*, from the Latin *speciosus*, 'showy' or 'beautiful,' refers more to the relatively large size of this moss than to its attractiveness—the capsules of showy bristle moss are less distinguished than those of most other species in the genus.

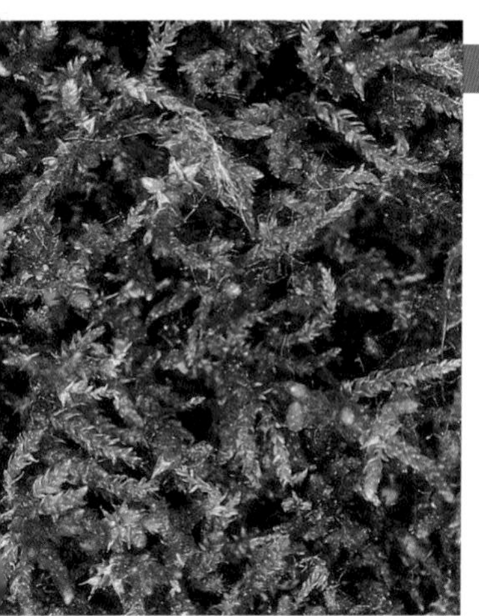

COMMON BEAKED MOSS • *Eurhynchium pulchellum*

GENERAL: Slender, shiny, **yellowish to dark green**; **forms extensive mats**; stems creeping, irregularly to regularly branched; branches flattened to rounded.

LEAVES: Stem leaves erect, about 1 mm long, egg-shaped, pointed; branch leaves **erect to wide-spreading**, slightly flattened, 0.5–1 mm long, narrowly oblong-egg-shaped, **bluntly pointed to rounded**, indistinctly ridged, **often twisted at tip, edges toothed all around**, midrib extends more than 3/4 length of leaf.

SPOROPHYTES: Single; stalk reddish, smooth, 1–2 cm long; capsules 1.5–2 mm long, oblong-cylindrical, strongly inclined to horizontal.

WHERE FOUND: Usually on rotting wood and bark at base of trees in both deciduous and coniferous forests; occasionally on rock; common at base of aspen trees; widespread across boreal forest and parkland; circumpolar.

NOTES: The small, egg-shaped, spreading, toothed and bluntly pointed leaves of common beaked moss are distinctive in comparison with other woodland mosses. • The genus name, *Eurhynchium*, from the Greek *eurys*, 'wide' or 'broad,' and *rhynchos*, 'a snout,' may refer to the blunt-tipped leaves and shoots. *Pulchellum*, is the diminutive form of *pulcher*, 'beautiful,' and means 'pretty little' moss. Both parts of this name are appropriate, but you'll probably need a microscope to appreciate the beauty of this small moss. Be sure to look at fresh plants. When this moss dries out it looks more dead than attractive.

GOLDEN RAGGED MOSS • *Brachythecium salebrosum*
SMOOTH-STALKED YELLOW FEATHER MOSS

GENERAL: Medium-sized to large moss in **loose, green to yellow-green mats**; stems upright or creeping along ground, irregularly branched.

LEAVES: Loosely erect or erect-spreading, sometimes curved and pointed in 1 direction, **somewhat to strongly pleated**, 2–3 mm long, lance-shaped to slightly broader, gradually pointed.

SPOROPHYTES: Single; stalk yellow to red, **smooth**, 1–3 cm long; capsule 2–3 mm long, oblong-cylindrical, asymmetrical, inclined or horizontal, **usually blackish**.

WHERE FOUND: Shady ground at tree bases; crawling over logs or humus; usually in rather dry or somewhat disturbed habitats; common and widespread across boreal forest; circumpolar.

NOTES: Golden ragged moss could be confused with sickle moss (p. 308), which also sometimes grows on the bases of tree trunks and has leaves that curve in 1 direction. However, the leaves of sickle moss are longer and much more strongly curled to one side. • The species name *salebrosum* comes from the Latin *salebra*, 'roughness,' perhaps in reference to the roughened appearance of the pleated leaves.

SMALL MOUSE-TAIL MOSS • *Myurella julacea*
ROUND MOSS

GENERAL: Small, yellowish to light green, shiny; **branches catkin-like**, blunt.

LEAVES: Crowded, overlapping wet or dry, faintly ridged when dry, 0.3–0.5 mm long, **rounded-egg-shaped**, concave, occasionally with short spine at tip, **edges toothed all around; midrib faint or lacking**.

SPOROPHYTES: Single; stalk orange-yellow, 6–15 mm long; capsule yellow-brown, 1–1.2 mm long, oblong-obovoid, **erect**, symmetrical, **smooth**.

WHERE FOUND: On damp, shaded, calcium-rich cliffs, soil or humus at base of trees, logs or stumps, or streambanks or pool edges; common in calcium-rich habitats; widespread across boreal forest; circumpolar.

NOTES: The branches of this distinctive moss have been said to look like worms or mouse tails. Small mouse-tail moss often grows with erect-fruited iris moss (p. 295) or slender-stemmed hair moss (p. 295). • The species name *julacea* refers to the 'julaceous' branches of this moss—shoots that appear smoothly cylindrical because of their closely and evenly overlapping leaves.

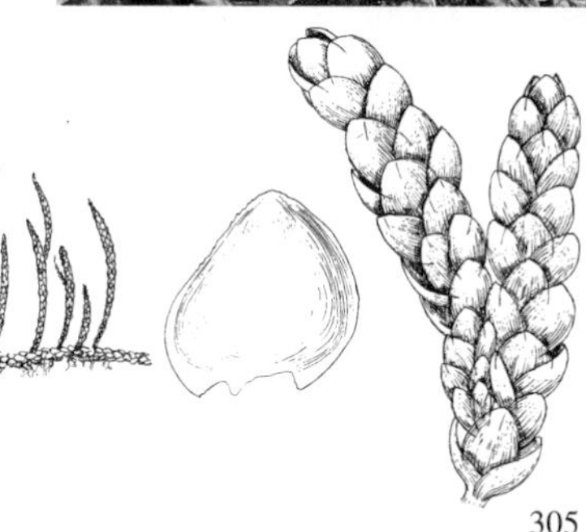

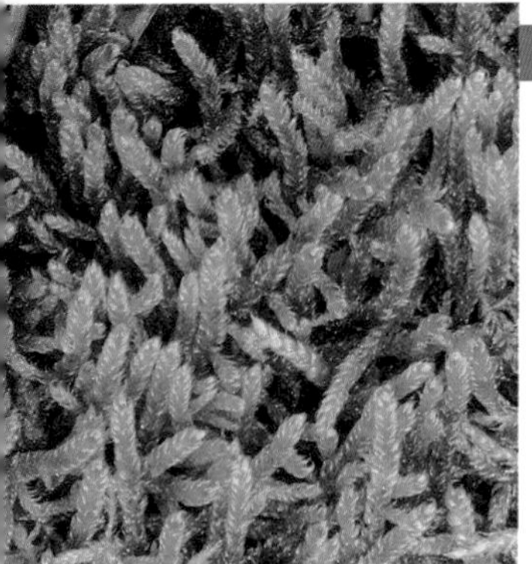

ROLLED-LEAF PIGTAIL MOSS • *Hypnum revolutum*

GENERAL: Golden to dark green or brownish; **forms mats**; stems crowded and ascending or loose and prostrate, irregularly to **regularly branched**.

LEAVES: Stem leaves moderately curved to point in 1 direction, concave, **strongly pleated when dry**, 1.2–2.2 mm long, oblong-lance-shaped, gradually pointed, **edges strongly curved backwards in lower half**, smooth-edged or **toothed near tip**, midrib short, double; branch leaves shorter and more distinctly toothed in upper part.

SPOROPHYTES: Single; stalk reddish, 10–15 mm long; capsule up to 3.5 mm long, cylindrical, curved, inclined.

WHERE FOUND: In thin or dense mats on rock, soil or humus in exposed places, and on logs or tree bases in forests; prefers calcium-rich substrates; widespread across boreal forest.

NOTES: Rolled-leaf pigtail moss is most abundant on calcium-rich rocks. It could be confused with **cypress pigtail moss** (*H. cupressiforme*) or **stump pigtail moss** (*H. pallescens*) but neither of these species has pleated leaves nor leaf edges that curve backwards. Stump pigtail moss is also smaller, its leaves are toothed almost all the way around and it commonly grows on tree bases. • The species name *revolutum*, from the Latin *re-*, 'back,' and *volvere*, 'to roll,' refers to the leaf edges.

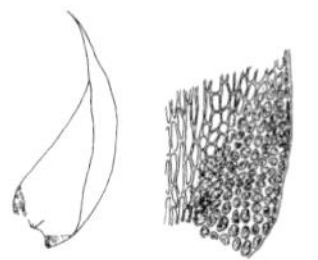

H. cupressiforme

H. pallescens

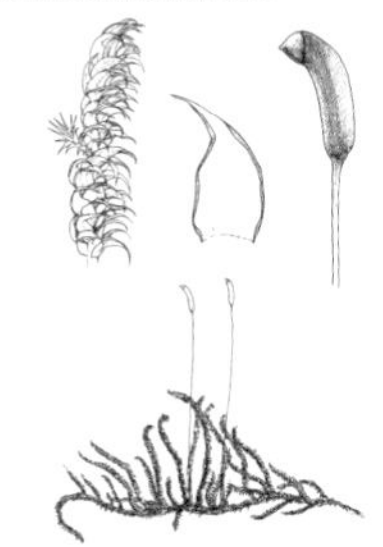

H. revolutum

STOCKING MOSS • *Pylaisiella polyantha*
ASPEN MOSS

GENERAL: Rather small and soft; **forms mats**; branches loosely **curled upwards** when dry.

LEAVES: Erect, 0.6–1.2 mm long, egg-lance-shaped, slenderly pointed, usually **smooth-edged, no midrib**.

SPOROPHYTES: Single; stalk smooth, 6–12 mm long; capsule 1.3–2 mm long, oblong-cylindrical, **erect, symmetrical**.

WHERE FOUND: Almost always on trunks of hardwoods, particularly poplars, at base as well as higher up trunk, occasionally on rotting wood, rarely on rock; common and widespread across boreal forest and parkland where poplars are common; circumpolar.

NOTES: Stocking moss could be confused with a similar moss, **common flat-brocade moss** (*Platygyrium repens*) but that species is darker, almost greasy in appearance, with conspicuous clusters of vegetative buds in the upper leaf axils. Common flat-brocade moss also grows on tree bases, but it usually prefers rotting logs, especially those that have lost their bark. • Stocking moss forms the characteristic 'stockings' on the bases of aspen trees. The habitat and the erect capsules (rare for a moss in this family) are good aids to identifying stocking moss. • The genus was named in honour of an early French bryologist, Bachelot de la Pylaie, who collected in Newfoundland.

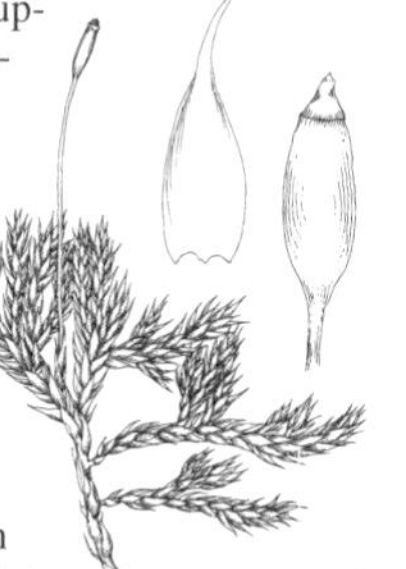

Pylaisiella polyantha

Platygyrium repens

COMMON HAIR-CAP • *Polytrichum commune*

GENERAL: **Dark green**, robust, unbranched, 4–15 cm tall or more; single-sexed, males have enlarged heads at plant tips, females produce sporophytes; lower portion covered by **grey rhizoids**.

LEAVES: 6–10 mm long, **lance-shaped, sharply pointed**; spread at right angles when moist, erect-flattened and rolled when dry; membranous, sheathing base; **edges coarsely toothed**; midrib covered on inner surface with 20–55 vertical tiers of cells (lamellae) 4–9 cells high, each tier ends with U-shaped cell (in cross-section).

SPOROPHYTES: Common, at plant tips; stalk wiry, very long; **capsules horizontal, 4-sided**; 64 short, rounded teeth and expanded central membrane around capsule mouth; **capsule hood has tuft of hairs at tip, covers entire capsule**.

WHERE FOUND: In moist coniferous forests; widespread across boreal forest; cosmopolitan.

NOTES: The hair-caps are the largest unbranched mosses in western Canada. • Common hair-cap could be mistaken for **alpine hair-cap** (*Pogonatum alpinum,* also called *Polytrichastrum alpinum*), but that moss has round (not square) capsules. • Tea made from common hair-cap was once taken to dissolve kidney and gall bladder stones. Based on the Doctrine of Signatures, this moss should be good for the hair, so a strong tea of common hair-cap was used as a rinse to 'strengthen and beautify' ladies' tresses. • The stems of common hair-cap often reach 30–45 cm or more in length. When the leaves are removed, the central stems form tough, pliable strands that have been used to make brooms and brushes or have been woven or plaited to make mats, rugs, baskets and hassocks. Apparently this was quite an art in ancient Europe. A hair-cap moss basket from the remains of an early Roman fort at Newstead, England, dates from 86 A.D.

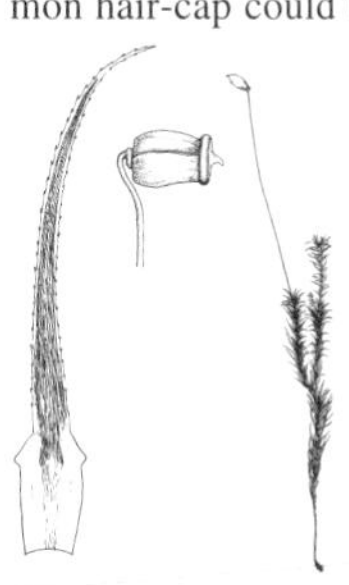

ELECTRIC EELS • *Dicranum polysetum*
WAVY DICRANUM

GENERAL: **Light green to yellow-green, large** (7 cm tall or more), covers large areas of ground; **stems covered with whitish, matted rhizoids**; good for sitting on.

LEAVES: To 1 cm long, **spread more or less at right angles from stem, edges wavy.**

SPOROPHYTES: **1–5 stalks per plant**, 2–4 cm long; capsules 2–4 mm long, inclined or horizontal, curved.

WHERE FOUND: Soil, rocks, decaying wood and humus in open, dry to moist forest; common and locally abundant (particularly in pine woods) across boreal forest north and west to interior Alaska.

NOTES: In most common *Dicranum* mosses (including electric eels), the male plants have been reduced to tiny buds on the leaves of female plants. This combines the advantages of having the sexes separate, to encourage outbreeding, with the convenience of having plants of the opposite sex nearby, to increase the chances of fertilization. Perhaps this is the reason that sporophytes are so common on the *Dicranum* mosses. • This species is often called 'electric eels' because the wavy leaves resemble miniature eels, and they stand out like they have been hit with an electric shock. The genus name *Dicranum* refers to the 2-forked teeth around the mouth of the capsule. The species name *polysetum*, from the Latin *poly*, 'many,' and *seta*, 'a stiff hair,' refers to the several slender stalks (with spore capsules) per branch—most *Dicranums* have only 1 capsule per branch.

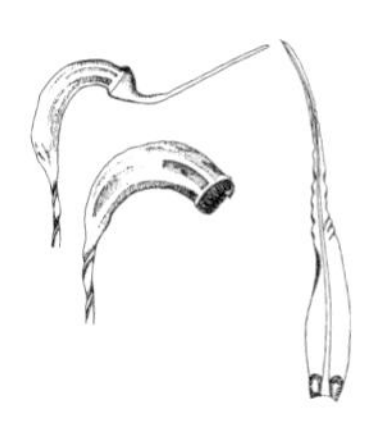

BROOM MOSS • *Dicranum scoparium*

GENERAL: Erect, little branched, densely matted rhizoids on lower stems; **forms large cushions** and sometimes mats, 2–8 cm high.

LEAVES: 5–12 mm long, **erect to curved, pointed in 1 direction** moist or dry; lance-shaped and sharply pointed; midrib single, ends in tip; **upper leaf cells longer than wide**, with irregularly thickened walls, become longer below; alar cells well developed, large, coloured, form well-marked group.

SPOROPHYTES: Often present, produced at plant tips; stalk single, straight, long; capsules curved, inclined, cylindrical, smooth; capsule teeth single.

WHERE FOUND: Tree bases, humus, rotting logs and rock outcrops; common across boreal forest; circumpolar.

NOTES: Also called *D. howellii*. • Cushion mosses (*Dicranum* spp.) are common throughout most of the boreal forest. They are identified by their medium to large size, erect stems, and leaves that curve to 1 side. Broom moss is one of the largest, most common species. The 4, well-developed, toothed ridges along the back of the midrib are characteristic of this species. At a more detailed level, the upper leaf cells of broom moss are longer than wide. Electric eels (p. 307) is the only other species to share this characteristic, and its leaves are distinctly wavy. • Broom moss is often used by florists to make banks of green in show windows. • The species name *scoparium*, from the Latin *scopae*, 'a broom,' and the common name 'broom moss,' both refer to the leaves of this moss, which look like they were swept to 1 side by someone sweeping the forest.

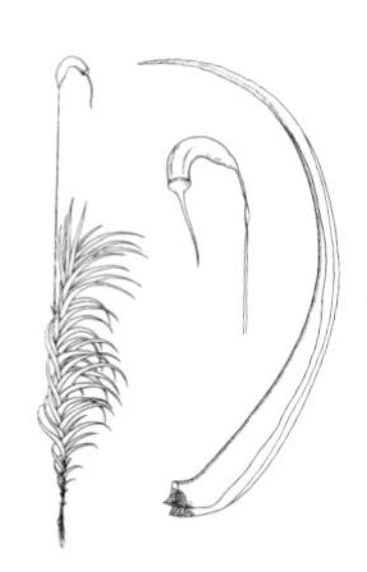

SICKLE MOSS • *Drepanocladus uncinatus*
HOOK MOSS

GENERAL: Yellow-brown to brown; stems slightly to irregularly **feathery branched**, 2–5 cm long; **forms tufts and mats**.

LEAVES: Long, narrow, **pleated lengthwise, all turned in same direction, almost curled around in crescent moon**.

SPOROPHYTES: Stalk red, 15–30 mm long, irregularly twisted; capsule brown, 2–3 mm long, strongly curved, almost horizontal.

WHERE FOUND: Soil, humus, rock, decayed wood, tree bases; wet to fairly dry sites, from gravel bars to peatlands (especially fens and swamps) to upland forests (usually at bases of trees or boulders); widespread across boreal forest; infrequent in favourable microsites in prairie and parkland; nearly cosmopolitan in cooler climates.

NOTES: The pleated leaves and sickle-shaped branch tips are distinctive of this species. • The name *Drepanocladus*, from the Latin *drepano*, 'curved,' and *clado*, 'a branch,' refers to the sickle-shaped leaves characteristic of the genus. The common name, 'sickle moss,' and the species name, *uncinatus*, from the Latin *uncio*, 'hooked,' both refer to the hook-shaped cluster of leaves at the tip of each stem.

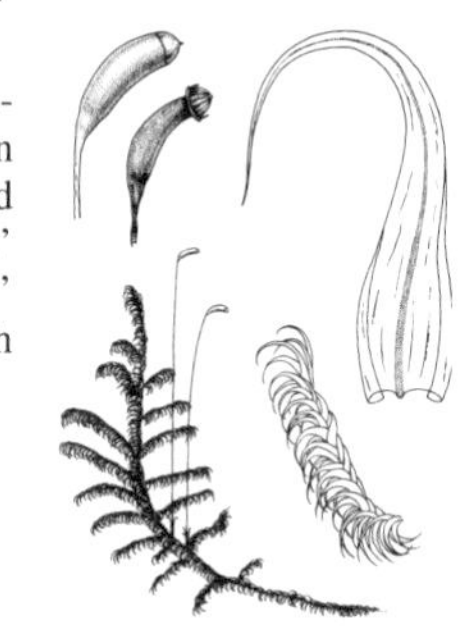

BIG RED STEM • *Pleurozium schreberi*
RED-STEMMED FEATHERMOSS

GENERAL: Stems orange to reddish, irregularly **pinnately branched**, ascending, 5–12 cm tall; forms light green to yellow-green **mats**.

LEAVES: Oblong to oval, rounded at tips, inrolled at sides.

SPOROPHYTES: Uncommon; stalks red to yellowish, 2–4 cm tall; capsules cylindrical, 2–3 mm long, horizontal.

WHERE FOUND: In most forested habitats; particularly abundant in dry forests; uncommon in exposed areas; widespread and abundant across boreal forest; circumpolar.

NOTES: Big red stem and stair-step moss (below) **are the 2 most common and widespread mosses in upland forests across the boreal region.** Big red stem is identified by its distinctive red stem, clearly visible through the pale, yellow-green leaves. Big red stem often grows as part of a continuous mat of mosses and lichens on the floor of mature coniferous forests. It usually grows with other feathermosses—especially stair-step moss and knight's plume (p. 310). • Big red stem was often used to chink log cabins. The Woods Cree placed handfuls of it at eye level to mark the location of rabbit snares or to flag trails through the woods. • The spore capsules of big red stem are rarely found, so sexual reproduction seems to be rare (possibly because of a lack of male plants), and there is no known mechanism of vegetative reproduction in this moss (other than stem fragmentation). It is interesting to speculate how a plant with almost no means of reproduction could become so common and widespread all around the world. • The genus name *Pleurozium* comes from the Latin *pleuro*, meaning 'ribs,' and presumably refers to the arrangement of the branches on either side of the stem that resembles a rib cage. This species is named in honour of J. C. Schreber, a German botanist.

STAIR-STEP MOSS • *Hylocomium splendens*

GENERAL: Olive green, yellowish or reddish green; stems creeping, 2–20 cm long, stems and branches reddish, often with branches on branches; current year's growth arises from near middle of previous year's branch, **producing feathery 'fronds' in step form**; forms springy mats.

LEAVES: 2–3 mm long, oval, smooth-edged, wide base, narrows abruptly to tip.

SPOROPHYTES: Uncommon; stalk red-brown, 1–3 cm long; capsules brown, inclined, 1.5–3 mm long, with long beak on lid.

WHERE FOUND: Soil, humus, decaying wood; in wide range of forest habitats; also in moist thickets and tundra; very common and widespread across boreal forest; circumpolar.

NOTES: Stair-step moss often grows in abundance with big red stem (above). Less vigourous stair-step moss specimens could be confused with big red stem, but stair-step moss is usually twice-branched, has many hair-like paraphyllia on its stems, and is not as shiny as big red stem. • Stair-step moss is the most common moss in the boreal forest. It is also our most northern feathermoss, extending well into the Arctic. • Those who know stair-step moss from warmer, moister climates may be surprised to discover the less robust, once- to twice-branched specimens of the boreal forest. • Stair-step moss is the only moss with a step-like arrangement of branch clusters. You can estimate the age of a plant by counting its 'steps'—a new level is produced each year. • Large quantities of this beautiful feathermoss have been used as green carpets in floral exhibitions. • The genus name *Hylocomium*, from the Greek *hyle*, 'wood' and *mnium*, an ancient name for a moss, means 'moss of the forest.' *Splendens* is the Latin word for 'shining'; although stair-step moss is seldom glossy, it is indeed splendid.

KNIGHT'S PLUME • *Ptilium crista-castrensis*

GENERAL: Green to golden-green, 3–12 cm long; branches somewhat upright to drooping, **symmetrical, feathery, tapered**; plumes **'well groomed'**; forms **decorative mats**.

LEAVES: Branch leaves pleated, egg-shaped, to 2 mm long, all curled towards branch below; stem leaves pleated, curled towards stem base.

SPOROPHYTES: Stalk reddish, 2–4.5 cm long; capsule chestnut brown, nearly horizontal, 2–3 mm long, curved.

WHERE FOUND: On humus, sometimes on logs, boulders and tree bases, in coniferous and mixed forest; common and often abundant across boreal forest; circumpolar.

NOTES: This is our most beautiful feathermoss, and though it is less common than the other feathermosses in our region (stair-step moss and big red stem), it often forms large patches in the moss carpets of boreal forest floors. Occasionally, knight's plume could be mistaken for sickle moss (p. 308), but that species is not so evenly branched, and it is less common. • The genus name *Ptilium* means 'plume-like.' The common name 'knight's plume' and the species name *crista-castrensis*, from the Latin *crista*, 'plume,' and *castrensis*, 'of the camp' or 'military,' both refer to the elegant, regularly branched plants, which were likened to the plume on a knight's helmet.

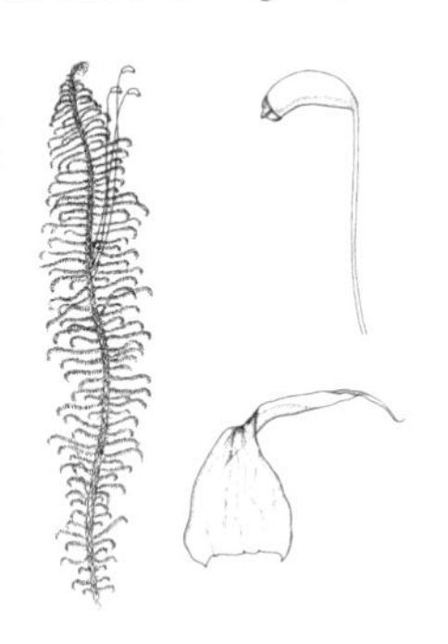

WIRY FERN MOSS • *Thuidium abietinum*

GENERAL: Yellow or dark green to brown; stems to 12 cm long, stiff; **branches simple** (not divided), **feathery** (somewhat like knight's plume), erect or ascending; with **abundant fuzz on stems**.

LEAVES: Furrowed, oval, 1.2–1.8 mm long.

SPOROPHYTES: Stalk reddish, 2–5 mm long, wavy; capsule 2–3 mm long, strongly curved and inclined.

WHERE FOUND: Dry, calcium-rich areas, exposed rocks, soil or sand, or humus in open, dry forests; locally abundant in suitable habitats across boreal forest, parkland and prairie; circumpolar.

NOTES: Also called *Abietinella abietina*. • This relatively large moss, with its dull and wiry appearance, is easily identified in the field. The dark green to brownish green, regularly pinnate plants that occur in drier sites are usually sufficient to distinguish this species. Wet plants, with their widely spreading leaves have a very different appearance from dry plants, in which the leaves are tightly flattened against the stem. Wiry fern moss rarely produces spores, especially in the eastern part of its range. • This genus was named *Thuidium* because the feathery branches were thought to resemble a redcedar (*Thuja* sp.). Similarly, the species name *abietinum* means 'like *Abies*,' a fir tree. Despite the similarities, it is relatively easy to distinguish wiry fern moss from the large coniferous trees.

HOOK-LEAF FERN MOSS • *Thuidium recognitum*

GENERAL: Light or yellow-green to yellow-brown; forms mats; stems 4–9 cm long, spreading, **2–3 times branched**; many thread-like scales, with **nipple-like bumps (papillae) mostly restricted to cell ends**.

LEAVES: Stem leaves about 1 mm long, broadly egg-shaped, abruptly pointed, **distinctly pleated**, toothed (especially toward tip), **midrib extends nearly to tip**; branch leaves shorter, with shorter midrib.

SPOROPHYTES: Single; stalk 2–4 cm long, reddish, smooth; capsule 2–3.5 mm long, cylindrical, curved and horizontal.

WHERE FOUND: On moist soil or humus, infrequently on logs or bark at bases of trees, in wooded areas; widespread, but not usually abundant, across boreal forest; circumpolar.

NOTES: Hook-leaf fern moss could be confused with wiry fern moss (p. 310) but it is smaller and more branched than that species and it prefers moister and more shaded habitats. • **Common fern moss** (*T. delicatulum*), is also very similar, but its leaves have a midrib that ends well below the tip and it has a long (1–2 mm long) beak on the lid covering the mouth of its capsule. • The species name *recognitum*, from the Latin *recognitus*, 'to recall to mind' or 'to recognize,' was given to this species because it was recognizable as separate from the widespread species common fern moss.

AQUATIC APPLE MOSS • *Philonotis fontana*
SWAMP MOSS

GENERAL: Yellow-green to powdery green, erect, simple, **in cushions or mats; stems red**; sometimes with several umbrella-like branches at tips of fertile plants; dull, but often glistening due to water droplets on surface, 1–10 cm high.

LEAVES: 1.5–2.0 mm long, loosely spreading when moist, erect-flattened when dry, lance- to egg-shaped, sharply pointed; leaf edges rolled under; midrib strong, single; leaf cells oblong, thick-walled throughout leaf, **each cell with 'bump'** (papilla) at either lower or upper end.

SPOROPHYTES: Uncommon, at plant tips, but sometimes seemingly from sides because of continued stem growth; stalk long, straight; capsules round, erect, smooth when young, become 16-ribbed, asymmetrical, and horizontal when old; 16 erect, double capsule teeth.

WHERE FOUND: In **seeps** and on **moist soil, banks and rock faces**; always associated with **calcium-rich water**; common along forestry roads and seismic lines, often in small ditches with seepy water; across boreal forest; circumpolar.

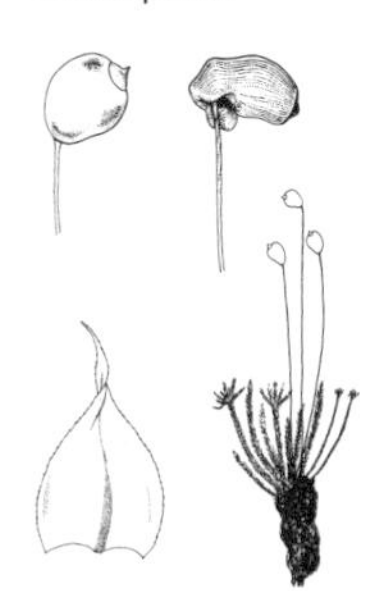

NOTES: The powdery whitish green colour and calcium-rich, seepy habitat are good identifying features of aquatic apple moss. • **Pale-leaved thread moss** (*Pohlia wahlenbergii*) has a similar dull, whitish green colour, and it often grows with aquatic apple moss. However, pale-leaved thread moss has broader leaves with smooth (not bumpy), thin-walled cells, and hanging capsules. • In most mosses, wart-like bumps grow on the sides of the cells, but in aquatic apple moss they project from the cell ends—from the lower end of cells on the outer surface of the leaf and from the upper end on the inner surface. This distinctive feature identifies aquatic apple moss. • Both the genus and species names refer to the wet, 'springy' habitats where this species grows.

TUFTED MOSS • *Aulacomnium palustre*
GLOW MOSS, RIBBED BOG MOSS

GENERAL: Yellow-green, drying brown, 3–9 cm tall; **stems erect, usually unbranched, with reddish brown, fuzzy covering**.

LEAVES: Lance- to egg-shaped, sharply pointed, 3–5 mm long, twisted when dry.

SPOROPHYTES: Stalk red-brown, to 4 cm tall, twisted when dry; **capsule** reddish brown, inclined to almost erect, to 4 mm long, curved, **strongly grooved**.

WHERE FOUND: Wet coniferous forests, swamps, fens and cold, wet, disturbed sites; very common and widespread in boreal forest; cosmopolitan.

NOTES: Tufted moss often grows mixed with golden fuzzy fen moss (p. 330) in treed fens. • It is hard to mistake the bright yellow-green patches of this moss in its cool, moist habitat. Sometimes the yellow is almost incandescent—hence the common name 'glow moss.' The reddish fuzzy stems contrasting with yellowish leaves, are characteristic of this species. • Tufted moss often reproduces asexually by producing small clusters of green, leafy-looking brood bodies on the tips of short green stalks. • The genus name *Aulacomnium*, from the Greek *aula*, 'furrowed,' and *mnium*, an ancient name for a moss, means 'furrowed moss,' and refers to the grooved capsules.

FERN MOSS • *Cratoneuron filicinum*

GENERAL: Medium-sized, yellowish, green or brownish; forms extensive cushions; stems up to 7 cm long, crowded, erect-ascending, usually **pinnately branched; scattered rhizoids in tufts along stem**.

LEAVES: Stem leaves **erect or erect-spreading** (sometimes curved to point in 1 direction), 1–1.8 mm long, egg-lance-shaped to broadly egg-shaped, abruptly pointed, **not pleated**, edges toothed all around, **midrib ends near tip**; **alar cells enlarged, colourless** (or yellowish), **in bulging group**; branch leaves shorter, narrower.

SPOROPHYTES: Single; stalk reddish, smooth, 2.5–3.5 cm long; capsule 2–2.5 mm long, curved-cylindrical, smooth, constricted below mouth when dry and empty, inclined to horizontal.

WHERE FOUND: On calcium-rich soil or rock, often near springs or along streams, frequently in drainage ditches; widespread across boreal forest; circumpolar.

NOTES: A related species, **hooked moss** (*C. commutatum*), has pleated leaves with the midrib extending beyond the tip, and it is more regularly branched. Because it grows in wet sites and has leaves that curve strongly to 1 side, hooked moss could be mistaken for a hook moss (*Drepanocladus,* p.308). However, the stems of hooked moss are covered with conspicuous hairs (visible with a hand lens) whereas those of *Drepanocladus* species are smooth. The habitat and the stems pointing upward like skinny fingers crooked at the tip make hooked moss fairly easy to recognize. • The name *Cratoneuron*, from the Greek *crates*, 'ruling' or 'powerful,' and *neuron*, 'a nerve,' refers to the strong midrib of the leaves of these mosses. The species name *filicinum*, from the Latin *filix*, 'fern,' refers to the fern-like appearance of this species.

WATERSIDE FEATHER MOSS • *Brachythecium rivulare*
COMMON VERDANT MOSS

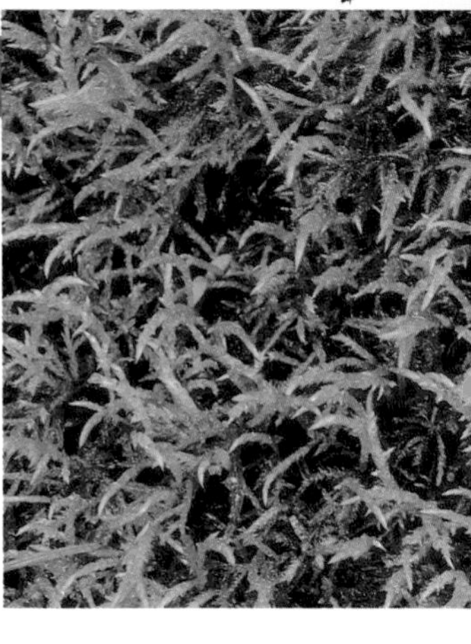

GENERAL: Robust, shiny, **pale yellow-green** to yellow-brown; forms extensive, loose mats or tufts; stems creeping to erect-ascending, irregularly **tree-like**.

LEAVES: Stem leaves broadly oblong-egg-shaped, gradually to abruptly pointed, 1.8–2.5 mm long, **alar cells noticeably enlarged**; branch leaves erect or erect-spreading, smooth or slightly pleated, 1–1.5 mm long, egg-lance-shaped to rounded-egg-shaped, pointed, **edges toothed upwards**, midrib about 2/3 length of leaf.

SPOROPHYTES: Single; stalk red-brown, **rough throughout**, 1.5–2.5 cm long; capsule 2–2.5 mm long, oblong-cylindrical, asymmetrical, strongly inclined to horizontal.

WHERE FOUND: On shaded rocks, soil and humus in woods around springs and in stream overflows; widespread across boreal forest; circumpolar.

NOTES: Waterside feather moss is recognized by its habitat, whitish green, soft, somewhat tree-like appearance, and highly differentiated alar cells, especially on the stem leaves. • The species name *rivulare*, from the Latin *rivus*, 'a brook' or 'a stream,' means 'growing beside a stream.'

COMMON TREE MOSS • *Climacium dendroides*

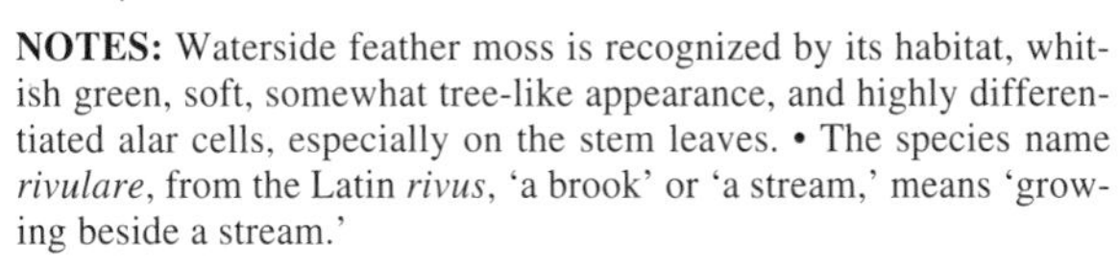

GENERAL: Large, robust, **tree-like**, about 2–10 cm high, from underground rhizome-like horizontal stems; stems branched, erect, with thread-like 'hairs.'

LEAVES: 2.5–3 mm long; midrib strong, single, long; cells thin-walled throughout; branch leaves loosely arranged, egg-shaped or narrower, sharp-pointed, pleated, toothed; stem leaves erect, pressed closely against stem, broader, with short, pointed tip, not pleated.

SPOROPHYTES: Rarely produced; when present, arise from top of erect stems; stalk long; capsules erect, smooth, dark, shortly cylindrical; capsule teeth double; capsule hood split.

WHERE FOUND: Most common in moist, humus-rich woods, peaty swamps, peatland edges, lake edges and in floodplain forests; widespread across boreal forest; circumpolar.

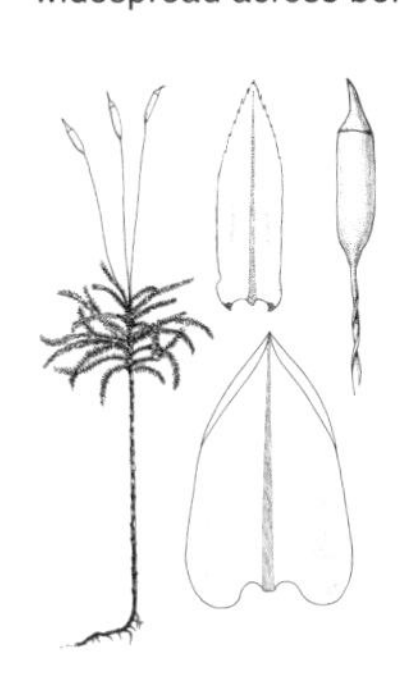

NOTES: Common tree moss has clusters of branches at the tips of its coarse, upright stems, giving each plant the appearance of a tiny tree. These 'trees' are interconnected by buried horizontal stems. This is the only common moss in the boreal forest that has a tree-like appearance. • The common name and the species name, *dendroides* (from the Greek *dendron*, 'a tree'), both refer to the tree-like appearance of this moss.

Peat Mosses

The genus *Sphagnum* (the peat mosses), with about 20 species in our region, is a common component of bogs, fens and some swamps. The peat mosses are easy to recognize as a group because of their distinctive morphology, but identification to species can be difficult and challenging.

Peat moss stems grow upright and have branches occurring in clusters (**fascicles**). The top portion of the stem, containing the clusters of young branches, is visible as a compact head (**capitulum**) at the tip of the plant. Some branches spread at right angles from the stem; the remainder hang down along it. The branches are covered by concave leaves without midribs. The **branch leaves** are 1 cell thick and consist of a net-like pattern of **small, green, living cells** without pores or strengthening ribs (**fibrils**), alternating with **large, clear, dead cells** with **pores** and **fibrils**. For each dead cell there are 2 small, green, living cells. The green cells have a variety of shapes when viewed in cross-section, and these are characteristic of groups of species. The stems also have **stem leaves** and these are often critical to the identification of many species.

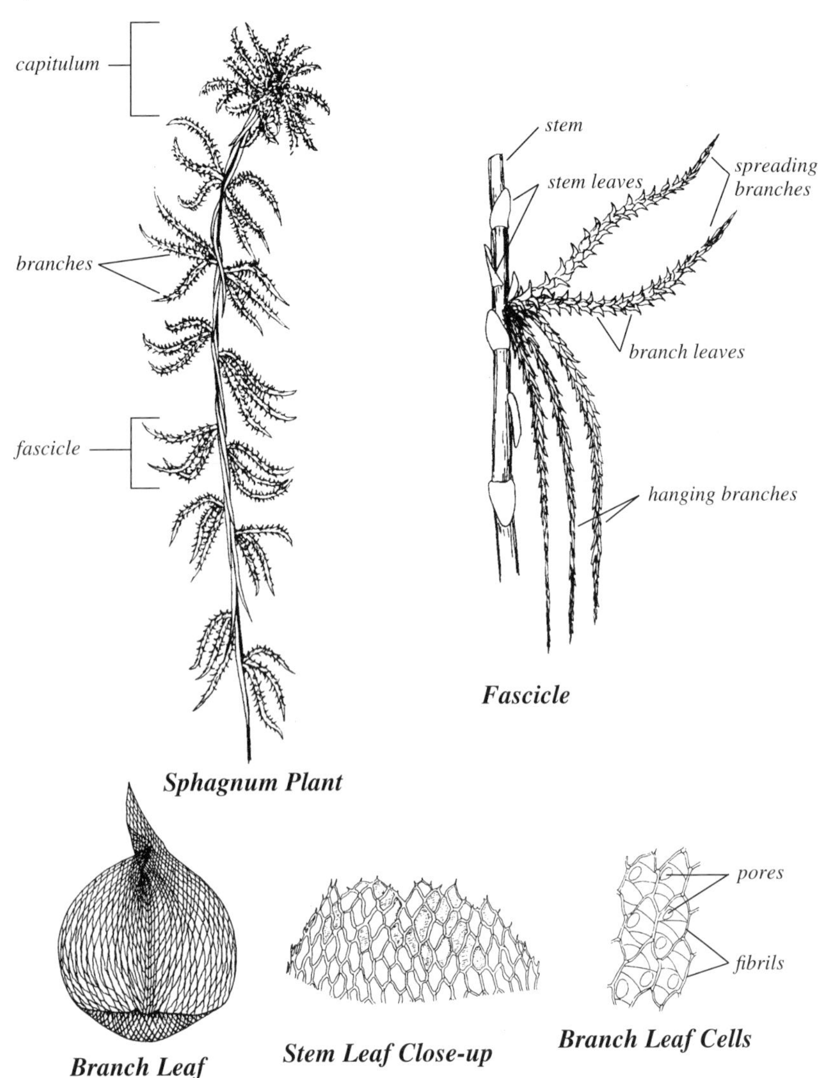

***Sphagnum* Plant**

Fascicle

Branch Leaf

Stem Leaf Close-up

Branch Leaf Cells

Key to the Peat Mosses (*Sphagnum*)

1a. Branch leaves quite wide (2–3 times as long as wide), strongly boat-shaped, hooded, blunt tip; stem leaves large (more than 1 mm long); plants pink to reddish (green in shade) .... ***Sphagnum magellanicum***

1b. Branch leaves narrow (3 or more times as long as wide), not strongly boat-shaped, minutely toothed, channelled tip; stem leaves often small (less than 1 mm long) 2

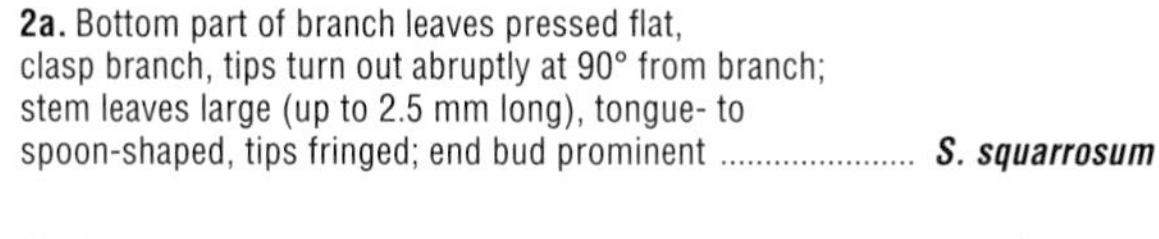

2a. Bottom part of branch leaves pressed flat, clasp branch, tips turn out abruptly at 90° from branch; stem leaves large (up to 2.5 mm long), tongue- to spoon-shaped, tips fringed; end bud prominent .... ***S. squarrosum***

2b. Branch leaves pressed to branch for most of their length; stem leaves smaller 3

3a. Young hanging branches (as seen between rays of head) paired, side by side; plants never red, purple or brown 4

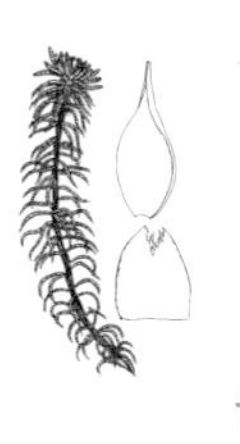

4a. End bud prominent; stem leaves have deep tear at tip; branch leaves egg-lance-shaped .... ***S. riparium***

4b. End bud not visible; stem leaves have blunt, smooth-edged or slightly toothed tip; branch leaves oblong or oblong-lance-shaped, wavy when dry; heads have dense 'pom pom' appearance .... ***S. angustifolium***

3b. Young hanging branches single; plants often have some red, purple or brown colouration 5

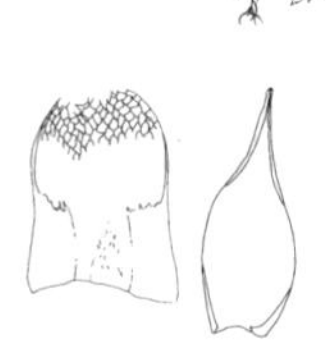

5a. Stem leaves fringed across straight top (suggestive of having been cut with pinking shears); plants green to yellowish green, without any red, purple or brown; heads flat .... ***S. girgensohnii***

5b. Stem leaves smooth-edged or slightly toothed at blunt tip; plants often have some red, purple or brown colouration 6

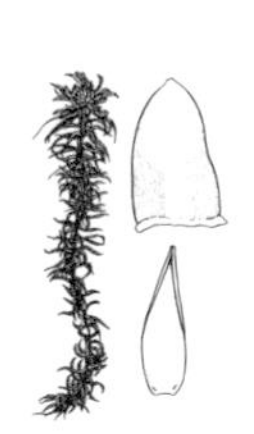

6a. Branch leaves 5-ranked, spread when dry; plants usually purple-red; stem green to red .... ***S. warnstorfii***

6b. Branch leaves never 5-ranked, overlap when dry; plants not purple-red 7

7a. Plants greenish brown to dark brown; stems dark brown; stem leaves about as long as wide; branch leaves less than 1 mm long; heads flat .... ***S. fuscum***

7b. Plants pink-red to green with reddish tinges; stems green; stem leaves longer than wide; branch leaves more than 1 mm long; heads rounded .... ***S. capillifolium***

MIDWAY PEAT MOSS • *Sphagnum magellanicum*

GENERAL: Medium-sized to large, **reddish to reddish purple**, sometimes light green when growing in shade; branches in clusters of 4–6, with 2–3 spreading away from the rest.

LEAVES: Stem leaves broadly tongue-shaped, **tip fringed, slightly toothed on sides**; branch leaves lie flat to somewhat spreading, broadly egg-shaped, toothed at tip.

SPOROPHYTES: Single; stalk short; capsule dark brown, spherical, erect, smooth.

WHERE FOUND: In open, relatively acidic habitats; especially noticeable on tops and sides of large hummocks in older, drier, more acidic parts of bogs, but also found in pioneering sites in wetter, more mineral-rich habitats; distribution closely parallels that of black spruce in our region; on all continents except Antarctica.

NOTES: Non-red specimens of midway peat moss are very similar to ***S. centrale***, and a microscope is needed to distinguish them. *S. centrale* also tends to occur in somewhat more mineral-rich sites. • Many peat mosses grow in very specific habitats. A sequence of species from poor fen peat moss (p. 317) in the hollows, to midway peat moss at mid-hummock (hence the common name), to rusty peat moss (p. 319) on the hummock top is commonly found throughout our region. • In horticulture, peat moss is added to mineral fertilizers as a filler or conditioner. It is also used as a potting medium for seedlings and cuttings and is added to soil as a source of organic material. When properly treated, peat soils are among the best agricultural soils. • The species name *magellanicum* refers to the Strait of Magellan region at the southern tip of South America, where this peat moss was first identified.

SQUARROSE PEAT MOSS • *Sphagnum squarrosum*
SPREADING-LEAVED PEAT MOSS

GENERAL: Large, **whitish to yellowish green, bristly**; head (capitulum) has large, **pinkish bud at tip**; stems **without fibrils or pores**.

LEAVES: Stem leaves oval, slightly fringed at broad tip; branch leaves 2–3.5 mm long, concave, spreading at right angles to bent backwards; tips toothed, not hooded, egg-shaped, narrow abruptly from broad base; green cells exposed on outer surface.

WHERE FOUND: Primarily **fen and swamp** species; found along streams, on peaty soils **in forests**, wet cliff shelves, and woodlands as well as in open sedge-dominated fens where it forms loose mats; **never in bogs and rarely in poor fens**; widespread and relatively common across boreal forest; circumpolar.

NOTES: Squarrose peat moss is easily identified by its widely spreading leaves that **bend out at right angles from the stem, with their tips turned backwards**. Occasionally, **thin-leafed peat moss** (*S. teres*) approaches squarrose peat moss in size and leaf orientation, but it is usually **smaller**. • When peat moss plants grow well, large amounts of dead material accumulate each year. As the lower layers are compressed by overlying mosses and water, peat is formed. In northern Europe, peat has been used as fuel for centuries, but in North America it has proved too costly to process. • The species name *squarrosum* is Latin for 'rough' or 'scabby,' and refers to the overlapping leaves and spreading leaf tips that give this moss a prickly appearance.

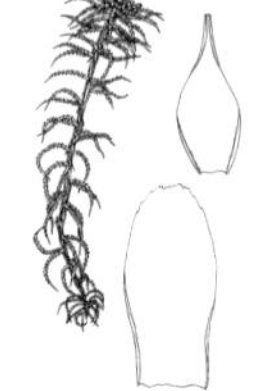

S. teres

S. squarrosum

SHORE-GROWING PEAT MOSS • *Sphagnum riparium*

GENERAL: Green, yellowish green or yellowish brown, medium-sized to large; forms extensive carpets; **prominent end buds**; branches in bunches of 4 or 5.

LEAVES: Stem leaves triangular, tip broad with single, deep tear; branch leaves egg-lance-shaped, overlapping to slightly spreading, weakly wavy, **abruptly curved backwards at tip when dry**.

WHERE FOUND: Depressions, hollows, roadside pools, cutlines through black spruce forest and disturbed areas; prefers slightly more mineral-rich sites than most other peat mosses; often dominant species in 'moats' surrounding peat plateaus and in permafrost collapse scars; widespread across our region; circumpolar.

NOTES: Greater peat moss (*S. majus*) is a large, shade-intolerant peat moss of open poor fens. The plants are usually dark green to almost black, especially at the branch tips, appearing as if they had been scorched by fire. The curved branch leaves often point in one direction. **Pendant branch peat moss** (*S. jensenii*) occurs in somewhat more mineral-rich sites than greater peat moss. It has very long stems (often submerged) and the plants often have an orangish brown hue. The stem leaves are sharply pointed and the branch leaves are curved but not all pointed in one direction. Both of these mosses are rare to uncommon in our region • Peat mosses control the drainage of vast sections of the boreal forest. They also control plant succession through their ability to hold water and to create an acidic environment. • Acidic peatlands are ideal for trapping and preserving plant and animal remains. Pollen preserved in peat is used to study prehistoric vegetation. • Peat moss spores are forced from their capsules under pressures 4–6 times atmospheric pressure—levels similar to those found in the tires of heavy trucks! These miniature explosions can be heard as tiny puffs, like the snapping of a wheat straw, and they hurl the spores 3 cm or more from the parent plant.

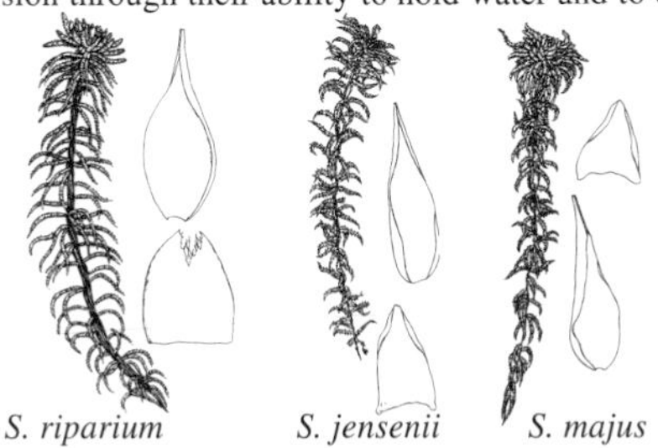

S. riparium *S. jensenii* *S. majus*

POOR FEN PEAT MOSS • *Sphagnum angustifolium*
YELLOW-GREEN PEAT MOSS

GENERAL: Small, **yellow-green**, slender; head (capitulum) bushy, rounded, not star-shaped; stems without fibrils or pores; **2 young, hanging branches** side by side.

LEAVES: Stem leaves 0.5–0.8 mm long, small, triangular, blunt, without pores and fibrils, widely spaced; branch leaves slender, toothed at tip, not hooded; most colourless cells have 1 large pore at tip and only 1–2 additional pores in corners of cell on outer surface; green cells triangular, exposed on outer surface only.

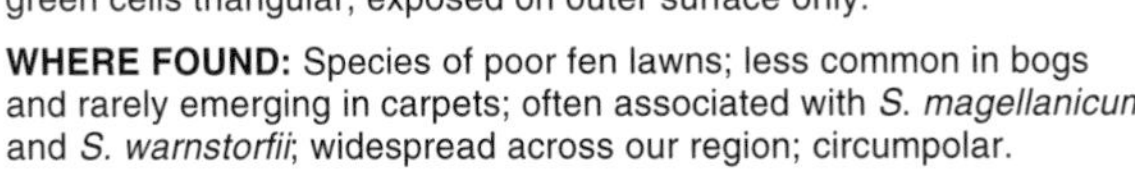

WHERE FOUND: Species of poor fen lawns; less common in bogs and rarely emerging in carpets; often associated with *S. magellanicum* and *S. warnstorfii*; widespread across our region; circumpolar.

NOTES: Also called *S. recurvum* var. *tenue*. • Poor fen peat moss is recognized by its yellow-green colour, by the pair of branches hanging side-by-side below each rather small 'pom pom' cluster of end branches, and by its small, triangular, blunt, stem leaves. It could be confused with ***S. fallax*** but that species has longer stem leaves (0.7–1.1 mm long) that are not toothed at the tip and its branch leaves are wavy with spreading tips when dry. *S. fallax* is rare in our region. • Poor fen peat moss has a much higher rate of water loss than many other plants. The inability of peat mosses to slow their water loss could be a key factor preventing their occurrence farther north. High Arctic regions are generally classified as polar deserts, and plants growing in these regions must conserve moisture. Many mosses and lichens have adapted to grow in extremely dry environments. They can shut down to withstand indefinite drying and resume metabolism when water becomes available.

GIRGENSOHN'S PEAT MOSS • *Sphagnum girgensohnii*
WHITE-TOOTHED PEAT MOSS

GENERAL: Medium-sized, **green**, robust, forms loose mats; head (capitulum) **star-shaped, flat-topped**; stems with **many pitted cells** in outer colourless, transparent layer; 1 young hanging branch.

LEAVES: Stem leaves broadly tongue-shaped, **torn along edge across flat top**; branch leaves similar to those of acute-leaved peat moss (below).

WHERE FOUND: Species of woodlands, and bog and fen edges across boreal forest; more common to north where it often forms low hummocks in shrubby tundra; circumpolar.

NOTES: Girgensohn's peat moss is distinguished by large, yet slender, **flat-topped, green** plants with **stem leaves** that look like they were cut with pinking shears. A similar species, **fringed bog moss** (*S. fimbriatum*) has stem leaves that are **torn all around** the edges. Acute-leaved peat moss (below), Warnstorf's peat moss (p. 319) and wide-tongued peat moss (below) differ from acute-leaved peat moss in that they all have some red colour. • Alaskan native peoples mixed peat moss with grease to make a salve for treating cuts. • Peat moss that had been burned to ashes or powder was used as a germicide in the early 1900s. • Peat water was said to be astringent and antiseptic and even the air near peatlands was considered extremely healthful. • Peat moss helps to prevent infection and it has often been used as an absorbent dressing for wounds. During World War I, peat moss was widely used for surgical dressings, freeing cotton for the manufacture of gunpowder. It was collected, picked clean, dried, and lightly packed in muslin bags. These were then sterilized and placed on wounds as dressings. Peat moss dressing could absorb more than 3–4 as much moisture as cotton dressings. It also retains liquids much better, reducing the number of time the dressing must be changed, and distributes liquids more uniformly throughout the dressing. Peat moss dressings are cooler, softer and less irritating than dressings stuffed with cotton, and they can be produced more quickly and cheaply in emergencies. • Sphagnol, an extract produced by distilling the tar from peat, was used to soothe and heal haemorrhoids, chilblains, and skin problems such as eczema, psoriasis, and acne.

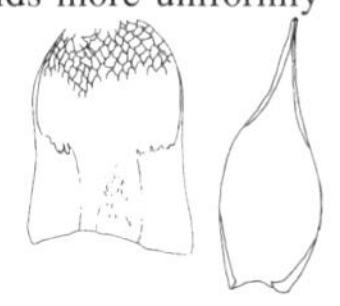

ACUTE-LEAVED PEAT MOSS • *Sphagnum capillifolium*
SMALL RED PEAT MOSS

GENERAL: Small, reddish, with **rounded, convex head** (capitulum); stems without pores or fibrils; **1 young, hanging branch**.

LEAVES: Stem leaves relatively long, broadly egg-shaped, narrower above than below, blunt, with at least a few fibrils and pores (or poorly formed gaps) in upper part; branch leaves slender, toothed at tip; colourless cells have many large pores on both surfaces; green cells triangular, exposed on inner surface only.

WHERE FOUND: Forms hummocks in wooded edges around bogs and poor fens, more rarely in open peatlands; widespread across boreal forest; circumpolar.

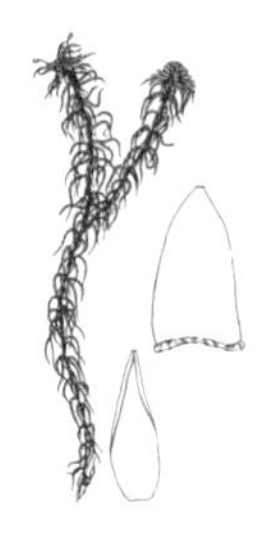

NOTES: Also called *S. nemoreum*. • **Wide-tongued peat moss** (*S. russowii*) is a similar species, but its outer stem cells have pores that are not found on the stems of acute-leaved peat moss. • The Chipewyan used acute-leaved peat moss as a dish scrubber. It was also used to insulate cabins and to fill the cracks between logs. Smudges of dried peat moss were used to keep flies and mosquitoes away. When smoke-curing hides or meat, the Chipewyan often burn damp peat moss with alder wood. The Woods Cree did not use red peat mosses in diapers, because they irritated the skin, but acute-leaved peat moss was used when it lacked its red colour. • The species name *capillifolium* means 'hair-leafed,' and refers to the relatively narrow leaves of this species.

RUSTY PEAT MOSS • *Sphagnum fuscum*
COMMON BROWN SPHAGNUM

GENERAL: Brown to greenish brown; stems slender, **brown**; individually less robust than other peat mosses, **in very compact hummocks; threadlike branches interwoven inside hummocks**.

LEAVES: Stem leaves tongue-shaped, blunt; branch leaves lance-shaped, pointed.

SPOROPHYTES: Stalk short (1–2 mm); capsule chocolate brown, 1–1.5 mm long.

WHERE FOUND: Caps hummocks in older or drier open bogs, forms extensive ground cover in black spruce bogs, or on isolated hummocks in fens; most common peat moss across our region; circumpolar.

NOTES: Rusty peat moss could be confused with unusually dark forms of acute-leaved peat moss (p. 318). The brown stems of rusty peat moss distinguish it from all other peat mosses in our region except **Lindberg's peat moss** (*S. lindbergii*) and, in the northwest, *S. lenense*. However, Lindberg's peat moss is much larger, has a prominent bud at its tip and its stem leaves have broad torn tips. It also prefers habitats that are slightly more mineral-rich. • Peat mosses absorb water better than a sponge. These plants are permeated with a system of minute tubes. Because of this sponge-like structure, peat moss absorbs water more quickly than cotton and can hold 3–4 times as much. • Peat moss has been used for many years in baby diapers and menstrual pads. Many babies spent the first 2 years of life in a 'moss bag,' made from moose hide and lined with rabbit skin. The baby was laid on the skin with dry moss and the bag was laced closed. Toddlers wore 'diapers' of soft skins packed with dry moss. The Chipewyan collected green, yellow, or white peat mosses from hummocks for use in diapers. Because of its absorptive and antiseptic qualities, the Woods Cree used rusty peat moss to treat diaper rash.

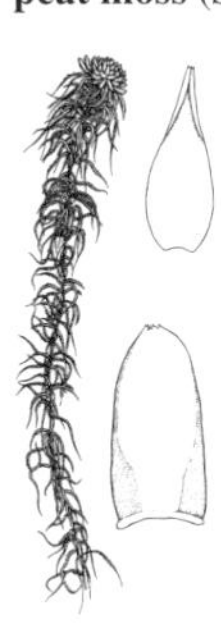

WARNSTORF'S PEAT MOSS • *Sphagnum warnstorfii*

GENERAL: Slender, small to medium-sized; **purplish red** growing in sun, greenish in shade; branches in clusters of 3–5 with 2 spreading away from the rest.

LEAVES: Stem leaves triangular to oblong tongue-shaped, tip blunt to broadly pointed, smooth-edged or weakly toothed; **branch leaves in 5 distinct rows**, spreading, egg-lance-shaped, tip narrow, curled inwards, toothed; upper half of young leaves (you'll need a microscope) have **very small, round, strongly ringed pores**, contrast with much larger pores in lower half of leaves.

SPOROPHYTES: Single; stalk short; capsule brownish, spherical, erect, smooth.

WHERE FOUND: Mineral-rich habitats across our region; often found with golden fuzzy fen moss (p. 330), forming small hummocks or lawns in rich fens; often occurs below rusty peat moss (above) on hummocks in treed fens; occasionally in white cedar swamps at southeastern edge of our region; circumpolar.

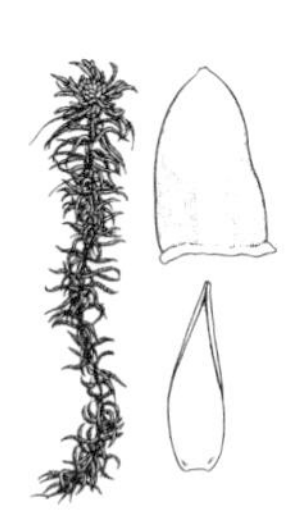

NOTES: Warnstorf's peat moss is the most common peat moss in Alberta, but it is less common farther north. • Peat mosses are the only mosses with considerable economic value. They are widely used as stable litter because of their ability to absorb and deodorize liquid manure, to reduce losses in available nitrogen, and to check insect pests. Peat moss has been used to produce many items, including ethyl and methyl alcohols, acetic and carbonic acids, ammonium compounds, nitrates, sugar, dyes, tanning materials, paper, woven fabrics, artificial wood, paraffin, naphtha, pitch, millboard, porous bricks, life preservers, gunpowder, fireworks, paint, paper, insulation, fabrics and charcoal.

SLENDER HAIR-CAP • *Polytrichum strictum*
BOG HAIR-CAP

GENERAL: Relatively slender, tall and stiff, 4–10 cm tall; grows singly or forms compact, deep tufts; stems dark brown, **conspicuously matted with whitish rhizoids well into leaves**.

LEAVES: Erect and straight when dry, **erect-spreading** when moist, 3.5–6 mm long, **not conspicuously awned**.

SPOROPHYTES: Single; stalk 2–6 cm long; capsule 2.5–3 mm long, erect to inclined or horizontal, **4–6 sided**.

WHERE FOUND: Closely associated with *Sphagnum* in open and wooded bogs and poor fens; typically found on top of old hummocks; common and widespread in peatlands across boreal forest; circumpolar.

NOTES: Also called *P. juniperinum* var. *affine*. • The presence of slender hair-cap often indicates the driest part of a *Sphagnum* hummock. In exposed areas, slender hair-cap stays within 1 cm above the surface of the peat moss. • The large capsules of hair-caps are eaten by many small mammals and birds. In Norway, they form an important part of the diet of grouse chicks. The capsules also resemble grains of wheat, hence the common name 'bird wheat.' They are edible and could be used as a survival food in an emergency. • The species name *strictum*, from the Latin *strictus*, 'close' or 'tight,' refers to the fact that the leaves of slender hair-cap do not spread widely from the stem when they are wet, as do those of most other hair-caps.

WAVY DICRANUM • *Dicranum undulatum*

GENERAL: In dense tufts, yellowish green or yellowish brown; stems 3–15 cm tall, densely matted with reddish brown rhizoids.

LEAVES: Commonly wrinkled and wavy, especially in upper part, straight or wavy when dry, 5–8 mm long, narrowly to broadly pointed, concave below, V-shaped above; smooth-edged below, toothed from about middle to tip.

SPOROPHYTES: Stalk yellow to yellowish brown, **single**, 2–4.5 cm long; capsule yellow to yellowish brown, inclined to horizontal, curved, 2–3 mm long, furrowed lengthwise when dry.

WHERE FOUND: Typical of wet areas in wooded peatlands across boreal forest; circumpolar.

NOTES: Wavy dicranum could be confused with electric eels (p. 307), but the stems of wavy dicranum are less woolly and its leaves are less pointed and tend to be erect rather than spreading at right angles to the stem. At a more detailed level, the upper leaf cells are short and there are no ridges on the back of the midrib. Usually these 2 species are most easily separated by habitat—wavy dicranum grows on organic soil, typically in boggy areas, whereas electric eels is found in upland forests. • **Sharp-leaved cushion moss** (*D. acutifolium*) is similar to these 2 species, but its leaves are more sharply pointed and less noticeably wavy. It is an upland species, more common farther north and at higher elevations. • The name *undulatum* refers to the wavy, undulating leaves of wavy dicranum.

TALL CLUSTERED THREAD MOSS • *Bryum pseudotriquetrum*

GENERAL: Greenish to reddish, grows in cushions, mats or as individuals among other mosses; **highly variable in form**; stems simple or rarely branched, **red, hairless**, about 2–8 cm high.

LEAVES: Up to 4 mm long, egg-shaped but variable, tip blunt to sharp; midrib ends near or extends just beyond tip; edges **smooth**, bordered by longer cells; **leaf bases extend downwards along stem**; leaf cells 6-sided, thin-walled, smooth.

B. pseudotriquetrum

SPOROPHYTES: Not common; stalk reddish to pale-green, erect, long; capsules hanging, smooth, cylindrical; capsule teeth well developed, double.

WHERE FOUND: Extremely variable, occurs in seeps, rich fens, swampy areas or on streambanks; always seems to be in areas receiving calcium-rich influence; common and widespread across boreal forest; circumpolar.

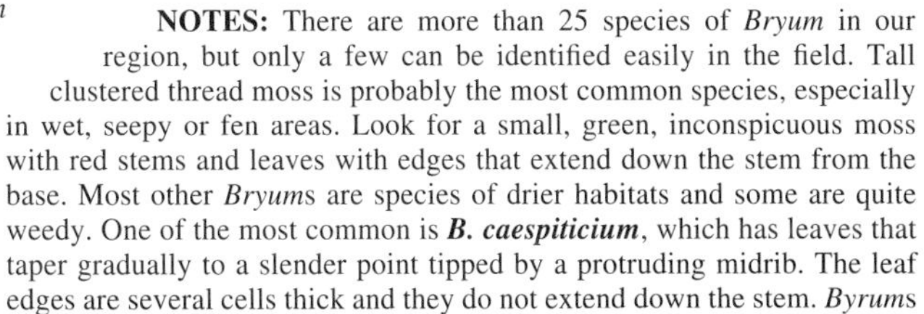

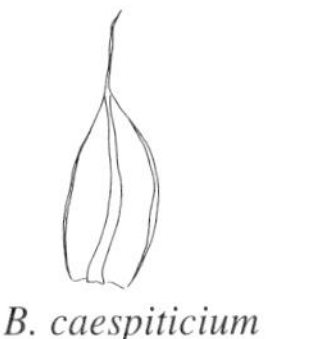

B. caespiticium

NOTES: There are more than 25 species of *Bryum* in our region, but only a few can be identified easily in the field. Tall clustered thread moss is probably the most common species, especially in wet, seepy or fen areas. Look for a small, green, inconspicuous moss with red stems and leaves with edges that extend down the stem from the base. Most other *Bryums* are species of drier habitats and some are quite weedy. One of the most common is ***B. caespiticium***, which has leaves that taper gradually to a slender point tipped by a protruding midrib. The leaf edges are several cells thick and they do not extend down the stem. *Byrums* could be confused with *Pohlia* species, but those mosses have toothed leaf edges and longer leaf cells. *Pohlias* also lack the distinctive row of long cells that border the leaves of *Bryum* species. • The name *Bryum* comes from the Greek word *bryon*, meaning 'moss.'

FELT ROUND MOSS • *Rhizomnium pseudopunctatum*

GENERAL: Dark green; forms loose tufts; stems 4–6 cm tall, with **thick rhizoids** towards base.

LEAVES: Erect or erect-spreading, moderately contorted when dry, **4.5–7 mm long, 6 mm wide**, rounded to broadly obovate, rounded or shallowly notched at tip, smooth-edged, bordered by 1–3 rows of linear cells **in double layer at leaf-base**.

SPOROPHYTES: Stalk **single**, 2–4.5 cm long, orange-brown or reddish; capsule pale brown, 2–2.5 mm long, narrowly elliptical, hanging.

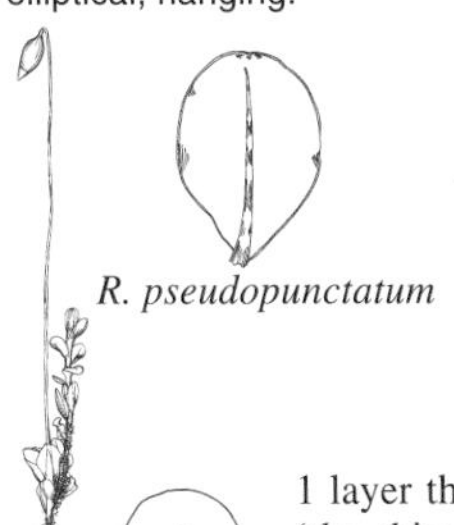

R. pseudopunctatum

R. gracile

WHERE FOUND: On peaty soil and wet humus, particularly in calcium-enriched habitats; common in fens and swamps across boreal forest; circumpolar.

NOTES: Also called *Mnium pseudopunctatum.* • Felt round moss could be confused with **slender round moss** (*R. gracile*) but the leaves of that species are smaller (up to 4 mm long and 3 mm wide) and are bordered by rows of cells that are only 1 layer thick at the leaf-base. • The name *Rhizomnium* means 'the rhizoid-bearing *Mnium*,' and refers to the thick, woolly layer of rhizoids on the stems of many species. The species name *pseudopunctatum* is derived from the Latin *pseudo*, 'false,' and *punctum*, 'a point.' The leaves of round mosses do not have points.

COMMON NORTHERN LANTERN MOSS • *Cinclidium stygium*

GENERAL: Green at tip, reddish brown to nearly black below; stems erect, simple, reddish brown, 2–10 cm tall.

LEAVES: Wavy and contorted when dry, 4–5 mm long, **obovate or rounded-egg-shaped** from narrow base, **tip broadly rounded** with **short, abrupt point, smooth-edged**, bordered.

SPOROPHYTES: Single; stalk yellow-brown, 3–7 cm long; capsule 3–4 mm long, brownish yellow, **egg-shaped, hanging, with short, distinct neck**.

WHERE FOUND: In rich fens across northern boreal forest, more common in Arctic; circumpolar.

NOTES: Common northern lantern moss is often found with three-ranked feather moss (*Calliergon trifarium*, p. 327) and golf club moss (*Catoscopium nigritum*, p. 324). The broad, blackish red leaves of common northern lantern moss, with their rolled edges and abrupt, sharp point are unmistakable. Lantern mosses (*Cinclidium* spp.) could be confused with round mosses (*Rhizomnium* spp.), but the round mosses have stronger borders on their leaves. • The species name *stygium* may come from the Latin *Stygius*, 'of the River Styx and the infernal regions,' either because the blackish lower stems and leaves look like they could have grown up from the nether world, or because the bryologist who described this moss felt it belonged there.

MARSH MAGNIFICENT MOSS • *Plagiomnium ellipticum*

GENERAL: Green or yellow-green, 3–9 cm high, forms loose to dense tufts; stems covered with rhizoids towards base; fertile stems erect; sterile stems bent to horizontal, often lengthened.

LEAVES: Small and widely spaced towards base, larger and crowded upwards; 5–8 mm long, elliptic to oval, rounded tip, abruptly tooth-pointed, **base does not extend as ridge down stem**, erect or spreading, strongly contorted when dry, **toothed nearly to base with short, blunt, 1–2 celled teeth** (sometimes nearly lacking), bordered by 2–4 rows of linear cells in 1 layer.

SPOROPHYTES: Stalk **single**, or rarely 2 or 3, yellowish towards top, reddish below, 2–5 cm long; capsule yellowish brown or brown, 3–5 mm long, oblong-cylindrical, hanging.

WHERE FOUND: In wet, nutrient-enriched habitats such as open, shrubby and treed fens, willow thickets along streams and in wooded swamps; common across boreal forest.

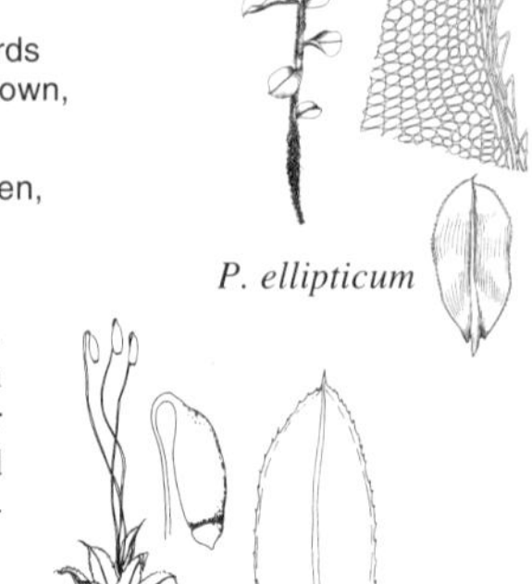

P. ellipticum

P. medium

NOTES: Also called *Mnium ellipticum* or *M. affine* var. *rugicum*. • Marsh magnificent moss could be confused with **common leafy moss** (*P. medium*) but that species has larger leaves with longer, sharper teeth, and its leaf-bases extend down the stem as a ridge. • The species name *ellipticum* refers to the oblong, rounded (elliptic) leaves.

THREE-ANGLED THREAD MOSS • *Meesia triquetra*

GENERAL: Plants 3–12 cm tall.

LEAVES: In 3 rows, spreading at right angles, often strongly contorted when dry, 2–4 mm long, egg-lance-shaped, **ridged above base**, narrow gradually to slender, pointed tip, **leaf-base extends down stem, edges distinctly toothed from middle to tip**.

SPOROPHYTES: Stalk single, 3–10 cm long; capsule 3–4.5 mm long, pear-shaped, curved above long, erect neck.

WHERE FOUND: Calcium-rich fens, where it can sometimes be abundant; widespread across boreal forest; circumpolar.

NOTES: Three-angled thread moss could be confused with **long-stalked thread moss** (*M. longiseta*), but that species has no teeth along the edges of its leaves. • Three-angled thread moss is relatively abundant in some rich fens. Look for it in pools and flarks with sausage moss (p. 329), red hook moss (p. 329), golf club moss (p. 324) and giant water moss (p. 326). • This genus was named in honour of David Meese, a Dutch gardener. The species name is derived from the Latin *tri*, 'three,' and *qued*, 'sharp,' and refers to the pointed leaves widely spread in 3 distinct rows.

CAPILLARY THREAD MOSS • *Meesia uliginosa*

GENERAL: Plants 1–2.5 cm tall.

LEAVES: Not in 3 rows, 2–4 mm long, **erect**, often contorted when dry, **narrowly lance- to linear tongue-shaped**, blunt-tipped, **not ridged, bases do not extend as ridge down stem, smooth-edged**.

SPOROPHYTES: Stalk 2–7 cm long, single; capsule 1.5–4 mm long, pear-shaped, curved above long, erect neck.

WHERE FOUND: On wet soil or peat, rarely on rotting wood, in calcium-rich fens and swamps; widespread, but never abundant, across boreal forest north to Arctic; circumpolar.

NOTES: The stalk of the spore capsule is exceptionally long in capillary thread moss, often more than double the height of the moss plant (the gametophyte). The erect, blunt leaves, with their extremely wide midrib, are unmistakable. • The species name *uliginosa* comes from the Latin *ulig*, 'moisture' or 'marshiness,' and refers to the wet habitats of this moss.

GOLF CLUB MOSS • *Catoscopium nigritum*
BLACK-FRUITED WEISSIA

GENERAL: Slender, 1.5–6 cm tall, **grows in very compact, green to blackish tufts**.

LEAVES: Erect-spreading, 0.8–1.5 mm long, somewhat contorted when dry, narrowly lance-shaped, slenderly pointed, ridged, smooth-edged.

SPOROPHYTES: Stalk single, erect, dark red, 8–25 mm long; **capsule very small**, egg-shaped, inclined or horizontal, **strongly asymmetrical, curved, mouth directed downwards at maturity**, 0.6–1 mm long, green when young, turns red-brown to **black with age**.

WHERE FOUND: On peaty humus, soil, or wet rocks, in calcium-rich fens and at edges of pools and lakes; widespread across boreal forest; circumpolar.

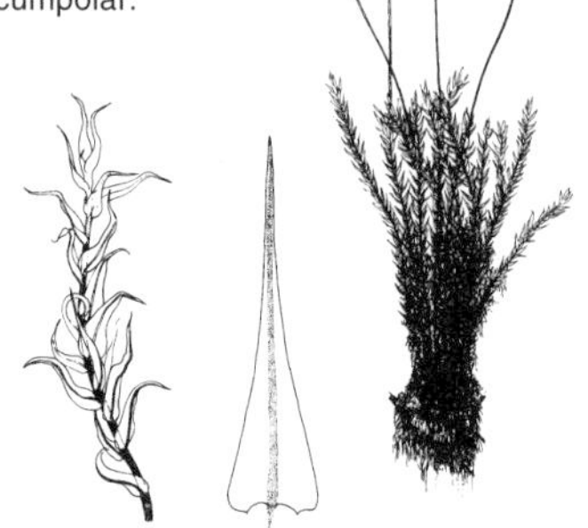

NOTES: The sporophytes of golf club moss, often produced in great numbers, resemble miniature golf clubs, and are always a delight to find. • The species name, from the Latin *niger*, 'black,' refers to the shiny black capsules, which are unusually dark for a moss. The genus name *Catoscopium* is derived from the Greek *kata*, 'down,' and *skopos*, 'a spy' or 'a watcher.' It refers to the capsule, which is directed downwards—hence 'looks' down. Crum (1973) says they remind him of someone looking back over his shoulder, bringing to mind some sound advice on the topic of staying young: Don't look back. Something may be gaining on you!

BLANDOW'S FEATHER MOSS • *Helodium blandowii*

GENERAL: Rather **robust**, 5–12 cm tall; **forms dense, soft, yellow to light green masses**; **stems red-brown**, erect, **closely once divided, branches nearly equal**, straight, somewhat tapered, **widely spreading**; many thread-like scales on stems and branches.

LEAVES: Stem leaves erect or erect-spreading, 1.2–1.5 mm long, **somewhat pleated**, broadly egg-shaped, gradually short-pointed, **leaf-base narrows abruptly, extends down stem**, hairy fringe near base, edges slightly toothed above, irregularly curled backwards often nearly to tip; branch leaves 0.5–0.8 mm long, egg-shaped, pointed.

SPOROPHYTES: Stalk single, reddish, 4–6 cm long; capsule 3–4 mm long.

WHERE FOUND: Prefers wet, calcium-rich habitats, fens, swamps and streambanks; widespread and relatively common across boreal forest; circumpolar.

NOTES: These large, soft, pale-coloured plants growing in fens are highly distinctive. Blandow's feather moss could be confused with wiry fern moss (p. 310), but that species grows on dry, calcium-rich rock outcrops and in dry upland forests, rather than peatlands, so the two are easily separated on the basis of habitat. • The genus name *Helodium*, from the Greek *helo*, 'marshy,' and *-odes*, '-like,' refers to the wet habitats preferred by these mosses. The species is named for Blandow, the botanist who first collected it in Germany.

Key to Common Rich Fen Mosses and Related Species

The leaves of mosses in this group are illustrated on p. 331.

1a. Leaves have strong midrib, ending at or near leaf tip 2

2a. Leaves pointed, erect-spreading to curved; found on calcium-rich streambanks and other wet places 3

3a. Leaves pleated, often coiled and crisped; plants irregularly to regularly branched; stems often have many, conspicuous paraphyllia (minute leaf- or thread-like structures) .... ***Cratoneuron commutatum*** (see *C. filicinum*)

3b. Leaves not pleated, erect-spreading to curved; plants irregularly branched; paraphyllia sparse and inconspicuous (if present) .... ***Cratoneuron filicinum***

2b. Leaves rounded at tip, blunt, erect; found in fens and peatland forests 4

4a. Stems more or less regularly pinnately branched; stem leaves nearly as wide as long .... ***Calliergon giganteum***

4b. Branches few and irregular; stem leaves usually much longer than wide .... ***Calliergon cordifolium*** (see *C. giganteum*)

1b. Leaves lack noticeable midrib, or midrib extends no more than 3/4 length of leaf 5

5a. Leaves wide-spreading from stem in star-like pattern, pointed, lack noticeable midrib; plants green to yellow-green 6

6a. Plants robust; leaves often over 2 mm long; found in fens .... ***Campylium stellatum***

6b. Plants small; leaves usually less than 1.5 mm long; found on rotting wood .... ***Campylium hispidulum*** (see *C. stellatum*)

5b. Leaves pointed at tip, usually pressed to stem or only slightly spreading 7

7a. Leaves sickle-shaped, curved, all point in 1 direction 8

8a. Leaves lack noticeable midrib; plants robust, reddish to brownish black, with wide, broadly pointed leaves, roughened when dry; found in rich fens .... ***Scorpidium scorpioides***

8b. Leaves have single midrib, usually reaches middle of leaf or beyond 9

9a. Leaves conspicuously pleated; found in moist, terrestrial habitats .... ***Drepanocladus uncinatus***

9b. Leaves smooth or with fine parallel ridges; aquatic, emergent or in peatlands 10

10a. Plants yellow-green, often regularly branched; found in fens .... ***Drepanocladus vernicosus*** (see *D. revolvens*)

10b. Plants green to reddish brown, irregularly branched; found in fens and ditches 11

11a. Leaf edge toothed at tip; alar cells enlarged; plants irregularly branched 12

12a. Midrib slender, reaches about middle of leaf; alar cells poorly differentiated .... ***Drepanocladus fluitans***

12b. Midrib stout, ends in upper part of leaf; alar cells well differentiated, form ear-like lobes at base of leaf .... ***Drepanocladus exannulatus*** (see *D. fluitans*)

11b. Leaves smooth-edged 13

13a. Plants robust, often reddish or blackish purple; alar cells little enlarged; found in fens .... ***Drepanocladus revolvens***

13b. Plants relatively slender, green; alar cells inflated; found in various wet habitats 14

14a. Lower leaf cells linear .... ***Drepanocladus aduncus***

14b. Lower leaf cells shorter and wider .... ***Drepanocladus polycarpus*** (see *D. aduncus*)

7b. Leaves rounded at tip, blunt or with tiny point, erect; found in fens and peatland forests 15

15a. Midrib weak, single or forked above leaf base, rarely reaches more than 1/3 length of leaf; leaf tips often have minute, channelled point .... ***Scorpidium turgescens*** (see *S. scorpioides*)

15b. Midrib strong, single, reaches 1/2 to 3/4 length of leaf 16

16a. Plants coarse, green, variously branched; leaves erect-spreading, broadly egg-shaped or heart-egg-shaped .... ***Calliergon richardsonii***

16b. Plants slender; stems little branched; leaves close and overlapping 17

17a. Plants yellow-green or straw-coloured throughout; leaves oblong ***Calliergon stramineum***

17b. Plants yellowish green above, become brown below; leaves egg-shaped to round .... ***Calliergon trifarium*** (see *C. stramineum*)

GIANT WATER MOSS • *Calliergon giganteum*
GIANT FEATHER MOSS

GENERAL: Yellow-green, green, or yellow-brown, up to 20 cm long or more, in deep, loose clumps; **feathery and more or less regularly branched**.

LEAVES: Stand well out from stem; stem leaves egg-shaped and blunt-tipped, **often as wide as long**; branch leaves rolled when dry; **midrib extends almost to tip of leaf** (use hand lens); **alar cells inflated, form well-defined regions**.

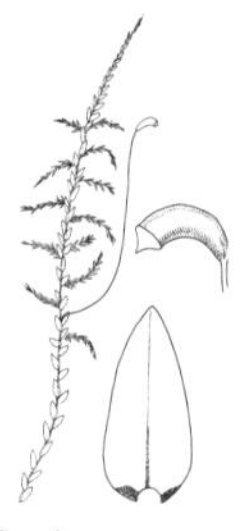

C. giganteum

SPOROPHYTES: Stalk cinnamon in colour, 4–7 cm long; capsules curved, 2–3 mm long.

WHERE FOUND: Marshes, swamps, fens, bog pools, alongside creeks; often floating or partially submerged; most common in fens; widespread and relatively common across boreal forest; circumpolar.

NOTES: Giant water moss could be confused with the rare **heart-leaved feather moss** (*C. cordifolium*), but that species has fewer and more irregular branches, and its stem leaves are usually noticeably longer than wide. At a more detailed level, the cells in the lower corners of the leaves of heart-leaved feather moss do not form well-defined regions. • The species name *giganteum* means 'giant,' an apt name for this moss, which can grow to over 20 cm in length.

C. cordifolium

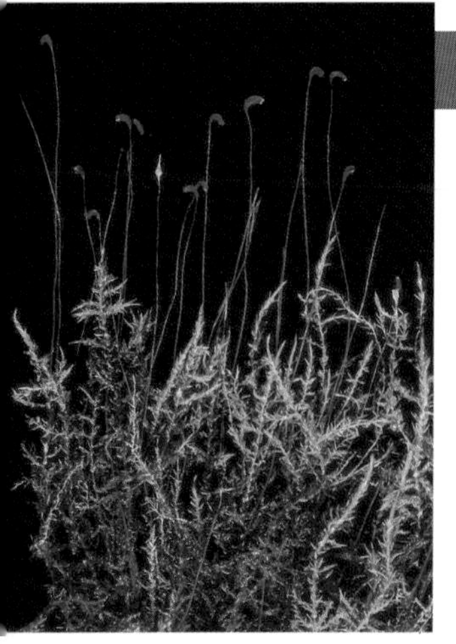

RICHARDSON'S WATER MOSS • *Calliergon richardsonii*
GOLD SPOON MOSS

GENERAL: Yellowish to green; forms soft, loose masses; stems up to 15 cm long, prostrate to ascending, **irregularly branched**.

LEAVES: Stem leaves **erect to wide-spreading**, 1.7–2 mm long, oblong or egg-heart-shaped, concave, rounded at tip, smooth-edged, **midrib 1/2–3/4 length of leaf**, usually forked at tip; **alar cells abruptly enlarged**, inflated, colourless, in large groups.

SPOROPHYTES: Single; stalk orange-red, 4.5–6 cm long; capsule 2.5–3 mm long, oblong-cylindrical, curved, horizontal.

WHERE FOUND: On wet soil or peat or in depressions; widely scattered in calcium-rich peatlands across boreal forest; circumpolar.

NOTES: The water mosses belong to a diverse group called the brown mosses, which grow in 'richer' (mineral-rich or calcium-rich) wetlands. The brown mosses include water mosses (*Calliergon* species), hook mosses (*Drepanocladus* species), golden fuzzy fen moss, sausage moss, clay pigtail moss, Blandow's feather moss, and yellow star moss. 'Poorer' wetlands are typically characterized by peat mosses. • The genus name *Calliergon*, from the Latin *callidus*, 'skillful' or 'crafty,' means 'skillfully made,' and refers to the 'well-constructed' appearance of water mosses. The species was named for Sir John Richardson, the naturalist on the Franklin expeditions.

STRAW-COLOURED WATER MOSS • *Calliergon stramineum*
STRAW-LIKE FEATHER MOSS

GENERAL: Slender, shiny, yellowish or straw-coloured to green; scattered among other mosses or forms loose to dense masses; stems erect-ascending, 5–12 cm tall, **sparsely branched**.

LEAVES: Small, crowded, overlapping, 1–1.2 mm long, oblong or oblong-egg-shaped, rounded and channelled at tip, smooth-edged, **midrib ends about 3/4 of way to tip**; **alar cells inflated**; often with many rhizoids near tips.

SPOROPHYTES: Very rare, single; stalk yellow-red to red, 3.5–5 cm long; capsule oblong-cylindrical, 2–3 mm long, curved, strongly inclined.

WHERE FOUND: Bogs and poor fens, rarely in open rich fens, among sedges at water's edge , or in 'moats' surrounding raised bogs; common and widespread across boreal forest; circumpolar.

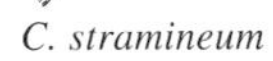

C. stramineum

C. trifarium

NOTES: Straw-coloured water moss is easily recognized by its slender, nearly simple stems, which end in narrow, rounded, shiny tips. It could be confused with **three-ranked feather moss** (*C. trifarium*), but that species is usually brownish in colour, does not have inflated alar cells, and grows in rich fens. • Straw-coloured water moss often forms mats or grows in single strands among peat moss in poorer sites. It almost never grows submerged with sausage moss in extremely rich fens—the common habitat of three-ranked feather moss. • The species name *stramineum*, from the Latin *stramen*, 'straw,' refers to the straw-coloured leaves and stems of this species, the palest of the water mosses.

YELLOW STAR MOSS • *Campylium stellatum*
YELLOW STARRY FEATHER MOSS

GENERAL: Moderately robust, **golden-yellow to green** or brownish; forms loose to dense mats; stems usually erect, **irregularly branched**.

LEAVES: Crowded, **wide-spreading from broad base** (giving moss bristly appearance), 1.2–2.6 mm long, **broadly egg-shaped to egg-heart-shaped**, narrow gradually to long, slender, channelled point, smooth-edged, **midrib short or lacking**, sometimes forked.

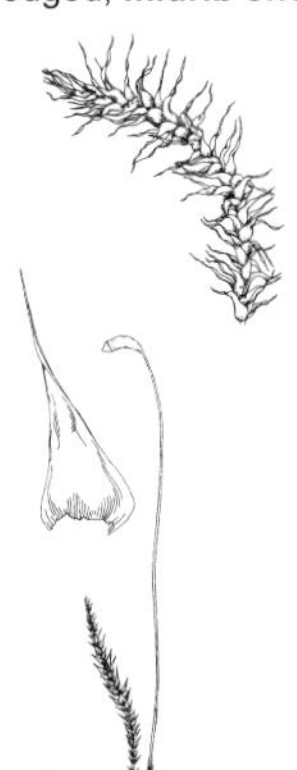

SPOROPHYTES: Rare, single; stalk yellow-red to red, 2–3.5 cm long; capsule 2–3 mm long, nearly cylindrical, curved, inclined to horizontal.

WHERE FOUND: On wet soil, humus or peat in open, calcium-rich wetlands, especially fens; common and widespread across boreal forest; circumpolar.

NOTES: False willow moss (*C. hispidulum*) is a smaller species that usually occurs on rotting wood in drier, upland sites in our region. It is recognized by its slender, creeping stems with leaves that seldom reach 1.5 mm in length. The leaves of yellow star moss are usually more than 2 mm long. • The slender, pointed leaves of yellow star moss spread out at right angles to the stem, giving the tip of each stem a star-like appearance—hence the species name *stellatum*, 'starred.' The genus name *Campylium*, from the Greek *campylos*, 'curved,' refers to the curved capsules of these mosses.

COMMON HOOK MOSS • *Drepanocladus aduncus*
CLAW-LEAVED FEATHER MOSS

GENERAL: Yellow-green to brownish; **forms soft, loose to dense mats**; stems spreading, irregularly to more or less regularly branched.

LEAVES: Curved to point in 1 direction, not pleated, egg-lance-shaped, narrow gradually to short or long, channelled tip, **smooth-edged, midrib weak, ends at or above leaf middle, alar cells clearly differentiated**.

SPOROPHYTES: Single; stalk 2–3 cm long; capsule 2–2.5 mm long, egg-shaped, curved, asymmetrical, inclined to horizontal.

WHERE FOUND: On soil or peat in wet habitats, including meadows, roadside ditches, marshes, swamps and fens; often submerged in shallow water; common and widespread across boreal forest; circumpolar.

NOTES: Because common hook moss prefers nutrient-rich, oxygen-poor water, it is often common in roadside ditches, disturbed pools and marshes. Another hook moss, ***D. polycarpus*** (sometimes considered a variety of *D. aduncus*), has shorter, wider cells near the leaf base and an area of enlarged cells in the lower corners of the leaves (alar cells) that extends nearly to the midrib. • The uppermost leaves of each stem of common hook moss grow together in a curved unit, so the growing stem tips have a hooked appearance. • The genus name *Drepanocladus* is from the Latin *drepano*, 'curved,' and *klados*, 'a branch,' and refers to the sickle-shaped shoots.

WATER HOOK MOSS • *Drepanocladus fluitans*
FLOATING FEATHER MOSS

GENERAL: Robust, **yellowish to reddish or brownish**; forms soft loose masses; **stems sparsely branched**.

LEAVES: Distant and spreading to crowded, curved and pointed in 1 direction, concave, **not ridged**, 2–4 mm long, narrowly lance-shaped, narrow gradually to long point, **toothed at tip, midrib slender, about 2/3 length of leaf, alar cells somewhat enlarged or inflated** but not abruptly differentiated or forming distinct ear-like lobes.

D. fluitans

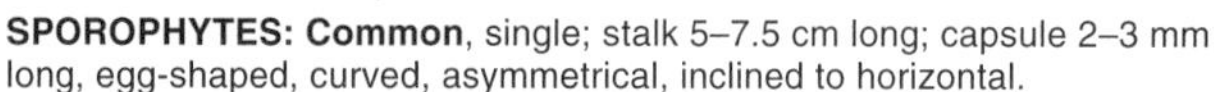

SPOROPHYTES: Common, single; stalk 5–7.5 cm long; capsule 2–3 mm long, egg-shaped, curved, asymmetrical, inclined to horizontal.

WHERE FOUND: Pools in poor fens and bogs; prefers acidic habitats; common and widespread across boreal forest; nearly cosmopolitan.

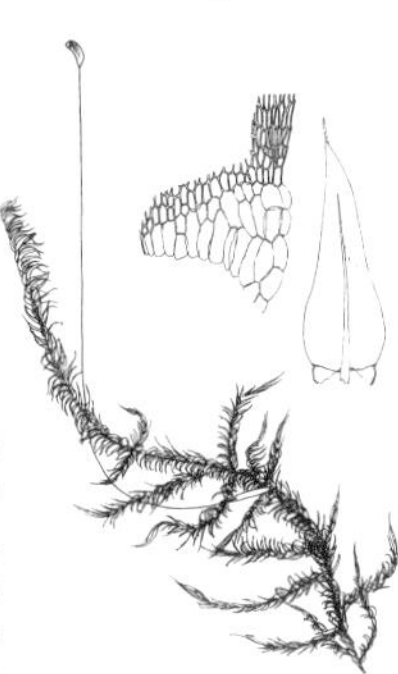

D. exannulatus

NOTES: Water hook moss could be confused with **marsh hook moss** (*D. exannulatus*), but that species is often wine-red to purplish. Also, it is firmer, less shiny and more regularly branched; its leaves are relatively flat, slightly ridged when dry, and have a broad midrib extending well into the tip; and it fruits less frequently. The alar cells are abruptly inflated in large groups that usually reach the midrib, and form distinct, ear-like lobes at the base of the leaf. Both of these species exhibit a confusing variety of growth forms caused by fluctuating water levels. • The species name *fluitans*, from the Latin *fluitare*, 'to float,' refers to the floating appearance of these plants. Water hook moss is actually attached to the substrate rather than free-floating.

RED HOOK MOSS • *Drepanocladus revolvens*
MAGNIFICENT HOOK MOSS

GENERAL: Robust, yellow-brown to **reddish or blackish purple**; forms dense, **glossy** tufts; stems erect-ascending, irregularly to somewhat regularly branched.

LEAVES: Leaves crowded, strongly curved to point in 1 direction, **nearly coiled into ring, smooth**, 3–4 mm long, concave, lance-shaped or egg-lance-shaped, **narrow gradually to long, thread-like point**, smooth-edged, **midrib about 3/4 length of leaf**.

SPOROPHYTES: Single; stalk red, 3–5.5 cm long; capsule 2–3 mm long, egg-shaped, curved, asymmetrical, inclined to horizontal.

WHERE FOUND: Calcium-rich fens; common and widespread across boreal forest; generally increases in abundance northwards; circumpolar.

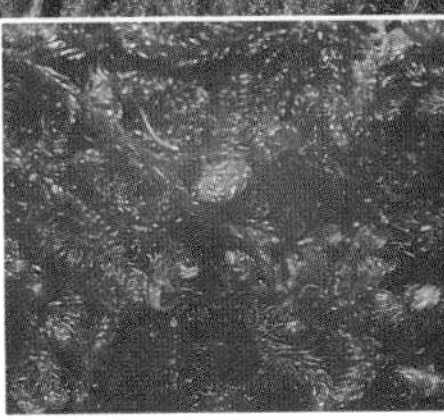

NOTES: Stick hook moss (*D. vernicosus*) is a similar species of calcium-rich fens. Usually it is easily distinguished by its green (rather than red to black) colour, but it could be confused with yellowish brown forms of red hook moss. However, stick hook moss has pleated leaves, whereas those of red hook moss are smooth. • The species name *revolvens*, from the Latin *revolvo*, 'to revolve,' 'to return' or 'to come round,' refers to the strongly curved leaves, the tips of which are often coiled almost into rings.

SAUSAGE MOSS • *Scorpidium scorpioides*
SCORPION FEATHER MOSS

S. scorpioides

S. turgescens

GENERAL: Robust, stiff, shiny; green, **reddish or brownish black**; stems up to 25 cm long and 4–6 mm wide, crowded and erect or loose and prostrate, **often encrusted with limy deposit**.

LEAVES: More or less **curved and pointed in 1 direction** (at least those at branch tips), **more or less ridged and obscurely wrinkled when dry**, concave, 2–3 mm long, elliptic to oblong-egg-shaped, blunt to pointed, **midrib weak, short and double, or none**.

SPOROPHYTES: Single; stalk long, red, irregularly twisted, 3–4.5 cm long; capsules 2–3 mm long, oblong-cylindrical, become curved, inclined to horizontal.

WHERE FOUND: Characteristic species of calcium-rich fens, usually submerged in shallow water, or floating in puddles among sedges or at pool edges; widespread across boreal forest; circumpolar.

NOTES: Yellow sausage moss (*S. turgescens*) is a related species with leaves that are usually straight, overlapping, and tipped by a short point. • Sausage moss is stiffer and larger than most hook mosses (*Drepanocladus* species) and its leaves are broader and shorter and do not have a midrib. • Sausage moss can be quite beautiful, with its robust growth form and shiny, deep reddish gold colour. However, in the rich fens where it usually grows, it is typically covered with limy deposits that give it a slimy and partially decomposed appearance. • Both the genus and species names are derived from the Greek *skorpios*, 'a scorpion,' because branches of these plants were thought to resemble the upward-turned tails of scorpions.

GOLDEN FUZZY FEN MOSS • *Tomenthypnum nitens*
GOLDEN MOSS

GENERAL: Yellow-green to golden brown, shiny when dry, 5–15 cm tall, **in upright stiff patches**; simple feathery branches, often curve downwards; **stems yellowish green to brown, covered with reddish brown felt**.

LEAVES: Long, tapered, **furrowed**, pointed, held upwards along stem.

SPOROPHYTES: Stalk reddish brown, 2–5 cm long; capsule similar in colour, inclined, 2–3 mm long.

WHERE FOUND: Wet, calcium-rich sites, open, shrubby and treed fens, swamps and tundra seepage; often forms hummocks in peatland sites; widespread and common in suitable habitats across boreal forest.

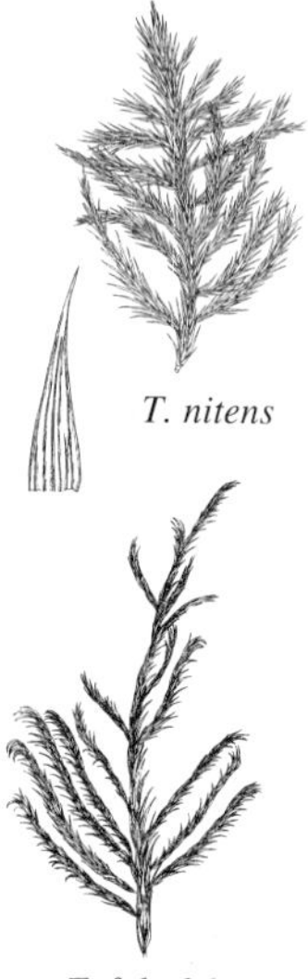

T. nitens

T. falcifolium

NOTES: The shining, golden-yellow, pleated leaves with midribs help to identify golden fuzzy fen moss. • The closely related species ***T. falcifolium*** has leaves that all curve in the same direction. It prefers the more acidic conditions found on the tops of hummocks in fens and swamps. • The golden mosses belong to a diverse group called the brown mosses, which grow in mineral-rich or calcium-rich wetlands (see C. richardsonii). • The name *Tomenthypnum*, from the Latin *tomentum*, 'a stuffing' (of wool) and *Hypnum*, a similar moss genus, refers to the fuzzy, felt-covered stems of this *Hypnum*-like moss. The species name *nitens* is Latin for 'bright' or 'shining,' and refers to the glossy, almost shining leaves of golden fuzzy fen moss.

THICK RAGGED MOSS • *Brachythecium turgidum*
THICK GRASS MOSS

GENERAL: Plants **robust**, shiny, green to golden-brown; **forms loose tufts**; stems **erect-ascending, sparsely branched**.

LEAVES: Loosely erect or erect-spreading, **pleated**, 2.5–3 mm long, egg-lance-shaped to oblong-egg-shaped, narrow abruptly to slender point, **usually smooth-edged**, midrib ends above middle of leaf.

SPOROPHYTES: Single; stalk reddish, **smooth**, 18–25 mm long; capsule **blackish**, about 2 mm long, oblong-cylindrical, asymmetrical, nearly horizontal.

WHERE FOUND: On wet soil or rocks, near streams, pools or waterfalls, or in wet depressions in peatlands, generally in calcium-rich habitats; scattered across boreal forest; circumpolar.

NOTES: Thick ragged moss is one of the larger *Brachythecium* mosses in the boreal forest. Unlike most other ragged mosses, it usually grows in erect tufts with its leaves directed towards the tips of its stems (rather than prostrate in mats with spreading or side-swept leaves). Some botanists classify thick ragged moss as a wet, calcium-rich habitat variant of golden ragged moss (p. 305). • The species name *turgidum*, from the Latin *turgidus*, 'swollen,' refers to the large, plump branches of this moss.

CLAY PIGTAIL MOSS • *Hypnum lindbergii*

GENERAL: Fairly robust, yellowish, light green or brownish; forms soft, loose, shiny mats; stems ascending, **irregularly branched**.

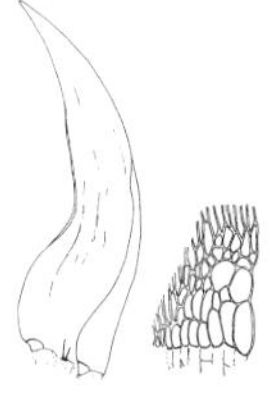

H. lindbergii

LEAVES: Erect-spreading, **flattened, tips strongly curved to point in 1 direction**, obscurely pleated when dry, **concave**, 1–2.5 mm long, oblong-egg-shaped, narrow gradually to point, smooth-edged or toothed at tip, **midrib short, double, alar cells abruptly inflated**.

SPOROPHYTES: Single; stalk smooth, 2.5–4 cm long; capsule 2–3 mm long, curved and asymmetrical, inclined to horizontal, **furrowed when dry and empty**.

WHERE FOUND: On wet soil, humus and peat, occasionally on rotten logs, in meadows, swamps and fens, especially at lake or pond edges; tends to favour calcium-rich habitats; fairly common and widespread across boreal forest; circumpolar.

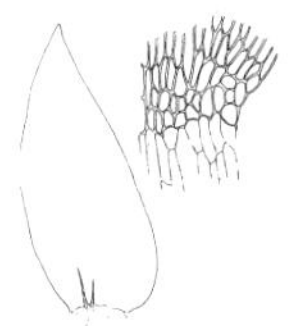

H. pratense

NOTES: Clay pigtail moss could be confused with **meadow pigtail moss** (*H. pratense*) but the leaves of that species are more clearly flattened and less strongly curved, and the alar cells are poorly developed. Meadow pigtail moss also strongly favours calcium-rich habitats. • The name *Hypnum*, from the Greek *hypnos*, 'to sleep,' apparently refers to the ancient use of some mosses as medicines. The species name *lindbergii* honours Sextus Lindberg.

Leaves of Brown Mosses and Related Species

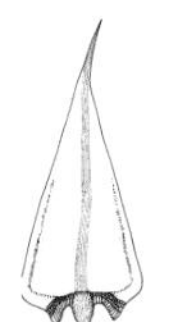

Cratoneuron commutatum
C. filicinum

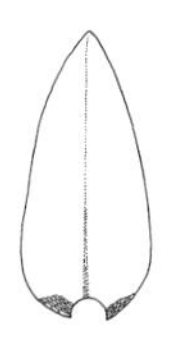

Calliergon giganteum

Calliergon cordifolium

Campylium stellatum
C. hispidulum

Scorpidium scorpoides

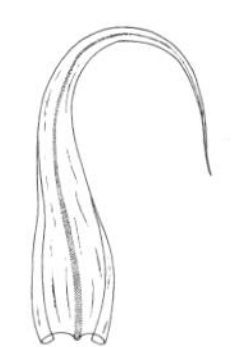

Drepanocladus uncinatus

Drepanocladus revolvens
D. vernicosus

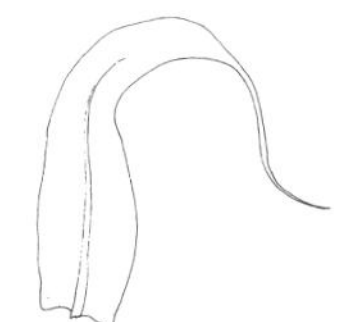

Drepanocladus fluitans

Drepanocladus exannulatus

Drepanocladus aduncus

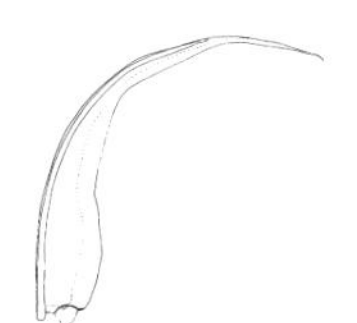

Drepanocladus polycarpus

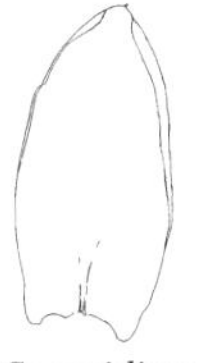

Scorpidium turgescens

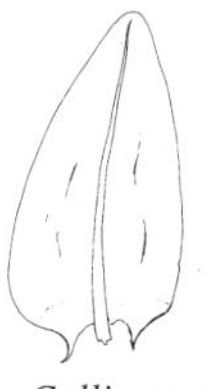

Calliergon richardsonii

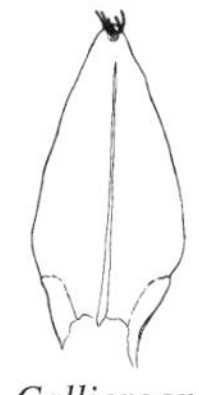

Calliergon straminium

Calliergon trifarium

Lichens

Well over 500 species of lichens make their home in forests of our region. Only a few hundred, however, are conspicuous and widespread, and many of these are discussed in this guide.

Someone once said that lichens are the banners of the fungal kingdom. Think of lichens as fungi that have discovered agriculture. Instead of invading or scavenging for a living like other fungi (moulds, mildews, mushrooms), lichen fungi cultivate algae within the fabric of fungal threads of which they themselves are composed. As photosynthesizers, algae are able to manufacture their own carbohydrates, proteins and vitamins. Some of these foodstuffs are 'harvested' by fungi for their own nourishment. In return, the fungus appears to provide the alga with protection from the elements. A lichen is simply the physical manifestation of this relationship, much as a gall is the manifestation of, for example, a larval insect feeding on a leaf.

Most lichen fungi cultivate green algae, though in some cases the photosynthesizing partner is a cyanobacterium (a blue-green 'alga'). In a few lichen species, both types of algae are present: the paler green alga scattered throughout, and the darker cyanobacterium localized in tiny colonies called **cephalodia**.

Lichens reproduce in several different ways. Sometimes the fungal partner produces saucer-like fruiting bodies (**apothecia**). In some species, the inner 'stuffing' (**medulla**) of the lichen becomes exposed at the surface as clusters of powdery dust (**soredia**). In other species, the upper surface bears tiny wart-like or coral-like outgrowths (**isidia**). When soredia and isidia are carried to new localities, as by birds, they may grow into new lichens.

Lichen Life Forms

Lichens come in many different shapes and sizes, but they never form leafy stems as mosses do. The species included in this guide are grouped in 6 growth forms: crust, scale, leaf, club, shrub and hair.

Crust: Closely attached to substrate; upper surface hard.

Scale: Tiny, shell-like lobes, usually with a cottony lower surface; forms overlapping colonies.

Leaf: Small to large leaf- or strap-like lobes, usually with rhizines and hard (non-cottony) lower surface.

Club: Unbranched or sparsely branched, cylindrical, usually upright stems.

Shrub: Much-branched cylindrical stems, usually tufted.

Hair: Intricately branched, tufted to more often hanging filaments.

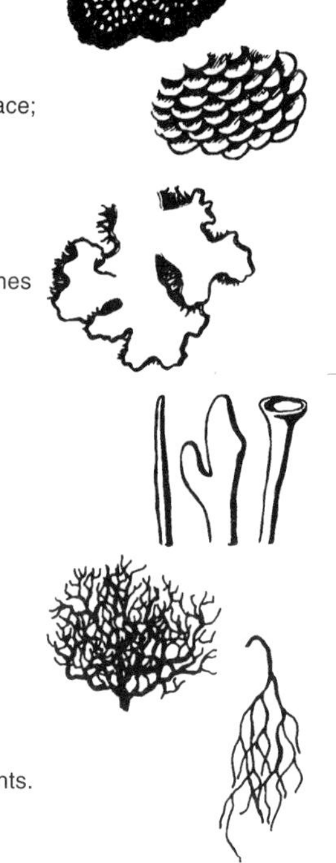

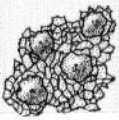

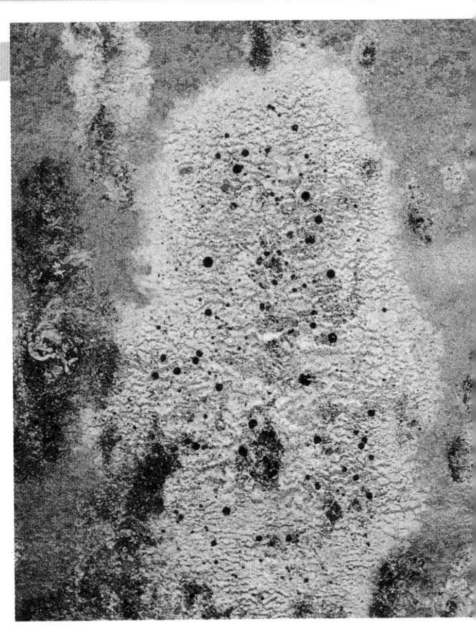

ASPEN COMMA • *Arthonia patellulata*

GENERAL: Crust lichen, thin, **pale grey-white**; **fruiting bodies** (apothecia) **appear as small black dots**, circular, about 0.5 mm wide, lacking a rim.

WHERE FOUND: On bark of deciduous trees, particularly **aspen**; common across southern boreal forest and parkland where aspen is common.

NOTES: Aspen comma often forms large, whitish, 'dusty' patches on smooth bark of **aspen trunks**. Take care not to confuse this lichen with the 'dust' that typically covers aspen trunks. • Most lichens in the boreal forest contain single-celled algae, of which the green alga *Trebouxia* is by far the most common. Dot lichen, however, contains a green alga of the genus *Trentepohlia*. In the free-living state, the cells of *Trentepohlia* grow in long, slender, thread-like strings. When growing in a lichen, however, the structure of its threads is very different, and consists of short, irregular strings, or sometimes single cells. Needless to say, the algal partner of a lichen cannot always be identified without first growing it on its own. • The species name *patellulata*, from the Latin *patella*, 'a small dish' or 'a plate,' refers to the round, flat to convex, fruiting bodies.

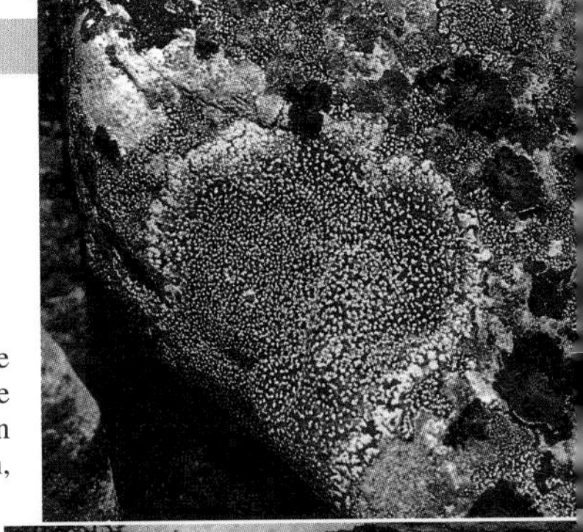

GREEN MAP LICHEN • *Rhizocarpon geographicum*

GENERAL: Crust lichen, consists of **mosaic of tiny yellowish green 'tiles'** (areoles) set against **black background**.

WHERE FOUND: Forms roughly circular colonies over **acidic, especially granitic, rocks** in open sites; common and widespread; most frequent in northern boreal forest; circumpolar.

NOTES: These 2 photos are of 2 different species in the *R. geographicum* complex. Green map lichen is only one of more than a dozen species of *Rhizocarpon* that occur in our region. Different species can be yellowish, green, greyish or rust red. • Most *Rhizocarpon*s are difficult to identify in the field, because the species are usually separated on the basis of spore features and chemical tests. • In arctic and alpine areas, map lichens grow from 0.02 to 0.5 mm a year. The latter rates are measured only on 'young' specimens, less than 300 years old. In our area, a patch 10 cm across could easily be about 1,000 years old. Once the growth rate is determined for an area, map lichen can be used to determine the age of the surface it has colonized. This technique (lichenometry) has been used to date moraines in the study of glacial history. Some map lichens are estimated to be more than 9,000 years old, placing them among the oldest living things on Earth.

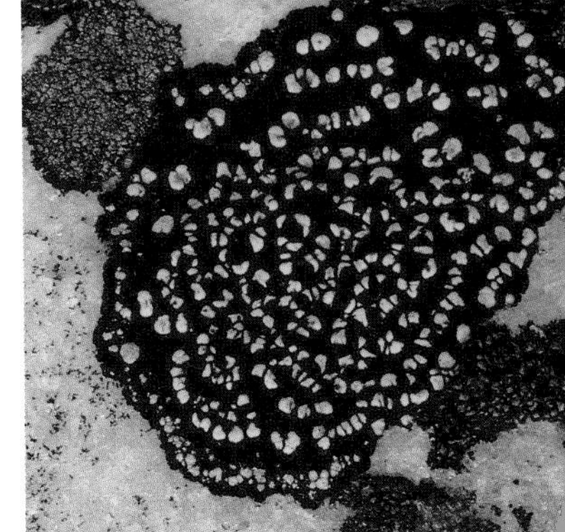

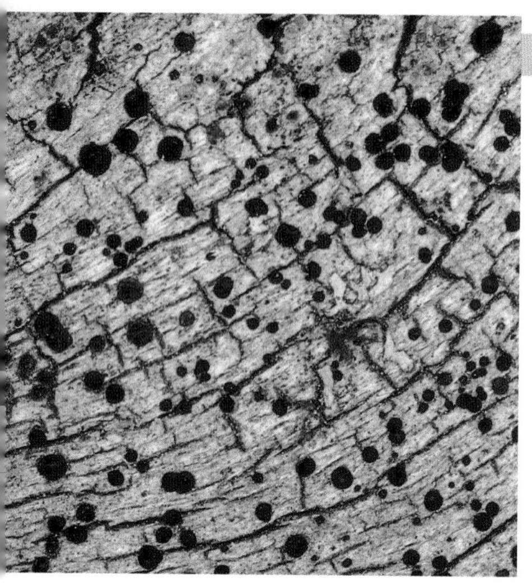

BUTTON LICHEN • *Buellia punctata*

GENERAL: Crust lichen, main body tile- or wart-like, or more often almost lacking, **ashy,** greenish or brownish; fruiting bodies (apothecia) **dull black**, up to 0.5 mm wide, flat, becoming convex, **with a thin rim**.

WHERE FOUND: Grows on **bark or old wood**; less often on rock or soil; widespread over southern boreal forest and parkland; circumpolar, primarily in boreal to temperate regions.

NOTES: *Buellia* is a relatively large genus of crust lichens, in which similar species are distinguished by spore characteristics. *Buellia* species grow both on rock and wood throughout our region. • Although button lichen is usually found on rock or wood, like many other calcium-loving lichens, it also grows on bone. It was one of a few species found colonizing old whale and caribou bones in the Arctic. In the days when the Doctrine of Signatures guided medical thought, lichens found growing on skulls were valued as a cure for epilepsy. Perhaps button lichen was one of these. • The species name *punctata* means 'dotted,' and refers to the blackish fruiting bodies scattered over the lighter-coloured crust.

DOT LICHEN • *Biatora vernalis*
SPRING DISC LICHEN

GENERAL: Crust lichen, thin to moderately thick, wart-like, warts well spaced or compacted, **olive green or paling to whitish or greyish**; fruiting bodies (apothecia) to 1 mm across, **convex** or hemispherical, whitish, **yellowish or reddish brown**, **lacking a rim**.

WHERE FOUND: Grows over mosses and on bark of both deciduous and coniferous trees; widespread across North America but especially common in boreal forest; circumpolar.

NOTES: Formerly called *Lecidea vernalis*. • The genus *Lecidea* (recently split into several genera, including *Biatora*) is represented by dozens of species across the boreal forest, many of which grow on rock. Some of the rock-growing species actually grow *within* their substrate. The surface 'skin' (cortex) of these lichens is absent. Instead, the algae and fungus are protected by a thin outer layer of rock crystals, often at a depth of 1 mm or more. Only the lichen fruiting bodies on the surface of the rock reveal the presence of the lichen within. • The species name *vernalis* is Latin for 'belonging to spring.'

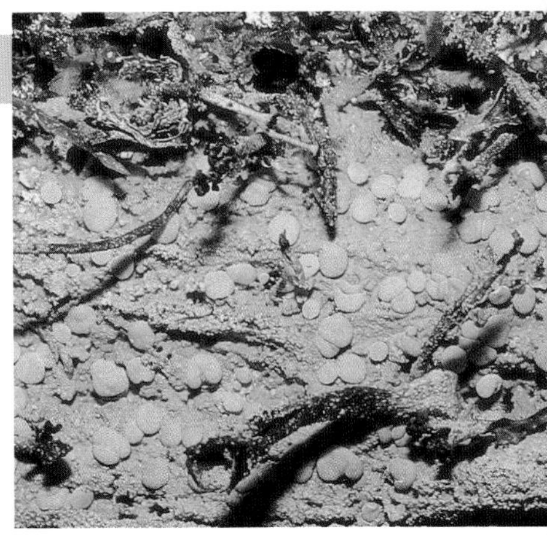

SPRAYPAINT • *Icmadophila ericetorum*
FAIRY PUKE

GENERAL: Crust lichen; upper surface **pale bluish green**, somewhat roughened, typically with small, **flesh-pink fruiting bodies** that lack stalks.

WHERE FOUND: Over decaying peat **moss hummocks and on rotting wood**, usually in cool, humid, coniferous forests and in bogs; common and widespread across boreal forest; circumpolar.

NOTES: At first glance, the main body of spraypaint lichen, a pale, whitish crust, could be mistaken for a mold. However, the pink, disc-like fruiting bodies are unmistakable and create a striking colour combination. • Spraypaint lichen is an unusual lichen because it is partly parasitic. Most lichen fungi draw their nourishment from their algal partner. However, in spraypaint lichen they supplement their diet with carbohydrates from the mosses on which they typically grow. • The genus name *Icmadophila*, from the Greek *ikmas*, 'juice' or 'moisture,' and *philos*, 'loving,' refers to the moist habitats preferred by this lichen. The species name *ericetorum*, from the Latin *erica*, 'heath,' refers to the bogs or other acidic habitats (heathlands) where it grows.

RIM LICHEN • *Lecanora circumpborealis*

GENERAL: Crust lichen, thin, smooth, **whitish**; **fruiting bodies** (apothecia), numerous, small, crowded, circular, up to 1 mm wide, **dark red-brown** to **occasionally blackish**, flat to somewhat convex with a thin rim, same colour as crust.

WHERE FOUND: Grows most abundantly on **conifers** but also on deciduous trees (especially birch) and shrubs (especially alder and willow) across boreal forest; circumpolar.

NOTES: Formerly called *L. coilocarpa* and *L. subfusca*. • Rim lichen is one of the most common crust lichens of trees in the boreal forest. • *Lecanora* is a very large genus with dozens of species found across our region on virtually all substrates. • In northern Africa and western Asia *Lecanora esculenta* and *L. affinis* have occasionally provided food for sheep and cattle. At various times, these lichens are said to have appeared suddenly, covering large areas with a layer of lichen sometimes 5-15 cm thick. This 'manna' was usually described as small, hard, irregular, greyish or whitish lumps, ranging in size from a pin head to a pea or small nut. The local people believed that it fell from heaven, and in times of famine they followed the example of their livestock and used it as food. It was said to have no odour and little flavour, and often it was made into bread. • These lichens do not 'rain from heaven,' but under dry conditions they can break into tiny fragments that are easily carried by the wind. So far as we know, no North American species has been put to similar use. • The species name *circumborealis* refers to the fact that this lichen is found in the boreal forest all around the world.

CRUSTED ORANGE LICHEN • *Caloplaca cerina*

GENERAL: Crust lichen, thallus smooth, uneven or thinly net-like, **ashy grey to dark greenish grey** or lacking; fruiting bodies (apothecia) up to 1.5(2) mm wide, concave to flat, yellow**, orange** or brownish yellow, **with a thick, pale grey to bluish grey rim**.

WHERE FOUND: Grows on wide variety of deciduous trees; also on soil or rock; especially common on **aspen**, balsam poplar and willow across southern boreal forest and parkland; circumpolar, primarily in temperate to boreal regions.

NOTES: Of the several species of *Caloplaca* that grow on deciduous trees in our region, the grey rim of the fruiting bodies is distinctive for crusted orange lichen. Another species commonly found on deciduous trees (and occasionally on conifers and rock), ***C. holocarpa***, has fruiting bodies with orange rims. Its distribution is similar and it is equally if not more common across our region. • A single tree provides many different environments. Light quality and humidity can change dramatically with height and distance from the outer canopy. The mosses and lichens at the base of a tree are usually very different from those in the canopy. Also, the composition of the plant 'communities' growing on a tree will change with time, as some species are replaced by others. A study of lichens growing on white ash (*Fraxinus americanus*) in Minnesota, showed crusted orange lichen invading branches of 9 years or older, and growing on most branches that had reached 15 years of age. • The species name *cerina*, is Latin for 'wax-coloured.'

COMMON SHINGLE • *Hypocenomyce scalaris*
COCKLESHELL LICHEN

GENERAL: Semi-erect **scale** lichen; lobes tiny, up to 1 mm long, broader than long, **shell-like**, loosely overlapping; upper surface pale greyish brown; **powdery dust (soredia) on lower surface of lobe edges**; **fruiting bodies** (apothecia) rare, black, flat.

WHERE FOUND: Forms loose, shingle-like colonies on **dead wood and charred stumps** (usually coniferous) in open forests; most common in boreal forests of Alberta and northeastern British Columbia; scattered occurrences to east and north, depending in part on fire history.

NOTES: Formerly called *Lecidea scalaris*. • Shingle lichen's selective choice of habitats, **weather-worn or charred wood**, makes it fairly easy to spot. It could be confused with wood-inhabiting species of *Cladonia,* but the scales of those lichens are usually longer than wide, and typically lack soredia. It also resembles ***H. friesii***, but that species often produces fruiting bodies and lacks the powdery 'dust' on its scales. • The genus name *Hypocenomyce* is from the Greek *hypo*, 'less than,' *kainos*, 'new' or 'recent,' and *mykes*, 'a mushroom.' *Scalaris* means 'scale-like' or 'overlapping,' and refers to the lichen's growth form.

CANDLEFLAME • *Candelaria concolor*
LEMON LICHEN

GENERAL: Leaf lichen, tiny, **bright yellow, centre often reduced to yellow granular crust, edges of lobes covered with granular or powdery soredia**; fruiting bodies (apothecia) rare, very small, 0.2–0.8 mm wide, scattered, circular, orange, yellow or yellowish brown, concave to flat, edge bright yellow.

WHERE FOUND: On coniferous and deciduous trees across southern boreal forest; circumpolar.

NOTES: Candleflame can be very abundant on the branches of conifers that are frequently used as bird perches, probably because of the nitrogen enrichment from the bird droppings. • When most lichens fruit, they produce disc-like structures called apothecia with spores contained in microscopic, long sacs called asci. Typically, the original cell undergoes 2 consecutive divisions, producing 8 spores in each ascus. However, the asci of candleflame usually contain 24 to 60 spores. • The species name *concolor* means 'of the same colour,' probably referring to the fact that the main lichen body and the fruiting bodies are the same colour.

POWDERED ORANGE LICHEN • *Xanthoria fallax*

GENERAL: Leaf lichen, small, yellow to **orange**, lobes rounded, to 2 mm wide, only slightly raised above substrate, **lower surface of margins with powdery soredia**; **fruiting bodies** (apothecia) **absent**.

WHERE FOUND: On trees (usually deciduous) and old wood; occasionally on rock; common and widespread across southern fringe of boreal forest and parkland; circumpolar.

NOTES: Powdered orange lichen could be confused with pincushion orange (*X. polycarpa*, p. 338), but that lichen commonly produces many fruiting bodies and lacks powdery soredia. These 2 lichens often grow together on the roofs of old barns and other buildings across the prairies, giving them a characteristic orange hue. They are also tolerant of pollution, and often grow on trees along city streets. • Another tiny orange lichen, ***X. candelaria***, has upright lobes that end in tiny, powdery soredia or wart-like isidia. • The species name *fallax* is Latin for 'deceitful' or 'deceptive,' perhaps because it is hiding something (the true colour and texture of the surface it covers?)—only the lichen knows for sure.

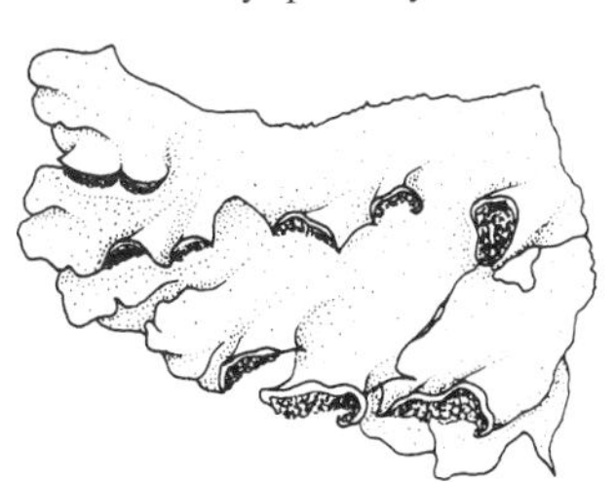

PINCUSHION ORANGE • *Xanthoria polycarpa*
CUSHION ORANGE

GENERAL: Leaf lichen, loosely flattened, forming small, loose, round **'pin cushions'**; lobes very narrow, about 0.2–0.3 mm wide; upper surface pale orange or **bright orange**, with **many orange fruiting bodies**.

WHERE FOUND: On **deciduous trees and shrubs** in open sites; occasionally on conifers; common and widespread across southern boreal forest and parkland.

NOTES: No other orange lichen shares the pincushion's copious fruiting bodies. • *Xanthorias* need large amounts of calcium and/or nitrogen to grow and reproduce, and they are often abundant in areas where bird droppings accumulate (e.g., near nests and roosting trees). • The name *Xanthoria*, from the Greek *xanthos*, 'yellow,' refers to the bright yellow colour of some species. The species name *polycarpa*, from the Greek *poly*, 'many' and *karpos*, 'fruits,' refers to the abundant fruiting bodies of this species.

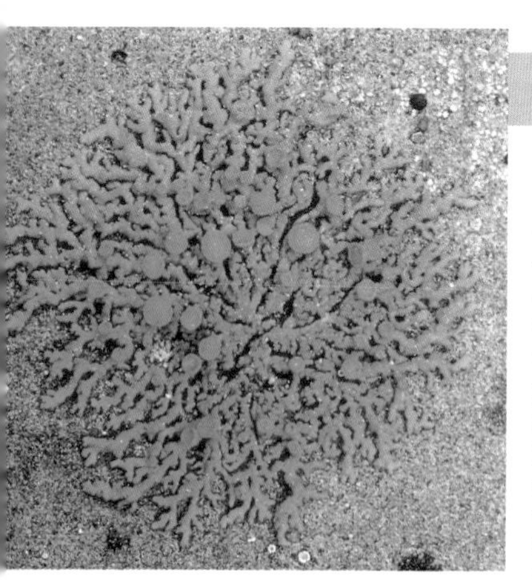

ELEGANT ORANGE LICHEN • *Xanthoria elegans*
ROCK ORANGE LICHEN

GENERAL: Flattened, **leaf** lichen, usually forms bright **orange**, circular patches on rock; lobes about 0.5 mm wide; **fruiting bodies** numerous, saucer-shaped, **orange**.

WHERE FOUND: On **calcium-rich rocks** (limestone, dolomite), acidic rocks enriched with calcium from bird droppings, old bones, and occasionally on old wood; common and widespread throughout the world.

NOTES: Lichens make it easy to distinguish acidic rocks, high in silicates, from basic rocks, high in carbonates. Map lichens (*Rhizocarpon* spp.) grow on the former; elegant orange lichen grows on the latter. • Because this lichen is so conspicuous and because it prefers sites rich in calcium and nitrogen, the bright splashes of orange created by its colonies are often used by biologists to locate the perches and nest sites of eagles, falcons and cliff-dwelling seabirds. Elegant orange lichen has been used in the same way as green map lichen (p. 334) to date surfaces. This lichen grows more quickly than map lichen, but its growth rate varies with habitat. Elegant orange lichen on the graves of crew members of Franklin's ill-fated 3rd expedition had reached only 4.4 cm in diameter after 102 years, suggesting an average growth of about 0.4 mm per year.

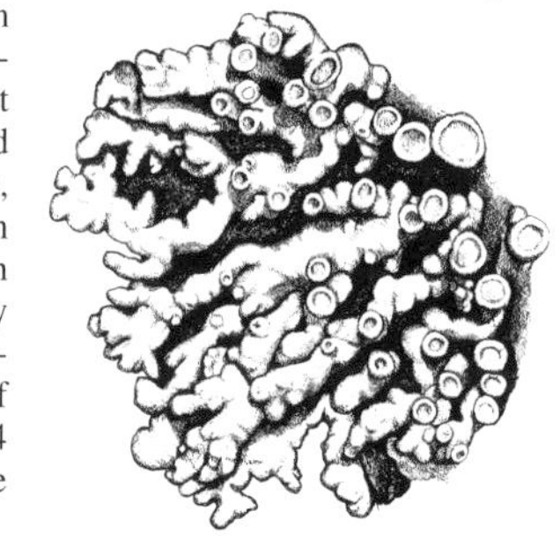

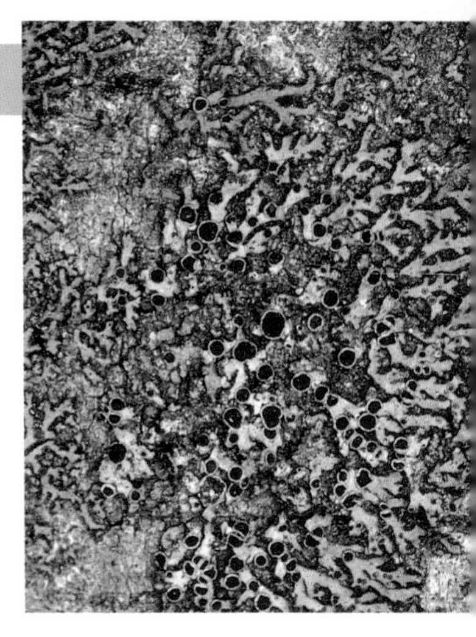

GRANULATED SHADOW • *Phaeophyscia orbicularis* WREATH LICHEN

GENERAL: Leaf lichen, 3–6 cm wide, **closely attached** to substrate; **upper surface** pale greenish to usually **brownish grey**, lobes 0.5-1.5 mm wide, **with tiny, head-shaped, powdery soredia**; lower surface black, densely covered with rhizines; fruiting bodies (apothecia) rare.

WHERE FOUND: Common on furrowed **bark of deciduous trees**, particularly aspen and balsam poplar; rarely on rock; widespread across southern boreal forest and parkland; circumpolar.

NOTES: Also called *Physcia orbicularis.* • At least 6 similar lichens (species of *Phaeophyscia*, *Physconia*, and *Physcia*) grow on trees in our region. Granulated shadow is the only one to produce tiny, head-shaped soredia on the upper surface. • Granulated shadow is relatively resistant to air pollution, and it flourishes on trees in many towns and cities. It is a common early colonizer of tree branches. In a study of lichen communities on white ash (*Fraxinus americana*) trees in Minnesota, granulated shadow was found on all branches over 10 years of age. • The species name *orbicularis* is Latin for 'round,' perhaps in reference to the rounded soralia (clusters of soredia).

HOODED ROSETTE • *Physcia adscendens* HOODED LICHEN

GENERAL: Leaf lichen, small, loosely flattened; lobes narrow, 0.5–1 mm wide, ending in **hood-shaped swellings,** the insides of which bear powdery **soredia** (use hand lens); upper surface **pale greyish**, dull, with small white spots; lobe edges fringed with **long, slender hairs** (cilia); lower surface white.

WHERE FOUND: Over **deciduous trees and shrubs** in open or somewhat shady sites; occasionally on bark and branches of conifers; common and widespread on poplars and willows in mixed wood boreal forest and parkland; circumpolar.

NOTES: Hooded rosette could be confused with the monk's hood lichen (*Hypogymnia physodes,* p. 342), but that species has hollow lobes that are black (rather than white) on the lower surface, and that lack hairs along the edges. ***P. tenella*** is very similar in size and colour, but it lacks the hood-shaped swellings at the tips of its lobes. • Occasionally, when a spore from the fungus of another lichen lands on or near a recently dislodged *Physcia* soredium, the germinating spore may commandeer its algae. This is one way that fungal spores can find appropriate algal partners with which to form a lichen. • The helmet-shaped lobe tips of hooded rosette lichen, with their tiny hairs, have been likened to miniature snakes, with heads raised and tongues out. • The genus name *Physcia* is derived from a Greek word meaning 'sausage' or 'large intestine,' probably in reference to the shape of the long lobes of some species. The species name *adscendens* is Latin for 'rising' or 'climbing,' and refers to the raised lobe tips.

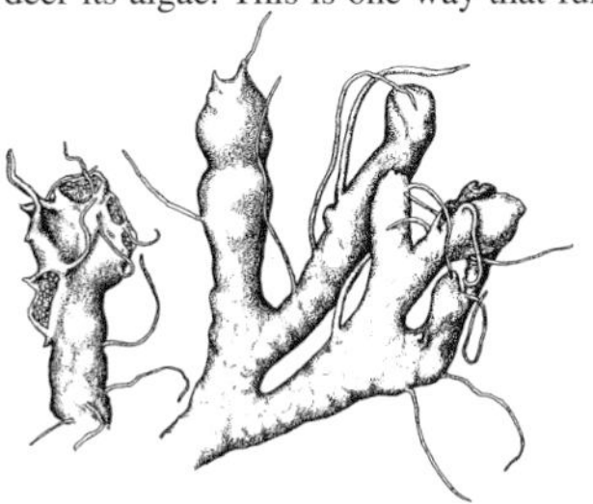

GREY-EYED ROSETTE • *Physcia aipolia*
HOARY ROSETTE

GENERAL: **Leaf** lichen, closely appressed; lobes narrow, 1–2 mm wide; upper surface **pale greyish**, with small white spots and **many fruiting bodies**, these usually with a **frosted appearance**; lower surface tan, with many pale rhizines.

WHERE FOUND: Forms rosettes (rounded patches) on the trunks and large branches of **deciduous trees** and shrubs, especially trembling aspen, balsam poplar, willow, and alder; common and widespread across boreal forest and parkland; circumpolar.

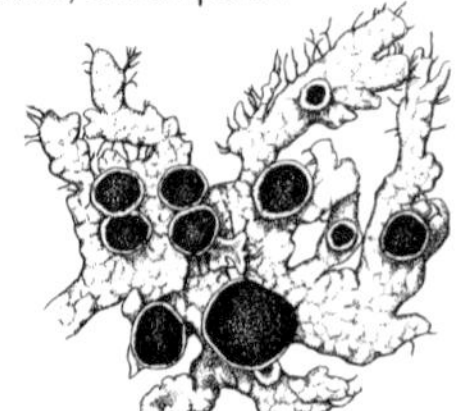

NOTES: Among tree-dwelling lichens, grey-eyed rosette is likely to be confused only with **star rosette** (*P. stellaris).* That species, however, lacks small white spots on its upper surface, and tends to grow on branches in the canopy. • The fruiting bodies of grey-eyed rosette get their frosted appearance from calcium oxalate crystals deposited at their surface. The crystals are formed using calcium taken from the bark surface. • Although star rosette lichen is one of the earliest colonizers of deciduous tree branches in many areas, it is very slow growing. Some specimens have been estimated to increase in diameter at a rate of 1.1 mm per year.

Key to *Cetraria*, *Parmelia* and Other Similar Lichens

1a. Lichen closely attached to substrate, small (usually less than 1 mm wide); usually on bark or wood 2

2a. Upper surface with soredia; underside black 3

3a. Upper surface and soredia pale greenish yellow ...... ***Parmeliopsis ambigua***

3b. Upper surface greyish white, with whitish soredia ...... ***Parmeliopsis hyperopta*** (see *P. ambigua*)

2b. Upper surface with isidia; underside pale ...... ***Imshaugia aleurites*** (see *Parmeliopsis ambigua*)

1b. Lichen loosely attached to substrate or erect, occasionally pressed flat, but, if so, not small (more than 1 mm wide); habitat various 4

4a. Lobes erect or loosely attached; fruiting bodies usually along lobe margins or at lobe tips 5

5a. Lichen brownish or dark olive-green 6

6a. Minute white patches present (check lower surface near lobe margins); lobes not ciliate; on ground 7

7a. White patches present only near lobe margins ..... ***Cetraria ericetorum*** (see *C. islandica*)

7b. White patches both at edge and sparsely scattered over the lower surface of lobes ...... ***Cetraria islandica***

6b. No minute white patches on thallus; lobes edges fringed with cilia; on trees ...... ***Tuckermannopsis americana***

5b. Lichen greenish yellow or yellowish white 8

8a. Soredia absent; medulla white; thallus pale yellowish or whitish green; on ground 9

9a. Lobes flattened; bases yellow to orange ...... ***Flavocetraria nivalis***

9b. Lobes somewhat tubular; bases purplish ...... ***Flavocetraria cucullata*** (see *F. nivalis*)

8b. Soredia present; medulla yellow; thallus lemon-yellow; on trees, shrubs and dead wood ...... ***Vulpicida pinastri***

4b. Lichen loosely attached to substrate to more or less flattened against it; fruiting bodies, if present, on upper surface of lobes 10

10a. Lobes hollow; rhizines (holdfasts) absent; lower surface black, shiny, wrinkled; soredia present 11

11a. Soredia located (at least in part) at lobe tips 12

12a. Soredia confined to the insides of burst lobe tips; upper surface grey .... ***Hypogymnia physodes***

12b. Soredia located on the upper surface of the lobes, usually near lobe tips; upper surface brown to brownish grey .... ***Hypogymnia bitteri*** (see *H. physodes*)

11b. Soredia located over central portions of upper surface .... ***Hypogymnia austerodes*** (see *H. physodes*)

10b. Lobes not hollow; rhizines present; lower surface pale or black, shiny or dull, smooth or wrinkled; soredia present or absent 13

13a. Upper surface yellowish green 14

14a. Soredia absent; on rock 15

15a. Underside whitish .... ***Arctoparmelia centrifuga***

15b. Underside dirty grey .... ***Arctoparmelia separata*** (see *A. centrifuga*)

14b. Soredia scattered over upper surface; usually on trees .... ***Flavopunctelia flaventior***

13b. Upper surface grey 16

16a. Upper surface strongly ridged 17

17a. Bearing soredia on ridges; on bark, rock or soil .... ***Parmelia sulcata***

17b. No soredia on ridges; on rock .... ***Parmelia omphalodes*** (see *P. sulcata*)

16b. No ridges on upper surface, bearing isidia: on rock . ***Parmelia saxatilis*** (see *P. sulcata*)

13c. Upper surface dark brown 18

18a. Soredia and isidia absent; fruiting bodies broadcast over upper surface (extending to lobe margins); lobes 1-3 mm wide; lichen small, never more than 5 cm across .... ***Melanelia septentrionalis***

18b. Soredia and/or isidia present 19

19a. Central upper surface covered with minute, solid isidia, which rub off to reveal faint, yellowish white soredia .... ***Melanelia subaurifera*** (see *M. exasperatula*)

19b. Central upper surface bearing inflated (i.e., hollow) isidia, these never asssociated with soredia 20

20a. Isidia club-shaped to barrel-shaped .... ***Melanelia exasperatula***

20b. Isidia distinctly cylindrical .... ***Melanelia elgantula*** (see *M. exasperatula*)

RIPPLED ROCKFROG • *Arctoparmelia centrifuga*
RING LICHEN

GENERAL: Leaf lichen, closely attached to substrate, often dies in centre and forms **concentric rings**, lobes branch irregularly; **upper side dull greenish yellow, lacking soredia and isidia**; **underside whitish** or pale brownish with brown to blackish rhizines; fruiting bodies (apothecia) rare.

WHERE FOUND: Grows on **acidic rocks** in well-lit locations; fairly common in boreal forest; circumpolar.

NOTES: Formerly called *Parmelia centrifuga* and *Xanthoparmelia centrifuga*. • The concentric growth pattern, when evident, helps to identify this lichen. • A less common species, ***A. separata***, has a dark underside and rarely grows in concentric rings. • Lichens have been in existence for hundreds of millions of years. Specimens of *Parmelia* (a closely related genus that once included *Arctoparmelia*) were found in the brown-coal formations of Saxony, dating from the Cenozoic era, 25–70 million years ago. • The species name *centrifuga*, from the Latin *centrum*, 'centre,' and *fugere*, 'to flee,' is a fanciful reference to the outer rings of lichen being forced outwards, away from the centre, as if they were spun in a centrifuge.

MONK'S HOOD LICHEN • *Hypogymnia physodes*
HOODED TUBE LICHEN

GENERAL: Leafy lichen, appressed; lobes narrow, 1–2 mm wide, **hollow**; **tiny, powdery soredia on insides of burst lobe tips**; upper surface pale greyish green; lower surface black, no rhizines.

WHERE FOUND: On numerous substrates, including trees, moss, boulders and soil; common and widespread across boreal forest; occasional in aspen parkland; circumpolar.

NOTES: Monk's hood lichen could be confused with *H. austerodes* and *H. bitteri*, but it is much less brown than either of those species, and it has soredia confined to the inside of burst lobe tips. In ***H. bitteri***, in contrast, the soredia are located primarily on the upper surface of the lobe tips, whereas in ***H. austerodes*** they are diffuse and mostly restricted to the central portions of the thallus. • Monk's hood lichen often grows together with waxpaper lichen (p. 343). At first glance these 2 lichens look alike, but monk's hood lichen has hollow, inflated lobes, while those of waxpaper lichen are solid. These 2 species are probably the most common lichens growing on trees in the northern half of the Northern Hemisphere. • Most lichens are very sensitive to air pollution, and soon disappear from industrialized areas. Monk's hood lichen is very tolerant of high levels of sulphur dioxide, and it is often found in city parks and along streets. • The name *Hypogymnia* is derived from the Greek *hypo*, 'under' and *gymnos*, 'naked,' referring to the lack of holdfasts (rhizines) on the lower surface of these lichens. The species name *physodes*, from the Greek *physa*, 'a bellows,' refers to the inflated or puffed out lobes of this species.

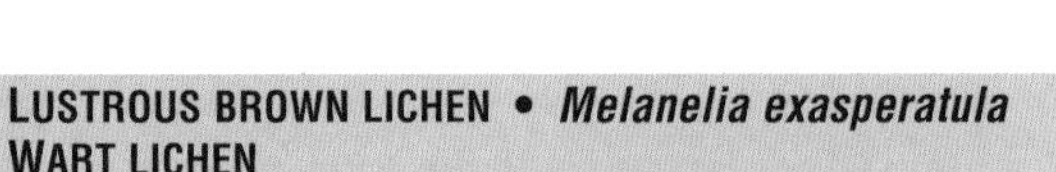

LUSTROUS BROWN LICHEN • *Melanelia exasperatula*
WART LICHEN

GENERAL: Leaf lichen, lobes 2–5 mm wide, pressed flat or sometimes raised at tips; upper surface **olive-green to olive-brown**, smooth to slightly wrinkled; small nipple-shaped bumps on upper surface enlarge to form **inflated, hollow, lustrous, club-shaped or barrel-shaped outgrowths** (isidia), to 2 mm long, simple or occasionally forked at tip; underside pale to dark tan, with many rhizines; fruiting bodies (apothecia) rare.

WHERE FOUND: Grows on **bark and twigs** of both deciduous and coniferous trees, rare on rocks; found across boreal forest, rare in the parkland; circumpolar.

NOTES: Also called *Parmelia exasperatula*. • Lustrous brown lichen is the most common of a group of brown bark lichens (including *M. subaurifera* and *M. elegantula*) that are distinguished from one another mainly by the characteristics of wart-like or grain-like isidia so minute you'll need a hand lens to see them. The isidia of lustrous brown lichen are club-shaped and hollow, whereas those of ***M. elegantula*** are cylindrical and solid. ***M. subaurifera*** (inset photo) has diffuse, powdery patches on its upper surface (absent in the other 2 species) that are mixed with tiny isidia. • The species name *exasperatula*, from the Latin *exaspero*, 'to roughen' or 'to provoke,' could refer to the way lustrous brown lichen roughens the appearance of the bark on which it grows. It could also reflect the frustration (exasperation) of the lichenologist who sorted out the taxonomy of this group.

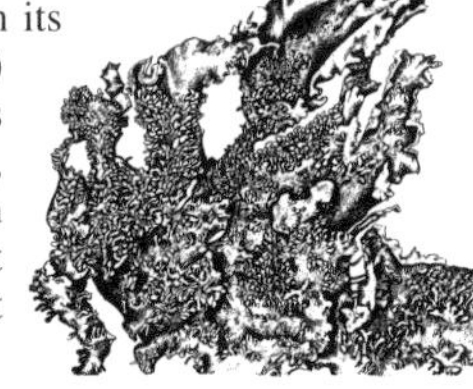

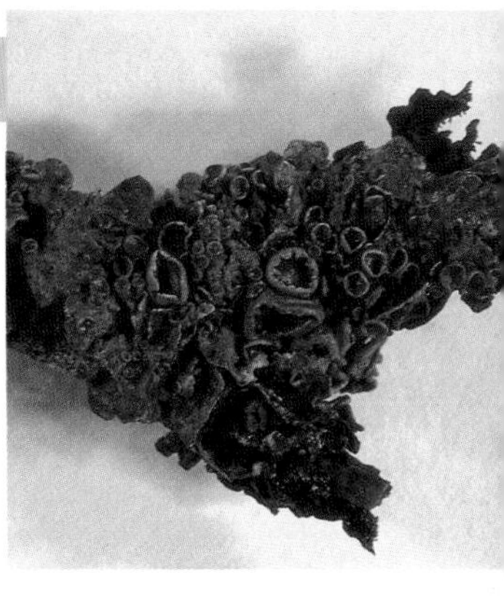

NORTHERN BROWN LICHEN • *Melanelia septentrionalis*
BROWN BARK LICHEN

GENERAL: Leaf lichen, lobes pressed flat, to 3–4 mm wide; upper surface dark to **olive-brown**, smooth to slightly roughened, dull or shiny along edges; lacking soredia and isidia; underside dark brown or black with many rhizines; **fruiting bodies** (apothecia) **common**, 1–3 mm wide, concave when young, convex with age, more or less **broadcast over the entire lichen**.

WHERE FOUND: Grows on **bark and twigs**; mainly across northern boreal forest north to low Arctic where it often occurs on dwarf birch; circumpolar.

NOTES: Also called *Parmelia septentrionalis*. • The species name *septentrionalis* means 'northern' or 'of the north.' It is from the Latin *septentriones* (originally *septem*, 'seven,' and *triones*, 'plow oxen'), the name for the 7 stars of the constellation Ursa Major (the Big Dipper), which is a northern constellation.

WAXPAPER LICHEN • *Parmelia sulcata*
POWDERED SHIELD

GENERAL: Leaf lichen, loosely appressed; lobes narrow, 1–4(5) mm wide; **upper surface** pale grey, **with powdery soredia in long narrow cracks**; **lower surface black**, with many black rhizines.

WHERE FOUND: On coniferous and deciduous trees and sometimes boulders or rotting wood in forests and openings; common and widespread across boreal forest; occasional in aspen parkland; circumpolar.

NOTES: Salted shield or rock shield (*P. saxatilis*) is similar to waxpaper lichen, but it has tiny club-shaped outgrowths (isidia) rather than powdery soredia on its upper surface. Also, it usually colonizes rock rather than trees or logs. Another rock-growing species, ***P. omphalodes***, bears neither isidia nor soredia. • Waxpaper lichen could also be mistaken for **monk's hood lichen** (*Hypogymnia physodes,* p. 342), which, however, has hollow, inflated lobes. • *Parmelia* lichens have long been used by northern residents around the world for dyeing. These lichens produce colours ranging from deep brown to dark tan, gold and even rusty brown. In Scotland, these 'crottle dyes' were said to bind well to fabric without mordants and to resist bleach. They were also said to give the wool a pleasant aroma and to render the cloth moth-proof, since caterpillars cannot eat the bitter lichen substances. The colour of a lichen and the colour it yields often have little in common. For example, *P. omphalodes* is dull grey, but it dyes wool golden- to chocolate-brown. • Waxpaper lichen is commonly used by rufous hummingbirds to decorate and camouflage their nests. • The species name *sulcata*, from the Latin *sulcus*, 'a furrow' or 'a groove,' refers to the network of cracks on the upper surface. *Saxatilis* is the Latin for 'growing among rocks,' from *saxum*, 'a rock.'

GREEN STARBURST • *Parmeliopsis ambigua*

GENERAL: Leaf lichen, **closely appressed**; **lobes very narrow**, 0.5–0.8 mm wide; upper surface pale **yellowish green,** with many **head-shaped clusters of powdery soredia**; lower surface blackened inwards of lobe tips, with sparse rhizines (use hand lens).

WHERE FOUND: Forms rosettes over **acid-barked trees and shrubs**; also over barkless logs; common and widespread across boreal forest north past treeline; circumpolar.

NOTES: Grey starburst (*P. hyperopta*) is identical to green starburst except in being greyish (usnic acid absent) rather than greenish (usnic acid present). A similar grey species, **floury starburst** (*Imshaugia aleurites*), has wart-like isidia on the upper surface and its underside is pale. Green starburst and grey starburst often grow together, and some taxonomists consider them chemical variants of the same species. However, they do have slightly different ecologies, grey starburst being more sensitive to prolonged dessication than green starburst. • The genus name *Parmeliopsis*, from *Parmelia*, another lichen genus, and the Latin *opsis*, 'a visual image,' means 'resembling *Parmelia*.' Quite so. The species name *ambigua* is Latin for 'doubtful' or 'unreliable'—this is a lichen of questionable character.

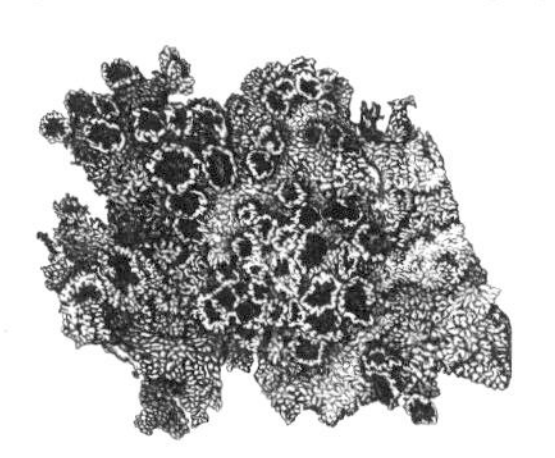

FRINGED RUFFLE • *Tuckermannopsis americana*
LEAFY BROWN BARK LICHEN

GENERAL: Leaf lichen, small; lobes 1–3 mm wide, **form loosely attached rosettes**; upper surface more or less wrinkled, **fringed with short hairs**, greenish brown when wet, dark brown when dry; underside usually brown with scattered rhizines; **fruiting bodies at** lobe **edges**, common, brown, to 3 mm wide, rim same colour as lobes.

WHERE FOUND: Grows **on coniferous trees** and old conifer wood, occasionally on poplar, birch and alder; widespread across boreal forest, with large gaps; occasional in parkland.

NOTES: Also called *Cetraria halei* or *Cetraria ciliaris* var. *halei*. • Fringed ruffle, with its marginal hairs (cilia), is unlikely to be confused with any other tree-dwelling lichen. It is often associated with horsehair lichens (*Bryoria* spp., p. 367), spruce moss (*Evernia mesomorpha*, p. 363) or old man's beard (*Usnea* spp., pp. 365-366) on the branches of conifers in our region. • The name *Tuckermannopsis* was meant to commemorate Edward Tuckermann, a 19th-century lichenologist. However, given that it literally means "resembling Tuckermann," the recipient of this honour might not have entirely approved.

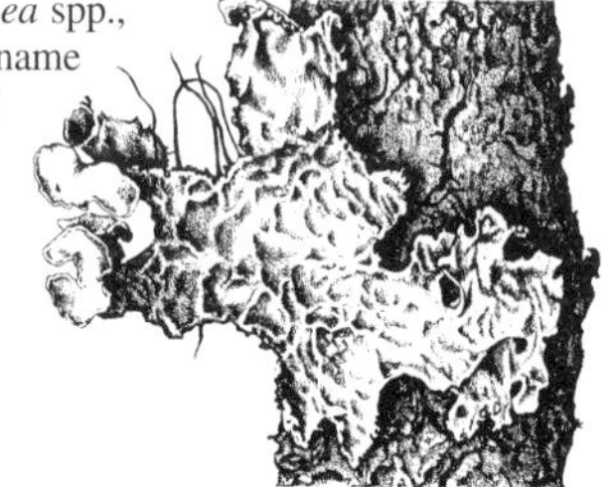

GREEN SPECKLEBACK • *Flavopunctelia flaventior*

GENERAL: Leaf lichen, 5–10 cm wide or larger, **yellow-green**, slightly wrinkled, lobe tips rounded; **upper surface and lobe margins bearing clusters of powdery soredia**; upper surface also bearing tiny round to irregular white 'pores' (check near lobe tips with a hand lens); lower surface tan, darkening to black in centre, with many rhizines; fruiting bodies very rare.

WHERE FOUND: Often forms large patches on **trunks and branches** of aspen, balsam poplar, birch, willow and conifers, especially spruce; widespread across southern boreal forest and northern parkland of prairie provinces.

NOTES: Also called *Parmelia flaventior*. • With its broad, sorediate, yellow-green lobes, green speckleback is unlikely to be confused with any other tree-dwelling lichen. In winter, the large patches of green speckleback on tree trunks provide a welcome splash of colour, sure to catch the eye of passing skiers or snowshoers. • The name *Flavopunctelia*, from the Latin *flavens*, 'yellow' or 'golden,' and *punctum*, 'a point' or 'a dot,' refers to the yellow lobes bearing many tiny pores. The species name *flaventior* also refers to the yellowish colour of this lichen.

POWDERED SUNSHINE • *Vulpicida pinastri*
MOONSHINE CETRARIA

GENERAL: Leaf lichen, small, **pale yellowish green**, loosely attached lobes relatively short, to 3–4 mm wide, with **bright yellow powdery soredia** along margins.

WHERE FOUND: On **bark near base of conifers**, dwarf birch, willow, alder, Labrador tea and Canada buffaloberry, where protected by snow in winter; common and widespread across boreal forest north past treeline to Arctic coast; circumpolar.

NOTES: Formerly called *Tuckermannopsis pinastri* or *Cetraria pinastri*. • Powdered sunshine is the only yellow bark-dwelling lichen with bright lemon-yellow soredia in our region. • The yellow colour derives from 2 poisonous lichen substances—pinastric and vulpinic acids. These are thought to deter grazing insects and other invertebrates • The name *vulpicida* is derived from the Latin *vulpes*, 'a fox,' and *caedere*, 'to kill,' referring to the former practice of using this lichen to poison wolves and foxes in northern Europe. The lichen was mixed with ground glass and nails and added to bait—apparently with lethal effect.

ICELANDMOSS • *Cetraria islandica*

GENERAL: Leaf lichen, **erect**, up to 10 cm tall, branched, with edges slightly to strongly rolled inwards, **lined with tiny, thorn-like spines**; upper surface olive-brown to dark or blackish **brown**, smooth, often turns red at base; underside lighter with tiny **scattered white patches** (pseudocyphellae) (use hand lens), especially along edges; fruiting bodies rare, at tips of lobes when present, reddish brown, smooth.

WHERE FOUND: On **soil or moss** in open sites, occasionally on old wood or on lower twigs of spruce trees (those covered by snow in winter); widespread across our region; circumpolar.

NOTES: In the related *C. ericetorum*, the white pores on the lower side of the lobes form a thin white line along the lobe margins. • The Icelandmosses contain large amounts of lichenin (a starch). When boiled they form a paste-like substance that has been widely used by folk healers to soothe irritated membranes associated with persistent problems of breathing (tuberculosis) and digestion (dysentery). It was usually taken in milk or cocoa. It also contains bitters, which can be used to stimulate appetite, especially in convalescence. Some Swiss companies still use Icelandmoss in throat lozenges, herbal sweets and herb tea. • Lapps, Icelanders and natives of northern Canada have used Icelandmoss for food. They got rid of its bitter, astringent taste by adding baking soda, lye, or wood ash to the cooking water and then rinsing it with cold water. The lichen was then dried and used to make jelly (500 g ground lichen:4 l milk or water, boiled, strained, and the liquid cooled), porridge (1 part lichen:3–4 parts water), or bread and biscuits (2–3 parts ground lichen:1–2 parts grain meal). The lichen residue left after making jelly was mixed with dressing and eaten as a salad. The jelly was often flavoured with sugar, lemon juice, wine or chocolate. Dried lichen keeps for many years and the foods made with Icelandmoss kept well and were less likely to be attacked by weevils during long voyages. • Icelandmoss is still sold in Iceland as a thickener for soups and stews. In Scandinavia and Iceland where it is harvested commercially, a site can be harvested every 3 years. • Scandinavians used Icelandmoss as fodder for livestock, and in Newfoundland it supplies up to 5% of the lichen diet of caribou.

FLATTENED SNOW LICHEN • *Flavocetraria nivalis*
RAGGED PAPERDOLL

GENERAL: Leaf lichen, 3–5 cm tall, **erect**, lobes **more or less flat**, usually forming small tussocks, irregularly branched; upper and lower surface straw-yellow, somewhat wrinkled, **base turning golden yellow** to brown when dying; fruiting bodies rare.

WHERE FOUND: Grows on **soil or rock**, or amongst mosses on soil; rarely on twigs below snow line; common and widespread across boreal forest; North American.

NOTES: Also called *Cetraria nivalis*. • **Curled snow lichen** (*F. cucullata*) (inset photo) could be confused with flattened snow lichen, but it has smooth, clearly channelled lobes (the edges are strongly curled inwards) and its base turns red with age. This reddish tinge is the result of the breakdown of protolichesterinic acid or usnic acid—two of about 500 compounds produced by lichens. • These 2 lichens have occasionally been used (like Icelandmoss) in foods, but they contain large amounts of usnic acid and are likely to cause indigestion. • The species name *nivalis* is Latin for 'snowy,' and refers to the areas of late-lying snow where this species is often found. The name *cucullata*, from the Latin *cucullus*, 'a hood' or 'a cowl,' refers to the wavy edges of the lobes, which are strongly curled inwards to form a pocket, like a hood.

Key to the *Peltigera* Lichens

1a. Thallus green when wet; warty 'freckles' (cephalodia) present over upper surface; fruiting bodies folded lengthwise, more or less vertical 2

2a. Underside darkening abruptly toward thallus centre; veins broad, usually indistinct .... ***Peltigera aphthosa***

2b. Underside darkening gradually toward thallus centre; veins broad or narrow, usually distinct .... ***Peltigera leucophlebia*** (see *P. aphthosa*)

1b. Thallus grey or greyish green; warty cephalodia absent; fruiting bodies folded lengthwise or not 3

3a. Small, rounded patches of soredia present on upper surface; thallus small, usually less than 3 cm across .... ***Peltigera didactyla***

3b. Soredia absent; thallus usually more than 3 cm across 4

4a. Thallus minutely hairy on upper surface (use hand lens), especially near lobe tips 5

5a. Hairs erect, especially near lobe tips; veins on underside broad and inconspicuous, forming a dark, thick, felty mat .... ***Peltigera malacea***

5b. Hairs appressed, branched 6

6a. Lobes upturned at the tip, mostly less than 1.5 cm wide .... ***Peltigera rufescens*** (see *P. canina*)

6b. Lobes turned down at the tip, broad, mostly more than 1.5 cm wide 7

7a. Veins narrow and somewhat raised; rhizines variously branched but never covered with minute, erect hairs; hairs on the upper surface typically extending to centre of thallus .... ***Peltigera canina***

7b. Lower surface with flat, soft veins, almost as wide as the spaces between them; rhizines mostly unbranched, densely covered with minute erect hairs; hairs on the upper surface thinning toward centre of thallus .... ***Peltigera retifoveata*** (see *P. canina*)

4b. Thallus not hairy 8

8a. Upper surface of thallus with dense, minute, wart-like roughenings .... ***Peltigera scabrosa*** (see *P. malacea*)

8b. Upper surface of thallus smooth 9

9a. Outermost rhizines aligned in one or more rows; fruiting bodies horizontal, as wide as long; regeneration lobules often developing along cracks in the thallus .... ***Peltigera elisabethae*** (see *P. neopolydactyla*)

9b. Outermost rhizines unaligned; fruiting bodies more or less vertical, longer than wide; regeneration lobules not produced 10

10a. Upper surface dull; rhizines 7-10 mm long; lobes to 3 cm wide; powdery coating on upper surface absent .... ***Peltigera neopolydactyla***

10b. Upper surface very shiny; rhizines to 5 mm long; lobes less than 2 cm wide; distinct powdery coating present on upper surface near margins .... ***Peltigera neckeri*** (see *P. neopolydactyla*)

FRECKLE PELT • *Peltigera aphthosa*
STUDDED LEATHER LICHEN

GENERAL: Leaf lichen, **loosely attached**; **lobes broad**, 2–5 cm wide, dull grey-green when dry, **bright green** when moist, bearing scattered '**warts**'; lower surface veinless or with broad, cottony, **inconspicuous veins**, that darken **abruptly inward from the lobe tips**; fruiting bodies large, reddish to blackish brown, on upper surface of extended lobes.

WHERE FOUND: On **moss, humus, decaying logs**, and occasionally rocks, typically in forested areas; very common and widespread across boreal forest north to Arctic; occasional in northern parkland; circumpolar.

NOTES: Freckle pelt is often confused with ***P. leucophlebia***, which has conspicuous, well-developed veins that darken only gradually towards the centre. Both lichens undergo a dramatic colour change, from grey-green to bright green, with wetting. • The brown to black dots or 'warts' on the upper surface of freckle pelt contain tiny colonies of cyanobacteria, which supply the lichen with nitrogen. These organisms can extract nitrogen from the air and supply this nutrient to the lichen fungus and its green algal partner. Freckle pelt is a symbiosis among representatives of 3 of the 5 kingdoms of life—Monera, Protista, and Fungi.• Mountain caribou forage for this lichen in winter. • The name *Peltigera* is from the Latin *pelta*, 'a light shield,' referring to the round shield-like apothecia. The species name, *aphthosa*, is from the Greek *aphthai*, 'an eruption or pustule,' referring to the scattered dark 'warts' on the upper surface of this lichen. • In the 1800s Swedish peasants believed that miliary fever was caused by the elf-mote or by meeting with elves, and they sought freckle pelt and dog pelt as a remedy. Freckle pelt lichen was also boiled in Scandinavia to make a wash for treating chapped skin on adults' feet and babies' bottoms.

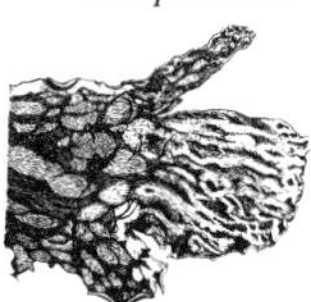

P. aphthosa

P. leucophlebia

TEMPORARY PELT • *Peltigera didactyla*
SMALL FELT LICHEN

GENERAL: Leaf lichen, small, brownish grey, loosely attached; lobes 3–7 cm wide; upper surface finely and minutely hairy (hand lens), with distinct **round,** grey, **powdery patches of soredia,** these disappearing with age; lower surface light tan, veins raised, pale or brownish, bearing flaring holdfasts (rhizines); fruiting bodies occasional, reddish brown, 3–5 mm wide, folded lengthwise, on tips of short lobes, erect, curved backwards.

WHERE FOUND: On **soil in woods or along road banks**; occasionally on rotting wood or mossy rocks; scattered across boreal forest.

NOTES: Previously called *P. spuria* • The conspicuous roundish powdery patches on the upper surface of young lobes are a distinguishing feature of temporary pelt. • This is one of the smaller species of *Peltigera* and although widespread it is often overlooked. It is rather transitory and colonies can appear and disappear in a few years. It inhabits freshly disturbed habitats such as cutbanks, where it is rapidly outcompeted by other, more aggressive colonizers. • The species name *didactyla*, from the Greek *di*, 'twice,' and *dactylos*, 'a finger,' probably refers to the short, little-divided lobes, or to the short-lobed fruiting bodies.

DOG PELT • *Peltigera canina*
DOG LICHEN

GENERAL: Leaf lichen, **loosely attached; lobes broad,** 1.5–2(3) cm wide; upper surface pale brown or grey, thinly **covered in minute appressed hairs,** especially near the lobe tips (use hand lens); lobe tips downturned; lower surface white, **cottony**, with narrow, **brownish veins** and dark, **flaring rhizines**; fruiting bodies common, large, light brown to chestnut- or blackish brown, folded lengthwise, vertical on extended lobes at edge of thallus.

WHERE FOUND: On **mineral soil, humus, decaying wood and moss** in open areas; very common and widespread across boreal forest; occasional in parkland; cosmopolitan.

NOTES: Dog pelt can be difficult to distinguish from other grey *Peltigeras* containing cyanobacteria rather than green algae and turning dull greyish rather than bright green when wet. Look for a thin, grey, felty covering on the upper surface (especially near the tips), downturned lobe tips, distinct veins and flaring holdfasts (rhizines). • **Rough pelt** (*P. scabrosa*) is similar, but it lacks hairs and has a roughened surface. Another similar species, **felt pelt** (*P. rufescens*) is more brittle, and its lobes are mostly less than 15 mm across with distinctly upturned tips. Specimens with very thick, very 'spongy' veins and holdfasts belong to the boreal species ***P. retifoveata***; its rhizines are densely covered with minute, erect hairs, and its upper surface is less hairy, especially towards the centre. *P. retifoveata* occurs most commonly on litter and moss in dry coniferous woods and in disturbed situations such as roadbanks from western Manitoba northwest to interior Alaska. • Herbalists once valued dog pelt as a cure for hydrophobia (fear of water) and a remedy for liver complaints (the plant being liver-shaped). The erect fruiting bodies were likened to dogs' teeth and in accordance with the Doctrine of Signatures dog pelt was recommended as a cure for rabies or the bite of a mad dog. This involved bleeding off 300 ml of blood and administering dog pelt tea to which black pepper had been added. The patient was also forced to take cold baths every morning for a month, and to go without breakfast—if the bite didn't drive you mad, the cure surely would. Better still, keep a piece of dog pelt in your shoe and avoid being bitten in the first place. • The species name *canina*, is Latin for 'of a dog.'

APPLE PELT • *Peltigera malacea*
BOXBOARD FELT LICHEN

GENERAL: Leaf lichen, loosely attached, forming patches to 30 cm in diameter; lobes ascending, 1–3 cm wide; upper surface greenish brown, bearing minute erect hairs near the lobe tips (use hand lens), **turning deep green when wet**; lower surface darkening towards centre; **veins very broad, barely distinguishable, holdfasts** (rhizines) **sparse**; fruiting bodies rare, erect.

WHERE FOUND: Common **on ground** in **open areas** in coniferous forests across boreal forest; circumpolar.

NOTES: Apple pelt is distinguished by its dull upper surface (with fine, minute, erect hairs near the edges) and lack of veins below. • **Freckle pelt** (*P. aphthosa*) (p. 348) is similar, but bears 'warts' over the upper surface and turns *bright* green when wet. Also similar is **rough pelt** (*P. scabrosa*), but that species has a roughened (scabrous) upper surface lacking hairs of any kind. It also has more distinct veins and it does not turn such a deep green when wet. • When this lichen dries, its stiff lobes feel like cardboard.

P. scabrosa

FROG PELT • *Peltigera neopolydactyla*
FINGER FELT LICHEN

GENERAL: Leaf lichen, large, **loosely appressed**; lobes broad, 10-25 mm wide; upper surface **hairless**, olive-green to pale or dark **bluish-grey**, lacking 'warts' (cephalodia), often bearing **brownish, tooth-like fruiting bodies** (apothecia) on raised lobes along the lobe margins; lower surface whitish, **cottony**, bearing low, broad, brownish or blackish veins and **long, slender holdfasts** (rhizines).

WHERE FOUND: On soil, moss, logs, rocks and bases of trees in moist woods; widespread across boreal forest; circumpolar.

NOTES: Frog pelt's broad lobes, naked upper surface, rather low veins and long, slender holdfasts (rhizines) distinguish this species from other *Peltigeras*. ***P. neckeri*** is somewhat similar but has a very shiny upper surface with a distinct powdery coating, rhizines that are only 5 mm long, and black fruiting bodies. Frog pelt could also be confused with ***P. elisabethae***, but that species frequently produces regeneration lobules along cracks in the thallus and it has horizontal, flattened fruiting bodies and outermost holdfasts that are concentrically arranged. *P. neckeri* occurs most commonly on moss or humus in mixed spruce-poplar stands in the boreal forest. It has a scattered distribution from Ontario westwards to the Yukon Territory • Studies of the physiology of frog pelts suggest that the algae (*Nostoc*) in these lichens photosynthesize most actively in boreal regions during late winter and early spring. • *Polydactyla*, from the Greek *poly*, 'many,' and *dactylos*, 'a finger,' probably refers to the narrow erect lobes of the numerous fruiting bodies.

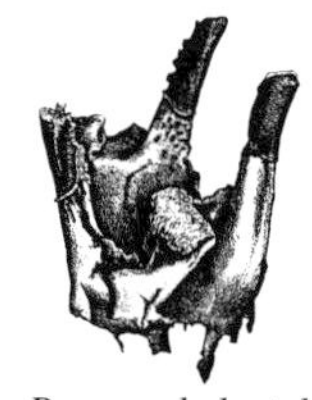

P. neopolydactyla

P. elisabethae

GREEN KIDNEY • *Nephroma arcticum*

GENERAL: Leaf lichen, up to 7-8 cm across; lobes very **broad, loosely appressed**; upper surface **yellowish to lime green**, lower surface pale, except blackening toward centre, dull, thinly covered in **fine hairs**; **fruiting bodies** common, to 2 cm across, positioned near the lobe margins on the **lower surface**.

WHERE FOUND: Over moss on rock outcrops, on moss **or** occasionally **rotting wood** in open, usually coniferous forests, in willow thickets, wet meadows and protected places in the tundra. Widespread across the northern boreal forest; circumpolar.

NOTES: Green kidney has the largest fruiting bodies of any of our lichens (regularly up to 3 cm across). Its yellowish-green color is distinctive. Two other widespread *Nephromas* in our region are greyish-brown. ***N. parile*** has soredia scattered over its upper surface; it grows on moist, shaded rocks and tree bases across the boreal forest. ***N. helveticum*** occurs in similar habitats, but it has short, flat teeth along its lobe edges. An unrelated leafy lichen, ***Leptogium saturninum***, is also common on tree bases (especially those of deciduous trees) and occasionally grows on rock. It is darker grey than the *Nephromas*, with small granular isidia on its upper surface, and whitish woolly hairs covering its pale grey underside. On the infrequent occasions when this species grows on rock it could be confused with some of the ***Umbilicarias***, but it is broadly attached to its substrate and lacks the umbilicus characteristic of the rocktripes.

BLISTERED ROCKTRIPE • *Umbilicaria hyperborea*

GENERAL: Leaf lichen, 1–4(5) cm across, roundish in outline, attached centrally by **single holdfast** (umbilicus); upper surface dark brown when moist, blackish when dry, **appears 'blistered'**; usually has **fruiting bodies**, these **concentrically ridged** (use hand lens); lower surface dark brown.

WHERE FOUND: On acidic rocks in open situations; common and widespread across boreal forest; most common of all rocktripe species in our region; circumpolar.

NOTES: Plated rocktripe (*U. muhlenbergii* or *Actinogyra muhlenbergii*) (inset photo) is similar, but its fruiting bodies have tiny ridges that radiate from their centres (rather than forming concentric rings as in other species). Plated rocktripe is less common than blistered rocktripe and is boreal to temperate in distribution, rarely occurring beyond treeline. ***U. proboscidea*** is also similar to blistered rocktribe but its lower surface has a conspicuous network of ridges, and it is generally more arctic-alpine. • The Chipewyan and Woods Cree boiled rocktripe with fish eggs, fish broth or caribou blood. They also added it to soup as a thickener and for its sour, mushroom-like flavour. These lichens contain very small amounts of minerals and vitamins. Their main food value appears to be as a filler in times of famine. Unless rocktripe is blanched or boiled in at least 1 change of water (preferably with baking soda or wood ashes added), its bitter lichen acids can cause severe upset in the stomach and intestines. Rocktripe is seldom used by people today, but musk oxen eat it during winter.

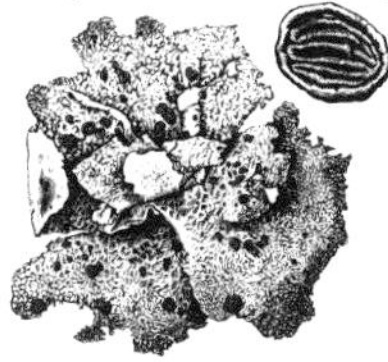

U. hyperborea

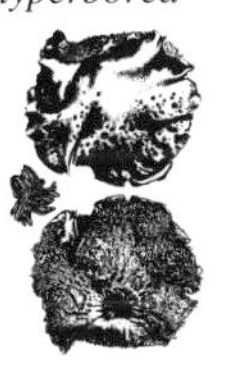

U. muhlenbergii

FROSTED ROCKTRIPE • *Umbilicaria vellea*

GENERAL: Leaf lichen, 10-15(25) cm across, attached centrally by single holdfast (umbilicus); **upper surface dull grey**, often covered with a whitish dust; underside black, covered with dense mat of black, **ball-tipped rhizines**; fruiting bodies rare, to 3 mm across, concentrically ridged (use hand lens).

WHERE FOUND: Open or shaded acid rocks, usually with some moisture seepage, in forests and on tundra; widespread across northern boreal forest; circumpolar.

NOTES: The thallus of this lichen is thick and brittle, resembling cardboard when dry. Recently, another rocktripe, *Umbilicaria americana*, was recognized as separate from *U. vellea*. These species are very similar in appearance, but *U. americana* is larger and tends to be cordilleran rather than arctic/boreal. • Other rocktripes growing in this region include ***U. polyphylla***, which has a smooth upper surface and sooty black lower surface; ***U. deusta*** which has minute granular isidia scattered over its upper surface; and ***U. torrefacta***, which has an upper surface punctured by copious minute perforations. • The Chipewyan burned rocktripe slightly in a pan, then mashed and boiled it to make a syrup to expel tapeworms. It was also chewed for the same purpose. Rocktripe soup was considered a nourishing, bland food for sick people. • In Scotland, rocktripe was used to make scarlet dye known as *corkir* for colouring tartan cloth. It also produces a rich purple dye when treated with ammonia, fermented and mixed with soda—the colour of the lichen has little to do with the colour of the dye produced.

Key to the *Cladina* and *Cladonia* Lichens

1a. Stems much branched; branches interwoven (=shrub lichens) 2

2a. Surface of branches with a cottony, spider-web-like texture; base scales lacking 3

3a. Stems branching into equal parts, forming dense, rounded heads .... ***Cladina stellaris***

3b. Stems do not branch into equal parts, having a main branch with smaller side branches 4

4a. Branches grey or whitish grey .... ***Cladina rangiferina***

4b. Branches yellow to greyish yellow 5

5a. Main stems tending to branch from the base; tips of branches often conspicuously browned ...... ***Cladina arbuscula*** (see *C. mitis*)

5b. Main stems distinguishable almost into the crown, not tending to branch from the base; tips of branches seldom conspicuously browned .... ***Cladina mitis***

2b. Surface of branches smooth, not cottony or felt-like; base scales usually present 6

6a. Branches often with small but distinct cups .... ***Cladonia amaurocraea*** (see *C. uncialis*)

6b. Branches lacking cups .... ***Cladonia uncialis***

1b. Stems little or not branched; awl- or cup-shaped (=club lichens) 7

7a. Stems terminating upward in cups 8

8a. Fruiting bodies red 9

9a. Stems lacking soredia; cups short and regular .... ***Cladonia borealis***

9b. Stems with soredia 10

10a. Clubs short (usually less than 3 cm tall) and broad, flaring quickly into deep cups .... ***Cladonia pleurota*** (see *C. borealis*)

10b. Clubs tall (2.5-8.5 cm) and slender, broadening gradually into shallow cups 11

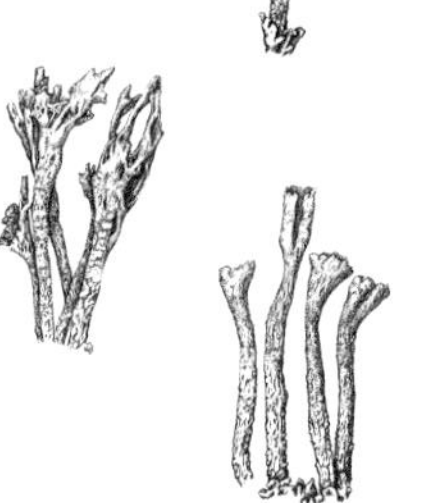

11a. Cups irregular, with split and torn sides; basal scales large, conspicuous .... ***Cladonia sulphurina***

11b. Cups more or less regular, sides not split or torn; basal scales small, inconspicuous .. ***Cladonia deformis*** (see *C. sulphurina*)

8b. Fruiting bodies brown 12

12a. Interior of cups closed, not opening into the hollow stem 13

13a. Stems with soredia 14

14a. Soredia granular; stems short and stout .... ***Cladonia chlorophaea***

14b. Soredia floury; stems taller and more slender .... ***Cladonia fimbriata*** (see *C. chlorophaea*)

13b. Stems without soredia 15

15a. Stems short, usually less than 1.5 cm tall; cups deep, goblet-shaped 16

16a. Base scales thin, ascending, not rosette-forming .... ***Cladonia pyxidata***

16b. Base scales thick, leathery, more or less appressed, forming rosettes around the stems .... ***Cladonia pocillum*** (see *C. pyxidata*)

15b. Stems taller (usually more than 2 cm tall); cups shallow, saucer-shaped 17

17a. Cups irregular; base of stem white-spotted .... ***Cladonia phyllophora*** (see *C. gracilis*)

17b. Cups regular, not split; base of stem usually not white-spotted 18

18a. Stems proliferating upward from the margins of the cups .... ***Cladonia gracilis* ssp. *turbinata***

18b. Stems proliferating upward from the centres of the cups .... ***Cladonia verticillata*** (see *C. gracilis*)

12b. Interior of the cups opening into the hollow stem 19

19a. Stems with floury soredia, forming narrow cups with inrolled edges .... ***Cladonia cenotea***

19b. Stems without soredia 20

20a. Interior of cups partly closed by torn or punctured membrane (sieve-like) .... ***Cladonia multiformis***

20b. Interior of cups entirely open to inside of club, no torn or punctured membrane .... ***Cladonia crispata*** (see *C. cenotea*)

7b. Stems not terminating upward in cups, instead awl-shaped with pointed tips or tipped with fruiting bodies that form heads 21

21a. Stems with soredia, seldom branched 22

22a. Soredia coarsely granular; lower parts of stems often covered with scales .... ***Cladonia scabriuscula***

22b. Soredia floury; lower parts of stems without scales 23

23a. Fruiting bodies red or pale brown; stems less than 2 cm tall, yellowish or grey 24

24a. Fruiting bodies red; stems yellowish or grey .... ***Cladonia bacillaris*** (see *C. coniocraea*)

24b. Fruiting bodies pale brown; stems yellowish .... ***Cladonia bacilliformis*** (see *C. coniocraea*)

23b. Fruiting bodies dark brown; stems often more than 2 cm tall, never yellowish 25

25a. Stems short, generally less than 3 cm tall, with most of outer surface covered in soredia .... ***Cladonia coniocraea***

25b. Stems tall, generally more than 3 cm; soredia scattered over surface or not 26

26a. Stems olive-green; soredia confined to rounded patches toward the upper part of stem .... ***Cladonia cornuta***

26b. Stems whitish or greyish; most of external surface covered in soredia ...... ***Cladonia subulata*** (see *C. cornuta*)

21b. Stems without soredia, upper parts distinctly branched 27

27a. Stems generally more than 3 cm tall, sometimes fissured; fruiting bodies very small, dark brown .... ***Cladonia furcata*** (see *C. multiformis*)

27b. Stems generally less than 2.5 cm tall; fruiting bodies larger, often conspicuous 28

28a. Fruiting bodies red .... ***Cladonia cristatella*** (see *C. borealis*)

28b. Fruiting bodies brown 29

29a. Fruiting bodies pale brown; stems yellowish; usually on rotten wood .... ***Cladonia botrytes***

29b. Fruiting bodies dark brown; stems whitish grey, sides torn and fissured; usually on soil .... ***Cladonia cariosa***

STUMP CLADONIA • *Cladonia botrytes*

GENERAL: Club lichen, upright, up to 1 cm tall, from very small basal scales (squamules); **clubs** (podetia) greyish yellow, branched towards top; **fruiting bodies** (apothecia) small, **pale brownish**, single or **grouped in small clusters** at branch tips.

WHERE FOUND: Characteristic of coniferous forests; grows on **rotting wood** or occasionally on soil rich in humus; widespread, but never abundant, across boreal forest; circumpolar.

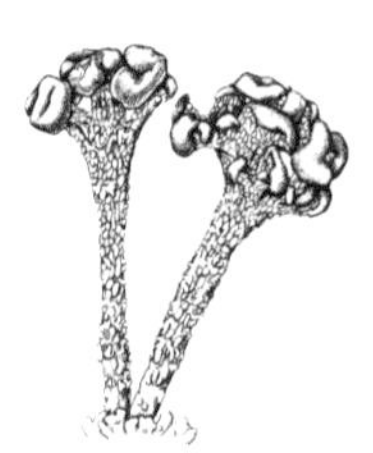

NOTES: Although it is one of the smaller species of its genus, stump cladonia is distinctive. It is the only *Cladonia* with large, tan to light-brown fruiting bodies on the tops of slender stalks. • *Cladonia* is a pre-Linnaean name that was proposed by John Hill in 1751 in *A History of Plants* to refer to all non-cup-bearing species of this group. The name was derived from the Greek *klados*, 'a branch' or 'a shoot,' and refers to the idea that these lichens look like little sticks, each topped with a cup or a cap. The species name *botrytes*, from the Greek *botrys*, 'a bunch of grapes,' refers to the clustered fruiting bodies at the ends of the branches.

RIBBED CLADONIA • *Cladonia cariosa*
TORN CLUB LICHEN

GENERAL: Club lichen, **upright**, 5-15(20) mm tall, arising from small basal scales (squamules); clubs (podetia) irregularly cylindrical, **whitish**, cupless, often somewhat branched towards top, sides markedly **fissured lengthwise**; fruiting bodies (apothecia) common, large, dark brown, at tips of clubs and branches.

WHERE FOUND: Forms colonies **over sandy mineral soil** in open forests; often best developed on disturbed sites; scattered across boreal forest.

NOTES: The whitish colour and the fissured and torn clubs give this lichen a distinctive 'half-dead' appearance. Also, the fruiting bodies are disproportionately large for the size of the clubs. • The species name *cariosa* is Latin for 'crumbling' or 'withered,' and probably refers to the warty, torn stalks that often appear to be falling apart.

RED PIXIE-CUP • *Cladonia borealis*

GENERAL: Club lichen, **upright**, 1–2 cm tall, **pale yellowish green**; clubs hollow, without soredia, **unbranched**, arising from colony of basal scales, terminating in flaring **cups**, these covered in **right-angle 'tiles'** (areoles) and often ringed with **red** fruiting bodies (apothecia).

WHERE FOUND: Forms colonies over **acidic soil and decaying conifer logs**; common across boreal forest but generally more abundant northwards to low Arctic; circumpolar.

C. borealis

C. cristatella

NOTES: Earlier books refer to this species as *C. coccifera*. These 2 species were recently separated on the basis of chemistry and slight differences in form. *C. borealis* is common across North America, whereas *C. coccifera* is more common in the east.• Of the more than 60 species of *Cladonia*s growing in the boreal forest, only a dozen or so have red fruiting bodies; the rest are brown-fruited. • ***C. pleurota*** is very similar to red pixie-cup, but it has a powdery (sorediate) surface. ***C. cristatella*** is another red-fruited species, but its fruiting bodies are borne at the tips of awl-shaped clubs, rather than on cups. • The main body of this lichen, the flattened, leaf-like base, has been observed to grow at a rate of 1.6–2 mm per year. Given the final size of the squamules (2–3 mm), this lichen appears to be adapted to completing its life cycle in a hurry • Red-fruited *Cladonia*s are often called British soldiers, because the tiny, red-capped figures scattered across the forest floor bring to mind the ranks of British red-coats that arrived in the forests of eastern North America over 200 years ago.

FALSE PIXIE-CUP • *Cladonia chlorophaea*

GENERAL: Club lichen, **upright**, 10–15 mm tall, arising from clusters of tiny grey basal scales; clubs unbranched, **pale greenish**, hollow, covered in **powdery or somewhat granular soredia**, terminating in **flaring cups**, these occasionally ringed with brown fruiting bodies.

WHERE FOUND: Over **acidic mineral soil**, on rotting wood, moss and humus, or over tree bases; especially in open sites; most common 'cup' *Cladonia*; widespread across boreal forest; circumpolar.

NOTES: This innocuous lichen has caused a lot of headaches among lichen taxonomists. It includes a large number of segregate species that generally look alike but have clearly defined differences in ecology, distribution and, above all, chemistry. For our purposes it is enough to treat this group as *C. chlorophaea*, in the broad sense. • False pixie-cup could be confused with brown pixie-cup (*C. pyxidata,* below), but that species lacks soredia. • ***C. fimbriata*** is another similar lichen commonly found in the boreal forest. It is covered with fine, floury soredia and its cups are taller, narrower and more trombone-shaped than the wider 'wine glasses' of false pixie-cup. • Most of the small cup lichens are commonly called pixie-cup lichens, because they resemble tiny goblets suitable for use by fairies or sprites. These cups can also be used to disperse the powdery soredia of the lichen, either by catching raindrops, which splash the soredia out of the cups, or by increasing exposure to air currents, which can carry the soredia to new sites.

BROWN PIXIE-CUP • *Cladonia pyxidata*

GENERAL: Club lichen, **upright,** 1–2 cm tall, greyish green; clubs hollow, **without soredia**, typically **unbranched**, arising from colonies of thin, erect base scales, terminating in **flaring cups**, these covered in **right-angle 'tiles'** (areoles) and sometimes ringed with **brown** fruiting bodies.

WHERE FOUND: On **acidic mineral soil** and humus in open areas; occasionally on rotting logs and at tree bases; common and widespread across boreal forest; cosmopolitan, found on all continents.

NOTES: Brown pixie-cup is often confused with false pixie-cup (*C. chlorophaea,* above). False pixie-cup has powdery soredia on its cups, but it is often difficult to tell the difference between these soredia and the small tile-like areoles on the cups of brown pixie-cup. Many specimens are intermediate between these 2 species. • A calcium-loving species, ***C. pocillum***, is also similar, but it has rosettes of thick, leathery scales (squamules) at the base of its clubs (podetia). The basal scales of brown pixie-cup are thin and do not form rosettes. • In Europe, brown pixie-cup was boiled to make a tea that was mixed with honey to make an expectorant for treating children's colds and whooping cough. • Brown pixie-cup was used to dye woollens red-purple and ashy-green colours. It took large numbers of these small plants to colour even small amounts of wool. • Lichens growing in sunny places can become very hot—often reaching 50–69°C (20–40°C above air temperature). When lichens are moist their tolerance of high temperatures is similar to that of most other plants. However, dry lichens can tolerate extremely high temperatures with minimal damage. In one study, the respiration of a dry brown pixie-cup, heated to 101°C for 30 minutes, was reduced by only 50%. • The species name *pyxidata* does not refer to pixies, fairies or sprites, but is derived from the Latin *pyxis*, 'a box,' perhaps because the tiny cups looked like miniature containers. A pyx is a box in a government mint in which sample coins are kept to be tested for purity or weight. Are the tiny scales in the cups of this lichen fairy coins?

SIEVE CLADONIA • *Cladonia multiformis*
SIEVE CUP LICHEN

GENERAL: Club lichen, **upright**, 3–5 cm tall, **abundantly branched**, greenish to brownish grey, smooth (no soredia), **clubs** (podetia) often bearing scattered **right-angle scales**; **cups** usually present, at least **partly perforated** with small holes; fruiting bodies rare, dark brown.

WHERE FOUND: Found on **soil, among mosses** and on rotting wood in moist habitats; common and widespread across boreal forest; occasional in northern parkland.

NOTES: The sieve-like holes in the cup lining are diagnostic. When lacking well-formed cups, sieve cladonia could be confused with **fork lichen** (*C. furcata*), which is cupless, well-branched and also lacks soredia. These two species often intergrade. In some cases, sieve cladonias with non-perforated cup linings could be confused with sterile forms of brown-foot cladonia (*C. gracilis,* below), though that species is coarser and much less branched. • The species name *multiformis*, from the Latin *multus*, 'many,' and *forma*, 'a shape,' refers to the many forms taken by this highly variable lichen.

BROWN-FOOT CLADONIA • *Cladonia gracilis* ssp. *turbinata*
SLENDER CUP LICHEN

GENERAL: Club lichen, **upright,** 3-5(7) cm tall, **brownish** grey, sparingly branched, lacking soredia; **cups distinct**, secondary cups **grow from cup edges**; fruiting bodies (apothecia) common, lining the cup edges, dark brown.

WHERE FOUND: On **soil, humus, mosses** and rotting wood; common and widespread across boreal forest; circumpolar, nearly cosmopolitan.

NOTES: Brown-foot cladonia is possibly the most common cup lichen in the boreal forest, especially in dry coniferous stands. • **Whorled cup lichen** (*C. verticillata*) is similar, but its secondary cups grow from the centres of the primary cups, rather than from the edges. • **Black-foot cladonia** (*C. phyllophora*) is also similar, but its cups are poorly developed and irregular, and the bases of the stems are black (with white spots), not brown. • In the north and west of our region, **orange-foot cladonia** (*C. ecmocyna*) is easily confused with slender cup lichen, but it tends to be pale greyish green with an orange base (rather than brownish) and it has narrower cups. These two species are most easily separated using chemical tests. • The species name *gracilis* is Latin for 'slender,' and the more northern variety of *C. gracilis*, with its slender stalks and narrow to non-existent cups is certainly that. The species name of whorled cup lichen, from the Latin *verticillus*, 'a whorl,' refers to the successive tiers of fruiting bodies on the rims of the cups within cups. The species name *phyllophora* is from the Greek *phyllon*, 'a leaf,' and *phoros*, 'to bear'; in some forms of this lichen, the stems bear numerous right-angle scales.

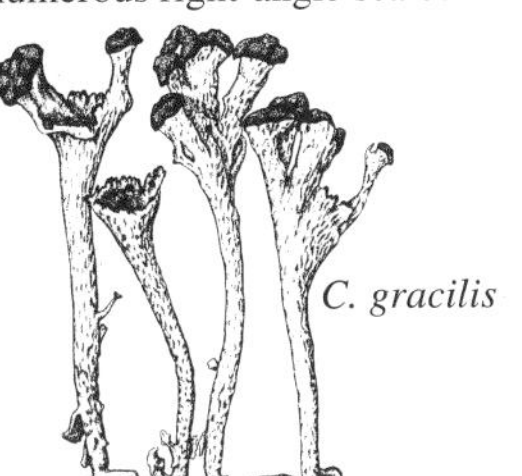

C. gracilis

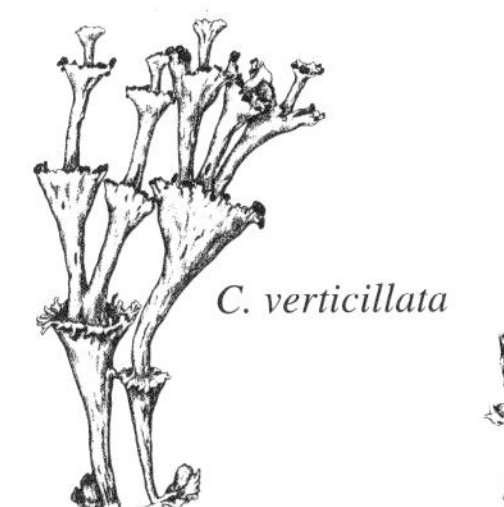

C. verticillata

C. phyllophora

SULPHUR CUP • *Cladonia sulphurina*
SULPHUR CLADONIA

GENERAL: Club lichen, **upright**, 2–5 cm tall (sometimes to 7 cm), **basal scales** (squamules) **large** (5–10 mm long), clubs (podetia) stout, **yellowish**, sides **fissured and split**, covered with **powdery soredia**; cups irregular, narrow; fruiting bodies (apothecia) ocassionally present, **red**.

C. sulphurina

WHERE FOUND: On humus, among mosses and especially on **rotting logs**; fairly common and widespread across boreal forest; circumpolar.

NOTES: Formerly called *C. gonecha.* • **Deformed cup** (*C. deformis*) is similar to sulphur cup, though in most cases, its base scales are much smaller and its cups are more regular. The main differences between these 2 species are those of chemistry. Sulphur cup fluoresces under ultraviolet light and deformed cup does not. Another cup lichen, ***C. pleurota***, also resembles these species, but it has much shorter, broader cups. • The species name refers to the pale yellow, sulphur-like colour of this lichen. The species name *deformis* is Latin for 'misshapen' or 'disfigured.' However, the clubs of this species are usually well-formed, and when the cups are edged with bright red fruiting bodies this is a very attractive lichen. The name 'deformed' would seem more appropriate for sulphur cup, which has split, irregular clubs. From this we may conclude that even lichen taxonomists (the people who give lichens their scientific names) aren't perfect.

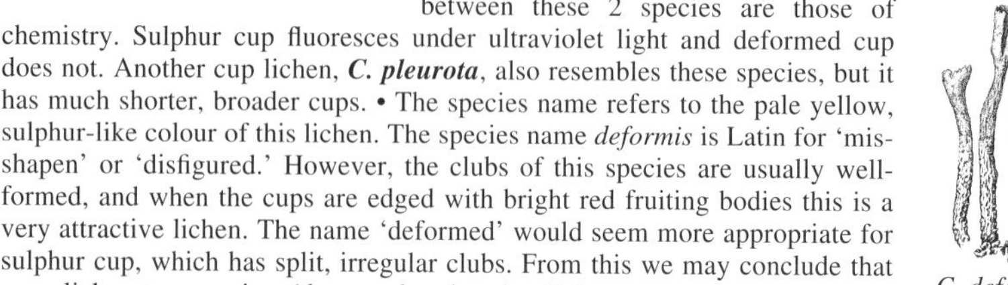

C. deformis

TINY TOOTHPICK CLADONIA • *Cladonia coniocraea*
AWL-SHAPED STUMP LICHEN

GENERAL: Club lichen, **upright**, 1–2.5 cm tall (sometimes to 3 cm), **whitish green**, hollow, unbranched, strongly tapered, **mostly pointed at tips**, covered in **floury soredia**; fruiting bodies rare, dark brown.

WHERE FOUND: On humus, **rotting logs and tree bases**; common and widespread across boreal forest; circumpolar.

NOTES: This is one of the most common 'pointed' or 'toothpick' lichens in our region. Look for large base lobes (squamules) with pale green, floury clubs (podetia) growing from their centres. Large specimens may be difficult to dintinguish from small specimens of horn cladonia (p. 359) but the soredia of that species form rounded patches on the upper part of the club. A less common species, ***C. bacilliformis,*** is almost identical, but its clubs are yellow. **Scarlet toothpick cladonia** (*C. bacillaris*) can only be distinguished by its red (rather than brown) apothecia or by chemical tests. • The *conio* in *coniocraea* is derived from the Greek *konis*, 'dust,' in reference to the powdery soredia that cover the clubs of this species.

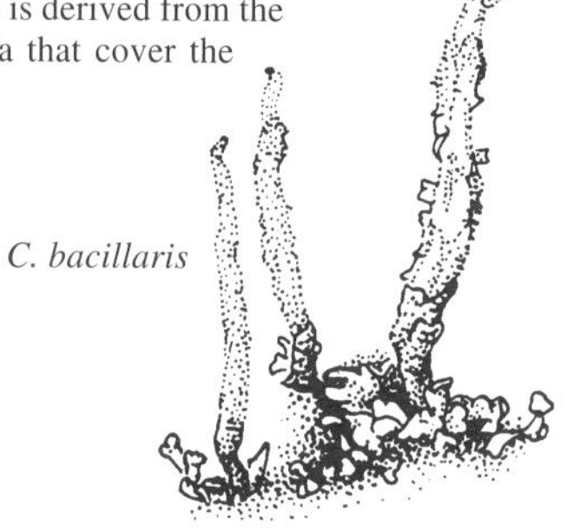

C. bacillaris

HORN CLADONIA • *Cladonia cornuta*
PIONEER CLADONIA

GENERAL: Club lichen, **upright,** 3–8 cm tall, greenish to **brownish grey**, hollow, generally **unbranched**, tapering to a **pointed tip, upper part covered in patches of powdery soredia** (hand lens), **lower half lacking soredia**; fruiting bodies rare, brown.

WHERE FOUND: Forms colonies **over humus and among mosses** in conifer forests and forest openings; occasionally on rotting wood; common and widespread across our region; circumpolar.

C. cornuta

C. subulata

NOTES: No other pointed, unbranched *Cladonia* has clubs with powdery soredia confined to rounded patches on their upper half. **Tall toothpick cladonia** (*C. subulata*) is very similar but it is generally whitish in colour with soredia covering the whole club (rather than occurring in patches and at the tip, as in horn cladonia). • Horn cladonias growing in full sunlight are darker than those of shaded habitats. • In many parts of the boreal forest, horn cladonia is an early successional species that enters an area soon after a fire but becomes scarce as the forest matures. However, it is often very common in open stands and tundra farther north. • The species name *cornuta*, from the Latin *cornu*, 'a horn,' refers to the horn-like appearance of the brown, pointed clubs of horn cladonia. The name *subulata* is taken from the Latin *subula*, 'an awl,' in reference to the shape of the slender, pointed clubs of tall toothpick cladonia.

POWDERED FUNNEL CLADONIA • *Cladonia cenotea*
FLOURY FUNNEL LICHEN

GENERAL: Club lichen, **upright,** 1-2.5 cm tall, arising from small basal scales (squamules); clubs narrow, generally **unbranched, whitish grey**, covered with **floury soredia**, terminating in **hollow cups**; fruiting bodies rare, pale to dark brown, on edges of cups.

WHERE FOUND: On **rotting wood and stumps**, soil rich in humus, and occasionally on thin humus over rocks; widespread across boreal forest; circumpolar.

NOTES: Sorediate, mostly unbranched clubs with irregular cups that open into hollow stalks are diagnostic of this species. **Shrub funnel cladonia** (*C. crispata*, p. 360) is very similar to this species, but it lacks soredia. Both of these common lichens apparently assist in the decay of dead wood. • The clubs of powdered funnel cladonia, with their flour-like covering of soredia, have been likened to the velvet-covered antlers of caribou. • The common name describes the small, sorediate cups that open into the hollow stalks. • The species name *cenotea* may be derived from the Greek *kenos*, 'empty'; the hollow cups are indeed empty.

SHRUB FUNNEL LICHEN • *Cladonia crispata*

GENERAL: Upright club lichen, 2–8 cm tall, up to 5 mm thick, **brownish mineral grey**, hollow, **branched, smooth** (no soredia), **cupless or with flaring cups**; secondary cups grow from cup edges; cups perforated, open into interior of club; fruiting bodies small, brown, at tips of branches.

WHERE FOUND: On soil, rotting wood or over mosses in generally more open areas; common and widespread across boreal forest; circumpolar.

NOTES: Shrub funnel lichen could be confused with sieve cladonia (p. 357), but that species has 'sieve-like' holes in its cups. • The species name *crispata*, from the Latin *crispus*, 'curled,' refers to the rippled edges of the cups.

SHINGLED CLADONIA • *Cladonia scabriuscula*
FORKED SHORE LICHEN

GENERAL: Club lichen, **upright**, 3–5(8) cm tall, pale greenish grey; clubs (podetia) sparsely to densely **covered (below) with right-angle scales, moderately covered (above) with granular soredia**; branches forking in pairs, usually pointed at tips; fruiting bodies rare, dark brown.

WHERE FOUND: On **moist soil** in woods, peatlands and old road banks; sometimes on thin soil over rock; scattered across boreal forest; circumpolar; found on every continent except Africa.

NOTES: Tall toothpick cladonia (*C. subulata,* p. 359) is similar to *C. scabriuscula*, but its soredia are fine and floury rather than granular and it lacks the right-angle scales of shingled cladonia. **Fork lichen** (*C. furcata*) is also similar, but it lacks both soredia and (for the most part) right-angle scales. • The species name *scabriuscula*, from the Latin *scabere*, 'to scratch,' means 'blotchy' or 'roughened with small knobs,' and refers to the rough appearance of the dense, flat scales covering the clubs.

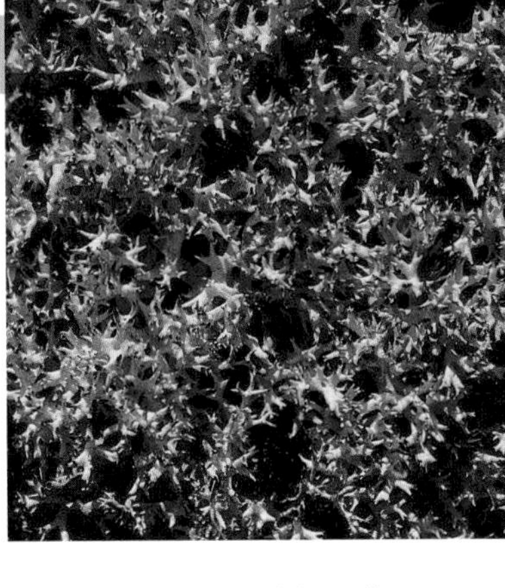

PRICKLE CLADONIA • *Cladonia uncialis*
SPIKE LICHEN

GENERAL: Shrub lichen, **upright**, 2–6 cm tall, 1–1.5 mm (sometimes to 4 mm) thick, yellowish green, forms **dense mats**, clubs (podetia) **freely branched**, final **branchlets spine-like**, very brittle, **shiny**, **cupless**; fruiting bodies rare, light brown.

WHERE FOUND: On soil (often sandy) and among mosses in open areas; widespread across boreal forest, more abundant northward; circumpolar.

NOTES: Another branching *Cladonia*, ***C. amaurocraea***, could be confused with prickle cladonia, but it is usually less densely branched and it has distinct cups at the tips of its branches. The smooth, hard, shiny surfaces of these 2 lichens distinguish them from the reindeer lichens (*Cladina* species), with which they often associate. • The species name *uncialis* is Latin for 'of an inch' or 'an inch high,' but this lichen is usually taller than its name suggests.

C. unicialis

C. amaurocraea

GREEN REINDEER LICHEN • *Cladina mitis*
YELLOW REINDEER LICHEN

GENERAL: Shrub lichen, **upright**, 4–7 cm (sometimes to 10 cm) tall, **pale yellowish green, intricately branching** from a main stem, not copiously fork-branching from the base; **branches hollow**, with dull, **appressed-cottony surface** (hand lens); end branchlets tending to point in 1 direction.

WHERE FOUND: Forms **mats on ground** in open coniferous forest; common and widespread across boreal forest; circumpolar.

NOTES: Tree reindeer lichen (*C. arbuscula*) (inset photo) is a similar, closely related species. Its branches tend to be coarser, more copiously branched from the base, and more strongly curved in 1 direction. The 2 species cannot be separated with certainty in the field, but they are easily identified using chemical tests. **Prickle cladonia** (*Cladonia uncialis,* above) is also similar at first glance, but its spreading branchlets and hard, shiny outer surface readily separate it from the *Cladina*s. • The Woods Cree boiled green reindeer lichen to make a medicinal tea for expelling intestinal worms. The powdered lichen was also taken in water for this purpose. • Reindeer lichens provide important ground cover in northern woodlands. Lichens are the principal winter food of many caribou, reindeer and musk-oxen. • Although hoofed mammals can survive for several months on a lichen diet, they usually lose body weight because of protein deficiency. The complex carbohydrates of the lichens are broken down by enzymes produced in the stomach of reindeer or caribou but in few other animals. The bacteria and protozoans in the rumen also help to break down lichen compounds into sugars that the animals can use. Humans cannot digest lichens and get little energy from eating these plants. However, partially digested lichen from the stomachs of freshly killed caribou was often eaten, and in this form it has much greater nutritional value. In some cases the rumen was mixed with blood and fat, plus meat scraps or liver, to make a pudding, highly esteemed by native people, but usually it was eaten fresh, warm and uncooked. It is said to taste like fresh lettuce salad.

GREY REINDEER LICHEN • *Cladina rangiferina*
TRUE REINDEER LICHEN

GENERAL: Shrub lichen, **upright**, 5–8(10) cm tall, **greyish white, intricately branching** from a main stem; **branches hollow**, with dull, **appressed cottony surface** (hand lens); many end branchlets pointing in 1 direction (**swept to 1 side**).

WHERE FOUND: Forms **extensive carpets over ground** in open coniferous forest (commonly on sandy soils) and in open sites, from lowland bogs on *Sphagnum* to arctic tundra; common and widespread across boreal forest; circumpolar.

NOTES: Green and grey reindeer lichens (right and left of upper photo), are most easily separated by color. • Northern native peoples used reindeer lichen in medicinal teas to treat colds, arthritis, fevers and other problems. Reindeer lichens were also used as a poultice to relieve the ache of arthritic joints. Reindeer lichens have been taken to treat fever, jaundice, constipation, convulsions, coughs, and tuberculosis. • Grey reindeer lichen is one of the lichens most frequently grazed by caribou and reindeer. In northern Europe it was collected as fodder for livestock, in the belief that milk from the cows would be creamier and their flesh would be fatter and sweeter. • Grey reindeer lichen is an excellent example of a plant that has adapted to surviving the severe conditions of the north. Stress studies showed that only boiling and radiation caused severe injury to these plants. With a 50% drop in water content, respiration slowed by only 20% and respiration continued until the lichens were almost completely dehydrated. • Reindeer lichens grow slowly, and mature clumps are often about 100 years old. These lichens generally produce a new branch each year, so the age of a clump can be estimated by counting back through the major branchings along a stem. Unfortunately, after about 20 years the lower parts start to decompose and eventually you must make an arbitrary decision as to what is living and what is dead.

NORTHERN REINDEER LICHEN • *Cladina stellaris*
STAR REINDEER LICHEN

GENERAL: Shrub lichen, **upright**, densely branched, usually with **no main stem**, forming **compact, rounded heads** (like cauliflower), 5–10 cm tall; heads grow singly, in small clumps, or forming billowing mats, yellowish white to pale yellowish green; branches hollow, dull with **appressed cottony surface**, uppermost branches usually **4–6 in a whorl**.

WHERE FOUND: On **soil, humus** or thin soil over rocks, in open coniferous forests, often common in lichen woodlands; widespread across boreal forest; circumpolar.

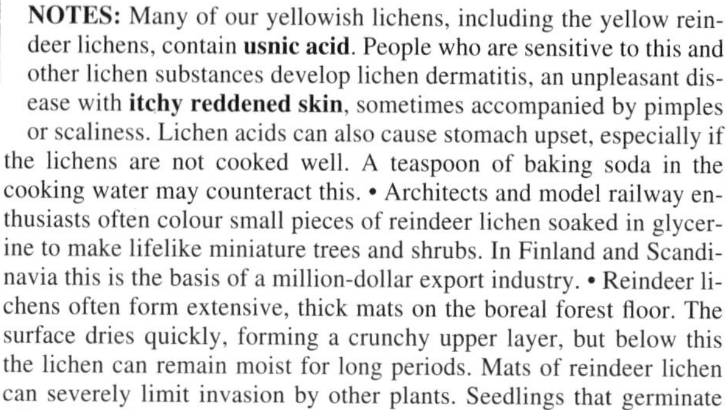

NOTES: Many of our yellowish lichens, including the yellow reindeer lichens, contain **usnic acid**. People who are sensitive to this and other lichen substances develop lichen dermatitis, an unpleasant disease with **itchy reddened skin**, sometimes accompanied by pimples or scaliness. Lichen acids can also cause stomach upset, especially if the lichens are not cooked well. A teaspoon of baking soda in the cooking water may counteract this. • Architects and model railway enthusiasts often colour small pieces of reindeer lichen soaked in glycerine to make lifelike miniature trees and shrubs. In Finland and Scandinavia this is the basis of a million-dollar export industry. • Reindeer lichens often form extensive, thick mats on the boreal forest floor. The surface dries quickly, forming a crunchy upper layer, but below this the lichen can remain moist for long periods. Mats of reindeer lichen can severely limit invasion by other plants. Seedlings that germinate on the lichen mat soon wither and die when the surface dries. Those that do manage to reach mineral soil often become enmeshed in lichen branches and are pulled out of the soil or snapped off by the repeated expansion and contraction of the lichens with changes in moisture.

WOOLLY CORAL • *Stereocaulon tomentosum*

GENERAL: Shrub lichen, **erect**, **ashy-grey**, 3–5(8) cm tall, much branched; branches solid, with dense, coral-like 'foliage', with a **distinct upper and lower surface**, the lower surface coated with a thick, spongy mat of hairs; fruiting bodies usually numerous, located towards ends of branches, 0.3–0.6 mm wide, dark brown.

WHERE FOUND: On soil in open places, occasionally on gravel; widespread across northern boreal forest; circumpolar.

NOTES: Woolly coral could be mistaken for ***S. paschale***, but that species lacks the thick woolly covering on its branches. The branches of all coral lichens bear small outgrowths called cephalodia, which contain secondary colonies of cyanobacteria. In woolly coral they appear as greyish lumps, whereas in woolly coral they take the form of tiny black tufts. • In some areas, *S. paschale* is one of the commonest species grazed by caribou and reindeer. When large areas are covered by this lichen in northern Sweden, it is usually the result of overgrazing of reindeer lichen (*Cladina* species) by reindeer. • The genus name *Stereocaulon*, from the Greek *stereos*, 'hard,' 'firm' or 'solid,' and *kaulos*, 'a stem,' refers to the solid stalks of these lichens, which become very firm and brittle when they dry. The species name *tomentosum*, from the Latin *tomentum*, 'a stuffing of hair or wool,' refers to the woolly covering on the lower surface of the stems.

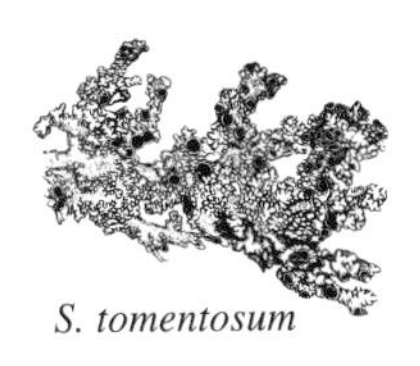

S. tomentosum

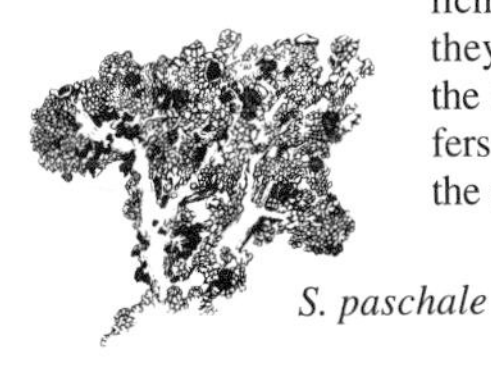

S. paschale

SPRUCE MOSS • *Evernia mesomorpha*
NORTHERN PERFUME

GENERAL: Shrub lichen, **semi-erect or hanging**, to 7 or 8 cm long, pale **yellowish green**, wrinkled, pliable; many branches somewhat **angular in cross-section,** with **coarse soredia** (hand lens).

WHERE FOUND: On **bark and twigs** of trees and shrubs, especially conifers, in open forests; occasionally on rocks or soil north of treeline; very common across boreal forest; rare on aspen and willow in parkland; circumpolar.

NOTES: The angular, green branches of spruce moss resemble those of some beard lichens (*Usnea* spp.), but those species always have a distinctive central cord in their branches. • The Chipewyan boiled spruce moss to make a strong tea that was cooled and dropped into the eyes to treat snow blindness. A related species, **oak moss** (*E. prunastri*), has been used in Europe since the 1600s to enhance and 'fix' the fragrances of perfumes, so that they last for hours rather than minutes. It was also collected and used for dyeing in Scandinavia. • The genus name *Evernia* may come from a Greek word meaning 'growing well,' but since the type species for the genus, (*E. prunastri*), grows only 2 mm per year (20 cm per 100 years), this seems inappropriate. Perhaps it refers to the abundance of this common lichen, rather than its growth rate.

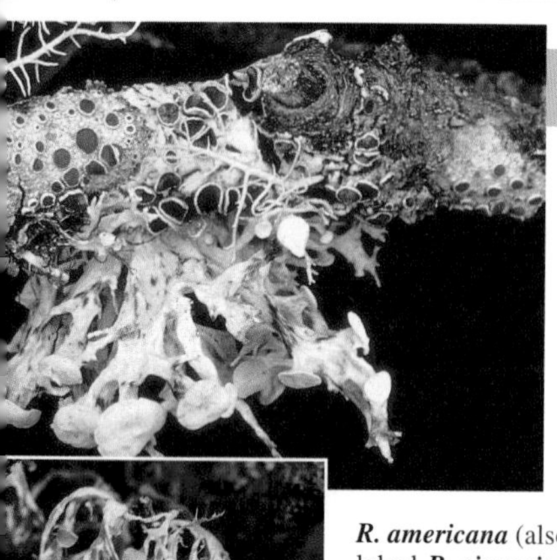

PUNCTURED GRISTLE • *Ramalina dilacerata*
CARTILAGE LICHEN

GENERAL: **Shrub** lichen, **tufted**, **greenish yellow**, less than 2 cm long; branches **hollow, perforated**, lacking soredia; **fruiting bodies common**, at end of branches.

WHERE FOUND: Stumps, **trunks and branches** of deciduous and coniferous trees and shrubs; widespread across boreal forest and northern parkland; total distribution incompletely known.

NOTES: **Also called** *R. minuscula* or *Fistulariella dilacerata* • The tufted habit and green, hollow, partly perforated branches separate punctured gristle from all other tree-dwelling lichens. The narrow-lobed ***R. americana*** (also called *R. fastigiata*) (inset photo) and the rather broad-lobed ***R. sinensis*** are both similar in general appearance, but they have solid branches. Of these 3 species, only *R. dilacerata* is common in our region. • Essential oils and other chemicals have been extracted from some cartilage lichens for use in the perfume industry, either as a pleasant scent for soaps or as an essence for perfumes. • The genus name *Ramalina* is from the Latin *ramalia*, 'twigs.'

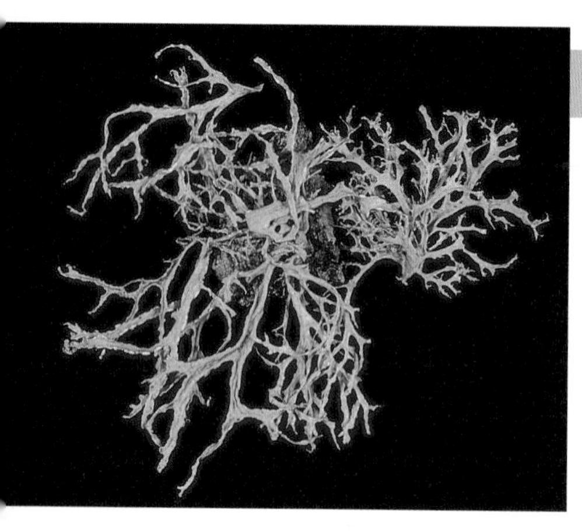

DUSTY GRISTLE • *Ramalina pollinaria*
DUSTY CARTILAGE LICHEN

GENERAL: **Shrub** lichen, **tufted**, much branched, **pale yellowish green**, 1–4 cm long, lobes flat to rounded, especially toward tip, surface smooth, **shining**, tips of branches often burst open; **granular soredia** present **on surface of lobes and at tips, not on edges**; **fruiting bodies very rare**, at lobe tips.

WHERE FOUND: On bark and twigs of deciduous and coniferous species in fairly sheltered areas, on dead wood and occasionally on rock; across southern boreal forest and northern parkland.

NOTES: Dusty gristle can be distinguished from most other tufted green shrub lichens by its shiny surface, solid branches, and soredia-bearing lobe tips. ***R. intermedia*** differs primarily in chemistry, though it almost invariably occurs on rock. ***R. farinacea*** is also similar, but its soredia are floury (not granular) and occur along the edges of the lobes, never on the lobe tips. • The name *pollinaria* derives from the Latin *pollen*, meaning 'fine flour,' an obvious reference to the coarse (not fine!) soredia of this species.

SUGARY BEARD • *Usnea hirta*
SHAGGY OLD MAN'S BEARD

GENERAL: Shrub or hair lichen, **tufted** to somewhat hanging, 3–5(6) cm long, pale **yellowish green throughout (including base)**, much branched; branches reinforced by a **tough white central cord**, **smooth**, bearing numerous lateral branchlets (fibrils), also bearing **copious isidia throughout**; fruiting bodies (apothecia) absent.

WHERE FOUND: On wide variety of substrates including **deciduous and coniferous trees**, dead wood, wooden fences, and rarely rocks; widespread across boreal forest; circumpolar.

NOTES: Sugary beard could be confused with ***U. subfloridana***, but that species has numerous minute, warty bumps on its branches and tiny peg-like isidia mixed in with its powdery soredia. • Although it is difficult to distinguish species of old man's beard, the genus *Usnea* is easily recognized by its characteristic tough, thread-like central cord. To see this, hold a strand between your finger and thumbnail and slowly pull lengthwise. This cord adds strength and resiliency, and it may also store energy-rich polysaccharides. • Lichen-covered branches make excellent kindling, but it seems a shame to burn these beautiful plants. Some songbirds incorporate hair-like lichens into their nests. Beard lichens also provide winter forage for deer and other hoofed mammals, when food on the ground may be hard to find. • The name *Usnea* appears to be derived from an Arabic word meaning 'moss.' The species name *hirta* is Latin for 'shaggy' or 'hairy,' which this lichen certainly is.

POWDERY BEARD • *Usnea lapponica*
POWDERY OLD MAN'S BEARD

GENERAL: Shrub or hair lichen, **tufted** to somewhat hanging, 3-5 cm long, pale **yellowish green,** but **blackening at base**, much branched; branches reinforced by a **tough white central cord**, **minutely bumpy** (check basal portions), bearing numerous side branchlets (fibrils); patches of **soredia** numerous, at first round in outline, but soon **encircling the branches**; fruiting bodies (apothecia) absent.

WHERE FOUND: Forms tufts **over conifers** and deciduous trees and shrubs; found across our region; circumpolar.

U. lapponica

U. subfloridana

NOTES: Also called *U. sorediifera*, *U. substerilis*, *U. fulvoreagens*. • Powdery beard is similar to another common boreal species, ***U. subfloridana***, but that species has clusters of peg-like isidia on its branches. Unfortunately, the isidia are not always obvious and the 2 species are easily confused. • The yellowish colour of these lichens reflects the presence of usnic acid. This acid has antibiotic properties (e.g., it is effective against *Mycobacterium tuberculosis*, which causes tuberculosis), and these lichens have been used to treat several skin diseases, ranging from diaper rash in babies to foot diseases in adults. • The species name *lapponica* means 'of Lapland,' presumably the homeland of the first specimens identified as this species.

PITTED BEARD • *Usnea cavernosa*
PITTED OLD MAN'S BEARD

GENERAL: Hair or shrub lichen, **hanging**, 10–25(40) cm long, soft, copiously branched, **straw yellow to pale green**, with a **white central cord**; branches to 1 mm thick, thicker branches distinctly **ridged and pitted**; fruiting bodies rare, small.

WHERE FOUND: Hangs from **conifer branches**, infrequently on deciduous trees; widespread across boreal forest; circumpolar.

NOTES: Pitted beard sometimes forms very large hanging masses on conifers. It is one of the most common *Usnea* species in our region, especially well developed under conditions of high humidity. The round, ridged and wrinkled main branches distinguish this from any other species of *Usnea*. ***Ramalina thrausta*** is similar, but it lacks the central cord characteristic of this genus. • Pitted beard is the largest of our lichens, with strands from 10–40 cm long. However, some *Usnea* species can reach 5 m in length. • The Woods Cree put fresh old man's beard into their nostrils to stop nose bleeds. In the 1400s, *Usnea* was prized as a treatment for lung diseases, a common complaint. According to the Doctrine of Signatures these lichens were effective against disorders of the hair and scalp. • The species name *cavernosa*, from the Latin *caverna*, 'a hollow' or 'a cave,' refers to the many hollows and pits on the branches of this species.

U. cavernosa

Ramalina thrausta

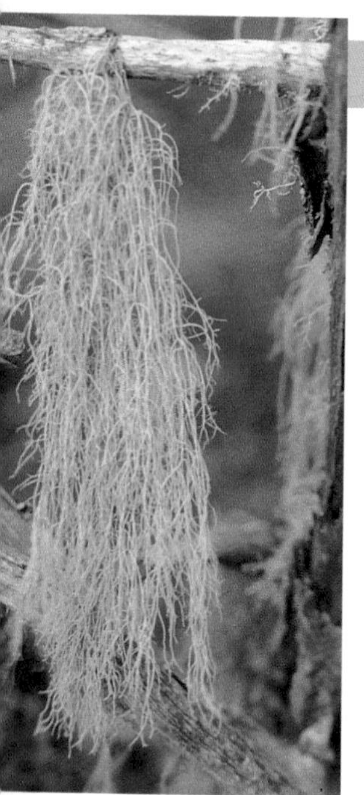

SCRUFFY BEARD • *Usnea scabrata*
SCRUFFY OLD MAN'S BEARD

GENERAL: Hair or shrub lichen, **hanging**, **10-20(30) cm** long, **pale yellowish green,** but **blackening at base**, much branched; branches reinforced by a **tough white central cord**, **minutely bumpy** (check basal portions), bearing copious side branchlets (fibrils), also bearing **scattered spine-like isidia**; fruiting bodies (apothecia) absent.

WHERE FOUND: On **conifers** and deciduous trees such as birch; scattered across our region.

NOTES: The pendent habit and scattered isidia are diagnostic of this species. • Some native groups reportedly used beard lichens as food by boiling them and eating them with fish, berries or grease. The usnic acid would make them very difficult to digest and they have very little food value for humans. Different species of beard lichen have been used in dyeing, and produce a tan colour. Scandinavians used beard lichens to make troll dolls—figures that were used to warn children to be good and kind to other people. • In Finland, usnic acid derivatives were found to be effective against a wide range of bacteria and also some fungi (e.g., athlete's foot and ringworm). Antibiotics from lichens are relatively non-toxic. • The species name *scabrata*, from the Latin *scaber*, 'rough' or 'scurfy,' refers to the many minute warty bumps and short branching spines that give the strands of this lichen a roughened appearance.

SPECKLED HORSEHAIR • *Bryoria fuscescens*

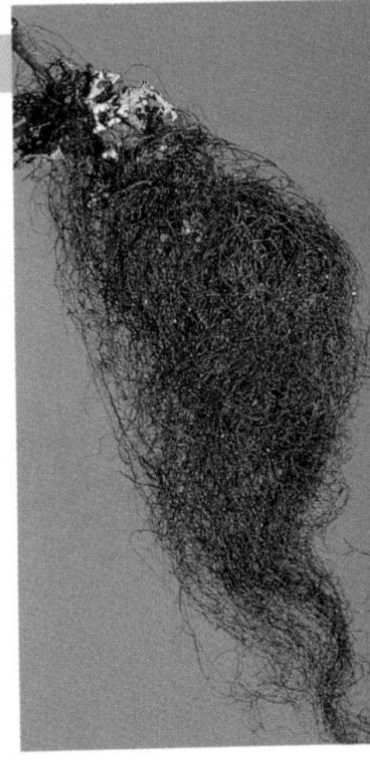

GENERAL: **Hair** lichen, usually hanging, 5–15 cm long, **dark medium-brown**, intricately branched; **branches even in width,** rather pliable, with **tiny whitish flecks of powdery soredia.**

ECOLOGY: Hanging from **branches of conifers**; common and widespread across boreal forest; circumpolar.

NOTES: This species could be confused with **brittle horsehair** (*B. lanestris*), which is distinguished (with practice) by its thinner (less than 0.2 mm), uneven, brittle branches. Another similar species, simple horeshair (below), is more tufted and has greenish black soredia. • Neither speckled nor brittle horsehair is known to produce fruiting bodies. Instead they reproduce entirely by vegetative means—either by breaking into pieces or by producing powdery, whitish soredia that are dispersed by the wind, and by birds, insects or other animals. • Horsehair lichens have been used as a source of yellow dye. • *Fuscescens* is Latin for 'becoming dark,' presumably referring to the tendency of this species to darken in sunny sites.

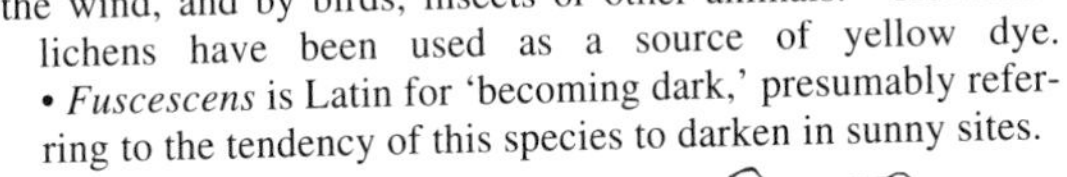

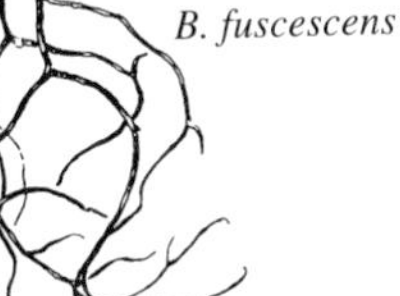

B. fuscescens

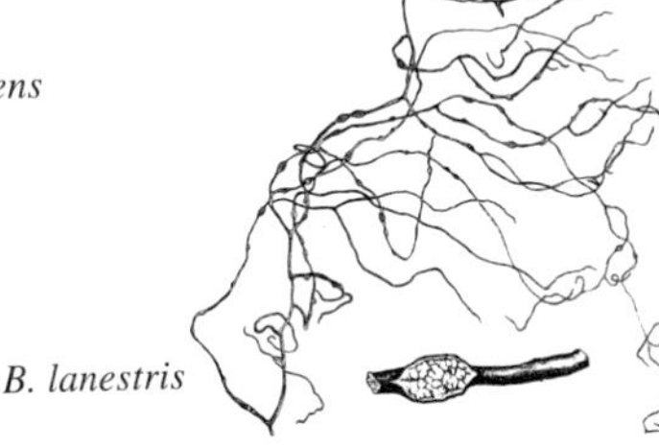

B. lanestris

SIMPLE HORSEHAIR • *Bryoria simplicior*

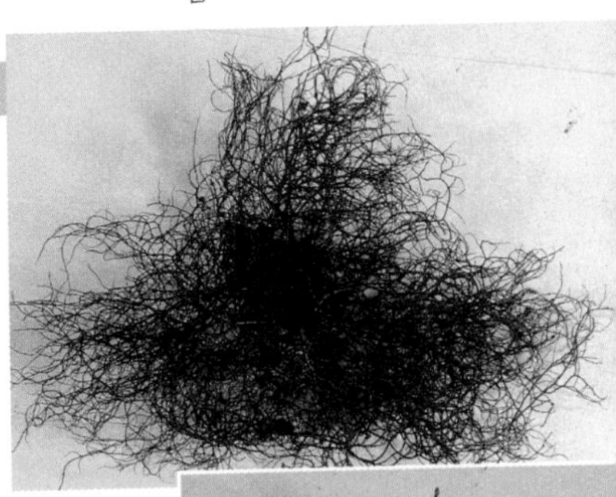

GENERAL: **Hair** lichen, **tufted**, up to 5 cm long, forming **small**, rather dense to slightly long **tufts, brownish black to black**, dull or slightly shiny; main branches to 0.3 mm thick, tips thinner, straight or slightly curved, **with tiny, round or oval clusters of powdery soredia containing greenish black dots**.

WHERE FOUND: On conifers and on birch, occasionally on old wood, rarely on rocks; characteristic of northern boreal forest; circumpolar.

NOTES: Simple horsehair often grows entangled with other horsehair lichens, but it is easily distinguished by the greenish black dots on its more even branches. ***B. furcellata*** differs in having white soralia that are narrower than the branches on which they occur, and that often bear characteristic tufts of spinules. This lichen is widespread on coniferous trees in our region, but it is more common in the eastern half on North America. • Northern native peoples heat horse hair lichen and make it into a powder for use on blisters and burns. • Horsehair lichens can be an important source of food for caribou and deer when these animals are unable to dig through deep or crusty snow to reach more desirable foods. In northern Saskatchewan, caribou soon strip the lichens from fallen trees. Some black spruce forests in Saskatchewan produce 359–572 kg of horsehair lichen per hectare. • The genus name *Bryoria* was derived from *Bryopogon* and *Alectoria*, 2 earlier names applied to these species.

Glossary

achene: a small, dry, 1-seeded, nut-like fruit (pp. 170, 227, 369)
acrocarpous: producing sporophytes from reproductive cells at the tip of a main stem or branch (p. 286)
adnate: attached
adventive: a foreign species growing away from its natural habitat; partly naturalized
alar cells: cells at the bottom corners of a moss leaf that attach the leaf to a stem
alkaline: containing unusually high levels of soluble mineral salts (usually chlorides, sulfates, carbonates, and bicarbonates of sodium, potassium, magnesium, and calcium)
alluvial: composed of deposits laid down by flowing water
alternate: situated singly at each node (p. 372); cf. opposite
amphibious: living both on land and in water; especially referring to plants that begin growing under water and continue on land after the water has receded or evaporated
annual: living for only one year; cf. biennial, perennial
anther: the pollen-bearing part of a stamen (pp. 43, 170, 369)
antheridium: male sex organ in bryophytes, a globose to shortly cylindric sac one cell layer thick
apical: at the apex or tip (p. 372)
apothecium: [apothecia] a disc-shaped or long fruiting body in which spores are produced in lichens (p. 374)
appressed: closely pressed against something, as leaves against a stem or as lichens adhering closely to the substrate.
aquatic: living or growing in water
archegonium: a female sex organ in bryophytes
areole: a small, discrete, greenish 'tile' on the surface of a lichen (e.g., *Rhizocarpon* spp., p. 333)
armed: with spines, prickles or thorns
ascending: growing obliquely upwards
ascus: a club-shaped, sac-like structure in which spores (ascospores) are formed by fungi belonging to the ascomycetes (p. 374)
auricle: a small projecting lobe or appendage, often at the base of an organ (pp. 254, 372)
awn: a slender bristle-shaped appendage (p. 254)
axil: the angle between the leaf and stem (p. 372)
axis: the central longitudinal line about which parts of the plant are arranged
babiche: rawhide thongs or lacings
banner: an upper, usually enlarged petal of a flower, as found in the pea family; also called a standard (p. 369)
beard: a clump of hairs (p. 371)
berry: a fruit, fleshy throughout, usually containing several or many seeds
biennial: living for 2 years, usually flowering and producing fruit in the second year
bilaterally symmetrical: (flowers) divisible vertically into equal halves (compare radially symmetrical)
bilobed: 2-lobed (p. 369)
biodiversity: the full range of life in all its forms—including genes, species and ecosystems—and the ecological processes that link them
biomass: total weight of living organisms in an area

Leaf Shapes

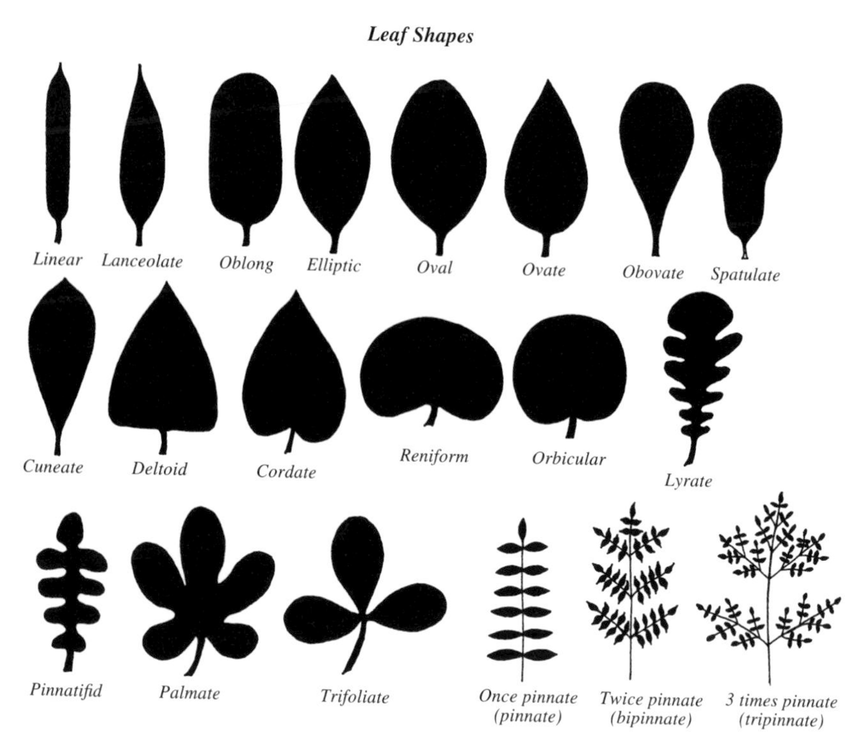

bipinnate: twice pinnate, leaflets again pinnate (p. 368)
blade: the broad, flat part of a leaf or petal (pp. 254, 369, 372)
bloom: a waxy powder covering a surface
bog: acidic peat-covered wetlands
brackish: of water, slightly salty
bract: a reduced or specialized leaf associated with, but not part of, a flower or flower cluster (pp. 43, 170); in conifers, an appendage of the central stalk of a cone
branchlet: a small branch
bryophyte: a plant of the group Bryophyta, a liverwort, moss or hornwort
bulb: a short, vertical, thickened underground stem with thickened leaves or leaf-bases (e.g., an onion)
bulbil: a small, bulb-like, reproductive structure, often located in a leaf axil or replacing a flower
calcareous: calcium-rich; soil rich in lime
callus: the firm, thickened base of a lemma in grasses
calyptra: [calyptrae] the thin covering or hood fitted over the upper part of a capsule in mosses (p. 292)
calyx: the sepals of a flower, collectively
canopy: the upper branches of the trees in a forest
capitulum: a tight, head-shaped cluster of branches (in sphagnum mosses), or flowers (in asters) (p. 314)
capsule: a dry fruit that splits open at maturity and is composed of more than 1 carpel (in seed plants) (p. 43); the spore-containing sac (in bryophytes) (pp. 286, 287, 292)
carpel: a fertile leaf bearing the undeveloped seed(s); 1 or more carpels join to form a pistil (p. 369)
catkin: a linear cluster of small flowers, usually of 1 sex, that lack petals but usually have surrounding bracts (p. 43)
caudex: the persistent, thickened stem of a perennial plant
cauline: on the stem
cephalodium: [cephalodia] a warty growth containing cyanobacteria (blue-green algae), in or on a lichen thallus (e.g., the black dots on *Peltigera aphthosa* p. 348)
ciliate: fringed with hairs or hair-like outgrowths
circumpolar: occurring in the polar region around the world
circumscissile: opening so that the top comes off as a lid
clambering: trailing over the ground
clasping: holding or surrounding tightly (p. 372)

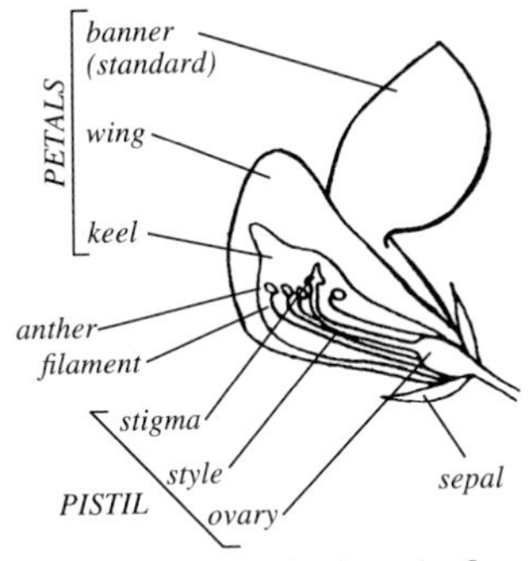

Section of an irregular flower with a superior ovary.

claw: stalk-like base of a petal or sepal (p. 369)
cleft: cut about halfway to the midrib or base, or a little deeper; deeply lobed (p. 372)
clonal: of a clone
clone: a group of plants that has originated by vegetative reproduction from a single individual
colluvial: debris of a variety of sizes, from large boulders to gravel, sand and silt, and deposited at the bottom of a slope by gravity
complanate: flattened together, compressed in 1 plane
complicate-bilobed: folded lengthwise and 2 lobed
compound: divided into smaller parts: leaves, divided into leaflets; flower clusters divided into smaller clusters (p. 372)
coniferous: bearing its reproductive organs in cones
cordate: heart-shaped, with the notch at the base (p. 368)
corm: a short, vertical, thickened underground stem without thickened leaves
corolla: the petals of a flower, collectively
cortex: the outermost layer of a stem or thallus (p. 375)
cosmopolitan: found all over the world
costa: the midrib or nerve of a moss leaf (p. 286)
creeping: growing along (or beneath) the surface of the ground and emitting roots at intervals, usually at the nodes
crisped: irregularly curled or rippled along the edges
cuneate: wedge-shaped or triangular, with the narrow end at the point of attachment (p. 368)
cyanobacteria: a group of organisms related to true bacteria; previously classified as Cyanophyta or blue-green algae because many contain chlorophyll mixed with a blue pigment (p. 374, 375)
cylindrical: shaped like a cylinder
cyme: a flower cluster in which the flowers at the branch tips bloom first (p. 370)

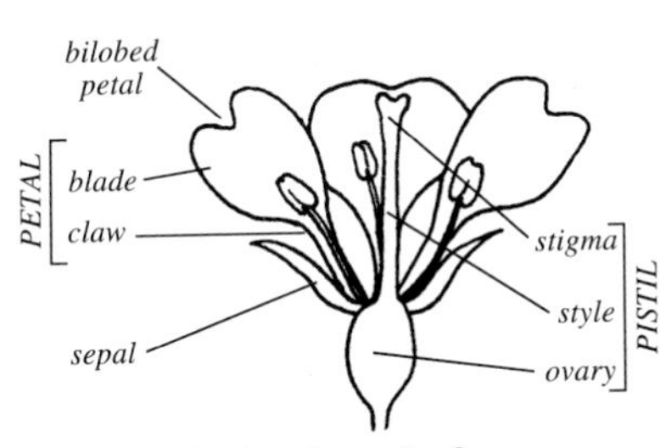

Section of a regular flower with an inferior ovary.

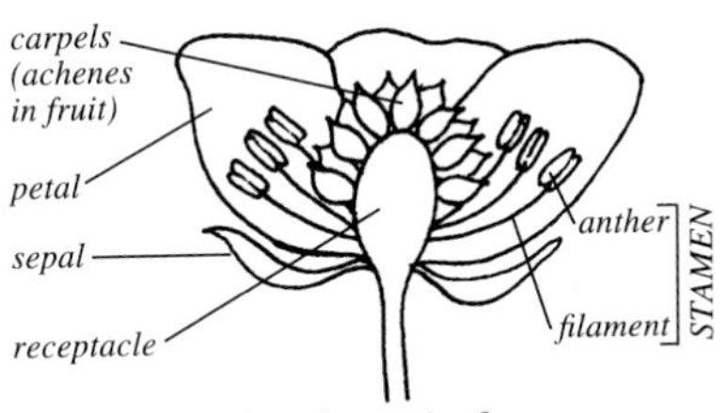

Section of a regular flower with numerous carpels.

deciduous: falling after completion of its normal function, often at the approach of a dormant season; cf. persistent

decoction: a solution prepared by boiling an animal or plant substance

decorticate: lacking a cortex

decumbent: reclining or lying flat on the ground, but with ascending tips

dehiscence: the act or method of opening, as in fruit, anthers or spore capsules

dehiscent: opening by a definite pore(s) or along a regular line(s) to discharge seeds or spores

deltoid: shaped like an equilateral triangle with one side at the base (p. 368)

dichotomous: branching into 2 equal parts, like a 'Y'

dicot: a seed plant bearing 2 cotyledons or seed leaves (includes most families of flowering plants in our region); most have leaves with net veins; cf. monocot

dimorphic: occurring in 2 forms, as in ferns with sterile and fertile leaves

dioecious: producing male and female flowers (or other reproductive structures) on separate individuals

disc flower: a flower with a tubular corolla; in the aster family, Asteraceae (p. 170)

disjunct: separated, referring to plant populations that are separated (distant) from all other populations of that species

dissected: deeply cut into segments or lobes

Doctrine of Signatures: a medieval theory that a plant could cure ailments of the part of the body that the plant resembled (e.g., thalloid liverworts were good for the liver)

dorsal: on the upper (outer) side, the side away from the stem; opposite to ventral (p. 372)

drupe: a fleshy or pulpy, 1-seeded fruit in which the seed has a stony covering (e.g., a cherry)

drupelet: a small drupe, usually in a cluster (e.g., a raspberry)

ecotone: the transition area between 2 plant communities

ellipsoid: a 3-dimensional form, in which every cross-section is shaped like an ellipse or a circle

elliptic: shaped like an ellipse, oval or oblong with the ends rounded and widest in the middle (p. 13)

emergent: coming out from; often referring to plants that are partly submerged in water

entire: without indentation or division (p. 372)

epiphyte: a plant growing on another plant

even-aged: all of similar age, referring to a population of plants (usually trees)

fascicle: a small bundle or cluster (p. 314)

fen: a mineral-rich wetland with slow-moving, often alkaline water with sedge and brown moss (not *Sphagnum*) peat

fibril: a thickening of a clear cell of a sphagnum moss that projects into the cells and forms oblique to transverse bars across the cell (p. 314)

fiddlehead: the young, coiled leaf of certain ferns

filament: the stalk of a stamen; supports the anther (pp. 43, 369)

filiform: thread-like (p. 292)

flark: a pool of water in a patterned fen that alternates with drier strings which form at right angles to the direction of water flow

fleshy: plump, firm and pulpy; succulent

flexuous: bent in a zig-zag manner

floodplain: a plain built up with stream deposits from repeated flooding

floret: a tiny flower, usually part of a cluster; usually applied to single flowers in the grass or aster family

follicle: a dry, single-carpel fruit that splits at maturity along 1 side only

forb: a broad-leaved, non-woody plant that dies back to the ground after each growing season

frond: a fern leaf (p. 274)

fruit: a ripened ovary, together with any other structures that ripen with it as a unit (p. 227)

gametophyte: the plant which produces sexual reproductive structures; the green leafy or thalloid plant of a moss or liverwort (pp. 286, 287)

gemma: [gemmae] a small body of a few cells serving in vegetative reproduction

gland: a bump, appendage or depression on a plant's surface that produces a sticky or greasy, viscous fluid

glandular: with glands

glandular-hairy: with glands and hairs

glaucous: covered with a white, waxy powder which can be rubbed off

globose: shaped like a sphere

glume: 1 of 2 empty bracts at the base of a grass spikelet

hairpoint: a thin, hair-like extension at the tip of a structure (p. 372)

herb: a plant without woody above-ground parts, the stems dying back to the ground each year

herbaceous: herb-like

humus: the organic constituent of soil, composed of decaying plant material

hybrid: an individual that is the offspring of parents of different kinds (usually, different species)

Types of Inflorescences

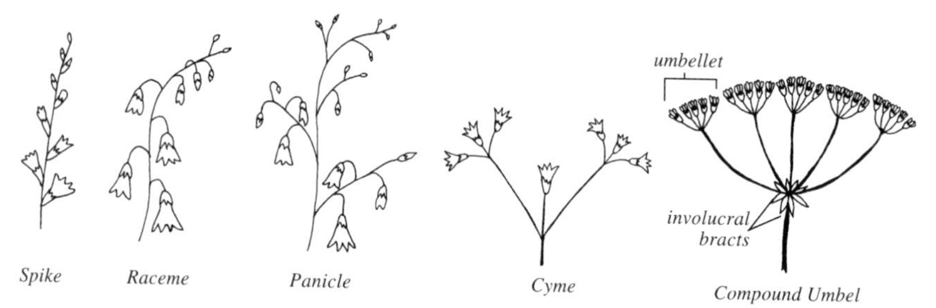

Spike *Raceme* *Panicle* *Cyme* *Compound Umbel*

hybridization: the process of creating a hybrid by cross-breeding different species
hypanthium: a ring or cup around the ovary, formed by the union of the lower parts of the sepals, petals and stamens
hypha: [hyphae] tiny, multicellular fungal threads making up the main body of a fungus or lichen (p. 375)
igneous rock: rock that has solidified from lava or magma
incubous: with leaves overlapping like shingles if the base of the plant is at the ridge of the roof and the tip is at the eaves; opposite of succubous
incurved: curved upwards and inwards
indusium: an outgrowth covering and protecting a spore cluster in ferns (p. 274)
inferior ovary: an ovary located below the point of attachment of the petals, sepals, and stamens (p. 369)
inflexed: incurved
inflorescence: a flower cluster (p. 254)
insectivorous: feeding on insects
internode: portion of a stem between 2 nodes (p. 254)
interrupted: discontinuous
introduced: a new element in the flora brought in from another region (e.g. Europe)
involucre: a set of bracts beneath an inflorescence (in Asteraceae and Apiaceae) (pp. 170, 370); a protective covering around the hood of a capsule (as in hepatics)
irregular flower: a flower with petals, or less often sepals, disimilar in form or orientation (p. 369)
isidium: [isidia] a tiny, wart-like outgrowth from the surface of a lichen, with algae and an outer cortex; functions as a vegetative reproductive propagule (p. 374)
keel: a sharp or conspicuous longitudinal ridge, like the keel of a boat; the 2 partly united lower petals of many species of the Fabaceae (p. 369)
lacerate: torn, or with an irregularly jagged margin
lamella: [lamellae] a green ridge or plate on the midrib of some moss leaves or on the undersurfaces of lichens (as in *Umbilicaria*)
lanceolate: lance-shaped, much longer than wide, widest below the middle and tapering to both ends (p. 368)
lateral: on the side of
layering: a form of vegetative reproduction in which branches droop to the ground and root
leader: the terminal shoot of a tree
leaflet: a single segment of a compound leaf
legume: the fruit of a pea family (Fabaceae) plant, composed of a single carpel, typically dry and splitting down both seams; a plant of the pea family (Fabaceae)
lemma: the lower of the 2 bracts immediately enclosing an individual grass flower (p. 254); cf. palea
lenticel: a slightly raised pore on root or stem bark
ligule: the flat, usually membranous projection from the top of the sheath of a grass; the strap-shaped corolla of a ray flower of a member of the aster family (p. 254)
linear: line-shaped; very long and narrow with essentially parallel sides (p. 368)
lip: a projection or expansion of something, such as the lower petal of an orchid or violet flower (pp. 87, 371)
loam: a loose-textured soil consisting of a mixture of sand, silt, clay and organic matter
lobe: a rounded or strap-shaped division of the thallus of a lichen or liverwort; or a rounded division of the leaf a liverwort or vascular plant (pp. 369, 372)
lobule: a small lobe
lyrate: pinnatifid with the terminal lobe largest and rounded (p. 368)
margin: an edge
marsh: a nutrient-rich wetland periodically inundated by slow-moving or standing water, characterized by emergent vegetation on mineral soils
medulla: inner part of a lichen thallus, usually consisting of loosely packed fungal hyphae (p. 375); cf. cortex
mericarp: an idividual carpel of a schizocarp
mesic: characterized by medium moisture supplies, neither very wet nor very dry
midrib: the central rib of a leaf (p. 292)
midvein: the central vein of a leaf (p. 372)
minerotrophic: nourished by mineral-rich waters, describing wetlands
monocot: a seed plant bearing one cotyledon (seed leaf) in the embryo (includes orchids, lilies, grasses and sedges); most have leaves with parallel veins; cf. dicot
monoculture: a single species comprising the vegetation a tract of land
mordant: a substance enabling a dye to become fixed in the fabric on which it is used, usually applied beforehand
mucilaginous: sticky, producing gummy or gelatinous substances
muskeg: a complex mosaic of boreal fens, bogs, swamps, pools and scrubby forest, increasingly common to the north
mycorrhiza: the symbiotic association of certain fungi with the roots of certain seed plants
naturalized: a foreign species that has adapted to the environment of a region
nectary: a gland that secretes nectar, usually associated with a flower (p. 43)
nerve: a prominent longitudinal line or vein in a leaf or other organ
net-veined: with a network of veins (p. 372); cf. parallel-veined
node: the place where a leaf or branch is attached (pp. 254, 372)
nut: a hard, dry, usually 1-seeded fruit that does not open at maturity; larger and thicker-walled than an achene
nutlet: a small nut; a very thick-walled achene

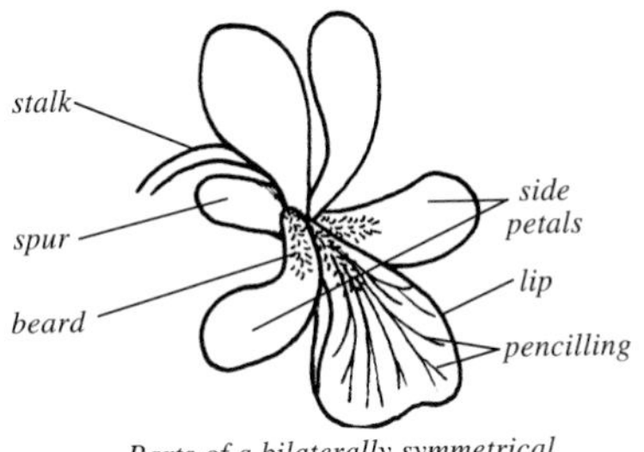

Parts of a bilaterally symmetrical (Viola) *flower*

oblanceolate: lanceolate with the broadest part above the middle
oblique: an angle of leaf attachment between 0° and 90° to the stem
oblong: shaped more or less like a rectangle (other than a square) with rounded corners (p. 368)
obovate: shaped like a long section through a hen's egg, broadest near the tip (describing 2-dimensional areas) (p. 368)
obovoid: egg-shaped, broadest near the tip (describing 3-dimensional structures)
offset: a short shoot arising near the base of a plant; usually propagative in function
operculum: the lid of a moss capsule (p. 286)
opposite: situated across from each other, not alternate or whorled; or, situated directly in front of organs of another kind (p. 372)
orbicular: essentially circular in outline (p. 368)
oval: broadly elliptic (p. 368)
ovary: the structure at the base of a pistil that contains the young, undeveloped seeds (pp. 87, 369)
ovate: shaped like a long section through a hen's egg, with the larger end toward the base (p. 368)
palea: the upper of the 2 bracts immediately enclosing an individual grass flower; cf. lemma
palmate: (leaves) divided into 3 or more lobes or leaflets diverging from a common point, like fingers of a hand (p. 368); cf. pinnate
panicle: a branched flower cluster blooming from the bottom up (p. 372)
papilla: [papillae] a minute, wart-like projection
papillose: with papillae
pappus: the modified hairs or bristles on the top of an achene (p. 170)
parallel-veined: with veins running parallel to one another, not branching to form a network (p. 372)
paraphyllium: [paraphyllia] minute leaf- or thread-like structure on the stems and among the leaves of some mosses
paraphysis: [paraphyses] a sterile hair growing between the asci in ascomycete fruiting bodies (p. 374), or between the sex organs of some mosses and liverworts
parasite: an organism that gets its food/water chiefly from a live organism to which it is attached
pencilled: marked with thin lines (p. 371)
perennial: growing for 3 or more years, usually flowering and producing fruit each year
perfect flower: a flower with both male and female functional reproductive organs
perianth: the sepals and petals of a flower, collectively; a tube of 2–3 fused leaves, surrounding the developing sporophyte of a liverwort (p. 287)
perichaetium: [perichaetia] (bryophytes) a special leaf surrounding the female organ (archegonium) at the base of a sporophyte stalk (seta)
periglacial: close to, or recently exposed by, retreating glaciers
perigynium: [perigynia] the inflated sac enclosing the achene of a sedge (*Carex*) (p. 227)
peristome: the fringe of teeth surrounding the mouth of a moss capsule (p. 286)
perithecium: a flask-shaped fruiting body of a fungus, immersed in a lichen thallus and with a dark and often protuding opening (p. 374)
permafrost: permanently frozen ground

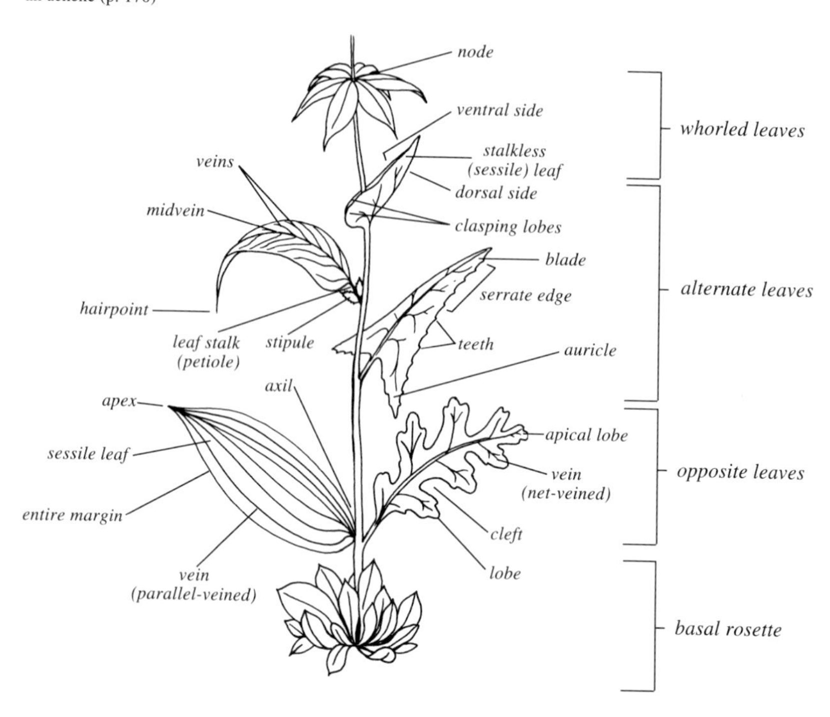

Leaf Characteristics

persistent: remaining attached after the normal function has been completed; cf. deciduous
petal: a member of the inside ring of modified flower leaves, usually white or brightly coloured (pp. 87, 170, 369, 371)
petiole: a leaf stalk (p. 372)
photobiont: the photosynthetically active component organism in a lichen, the alga or cyanobacterium
photosynthesis: the process by which green plants produce their food (carbohydrates) from water, carbon dioxide and minerals, using the sun's energy
phytogeography: study of the geographical distribution and relationships of plants
pingo: a low, conical or dome-shaped mound consisting of a layer of soil over a large core of ice, formed in areas with thin or interrupted permafrost
pinna: [pinnae] the primary division of a pinnate leaf or frond (p. 274)
pinnate: feather-formed; of a compound leaf in which the leaflets are placed on each side of the common axis (p. 368)
pinnatifid: pinnately cleft halfway to the middle (p. 368)
pinnule: the secondary or ultimate leaflet or division of a pinnately compound leaf or frond (p. 274)
pioneer: a plant (or animal) that establishes itself in a recently cleared or otherwise unoccupied area
pistil: the female organ of a flower, usually consisting of an ovary, style and stigma (p. 369)
pistillate: with a pistil; usually applied to flowers with pistils only (lacking stamens)
pith: the soft, spongy tissue in the centre of the stems and branches of certain plants
pleated: having parallel folds or ridges, like pleats in a skirt
pleurocarpous: producing sporophytes from the side of the stem or on short, specialized branches, not at the stem tip.
plicate: folded like a fan
pod: a dry fruit which opens to release its seeds
podetium: [podetia] a stalk-like growth of a lichen, usually supporting 1 or more fruiting bodies (e.g., the clubs or cups of some *Cladonia* lichens (pp. 354-361)
pollinium: [pollinia] a waxy mass of many pollen grains transported as a unit during pollination, as in orchids
pome: a fruit with a core (e.g., an apple)
potherb: an herb cooked as a vegetable
poultice: a moist mass of plant and/or animal material applied to a sore or inflamed part of the body
prostrate: growing flat along the ground
pseudocyphellum: [pseudocyphellae] a white patch, dot, or line on the surface of some lichens, caused by a break in the cortex and the extension of the inner medulla to the surface
pseudopodium: [pseudopodia] a stalk produced by the gametophyte that functions as a sporophyte stalk in the moss genus *Sphagnum*
pycnidium: [pycnidia] a tiny, flask-shaped structure in the thallus of a lichen, in which minute spore-like reproductive cells (pycnoconidia) are produced (p. 374)
quadrangular: 4-sided in cross-section
raceme: an unbranched cluster of stalked flowers on a common, elongated central stalk, blooming from the bottom up (p. 372)
rachis: the main axis (as in a compound leaf) (p. 274)
radially symmetrical: developing uniformly on all sides, like spokes on a wheel
rank: row, as a vertical row of leaves on a stem
ray flower: a flower with a flattened, strap-like corolla, found in flower clusters of the aster family, often radiating from the edge of the head (p. 170)
receptacle: the end of a stem to which the flower parts (or, in Asteraceae, the flowers) are attached (pp. 170, 369); in some liverworts, the umbrella-like structure to which the sporophytes are attached (p. 287)
recurved: curved under (usually referring to leaf margins) (p. 292)
reflexed: abruptly bent or turned back or down
regular flower: a flower in which the members of each circle of parts (or at least the sepals and petals) are similar in size, shape and orientation; a radially symmetrical flower (p. 369)
reniform: kidney-shaped (p. 368)
retort cells: (*Sphagnum*) cells with a short projecting neck at the upper end, terminating in a pore, found in the cortex of branches
rhizine: thread-like, branched, unbranched, tufted or brush-like organ of attachment for lichens, composed of fungal hyphae (p. 375)
rhizoid: a structure of root-like form and function, but of simpler anatomy, found in bryophytes (pp. 286, 287)
rhizomatous: with rhizomes
rhizome: an underground, often lengthened stem; distinguished from a root by the presence of nodes and buds or scale-like leaves
riparian: of or pertaining to a river
rosette: a cluster of organs (usually leaves) arranged in a circle or disk, often at the base of a plant (p. 372)
runner: a slender stolon
samara: a dry, usually 1-seeded, winged fruit (e.g., a maple fruit)
saprophyte: a plant that lives on dead organic matter, neither parasitic nor making its own food
sapwood: the new wood immediately below the bark of a tree
scale: any small, thin or flat structure (p. 274)
scape: a leafless flower stalk arising from near the ground
schizocarp: a fruit that splits into several parts
scree: rock debris at the foot of a rock wall; talus
scurvy: a disease caused by the lack of vitamin C in the diet, characterized by swollen and bleeding gums
sepal: a member of the outermost ring of modified flower leaves; usually green and more or less leafy in texture (pp. 87, 369)
serotinous: appearing, blooming or producing leaves late in the season; often applied to the cones of evergreens that remain on the tree unopened until high temperatures melt their resins, opening the scales
serrate: saw-toothed, having sharp, forward-pointing teeth (p. 372)
sessile: without a stalk
seta: [setae] a stalk supporting a bryophyte capsule (pp. 286, 287)
sheath: an organ which partly or completely surrounds another organ, as the sheath of a grass leaf surrounds the stem (p. 254)

sheathing: partly or wholly surrounding an organ
silicle: a short silique, not much longer than wide
silique: a pod-like fruit of certain members of the mustard family, much longer than wide
simple: not divided or subdivided
site: a plot of ground or place of location
slough: a wet depression or pond
snag: a short, uneven stump left standing out from a tree trunk or branch; a standing dead tree
soralium: [soralia] a localized clump of soredia on the surface or along the edge of a lichen
soredium: [soredia] (lichens) a microscopic clump of several algal cells surrounded by fungal hyphae; soredia are produced as a means of vegetative reproduction, and break through the outer surface (cortex) to appear as a powder on the surface of a lichen (p. 374)
sorus: [sori] a cluster of small spore cases (sporangia) on the underside of a fern leaf (p. 274)

spadix: a spike with small, crowded flowers on a thickened, fleshy axis (used only among monocotyledons)
spathe: a large, usually solitary, bract below and often enclosing a flower cluster (usually a spadix); applied to monocots only
spatulate: shaped like a spatula, or like a long section through a pear; rounded at the tip, rounded, but narrower at the base (p. 368)
spike: a more or less elongate inflorescence with non-stalked flowers (p. 370)
spikelet: a small or secondary spike in a panicle; the floral unit, or ultimate cluster, of a grass flower cluster (p. 254)
spinulose: covered by tiny spines
sporangium: [sporangia] a spore case
spore: a 1- to several-celled reproductive body produced in an ascus (lichens) (p. 374), capsule (mosses and hepatics) (p. 286), or sporangium (ferns and allies); capable of giving rise to a new plant
sporophyll: a spore-bearing leaf
sporophyte: the spore-bearing part or phase of a plant; in bryophytes composed of a foot, stalk and capsule; in vascular plants, the leafy plant (pp. 287, 287); cf. gametophyte
spur: a hollow appendage on a petal or sepal, usually functioning as a nectary (pp. 87, 371)
spurred: with a spur
squamule: a small scale-like lobe of a lichen (see shingle lichen, p. 336), often near the bases of *Cladonia* species (pp. 354-61)
stamen: the pollen-bearing (male) organ of a flower consisting of an anther and a filament (pp. 43, 369)
staminate: with stamens; usually applied to flowers with stamens only (lacking pistils)

Cross-sections of Lichen Reproductive Structures

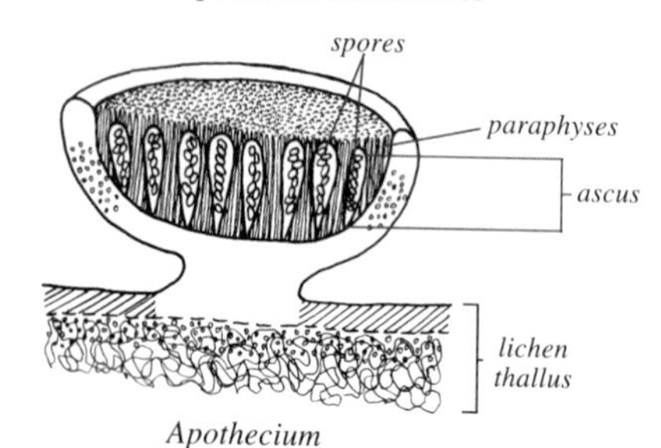

Apothecium

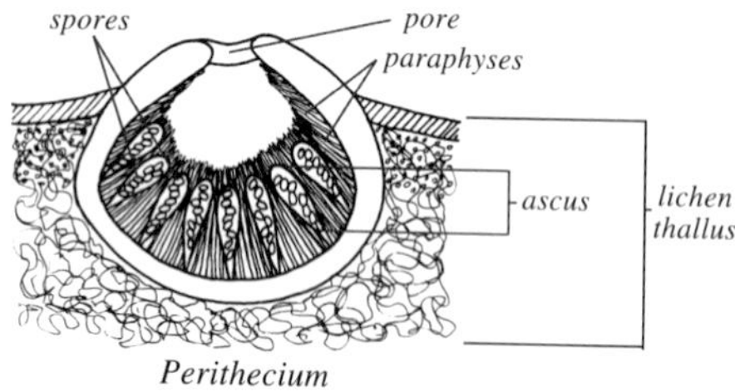

Perithecium

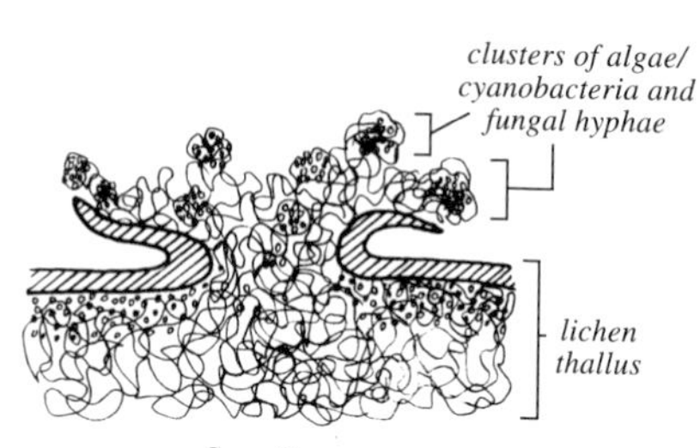

Soredium

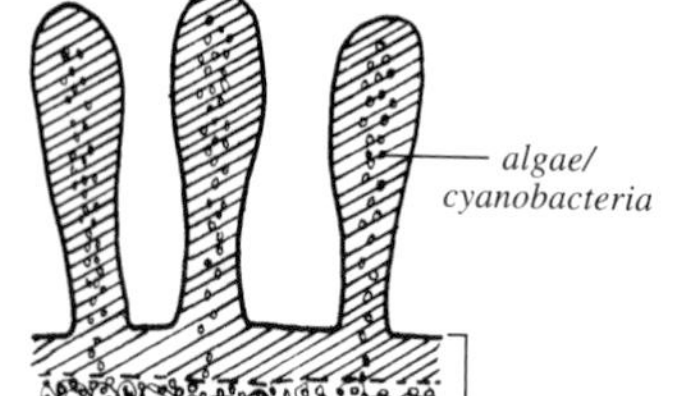

Isidia

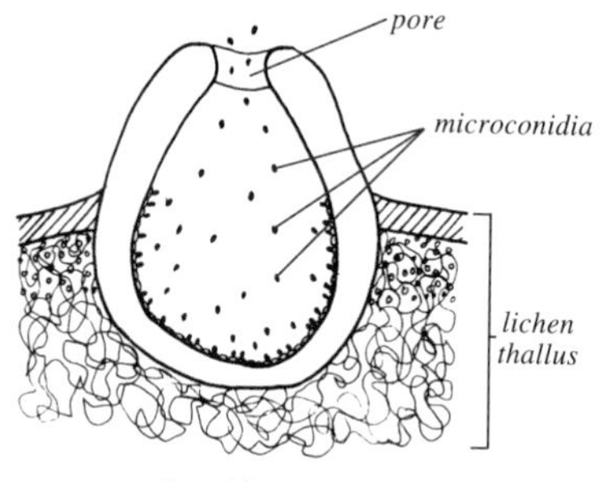

Pycnidium

stand: a group of growing plants (especially trees) of the same type (e.g., a white spruce stand)
standard: the upper, usually enlarged petal of a flower, as found in the pea family; also called a banner
stigma: the tip of the female organ (pistil) in plants, where the pollen lands (pp. 43, 170, 227, 369)
stipe: a stalk-like support (p. 274)
stipule: a leaf-like appendage at the base of a leaf stalk (p. 372)
stolon: a horizontally-spreading stem or runner on the ground, usually rooting at nodes or tips
stoloniferous: with stolons
stoma, stomate: [stomata] a tiny opening in the epidermis of plants, bounded by a pair of guard cells which can close off the opening by changing shape
striate: marked with fine, parallel ridges
strobilus: [strobili] a cone-like cluster of sporophylls
style: the stalk, or middle part, of a plant's female organ, connecting the stigma and ovary (p. 369)
sub-: almost
substrate: a surface on which something grows
subtend: to be directly below and close to
succession: the process of change in plant community composition and structure over time
succubous: with leaves overlapping like shingles on a roof if the base of the plant is the 'eaves' and the tip is the 'ridge'; the opposite of incubous
succulent: fleshy and juicy; a plant that stores reserves of water in the fleshy stems or leaves
superior ovary: an ovary located above the point of attachment of the petals, sepals, and stamens of the flower (p. 369)
suture: a seam or line of fusion; usually applied to the vertical lines along which a fruit splits open
swamp: a wetland with permanently waterlogged subsurface, periodically innundated by standing or gently moving water, nutritionally intermediate between a bog and a fen
talus: rocky debris at the foot of a rock wall; scree
tannin: a brownish white, astringent compound extracted from plants and used in the manufacture of leather
taproot: a primary, descending root
tepal: a sepal or petal, when these structures cannot be distinguished
terminal: at the end, or top, of
terrestrial: living on or growing in the earth
thalloid: with a thallus (p. 287)
thallus: a main plant body, not differentiated into stems and leaves, in lichens and some liverworts (pp. 287, 374, 375)
throat: the opening of a corolla or calyx of fused petals or petal-like sepals; in grasses, the upper part of a leaf sheath
trailing: flat on the ground, but not rooting
trifoliate: 3-leaved (p. 368)
tripinnate: 3 times pinnate, leaflets twice divided again (p. 368)
tuber: a thickening, usually at the end of a rhizome, serving in food storage and often also in reproduction; sometimes loosely applied to tuberous roots
tubercle: a small swelling or projection on an organ
tuberous: thickened like a tuber
tussocks: compact tufts of grasses or sedges
umbel: an often flat-topped flower cluster in which the flower stalks arise from a common point, much like the stays of a reversed umbrella (p. 372)
umbellet: 1 of the umbel clusters in a branched (compound) umbel (p. 372)
umbelliferous: bearing umbels
umbilicus: a depression, or 'belly-button,' on the upper surface of a lichen corresponding with the point of attachment on the lower surface (as in *Umbilicaria*, p. 351)
understory: the vegetation growing below the main canopy of a forest
utricles: small, thin-walled, 1-seeded, inflated fruits
valve: one of the pieces into which a pod or capsule splits
vegetative reproduction: producing new plants from asexual parts (e.g., rhizomes, leaves, bulbils); offspring are genetically identical to the parent plant
vein: a strand of conducting tubes (a vascular bundle), especially if visible externally, as in a leaf (pp. 274, 372)
ventral: on the lower (inner) side, closest to the stem; opposite to dorsal (p. 372)
vermifuge: a remedy that destroys intestinal worms
weed: a common, undesirable or troublesome plant that grows in abundance, especially on cultivated or waste ground; many of our weeds have been introduced from Europe
whorl: a ring of 3 or more similar structures (e.g., leaves around a node on a stem) (p. 372)
whorled: arranged in whorls (p. 372)
wing: a thin, flat extension or projection from the side or tip; one of the 2 side petals in a flower of the pea family (Fabaceae) (p. 369)

Cross-section of a Stratified (Layered) Leafy Lichen Thallus

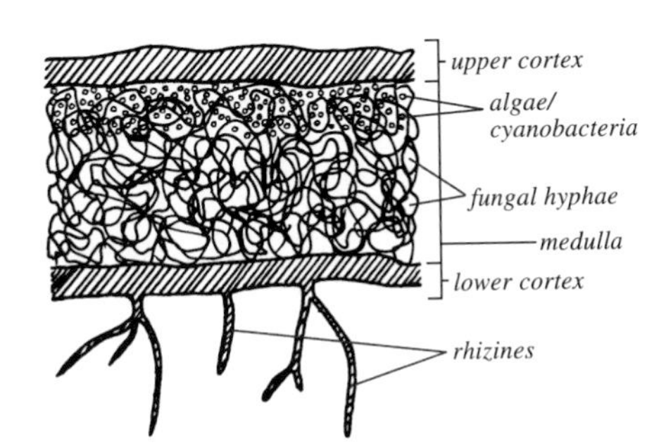

Photo and Illustration Credits

Photo Credits

Andrusek, Blaine: 25a, 25b, 28a, 28b, 28c, 29b, 30a, 30b, 32b, 32c, 40c, 46a, 47a, 61b, 153a, 163a, 185a, 193a

Antos, Joe and Gerry Allen: 126b, 133b, 149b, 174a, 204c, 208b, 220c

Boas, Frank: 41a, 51b, 55c, 62a, 63b, 63c, 66a, 67c, 67d, 68c, 69a, 70a, 70b, 72a, 72b, 73a, 73c, 83a, 83c, 86a, 86b, 92a, 96b, 96c, 97a, 97b, 97c, 98, 102, 112a, 112b, 114b, 114c, 123a, 124a, 125a, 132b, 135c, 136a, 141a, 145b, 147a, 148a, 150a, 151a, 156a, 157b, 159b, 161b, 161c, 167c, 168b, 169a, 179c, 186a, 194b, 196a, 197b, 200c, 202c, 203a, 204a, 208a, 231b, 231c, 236a, 236b, 248c, 276c, 276d, 277c, 277d, 291a, 293c, 301b, 302b, 307b, 307c, 308b, 308c, 310a, 310b, 310c, 311b, 312b, 316c, 319b, 321a, 356b, 362b

Bovey, Robin: 287, 288a, 288b, 290a, 290b, 294a, 294b, 295a, 295b, 296a, 300a, 302c, 303b, 304a, 304b, 305a, 305b, 306a, 306c, 311a, 312c, 313a, 319c, 320b, 322a, 323a, 323b, 324a, 324b, 326a, 327a, 327b, 328a, 329c, 331, 333a, 333b, 334a, 336b, 337b, 338a, 338b, 339a, 339b, 340, 342a, 342b, 343b, 344a, 345c, 348b, 349b, 355a, 357b, 359a, 359b, 360a, 360b, 361a, 363d, 364a, 365a, 365b

Ceska, Adolf: 53b 159a, 221a, 224b, 242a, 252b, 261c

Dickens, R. Blake: 33a, 33b, 33c, 33d, 158c

Frisch, Robert: 58a, 145a, 192b, 195b, 282b, 346c, 355b

Hrapko, Julie: 24a, 26a, 26c, 32a, 54c, 58b, 65a, 65b, 66c, 75c, 83b, 216d, 217b, 285a

Johnson, Derek: 11, 12, 14a, 14b, 14c, 16a, 16b, 16c, 16d, 16e, 17a, 17b, 17c, 24b, 25c, 26b, 36a, 48a, 53a, 54b, 55a, 55b, 56a, 57d, 61c, 64a, 65c, 66b, 67a, 67b, 68a, 68b, 69c, 70c, 71a, 71b, 75b, 82, 85b, 86d, 90b, 90c, 90d, 91b, 91c, 92d, 94b, 94c, 95b, 95c, 99a, 100a, 100b, 100c, 101a, 101b, 104c, 107, 108b, 108c, 109a, 111a, 111c, 114a, 115d, 119a, 119b, 119d, 120a, 120c, 121a, 121c, 122a, 122b, 123b, 127c, 128c, 128d, 134d, 135b, 135d, 136b, 136c, 137d, 140c, 141c, 142b, 143a, 144c, 146b, 146c, 147b, 151b, 151c, 152, 154a, 155a, 159c, 160a, 160b, 162a, 163d, 164b, 164c, 165c, 166b, 167b, 167d, 174c, 175b, 179a, 181b, 182c, 183b, 184a, 184b, 185b, 185c, 185d, 187b, 187c, 189c, 190c, 191a, 195a, 195c, 195d, 196b, 198c, 199a, 199b, 199c, 200a, 200b, 202b, 205a, 205b, 209c, 209d, 210a, 210b, 210d, 211b, 216c, 218c, 219b, 221b, 223a, 226a, 232a, 240b, 241d, 243a, 243b, 245b, 252a, 263c, 263d, 268b, 269b, 270b, 272b, 273c, 278b, 298a, 299c, 306b, 316a, 316b, 317a, 317b, 318c, 319a, 320a, 335a, 335b, 336a, 337a, 341, 342c, 34b, 351b, 351c, 354, 356c, 357a, 362d, 363c, 364b, 366c

Kershaw, Linda: 27a, 29a, 29c, 31a, 31b, 34c, 35b, 35c, 36c, 40a, 40b, 41b, 49b, 50a, 50b, 52a, 52b, 54a, 55d, 56b, 57b, 60b, 61a, 62b, 62c, 68d, 73b, 74a, 75a, 84a, 84b, 85a, 85c, 86c, 90a, 91a, 92c, 93a, 93d, 94a, 96a, 96d, 100d, 103b, 104a, 104b, 105, 109b, 110a, 111b, 113b, 115b, 115c, 120b, 121b, 125d, 126c, 126d, 127a, 127b, 128a, 128b, 129, 132c, 134a, 134b, 134c, 135a, 137a, 137b, 137c, 140a, 140b, 141b, 142a, 143b, 144a, 144b, 148c, 150c, 155b, 158b, 159d, 162b, 163c, 165a, 165b, 166a, 174d, 175a, 176a, 176b, 177c, 178b, 181a, 186b, 187d, 189a, 189b, 191b, 192c, 193b, 193c, 194c, 197a, 198b, 201c, 203c, 204b, 206a, 206b, 211a, 211d, 215c, 216b, 218a, 219a, 219c, 220a, 222a, 222b, 223c, 224c, 225c, 226b, 226c, 231a, 237b, 240a, 241b, 244a, 245a, 246b, 248a, 249b, 253a, 260a, 261b, 262b, 262c, 263b, 265b, 268a, 270a, 273b, 275a, 275b, 276b, 277b, 278a, 278c, 279a, 281a, 282c, 285b, 289b, 291c, 300b, 301a, 303a, 309b, 321b, 321c, 322b, 326b, 328b, 329a, 329b, 330b, 330c, 330d, 333c, 334b, 334c, 343a, 344b, 346b, 346d, 348a, 349a, 361b, 361c, 362a, 362c, 363a, 364c, 366b, 367b, 367c

Merilees, Bill: 27b, 42c, 99b, 103a, 145c, 149c, 169b, 174b, 178a, 182a, 192a, 209a, 210c, 218b, 225a, 225b

Norton, Robert: 59d, 72d, 93c, 119c, 148b, 153b, 156b, 157a, 168c, 169c, 179b, 202a, 206c, 219d, 220b, 234a, 235d, 236d, 239b, 242b, 262a, 267b, 267c, 267d

Pojar, Jim: 24c, 24d, 27c, 34d, 49c, 49d, 56d, 57a, 59b, 60a, 63a, 64b, 64c, 69b, 72c, 73d, 74b, 95a, 108a, 113a, 113c, 113d, 115a, 124b, 125b, 125c, 132a, 133a, 145d, 146a, 149a, 154b, 158a, 164a, 167a, 168a, 177a, 177b, 182b, 187a, 188, 190a, 190b, 194a, 196c, 196d, 197c, 201a, 201b, 209b, 215a, 216a, 217a, 220d, 237a, 243c, 244b, 248b, 251a, 251b, 261a, 264b, 265c, 268c, 269a, 276a, 277a, 279b, 281b, 282a, 283b, 283c, 283d, 284, 318a, 318b, 330a

Roberts, Anna: 46b, 46c, 48b, 49a, 51a, 59c, 109c, 183a, 184c, 215b, 231d, 232b, 233b, 234b, 234c, 235a, 235b, 235c, 236c, 238a, 238b, 239a, 239c, 241a, 241c, 242c, 253b, 260b, 263a, 264a, 265a,266, 267a, 271a, 271c, 272a, 273a

Steen, Trygve: 150b, 163b, 203b, 211c, 224a, 226d, 246a,247, 271b

Vitt, Dale: 289a, 291b, 293a, 293b, 296b, 297a, 297b, 299a, 299b, 302a, 307a, 308a, 309a, 309c, 311c, 312a, 313b, 313c, 345a, 345b, 358a, 358b

Williams, Dave: 350, 351a, 356a, 356d, 367a

Additional Photographers: B.C. Ministry of Forests: 35a, 56c, 126a, 223b; **Corns, Ian:** 15, 35d; **Goward, Trevor:** 346a; **Long, Ron:** 83d, 110b, 161a, 198a; **Meidinger, Del:** 283a; **Naga, M.:** 36b; **Otto, George:** 363b; **Scagel, Rob:** 59a; **Thomas, Don:** 249a, 366a; **Thormin, Terry:** 92b, 93b; **Turner, Nancy and Robert:** 269c; **Wallace, Cliff:** 57c; **Wershler, Cleve:** 34a; **Wilkinson, Kathleen:** 39, 47b; **Woollett, J.M.:** 42a, 42b

Illustration Credits

Brayshaw, T.C. 1976. *Catkin Bearing Plants (Amentiferae) of British Columbia*. Occasional Papers of the British Columbia Provincial Museum, no. 18. British Columbia Provincial Museum, Victoria: 40a, 40b, 49b, 50a, 51a, 52a, 53a

Brayshaw, T.C. 1985. *Pondweeds and Bur-reeds, and Their Relatives, of British Columbia*. Occasional Papers of the British Columbia Provincial Museum, no. 26. British Columbia Provincial Museum, Victoria: 218b

British Columbia Forest Service, Research Branch (original line drawings): 127a, 201a

Britton, N.L., and A. Brown. [1913] 1970. *An Illustrated Flora of the Northern United States and Canada*. Vol. 1: *Ferns to Buckwheat*. Reprint of 2nd ed. Dover Publications, New York: 24, 25, 26, 28, 29, 30, 31, 36, 41a, 93a, 101, 272a

Britton, N.L., and A. Brown. [1913] 1970. *An Illustrated Flora of the Northern United States and Canada*. Vol. 2: *Amaranth to Polypremum*. Reprint of 2nd ed. Dover Publications, New York: 32, 33, 55b, 56a, 61b, 62b, 64a, 68b, 71b, 75b, 115b, 120a, 121a, 121b, 124b, 134b, 140b, 144a, 158b, 160b, 197a, 210a, 224b

Britton, N.L., and A. Brown. [1913] 1970. *An Illustrated Flora of the Northern United States and Canada*. Vol. 3: *Gentian to Thistle*. Reprint of 2nd ed. Dover Publications, New York: 65a, 65b, 67a, 68a, 183b, 199a, 202a, 203a, 205a, 210b, 211a

Crum, H. 1973. *Mosses of the Great Lakes Forest*. Contributions from the University of Michigan Herbarium, vol. 10. University of Michigan Press, Ann Arbor: 298b, 301a, 302b, 303a, 303b, 305a, 306a, 306b, 309b, 310b, 311a, 313a, 321a, 322a, 322b, 323b, 324b, 326b, 327a, 328a, 328b, 329b, 331a

Crum, H. 1988. *A Focus on Peatlands and Peat Mosses*. University of Michigan Press, Ann Arbor: 316a, 316b, 317a, 317b, 318b, 319a, 319b

Crum, H.A., and L.E. Anderson. 1981. *Mosses of Eastern North America*. 2 vols. Columbia University Press, New York: 293b, 296a, 298a, 299a, 299b, 301a, 302b, 304a, 305b, 306b, 307b, 311a, 318a, 320a, 320b, 321b, 322a, 322b, 326a, 326b, 327b, 328a, 329a, 330a, 330b

Frye, T.C., and L. Clark. 1937. *Hepaticae of North America*. University of Washington Publications in Biology, vol. 6, nos. 1–5. University of Washington Press, Seattle: 287b, 288a, 290b

Goward, Trevor (original line drawings): 338a, 342a, 356a, 356b, 367a

Hale, M.E. 1979. *How to know the Lichens*. 2nd ed. Wm. C. Brown Company Publishers, Dubuque, Iowa: 337b, 355b, 364a

Hitchcock, C.L., A. Cronquist, M. Ownbey and J.W. Thompson. 1955. *Vascular Plants of the Pacific Northwest*. Part 5: *Compositae*. University of Washington Publications in Bilology, vol. 17. University of Washington Press, Seattle: 174b, 175a, 175b, 176a, 177a, 177b, 178a, 179b, 181a, 181b, 182a, 182b, 183a, 184b, 185a, 185b, 186b, 187a, 187b, 188b, 189b, 190a, 191a, 191b

Hitchcock, C.L., A. Cronquist, M. Ownbey and J.W. Thompson. 1959. *Vascular Plants of the Pacific Northwest*. Part 4: *Ericaceae through Campanulaceae*. University of Washington Publications in Bilology, vol. 17. University of Washington Press, Seattle: 64b, 66a, 66b, 67b, 69a, 70, 71a, 72a, 72b, 73b, 156a, 157a, 157b, 159a, 159b, 160a, 161a, 161b, 162b, 163a, 163b, 164a, 164b, 165a, 167b, 169a, 196b, 198a, 199b, 200a, 200b, 201b, 203a, 203b, 204b, 205b, 211b, 219b

Hitchcock, C.L., A. Cronquist, M. Ownbey and J.W. Thompson. 1961. *Vascular Plants of the Pacific Northwest*. Part 3: *Saxifragaceae to Ericaceae*. University of Washington Publications in Bilology, vol. 17. University of Washington Press, Seattle: 54b, 55a, 56b, 57a, 57b, 58, 61a, 62a, 63a, 63b, 74a, 114a, 114b, 115a, 115b, 133b, 134a, 135a, 136b, 137a, 137b, 140a, 141b, 142a, 143a, 144a, 144b, 145a, 146a, 147a, 147b, 148a, 148b, 150a, 150b, 151a, 151b, 152b, 153a, 154a, 154b, 155b, 194b, 196a, 220a, 223a, 223b

Hitchcock, C.L., A. Cronquist, M. Ownbey and J.W. Thompson. 1964. *Vascular Plants of the Pacific Northwest*. Part 2: *Salicaceae to Saxifragaceae*. University of Washington Publications in Bilology, vol. 17. University of Washington Press, Seattle: 34, 39b, 41b, 42a, 42b, 46a, 46b, 47b, 48b, 49a, 50b, 51b, 52b, 53b, 98b, 99b, 100a, 100b, 101, 102b, 103a, 103b, 104a, 104b, 105a, 105b, 107b, 108a, 108b, 109a, 109b, 110a, 110b, 111a, 111b, 112a, 112b, 113a, 116, 119a, 119b, 122a, 122b, 123a, 123b, 124a, 125a, 125b, 128a, 128b, 129a, 192a, 192b, 193b, 195a, 208a, 209a, 209b, 220b, 224a, 225a, 225b

Hitchcock, C.L., A. Cronquist, M. Ownbey and J.W. Thompson. 1969. *Vascular Plants of the Pacific Northwest*. Part 1: *Vascular Cryptogams, Gymnosperms, and Monocotyledons*. University of Washington Publications in Bilology, vol. 17. University of Washington Press, Seattle: 27, 75a, 83a, 83b, 85a, 85b, 86a, 92b, 93b, 94a, 94b, 95b, 96a, 96b, 97a, 97b, 206a, 206b, 215a, 215b, 216a, 216b, 217b, 219a, 221a, 221b, 222a, 222b, 226b, 231a, 231b, 232a, 232b, 233b, 235a, 235b, 236a, 236b, 237a, 238b, 239a, 239b, 240a, 241a, 241b, 242a, 242b, 243a, 243b, 244b, 245, 246b, 247a, 248a, 248b, 249a, 249b, 251a, 251b, 252a, 252b, 253b, 260a, 260b, 261a, 261b, 262a, 262b, 263a, 263b, 264a, 264b, 265a, 265b, 266a, 266b, 267a, 267b, 268a, 268b, 269a, 269b, 270a, 270b, 271a, 271b, 272b, 273a, 273b, 276b, 277a, 279a, 279b, 281a, 281b, 282a, 282b, 283a, 283b, 284b, 285a, 285b

Hulten, E. 1968. *Flora of Alaska and Neighboring Territories: A Manual of the Vascular Plants*. Stanford University Press, Stanford, Calif.: 178b, 245

Maywood, John (original line drawings): 291a, 291b

Porsild, A.E., and W.J. Cody. 1980. *Vascular Plants of Continental Northwest Territories, Canada*. National Museums of Canada, Ottawa: 47a, 48a, 56a, 84a, 84b, 124b, 154b, 155a, 165b, 186a, 198b, 276a, 278a, 278b

Royal British Columbia Museum (original line drawings): 179a, 186a

Salkeld, Shirley (original line drawings): 234a, 236b, 237b, 253a, 261b, 262a, 264a, 265a, 265b, 266a, 267a, 268b, 269a, 269b, 271a, 271b

Schofield, W.B. 1992. *Some Common Mosses of British Columbia*. 2nd ed. Royal British Columbia Museum Handbook no. 28. Royal British Columbia Museum, Victoria: 293a, 293b, 294a, 294b, 295a, 295b, 296a, 296b, 297a, 297b, 298a, 299a, 299b, 300a, 300b, 301b, 302a, 304b, 306a, 307a, 308a, 308b, 309a, 310a, 311b, 312a, 312b, 313b, 323a, 324a, 326a, 328b

Schuster, R.M. 1969. *The Hepaticae and Anthocerotae of North America East of the Hundredth Meridian*. Vol. 2. Columbia University Press, New York: 288a, 288b, 289a, 289b, 290a

Szczawinski, A.F. 1962. *The Heather Family of British Columbia*. British Columbia Provincial Museum Handbook no. 19. British Columbia Provincial Museum, Victoria: 69b, 70, 208b

Taylor, T.M.C. 1963. *The Ferns and Fern-allies of British Columbia*. 2nd ed. British Columbia Provincial Museum Handbook no. 12. British Columbia Provincial Museum, Victoria: 277b

Taylor, T.M.C. 1973. *The Rose Family of British Columbia*. British Columbia Provincial Museum Handbook no. 30. British Columbia Provincial Museum, Victoria: 59a, 59b, 60a, 132a, 132b, 133a, 135b

Taylor, T.M.C. 1974. *The Pea Family of British Columbia*. British Columbia Provincial Museum Handbook no. 32. British Columbia Provincial Museum, Victoria: 141a, 142b, 143b, 145b, 146b

Taylor, T.M.C. 1974. *The Figwort Family of British Columbia*. British Columbia Provincial Museum Handbook no. 33. British Columbia Provincial Museum, Victoria: 166b, 167a, 168a, 168b, 169b

Taylor, T.M.C. 1983. *The Sedge Family of British Columbia*. British Columbia Provincial Museum Handbook no. 43. British Columbia Provincial Museum, Victoria: 233a, 234a, 234b, 238a, 238b

Thompson, J.W. 1984. *American Arctic Lichens: 1. The Macrolichens*. Columbia University Press, New York: 338b, 339b, 340a, 342b, 343a, 343b, 344a, 344b, 345b, 348a, 349b, 350a, 351a, 354b, 355a, 355b, 356a, 357a, 357b, 358a, 358b, 359a, 359b, 360a, 361a, 363a, 363b, 364b, 365a, 365b, 366a, 367a, 367b

References

Abbe, E. 1981. *The Fern Herbal*. Comstock Publishing Associates, Cornell University Press, Ithaca, New York.

Alberta Provincial Museum. 1988. *Collection of Plants Annotated with Historical Uses*. Edmonton, Alberta.

Alcock, R.H. 1971. *Botanical Names for English Readers*. Grand River Books, Detroit.

Allen, O.N., and E.K. Allen. 1981. *The Leguminosae*. University of Wisconsin Press, Madison.

Anderson, A. 1977. *Some Native Herbal Remedies, as Told to Anne Anderson by Luke Chalifoux*. Friends of the Botanic Garden of the University of Alberta Publication no.8. University of Alberta, Edmonton.

Angier, B. 1978. *Field Guide to Medicinal Wild Plants*. Stackpole Books, Harrisburg, Pennsylvania.

Argus, G.W. 1973. *The Genus* Salix *in Alaska and the Yukon*. National Museum of Natural Sciences Publications in Botany no. 2. National Museums of Canada, Ottawa.

Bailey, R.G. 1994. *The United States ecoregions map*. Revised @ 1:7,500,000. U.S.D.A. Forest Service, Fort Collins, Colorado.

Baldwin, K.A., and R.A. Sims. 1989. *Field Guide to the Common Forest Plants in Northwestern Ontario*. Forestry Canada and Ontario Ministry of Natural Resources, Thunder Bay, Ontario.

Bartlett, M. 1975. *Gentians*. Blandford Press, Poole, Dorset.

Bews, J.W. 1929. *The World's Grasses: Their Differentiation, Distribution, Economics and Ecology*. Longmans, Green & Co., Toronto.

Bischler, H. 1984. Marchantia *L. The New World Species*. Bryophytorum Bibliotheca, Band 26. A.R. Gantner Verlag K.G., Vanduz, Germany.

Bliss, A. 1978. *Weeds: A Guide for Dyers and Herbalists*. Juniper House, Boulder, Colorado.

Bliss, L.C. 1971. Arctic and alpine plant life cycles. In *Annual Review of Ecology and Systematics*, vol. 2. Annual Reviews, Palo Alto, California.

Boivin, B. 1967–69. Flora of the Prairie Provinces: A handbook to the flora of the provinces of Manitoba. Saskatchewan, and Alberta. *Phytologia* 15:121–159, 329–446; 16:1–47, 219–261, 265–339; 17: 58–112; 18:281–293.

Brayshaw, T.C. 1976. *Catkin Bearing Plants (Amentiferae) of British Columbia*. Occasional Papers of the British Columbia Provincial Museum, no. 18. Victoria.

———. 1985. *Pondweeds and Bur-reeds, and Their Relatives, of British Columbia*. Occasional Papers of the British Columbia Provincial Museum, no. 26. Victoria.

———. 1989. *Buttercups, Waterlilies and Their Relatives in British Columbia*. Royal British Columbia Museum Memoir no. 1. Victoria.

Bristow, A. 1978. *The Sex Life of Plants*. Barrie & Jenkins, London.

Britton, N., and A. Brown. [1913] 1970. *An Illustrated Flora of the Northern United States and Canada*. 3 vols. Reprint of 2nd ed. Dover Publications, New York.

Brodo, I.M., and D.L. Hawksworth. 1977. Alectoria *and allied genera in North America*. National Museum of Natural Sciences Publications in Botany, no. 6. National Museums of Canada, Ottawa.

Brown, A. 1970. *Old Man's Garden*. Gray's Publishing, Sidney, British Columbia.

Bush, C.D. 1990. *A Compact Guide to Wildflowers of the Rockies*. Lone Pine Publishing, Edmonton, Alberta.

Chinnappa, C.C., and J.K. Morton. 1991. Studies on the *Stellaria longipes* complex (Caryophyllaceae): Taxonomy. *Rhodora* 93:129–135.

Clark, L.J. 1976. *Wild Flowers of the Pacific Northwest*. Gray's Publishing, Sidney, British Columbia.

Cody, W.J. 1988. *Plants of Riding Mountain National Park, Manitoba*. Publication 1818/E. Research Branch, Agriculture Canada, Ottawa.

Cody, W.J., and D.M. Britton. 1989. *Ferns and Fern Allies of Canada*. Publication 1829/E. Research Branch, Agriculture Canada, Ottawa.

Conard, H.S., and P.L. Redfearn. 1979. *How to Know the Mosses and Liverworts*. Wm. C. Brown Co. Publishers, Dubuque, Iowa.

Coombes, A.J. 1985. *Dictionary of Plant Names*. Timber Press, Portland, Oregon.

Coon, N. 1963. *Using Plants for Healing*. Hearthside Press, U.S.A.

Cormack, R.G.H. 1977. *Wild Flowers of Alberta*. Hurtig Publishers, Edmonton, Alberta.

Corns, I.G., and R.M. Annas. 1986. *Field Guide to Forest Ecosystems of West-central Alberta*. Northern Forestry Centre, Canadian Forest Service, Edmonton, Alberta.

Correll, D.S., and H.B. Correll. 1972. *Aquatic and Wetland Plants of Southwestern United States*. 2 vols. Stanford University Press, Stanford, California.

Council for Yukon Indians. 1993. *Land of My Ancestors: Plants As Food and Medicine*. Council for Yukon Indians, Whitehorse.

Crum, H.A. 1973. *Mosses of the Great Lakes Forest*. Contributions from the University of Michigan Herbarium, vol. 10. Ann Arbor.

———. 1976. *Mosses of the Great Lakes Forest*. Rev. ed. Contributions from the University of Michigan Herbarium, vol. 10. Ann Arbor.

———. 1988. *A Focus on Peatlands and Peat Mosses*. University of Michigan Press, Ann Arbor.

Crum. H.A., and L.E. Anderson. 1981. *Mosses of Eastern North America*. 2 vols. Columbia University Press, New York.

Currah, R.S., S. Hambleton and E.A. Smreciu. 1986. *Orchids of Alberta: Studies of Their Microhabitat, Mycotrophy and Micropropagation*. Devonian Botanic Garden, University of Alberta, Edmonton, Alberta.

Demarchi, D.A., R.D. Marsh, A.P. Harcombe and E.C. Lea. 1990. The environment. In *The Birds of British Columbia*, ed. R.W. Campbell, N.K. Dawe, I. McTaggart-Cowan, J.M. Cooper, G.W. Kaiser and M.C.E. McNall, vol. 1:55–144. Royal British Columbia Museum and Canadian Wildlife Service, Victoria.

Densmore, F. 1974. *How Indians Use Wild Plants*. Dover Publications, New York.

Dore, W.G., and J. McNeill. 1980. *Grasses of Ontario*. Monograph 26. Biosystematics Research Institute, Research Branch, Agriculture Canada, Ottawa.
Douglas, G.W. 1982. *The Sunflower Family (Asteraceae) of British Columbia. Volume 1—Senecioneae*. Occasional Papers of the British Columbia Provincial Museum, no. 23. Victoria.
———. 1995. *The Sunflower Family (Asteraceae) of British Columbia. Volume 2—Asteraceae, Anthemideae, Eupatorieae and Inuleae*. Royal British Columbia Museum, Victoria.
Downie, S.R. 1987. *A Biosystematic Study of* Arnica, *Subgenus* Arctica. PhD Thesis. University of Alberta, Edmonton, Alberta.
Duke, J.A. 1986. *Handbook of Northeastern Indian Medicinal Plants*. Quarterman Publications, Lincoln, Massachusetts.
Duke, J.A., and E.S. Ayensu. 1985. *Medicinal Plants of China*. Reference Publications, Algonac, Michigan.
Dunn, D.B., and J.M. Gillett. 1966. *The Lupines of Canada and Alaska*. Monograph No. 2. Research Branch, Agriculture Canada, Ottawa.
Ealey, D.M., ed. 1992. *Alberta Plants and Fungi: Master Species List and Species Group Checklists*. Alberta Energy, Forestry, Lands and Wildlife, Edmonton, Alberta.
Ecoregions Working Group. 1989. *Ecoclimatic regions of Canada, first approximation*. Ecological Land Classification Series no. 23. Canada Committee on Ecological Land Classification, Environment Canada, Ottawa.
Egan, R.S. 1987. A fifth checklist of the lichen-forming, lichenicolous and allied fungi of the continental United States and Canada. *The Bryologist* 90:77–173.
Elliot-Fisk, D.L. 1988. The boreal forest. In *North American Terrestrial Vegetation*, ed. M.G. Barbour and W.D. Billings, 33–62. Cambridge University Press, Cambridge.
Emery, R.J.N., and C.C. Chinnappa. 1994. Morphological variation among members of the *Stellaria longipes* complex: *S. longipes*, *S. longifolia*, and *S. porsildii* (Caryophyllaceae). *Plant Systematics and Evolution* 190:69–78.
Erichsen-Brown, C. 1979. *Use of Plants for the Past 500 Years*. Breezy Creeks Press, Aurora, Ontario.
Featherly, H.I. 1954. *Taxonomic Terminology of the Higher Plants*. Iowa State College Press, Ames.
Ferguson, M., and R.M. Saunders. 1976. *Canadian Wildflowers*. Van Nostrand Reinhold, Toronto.
Fernald, M.L., and A.C. Kinsey. 1943. *Edible Wild Plants of Eastern North America*. Idlewild Press, Cornwall-on-Hudson, New York.
Fielder, M. 1975. *Plant Medicine and Folklore*. Winchester Press, New York.
Flint, C.L. 1888. *Grasses and Forage Plants: A Practical Treatise*. Lee and Shepard Publishers, Boston.
Foster, S., and J.A. Duke. 1990. *A Field Guide to Medicinal Plants: Eastern and Central North America*. Houghton Mifflin Co., Boston.
Fowells, H.A. 1965. *Silvics of Forest Trees of the United States*. Agriculture Handbook No. 271. U.S.D.A. Forest Service, Washington, D.C.
Frye, T.C., and L. Clark. 1937. *Hepaticae of North America*. University of Washington Publications in Biology, vol. 6, nos. 1–5. University of Washington Press, Seattle.
Gallant, A.L., E.F. Binnian, J.M. Omernik and M.B. Shasby. 1994. *Ecoregions of Alaska*. U.S.G.S. Professional Paper, U.S. Geological Survey, U.S. Environmental Protection Agency and Colorado State University (in press).
Genders, R. 1988. *Edible Wild Plants: A Guide to Natural Foods*. van der Marck Editions, New York.
———. 1977. *Scented Flora of the World*. St. Martin's Press, New York.
Gibbons, B. 1984. *How Flowers Work*. Blandford Press, Poole, Dorset.
Gleason, H.A. 1952. *The New Britton and Brown Illustrated Flora of the Northeastern United States and Adjacent Canada*. 3 vols. Macmillan Co., New York.
Gleason, H.A., and A. Cronquist. 1963. *Manual of the Vascular Plants of Northeastern United States and Adjacent Canada*. Van Nostrand, Princeton, New Jersey.
Glime, J.M. 1989. Should mosses have common names? *Evansia* 6(1):1–6.
———. 1991. Should mosses have common names? Part 5. The common names of Pottiales. *Evansia* 8(2):32–37.
———. 1992. Should mosses have common names? Part 7. The common names of Funariales. *Evansia* 9(2):45–48.
———. 1993a. Should mosses have common names? Part 8. The common names of Bryales. *Evansia* 10(1):1–8.
———. 1993b. Should mosses have common names? Part 9. The common names of Orthotrichales. *Evansia* 10(2):64–67.
———. 1994a. Should mosses have common names? Part 10. The common names of Leucodontales. *Evansia* 11(1):30–34.
———. 1994b. Should mosses have common names? Part 11. The common names of Tetraphidales. *Evansia* 11(3):93–94.
———. 1994c. Should mosses have common names? Part 12. The common names of Hypnales. *Evansia* 11(3):102–114.
Glime, J.M., and J. Zhang. 1990. Should mosses have common names? Part 4. The common names of Dicranales. *Evansia* 7(3):41–46.
Goffinet, B., and R.I. Hastings. 1994. *The Lichen Genus* Peltigera *(Lichenized Ascomycetes) in Alberta*. Natural History Occasional Paper no. 21. Provincial Museum of Alberta, Edmonton, Alberta.
Gordon, L. 1977. *Green Magic: Flowers, Plants and Herbs in Lore and Legend*. Ebury Press, Webb and Bower, Exeter, England.
Gould, F.W. 1968. *Grass Systematics*. McGraw-Hill Book Co., Toronto.
Goward, T., B. McCune and D. Meidinger. 1994. *The Lichens of British Columbia. Illustrated Keys. Part 1—Foliose and Squamulose Species*. British Columbia Ministry of Forests, Victoria.
Graham, F.K. 1985. *Plant Lore of an Alaskan Island*. Alaska Northwest Publishing Co., Anchorage.

Grant, V.E., and K.A. Grant. 1965. *Flower Pollination in the Phlox Family*. Columbia University Press, New York.

Grieve, M. [1931] 1976. *A Modern Herbal*. Reprint. Penguin Books, Harmondsworth, Middlesex, England.

Hale, M.E. 1967. *The Biology of Lichens*. Edward Arnold, London.

———. 1979. *How to Know the Lichens*. 2nd ed. Wm. C. Brown Co. Publishers, Dubuque, Iowa.

Hardin, J.W., and J.M. Arena. 1974. *Human Poisoning From Native and Cultivated Plants*. 2nd ed. Duke University Press, Durham, North Carolina.

Harding, W. 1981. *Saxifrages: The Genus* Saxifraga *in the Wild and in Cultivation*. The Alpine Garden Society, Woking, Surrey, England.

Harrington, H.D. 1968. *Edible Native Plants of the Rocky Mountains*. University of New Mexico Press, Albuquerque, New Mexico.

Harris, B.C. 1972. *The Compleat Herbal*. Barre Publishers, Barre, Massachusetts.

Hermann, F.J. 1970. *Manual of the Carices of the Rocky Mountains and Colorado Basin*. Agriculture Handbook No. 374. U.S.D.A. Forest Service, Washington, D.C.

Hitchcock, C.L., A. Cronquist, M. Ownbey and J.W. Thompson. 1955–69. *Vascular Plants of the Pacific Northwest*. 5 vols. University of Washington Publications in Biology, vol. 17. University of Washington Press, Seattle.

Hogg, E.H. 1994. Climate and the southern limit of the western Canadian boreal forest. *Canadian Journal of Forest Research* 24:1835–1845.

Hosie, R.C. 1969. *Native Trees of Canada*. Canadian Forest Service, Ottawa.

Hubbard, C.E. 1954. *Grasses: A Guide to Their Structure, Identification, Uses, and Distribution in the British Isles*. Penguin Books, Harmondsworth, Middlesex, England.

Hudson, J.H. 1977. Carex *in Saskatchewan*. Bison Publishing House, Saskatoon, Saskatchewan.

Hultén, E. 1968. *Flora of Alaska and Neighboring Territories*. Stanford University Press, Stanford, California.

Ireland, R.R. 1982. *Moss Flora of the Maritime Provinces*. National Museum of Natural Sciences, Ottawa.

Ireland, R.R., G.R. Brassard, W.B. Schofield and D.H. Vitt. 1987. Checklist of mosses of Canada 2. *Lindbergia* 13:1–62.

Janzen, D.H. 1971. Seed predation by animals. In *Annual Review of Ecology and Biosystematics*, vol. 2. Annual Reviews, Palo Alto, California.

Johnson, K.L. 1987. *Wildflowers of Churchill and the Hudson Bay Region*. Manitoba Museum of Man and Nature, Winnipeg.

Johnston, A. 1987. *Plants and the Blackfoot*. Lethbridge Historical Society Occasional Paper no. 15. Lethbridge, Alberta.

Jones, D.L. 1987. *Encyclopedia of Ferns*. Lothian Publishing Co., Sydney, Australia.

Kari, P.R. 1987. *Tanaina Plantlore: Dena'na K'et'una*. National Park Service, Alaska Region.

Kehoe, A.B. 1992. *North American Indians: A Comprehensive Account*. Prentice Hall, Englewood Cliffs, New Jersey.

Kerik, J. 1985. *Living With The Land: Use of Plants by the Native People of Alberta*. Provincial Museum of Alberta, Edmonton, Alberta.

Kindscher, K. 1987. *Edible Wild Plants of the Prairie: An Ethnobotanical Guide*. University Press of Kansas, Lawrence.

Kingsbury, J.M. 1964. *Poisonous Plants of the United States and Canada*. Prentice-Hall, Engelwood Cliffs, New Jersey.

Kirk, D.R. 1975. *Wild Edible Plants of Western North America*. Naturegraphic Publishers, Happy Camp, California.

Knap, A.H. 1975. *Wild Harvest*. Pagurian Press, Toronto.

Kowalchik, C., and W.H. Hylton, eds. 1987. *Rodale's Illustrated Encyclopedia of Herbs*. Rodale Press, Emmaus, Pennsylvania.

Kuijt, J. 1969. *The Biology of Parasitic Flowering Plants*. University of California Press, Berkeley.

Kunkel, G. 1984. *Plants For Human Consumption: An Annotated Checklist of the Edible Phanerogams and Ferns*. Koeltz Scientific Books, Koenigstein, Germany.

Langshaw, R. 1983. *Naturally: Medicinal Herbs and Edible Plants of the Canadian Rockies*. Summerthought Publications, Banff, Alberta.

Larsen, J.A. 1980. *The Boreal Ecosystem*. Academic Press, New York.

Lawler, L.J. 1984. Ethnobotany of the Orchidaceae. In *Orchid Biology Reviews and Perspectives, 3*. Cornell University Press, Ithaca, New York.

Lawton, E. 1971. *Moss Flora of the Pacific Northwest*. Hattori Botanical Laboratory, Japan.

Leacock, E.B., and N.O. Lurie, eds. 1971. *North American Indians in Historical Perspective*. Random House, New York.

Leighton, A.L. 1985. *Wild Plant Use By the Woods Cree (Nihithawak) of East-central Saskatchewan*. National Museum of Man, Mercury Series Paper no. 101. National Museums of Canada, Ottawa.

Lewis, W.H., and M.P.F. Elvin-Lewis. 1977. *Medical Botany: Plants Affecting Man's Health*. John Wiley & Sons, Toronto.

Li, Y. and J.M. Glime. 1989. Should mosses have common names? Part 2. The common names of *Sphagnum*. *Evansia* 6(2):25–27.

Line, L., and W.H. Hodge. 1978. *The Audubon Society Book of Wildflowers*. Harry N. Abrams Publishers, New York.

Little, E.L. 1980. *The Audobon Society Field Guide to North American Trees: Eastern Region*. Alfred A. Knopf, New York.

Long, J.C. 1965. *Native Orchids of Colorado*. Museum Pictorial no. 16. Denver Museum of Natural History, Denver.
Looman, J. 1982. *Prairie Grasses Identified and Described by Vegetative Characters*. Publication 1413, Agriculture Canada, Ottawa.
Looman, J., and K.F. Best. 1979. *Budd's Flora of the Canadian Prairie Provinces*. Publication 1662. Research Branch, Agriculture Canada, Ottawa.
MacKinnon, A., J. Pojar and R. Coupé, eds. 1992. *Plants of Northern British Columbia*. Lone Pine Publishing, Edmonton, Alberta.
Marles, R.J. 1984. *The Ethnobotany of the Chipewyan of Northern Saskatchewan*. M.Sc. Thesis. University of Saskatchewan, Saskatoon.
McGuffin, N.J., ed. 1986. *Dye Plants of Ontario*. Compiled by Burr House Spinners and Weavers Guild, Richmond Hill, Ontario. Publishing and Printing Services, Concord, Ontario.
Medsger, O.P. 1966. *Edible Wild Plants*. Collier Books, New York.
Meeuse, B., and S. Morris. 1984. *The Sex Life of Flowers*. Facts on File Publications, New York.
Meidinger, D.V. and J. Pojar. 1991. *Ecosystems of British Columbia*. Special Report Series no. 6. Research Branch, British Columbia Ministry of Forests, Victoria.
Mercatante, A.S. 1976. *The Magic Garden: The Myth and Folklore of Flowers, Plants, Trees, and Herbs*. Harper & Row, New York.
Michael, P. 1980. *All Good Things Around Us*. Holt, Rinehart & Winston, New York.
Millspaugh, C.F. 1892. *American Medicinal Plants*. John C. Yorston & Co., Philadelphia.
Moerman, D.E. 1986. *Medicinal Plants of Native America*. Vol. 1. Technical Report no. 19. University of Michigan Museum of Anthropology, Ann Arbor.
Moore, M. 1979. *Medicinal Plants of the Mountain West*. Museum of New Mexico Press, Santa Fe.
Moss, E.H. 1983. *Flora of Alberta*. 2nd ed. Revised by J.G. Packer. University of Toronto Press, Toronto.
Mulligan, G.A., and D.B. Munro. 1990. *Poisonous Plants of Canada*. Publication 1842/E. Agriculture Canada, Ottawa.
Oswald, E.T., and J.P. Senyk. 1977. *Ecoregions of the Yukon Territory*. Environment Canada, Canadian Forestry Service, Victoria, B.C.
Pojar, J., and A. MacKinnon, eds. 1994. *Plants of Coastal British Columbia including Washington, Oregon & Alaska*. Lone Pine Publishing, Edmonton, Alberta.
Pond, B. 1974. *A Sampler of Wayside Herbs, Discovering Old Uses for Familiar Wild Plants*. The Chathan Press, Riverside, Connecticut.
Porsild, A.E. 1951. Plant life in the Arctic. *Canadian Geographical Journal* (March).
Porsild, A.E., and W.J. Cody. 1980. *Vascular Plants of Continental Northwest Territories, Canada*. National Museums of Canada, Ottawa.
Reader's Digest General Books. 1986. *Magic and Medicine of Plants*. Reader's Digest, Montreal.
Richardson, D. 1975. *The Vanishing Lichens: Their History, Biology and Importance*. David & Charles, Vancouver.
Robertson, A. 1984. Carex *of Newfoundland*. Newfoundland Forest Research Centre, Canadian Forestry Service, St. John's, Newfoundland.
Rogers, R.D. 1992. *Sundew, Moonwort and More Medicinal Plants of Alberta*. Published by the author, Edmonton, Alberta.
Rosentreter, R., L.C. Smithman, and A. DeBolt. 1993. Vernacular lichen names: Swedish names translated to English. *Evansia* 10(3):104–111.
Rowe, J.S. 1972. *Forest Regions of Canada*. Publication 1300. Canadian Forestry Service, Ottawa.
Savile, D.B.O. 1972. *Arctic Adaptations in Plants*. Monograph no.6. Research Branch, Agriculture Canada, Ottawa.
Schofield, W.B. 1992. *Some Common Mosses of British Columbia*. 2nd ed. Royal British Columbia Museum Handbook no. 28. Victoria.
Schuster, R.M. 1969. *The Hepaticae and Anthocerotae of North America East of the Hundredth Meridian*. Vol. 2. Columbia University Press, New York.
Scoggan, H.J. 1957. *Flora of Manitoba*. National Museum of Canada Bulletin no. 140. Ottawa.
———. 1978–79. *The Flora of Canada*. 4 vols. National Museum of Natural Sciences Publications in Biology, no. 7. National Museums of Canada, Ottawa.
Scotter, G.W., and H. Flygare. 1986. *Wildflowers of the Canadian Rockies*. Hurtig Publishers, Edmonton, Alberta.
Simmons, H., and S. Miller. 1982. *Notes on the Vascular Plants of the Mackenzie Mountain Barrens and Surrounding Area*. Information Report no. 3. Wildlife Service, Northwest Territories Renewable Resources, Yellowknife.
Skinner, C.M. 1911. *Myths and Legends of Flowers, Trees, Fruits, and Plants*. J.P. Lippincott Co., Philadelphia.
Slack, A. 1979. *Carnivorous Plants*. M.I.T. Press, Cambridge, Massachusetts.
Soper, J.H., and M.L. Heimburger. 1982. *Shrubs of Ontario*. Royal Ontario Museum, Toronto.
Spencer, J.D., E. Johnson, A.R. King, R.F. Spencer, T.Stern, K.M. Stewart and W.J. Wallace. 1977. *The Native Americans*. 2nd ed. Harper & Row, New York.
Stebbins, G.L. 1971. Adaptive radiation of reproductive characteristics in angiosperms, 2: Seeds and seedlings. In *Annual Review of Ecology and Systematics*, vol. 2. Annual Reviews, Palo Alto, California.
Stephens, H.A. 1973. *Woody Plants of the North Central Plains*. University Press of Kansas, Lawrence.
Stokes, D.W. 1981. *The Natural History of Wild Shrubs and Vines—Eastern and Central North America*. Harper & Row, New York.
Stokoe, W.J., and A.L. Wells. 1962. *The Observer's Book of Grasses, Sedges and Rushes*. Frederick Warne & Co., New York.

Stotler, R., and B. Crandall-Stotler. 1977. A checklist of the liverworts and hornworts of North America. *The Bryologist* 80:405–428.
Strong, D.R., and K.R. Leggat. 1981. *Ecoregions of Alberta*. ENR Technical Report no. T/4. Alberta Energy and Natural Resources, Edmonton, Alberta.
Stubbendieck, J., S.L. Hatch and K.J. Kjar. 1982. *North American Range Plants*. University of Nebraska Press, London.
Szczawinski, A.F. 1962. *The Heather Family (Ericaceae) of British Columbia*. British Columbia Provincial Museum Handbook no. 19. Victoria.
Szczawinski, A.F., and N.J. Turner. 1978. *Edible Garden Weeds of Canada*. Edible Wild Plants of Canada Series no. 1. National Museums of Canada, Ottawa.
———. 1980. *Wild Green Vegetables of Canada*. Edible Wild Plants of Canada Series no. 4. National Museums of Canada, Ottawa.
Tanaka, T. 1976. *Tanaka's Cyclopedia of Edible Plants of the World*. Keigaki Publishing Co., Tokyo.
Taylor, T.M.C. 1963. *The Ferns and Fern Allies of British Columbia*. 2nd. ed. British Columbia Provincial Museum Handbook no. 12. Victoria.
———. 1966. *The Lily Family (Liliaceae) of British Columbia*. British Columbia Provincial Museum Handbook no. 25. Victoria.
———. 1970. *Pacific Northwest Ferns and Their Allies*. University of Toronto Press, Toronto.
———. 1973. *The Rose Family (Rosaceae) of British Columbia*. British Columbia Provincial Museum Handbook no. 30. Victoria.
———. 1974a. *The Pea Family (Leguminosae) of British Columbia*. British Columbia Provincial Museum Handbook no. 32. Victoria.
———. 1974b. *The Figwort Family (Scrophulariaceae) of British Columbia*. British Columbia Provincial Museum Handbook no. 33. Victoria.
———. 1983. *The Sedge Family (Cyperaceae) of British Columbia*. British Columbia Provincial Museum Handbook no. 43. Victoria.
Thieret, J.W. 1956. Bryophytes as economic plants. *Economic Botany* 10:75–91.
Thiselton-Dyer, T.F. [1889] 1968. *The Folk-lore of Plants*. Reprint. Singing Tree Press, Detroit.
Thomson, J.W. 1967. *The Lichen Genus* Cladonia *in North America*. University of Toronto Press, Toronto.
———. 1979. *Lichens of the Alaskan Arctic Slope*. University of Toronto Press, Toronto.
———. 1984. *American Arctic Lichens: 1. The Macrolichens*. Columbia University Press, New York.
Tsvelev, N.N. 1983. *Grasses of the Soviet Union*. 2 vols. Oxonian Press, New Delhi, India.
Turner, G. 1979. *Indians of North America*. Blandford Press, Poole, Dorset.
Turner, N.J. 1975. *Food Plants of British Columbia Indians. Part 1: Coastal Peoples*. British Columbia Provincial Museum Handbook no. 34. Victoria.
———. 1978. *Food Plants of British Columbia Indians. Part 2: Interior Peoples*. British Columbia Provincial Museum Handbook no. 36. Victoria.
Turner, N.J., and A.F. Szczawinski. 1978. *Wild Coffee and Tea Substitutes of Canada*. Edible Wild Plants of Canada Series no. 2. National Museums of Canada, Ottawa.
Tyler, V.E., L.R. Brady and J.E. Robbers. 1988. *Pharmacognosy*. 9th ed. Lea & Febiger, Philadelphia.
van der Kloet, S.P. 1988. *The Genus* Vaccinium *in North America*. Publication 1828. Research Branch, Agriculture Canada, Ottawa.
van der Pijl, L., and C.H. Dodson. 1966. *Orchid Flowers: Their Pollination and Evolution*. University of Miami Press, Coral Gables, Florida.
Vance, F.R., J.R. Jowsey and J.S. McLean. 1984. *Wildflowers Across the Prairies*. Western Producer Prairie Books, Saskatoon, Saskatchewan.
Viereck, E.G. 1987. *Alaska's Wilderness Medicines: Healthful Plants of the Far North*. Alaska Northwest Publishing Co., Anchorage.
Viereck, L.A. 1971. *Alaska Vegetation Types*. Map @ 1:5,000,000. Compiled for *Alaska Trees and Shrubs*, L.A. Viereck and E.L. Little, 1972. Agriculture Handbook 410. U.S.D.A. Forest Service
Vitt, D.H., J.E. Marsh and R.B. Bovey. 1988. *Mosses, Lichens and Ferns of Northwestern North America*. Lone Pine Publishing, Edmonton, Alberta.
von Marilaun, A.K., and F.W. Oliver. 1902. *The Natural History of Plants*. Vols. 1 & 2. Blackie & Son, Glasgow, Scotland.
Walker, M. 1984. *Harvesting the Northern Wild*. Outcrop, Yellowknife, N.W.T.
Wallis, T.E. 1967. *Textbook of Pharmacognosy*. 5th ed. J.A. Churchill, London.
Ward-Harris, J. 1983. *More Than Meets The Eye: The Life and Lore of Western Wildflowers*. Oxford University Press, Toronto.
Weiner, M.A. 1972. *Earth Medicine—Earth Foods*. Collier Books, Collier Macmillan Publishers, New York.
Welsh, S.L. 1974. *Anderson's Flora of Alaska and Adjacent Parts of Canada*. Brigham Young University Press, Provo, Utah.
Wilkinson, K. 1990. *Trees and Shrubs of Alberta*. Lone Pine Publishing, Edmonton, Alberta.
Willard, T. 1992. *Edible and Medicinal Plants of the Rocky Mountains and Neighbouring Territories*. Wild Rose College of Natural Healing, Calgary, Alberta.
Yanovsky, E. 1936. *Food Plants of the North American Indians*. Miscellaneous Pulbication no. 237. U.S. Department of Agriculture, Washington, D.C.
Zasada, J.C., and E.C. Packee. 1994. The Alaska region. In *Regional Silviculture of the United States*. Wiley-Interscience (in press).

Index to Common and Scientific Names

D

E

N–O

P

Q–R

S

T

U

V

W

X–Y–Z

Authors and Contributors

Derek Johnson
Canadian Forest Service
Edmonton, Alberta

Andy MacKinnon
B.C. Forest Service, Research
Victoria B.C.

Trevor Goward
Naturalist, Lichenologist
Clearwater, B.C.

Linda Kershaw
Botanist, Naturalist
Sherwood Park, Alberta

Jim Pojar
B.C. Forest Service, Research
Smithers, B.C.

Dale Vitt
University of Alberta
Edmonton, Alberta